Living with Wildlife in the Pacific Northwest

Living with
Wildlife
IN THE PACIFIC NORTHWEST

RUSSELL LINK

UNIVERSITY OF WASHINGTON PRESS

Seattle and London

in association with the

WASHINGTON DEPARTMENT OF FISH AND WILDLIFE

Living with Wildlife in the Pacific Northwest has in part been made possible by a grant from the U.S. Fish and Wildlife Service.

University of Washington Press
PO Box 50096, Seattle, WA 98145
www.washington.edu/uwpress

Library of Congress Cataloging-in-Publication Data
Link, Russell.
    Living with wildlife in the Pacific Northwest / Russell Link.
        p. cm.
    Includes index.
    "Published in association with the Washington Department of Fish and Wildlife."
    ISBN 0-295-98386-8 (alk. paper)
    1. Animals—Northwest, Pacific. 2. Human-animal relationships, Northwest, Pacific. I. Title.
    QL151.L56 2004
    519.9795—dc22   2004041961

The paper used in this publication is acid-free and recycled from 20 percent post-consumer and at least 50 percent pre-consumer waste. It meets the minimum requirements of American National Standard for Information Sciences—Permanence of Paper for Printed Library Materials, ANSI Z39.48-1984.

Design and layout: Magrit Baurecht Design
Cover design: Magrit Baurecht Design
Editorial review and index: Sigrid Asmus
Word processing: Renee Blough, Kelly Smith
Illustrations: As credited
Front cover photographer: Don R. Sinclair
Back cover photos/Photographer (starting from upper left):

Townsend mole/Jim Pruske	Big brown bat/Ty Smedes
Red-eared slider/Idie Ulsh	Yellow-bellied marmot/Alan Bauer
Steller's jay/Jim Pruske	Canada goose/Ty Smedes
Pacific treefrog/Alan Baurer	Common garter snake/Jon McGinnis
Northern flicker/Jim Pruske	Douglas squirrel (Chickaree)/Jim Pruske
White-tailed deer twins/Bob Griffith	Coyote/Ty Smedes
Sharp-shinned hawk/Kim Chandler	Golden-mantled ground squirrel/Joy Spurr
Opossum/Jim Pruske	Mountain lion (Cougar)/Ty Smedes
Pocket gopher/Ty Smedes	Eastern cottontail rabbit/Jim Pruske

CONTENTS

Introduction

Living with Wildlife in the Pacific Northwest was written for people interested in knowing more about the wildlife occupying areas around their homes or property.

From bats to woodpeckers, the animal species covered in the book were selected after surveying agencies and organizations that receive the most calls from the public about local wildlife. While many people call for general information about wildlife, in many cases the calls are from people who are—or think they are—experiencing a conflict with a wild animal and are seeking a way to remedy the problem. These conflicts can be classified as either *destructive* (damage to structures, landscapes, crops, etc.), *aggressive* (probable harm to people or pets), or as *health risks* (potential transfer of disease/parasites to people and their pets).

While the book includes information that will be helpful to anyone interested in local wildlife, it was written primarily for people seeking solutions to these human/wildlife conflicts.

It's important to note that not all wildlife create conflicts. Although it might not appear so at the time, the animals, which are often referred to as nuisance or problem animals, are innocent. When a conflict exists between humans and animals it is usually because the animal is only doing what it needs to do to survive. It is simply following its own instincts, and intends no harm or discomfort.

Dealing with a conflict can be difficult because it is often a community issue. Some people habitually feed and perhaps inadvertently shelter wildlife, while their neighbors may not want wildlife around at all. "One person's nuisance is another person's joy," as the saying goes. This scenario can create undesirable situations for people, pets, and the animals themselves. Raccoons, coyotes, and squirrels that are fed by people often lose their fear of humans and may become aggressive when not fed as expected. These hungry visitors might approach a neighbor who might choose to remove these animals, or have them removed.

A conflict can also quickly alter a wildlife lover's perception about a certain species, especially when the situation exceeds his/her current level of tolerance. Such is the paradox that wildlife around homes and property present: We want them and we don't want them, depending on what they are doing at any given moment.

This book encompasses four parts. The first three parts include individual chapters on the mammals, birds, reptiles, and amphibians that are likely to be encountered in or around homes, properties, or neighborhoods. Each chapter begins with a description of one or more species followed by details on feeding behavior, reproduction, and other biological information.

Next, for people interested in safely viewing these animals, tips are provided on where to look, when to look, and what to look for. For people needing to learn more about an animal to help solve a conflict, details on tracks, burrows, nest sites, and other clues are provided. Finally, for those interested in attracting the animal, tips for attracting and maintaining its habitat are provided.

The next section in each chapter is titled "Preventing Conflicts." Here is where common encounters between humans and the wildlife that are found in yards, gardens, and structures are described. This section provides several methods for resolving human/wildlife conflicts, including changing human behavior or perceptions so that people are willing to tolerate some damage.

The best approach is always to use knowledge of the biology and behavior of the species to reduce or eliminate conflicts and to do this in a way that is humane, environmentally benign, and socially acceptable.

Next, public health concerns for each species are described, followed by the animals' legal status. Before taking any action to remedy a conflict associated with wildlife, its legal status must be determined. Most mammals, birds, reptiles, and amphibians that occur in the wild in Oregon, Washington, and British Columbia are protected or regulated by state, provincial, and/or federal laws. These laws not only pertain to the killing of regulated species but may also prohibit live trapping and relocation, harassment, and possession of the animal dead or alive.

Part four consists of appendices, including detailed information referenced in the first three parts, and helpful options on where to go for further information.

The book concludes with an index that will help you find out more about the wildlife in your area.

Acknowledgments

I offer my heartfelt gratitude to the many people who have participated in the creation of this book. For their continuous support throughout the project, I thank Dave Brittell and Lora Leschner from the Washington Department of Fish and Wildlife. For help with the manuscript from beginning to end, I thank biologists Mike Badry and Ray Rainbolt. A special thank you goes to Flora Johnson, who donated so much of her time and talent to this undertaking.

For reviewing the technical information about the wildlife species included in the book, I thank Dana Base, Rich Beausoleil, Brian Bell, Georgia Conti, Caren Crandell, Janean Creighton, Fred Dobler, Howard Ferguson, Margaret Gaspari, Mark Goldsmith, Min Huang, Melissa Keigley, Steve Layman, Eric Larsen, Jeff Lewis, Mary Linders, Kelly McAllister, Scott McCorquodale, Hal Michael, Klaus Richter, Laura Simon, Rocky Spencer, Derek Stinson, Larry Sullivan, Jim Tabor, Patricia Thompson, Susan Van Leuven, Diane West, Mike West, and Dr. Eric Yensen. Any errors that survived are my fault entirely.

For sharing their experience with human/wildlife conflicts, I thank Wendy M. Arjo, Karen Awrylo, John Consolini, Veda DePaepe, Tami DuBow, Paul Dye, Fred Goodman, Bill Hebner, Ginger Holser, Jake Jacobson, Kevin Mack, Rex E. Marsh, Mike Marshall, Troy McCormick, Sue Murphy, Kip Parker, Dave Pehling, David Powell, Carol Powers, Roger Woodruff, and Pete Wise.

For reviewing the sections on public health concerns, I thank John Grendon and Briggs Hall.

For reviewing the sections on legal status, I thank Steve Dauma, Sean Carrell, and Evan Jacoby.

For their editing and technical writing assistance, I thank Renee Blough, Donna Gleisner, Mike O'Malley, Kelly Smith, and Bill Weiler.

For assisting with artwork acquisition I thank David Hutchinson and Kay Barton. For kindly agreeing to include their artwork, I thank Elva Hamerstrom Paulson, Kim A. Cabrera, the Oregon Department of Fish and Wildlife, and the University of California Press. A special thank you goes to James R. Christensen, who allowed me to include his incredible illustrations throughout Part One.

Thanks to Naomi Pascal, Mary Ribesky, Marilyn Trueblood, John Stevenson, and all those at the University of Washington Press whose insight, enthusiasm, and patience made this book possible.

Once again, it has been an honor working with artist Jenifer Rees, designer Magrit Baurecht, and editor Sigrid Asmus, and I deeply appreciate their willingness to see this project through to the end.

Finally, I thank Kathy and Vanessa, who yielded family time while I completed this book.

Note: The tips and information provided here are not intended as replacements for professional advice. Some of the procedures described herein are best performed by a professional. The publisher and author cannot assume any responsibilities for

PART 1

Mammals

CHAPTER 1

Badgers

The badger *(Taxidea taxus)* is a stocky member of the weasel family with a broad head, a short thick neck, short bowed legs, and a short bushy tail (Fig. 1). Its front legs are stout and muscular and its front claws are long, heavy, and designed for digging (Fig. 5). Badgers measure 23 to 30 inches (58–76 cm) in length, including a 5-inch (13 cm) long tail, and are powerful animals for their size.

Badgers occur east of the Cascade Range in Oregon and Washington, and in the central and southern interior of British Columbia as far north as Williams Lake. They occupy grasslands, farm and agricultural habitats, open fir and pine forests, shrub-steppe ecosystems, and other treeless areas with loose soil and an abundance of prey.

There are no documented cases of badgers attacking people; however, they will hiss, growl, and snarl when frightened or cornered, and will make short charges at any intruder. Although badgers pose no threat to humans unless provoked, the presence of aggressive dogs can agitate badgers into fighting back.

Badgers are beneficial to humans. They cause little damage and perform a significant service by limiting rodent, snake, and insect populations. In addition, foxes, burrowing owls, and other wildlife use badger burrows as nest sites. In some areas, badger burrows are the only holes of the right size available for other species of wildlife.

Badgers have a very loose, tough skin but can't, as fable would have it, "turn around in their own skin."

Facts about Badgers
Food and Feeding Habits

- Badgers dig with their forefeet and push the excavated soil to the rear with their hind feet, tunneling with amazing speed in search of marmots, mice, ground squirrels, pocket gophers, and other prey that live underground.
- Although small burrowing mammals make up most of their diet, badgers also eat birds, reptiles (including rattlesnakes), insects, and carrion (animal carcasses).

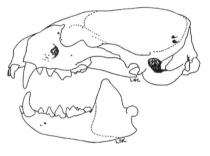

Figure 2. A badger's jaws are extremely strong, with well-developed canine and shearing teeth that can penetrate the skull of their prey with a single bite. (From Verts and Carraway, Land Mammals of Oregon.*)*

Figure 1. The badger is low and wide in general appearance. The shaggy fur on its back and sides ranges from gray to red. The triangular, black "badges" on each cheek are the basis for the badger's name. A white stripe extends back over the head from the nose. (From Larrison, Mammals of the Northwest.*)*

Figure 3. When threatened, a badger will quickly dig a hole, then turn around and face the threat. (From Christensen and Larrison, Mammals of the Pacific Northwest: A Pictorial Introduction.)

Den Sites

- Badger burrows are constructed mainly from tunnels they create in the pursuit of prey.
- Burrows are up to 30 feet (9 m) long, 10 feet (3 m) below ground, and are used for concealment, protection from weather, and as nurseries.
- During the summer, adult badgers without young will visit a new den daily. In winter, a single den may be used for longer periods.
- The nursery den is dug by the female and is used for extended periods. These dens contain a branched main tunnel to allow animals to pass one another, pockets and side-tunnels for disposal of droppings, and a grass-lined bedding chamber.
- Females change dens to forage in new areas close to the nursery. The female will carry the young in her teeth if the young are too small to follow on their own.

Reproduction and Family Structure

- Badgers mate in summer or early fall, and both sexes mate with multiple partners.
- Badgers experience delayed implantation. This allows badgers, a species that needs only five or six weeks for fetal development, to breed when adults are most active and likely to find each other. It also allows them to raise young in spring when food is most abundant, without high fetal development costs during the winter when food supplies are low.
- One to four kits are born from late March to early May.
- Kits first emerge from the den on their own when six to seven weeks of age.
- Juveniles approach adult size by late summer and early fall, when they leave the den to establish their own territories.

- Male badgers are solitary except during breeding season; females are also solitary except during the breeding season and when rearing young.

Mortality and Longevity

- Badgers have few natural enemies. Golden eagles, coyotes, and bobcats are potential predators of young badgers. Cougars, bears, and wolves are able to kill adults.
- The most long-lasting negative impacts on badger populations result from loss of habitat and food sources, indiscriminate shooting, and collisions with vehicles and farm machinery.
- Average longevity for a badger in the wild is six to eight years.

Viewing Badgers

Badgers are primarily active at night; however, juveniles and females with kits are seen during the day. Badgers are active at temperatures well below freezing and in fairly deep snow. Although not true hibernators, badgers are able to remain inactive in their burrows for a month or more during periods of severe weather.

When seen at a distance, a badger can be confused with the similarly sized yellow-bellied marmot (see Chapter 11). However, the badger runs with a steady trot, while the marmot runs with a loping gallop. To get a closer look and avoid disturbing a badger, use binoculars—or, better yet, a spotting scope.

When threatened or attacked, a badger will quickly dig a hole, then turn around and face the threat (Fig. 3). In this position the formidable badger becomes nearly unbeatable. If the threat or attack is not immediate the badger will continue to dig until it is completely underground.

Den Sites

The most identifiable badger sign is its den—or the excavated areas where badgers dig for prey, which may be one and the same.

The entrance to the den is roughly elliptical, as one would expect from such a flattened animal. Entry size varies with the badger, but averages 8 to 12 inches (20–30 cm) wide.

More so than foxes and other borrowing animals, badgers tend to pile excavated earth in a single mound in front of their den entrance. There usually will be recognizable tracks in fresh, loose soil.

Scattered food remains found around the entrance may include snake rattles, bones, fur, and broken eggshells.

Tracks and Trails

A badger's low-slung body and waddling gait leave an easily identified, trough-shaped trail through thick grass. In snow, the badger leaves a trail similar to that of the porcupine, but one that lacks the broomlike brush marks made by a porcupine's swinging tail.

Badger tracks may appear similar to coyote tracks; however, the front tracks have a distinct pigeon-toed appearance (Fig. 4).

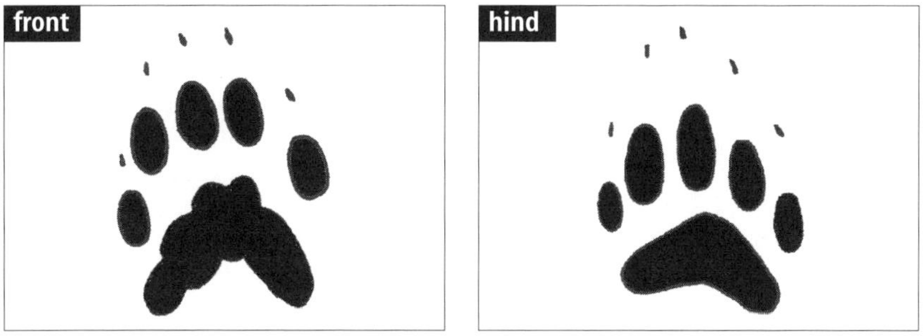

Protecting Badgers

The badger is at risk in the Pacific Northwest because the amount of habitat suitable for them is small. The little that remains—grasslands and open pine or fir forests along the major valleys—has been greatly modified by the development of towns, rural subdivisions, ranches, agriculture, and highways.

Although damage by badgers is usually not serious, it has resulted in considerable persecution over the years. Early rodent-control programs, where ground squirrels and pocket gophers were poisoned, probably reduced the food supply for badgers and resulted in deaths of some that ate poisoned rodents. These historical factors, in combination, are believed to have seriously reduced badger populations.

Based on current research, management practices to protect badgers should include:

- Identify and conserve suitable badger habitat.
- Increase the use of culverts under highways that cross high-quality badger habitat to reduce road kills.
- Protect badger prey populations, especially ground squirrel colonies.
- Discourage indiscriminate shooting of badgers and their prey.
- Restore or create badger habitat by reestablishing historic fire regimes to create continuous, nonforested habitat.
- Continue to dispel badger myths and encourage ecological understanding, particularly when badgers are present in park campgrounds.
- Ranchers, farmers, and golf course owners should be encouraged to tolerate minor amounts of digging by badgers and ground squirrels.

Figure 4. Measured from claw tips to heel pad, front tracks are 2½ to 3 inches (6.5–7.5 cm) long and about 2 inches (5 cm) wide. The toes of the front feet are nearly twice as long and wide as those of the hind feet. The claws of the front feet are exceptionally long—up to 1 inch (2.5 cm) or more for the three center toes—and distinctly angled toward the inside, or smallest, toe. The claws on the hind feet are usually less than ½ inch (12 mm) long. (Drawing by Kim A. Cabrera.)

Figure 5. The claws on a badger's front feet are exceptionally long and designed for digging. (From Christensen and Larrison, Mammals of the Pacific Northwest: A Pictorial Introduction.*)*

Droppings

Badger droppings are identical to those of other carnivores with similar diets and body weights. However, unlike bobcats, coyotes, and foxes, badgers are meticulous about burying their droppings.

Preventing Conflicts

Although occasions are rare, the burrowing activity of badgers can bring them into conflict with humans. It is perceived that livestock and horses can break legs when stepping in badger burrows. Badgers digging in crop fields may slow harvesting or cause damage to machinery. Badgers are known to kill domestic fowl and young

livestock, however these are rare occurrences.

Badgers may burrow in dams, dikes, and irrigation canal banks; these are likely the only badger activities that cause significant damage. For information on how to prevent this damage, see "Preventing Conflicts" in Chapter 15.

To prevent conflicts or remedy existing problems:

Never approach a badger's natal den. A badger mother's protective instincts make her dangerous if she has young in the den. Observe these dens with binoculars or a spotting scope.

Keep dogs contained, especially from dusk to dawn. Outside at night, dogs may confront a badg-

er and suffer injury as a result of the conflict (see Appendix D).

Enclose ducks, chickens, and other small livestock in secure coops or pens at night. To prevent badgers from tunneling into structures housing domestic animals, line vulnerable areas with 1-inch (2.5 cm) mesh hardware cloth that extends at least 12 inches (30 cm) underground. Alternatively, fit fencing tightly to the ground with an L extension out on the surface (see "Preventing Conflicts" in Chapter 23 for diagrams).

A single electrified wire, strung 6 to 8 inches (15–20 cm) above the ground and 6 inches (15 cm) out from the structure will also prevent badgers from digging into the area. The wire needs to be turned on only at night. (See "Electric Fences" in Chapter 7 for additional information.)

Badgers may be **temporarily** discouraged from using an area by the use of bright lights at night.

Encourage natural control. A long-term approach to prevent conflicts with badgers is to reduce their natural food supply around homes and properties. Installing nest boxes and perching poles for owls and hawks may help keep snake and rodent populations at a level that forces badgers to move on and hunt elsewhere (see "Maintaining Hawk Habitat" in Chapter 33 for information).

Trapping and Lethal Control

If conflicts with a badger continue, the animal may have to be trapped. Trapping a badger should be a last resort and can never be justified without first applying the above-described prevention measures. Trapping is also rarely a permanent solution since other badgers are likely to move into the area if attractive habitat is still available.

Biologists might want to trap and release badgers if there was a need for population recovery in an area. Such a project would have to be under the supervision of a federal, state, or provincial wildlife biologist.

Because of the strength and potentially aggressive nature of badgers, a professional trapper should be hired to trap them. (See Appendix A, "Trapping Wildlife," Appendix C, "Hiring a Wildlife Damage Control Company," and "Legal Status," in this chapter.)

Public Health Concerns

The diseases or parasites associated with badgers in the Pacific Northwest are rarely a risk to humans.

Canine distemper, a disease that affects domestic dogs, is found in our badger populations (see "Public Heath Concerns" in Chapter 21). Have your dogs vaccinated for canine distemper to prevent them from contracting the disease.

Legal Status

Because badgers' legal status, trapping restrictions, and other information change, contact your local, state, or provincial wildlife office or visit their Web site for updates. (See Appendix E for contact information and where to access the state and provincial laws mentioned below.)

Oregon: Although not classified as furbearers, badgers are considered mammals with commercial fur value and a trapping or hunting license is required while someone is involved in these activities. Otherwise, badgers are not protected and can be killed without restrictions.

Washington: The badger is classified as a furbearer (WAC 232-12-007). A trapping license and open season are required to trap badgers.

A property owner or the owner's immediate family, employee, or tenant may kill or trap a badger on that property if it is damaging crops or domestic animals (RCW 77.36.030). In such cases, no special trapping permit is necessary for the use of live traps. However, a special trapping permit is required for the use of all traps other than live traps (RCW 77.15.192, 77.15.194; WAC 232-12-142). There are no exceptions for emergencies and no provisions for verbal approval. All special trapping permit applications must be in writing on a form available from the Department of Fish and Wildlife (WDFW). WDFW must be notified immediately after taking a badger in all situations.

British Columbia: The badger is not designated as a furbearing animal and there are no open hunting or trapping seasons for this species. Badgers are protected under the Wildlife Act, Designation and Exemption Regulation, Schedule A (B.C. Reg. 168/90), and so a permit is required if anyone needs to remove a problem badger from their private property, or wishes to import or export badgers to or from the province.

The *T. t. jeffersonii* subspecies is Red-Listed by the B.C. Conservation Data Center.

Species Ranking in British Columbia

Red-Listed Species include any indigenous species or subspecies that have, or are candidates for Extirpated, Endangered, or Threatened status in British Columbia. Extirpated species no longer exist in the wild in British Columbia, but do occur elsewhere. Endangered species are facing imminent extirpation or extinction. Threatened species are likely to become endangered if limiting factors are not reversed. Not all Red-Listed species will necessarily become formally designated. Placing species on these lists flags them as being at risk and requiring investigation.

Blue-Listed Species include any indigenous species or subspecies considered to be Vulnerable in British Columbia. Vulnerable species are of special concern because of characteristics that make them particularly sensitive to human activities or natural events. Blue-Listed species are at risk, but are not Extirpated, Endangered, or Threatened.

Yellow-Listed Species include uncommon, common, declining, and increasing species—all species not included on the Red or Blue lists.

For additional information contact your local wildlife office (see Appendix E), or see http://srmwww.gov.bc.ca/atrisk/index.

CHAPTER 2

Bats

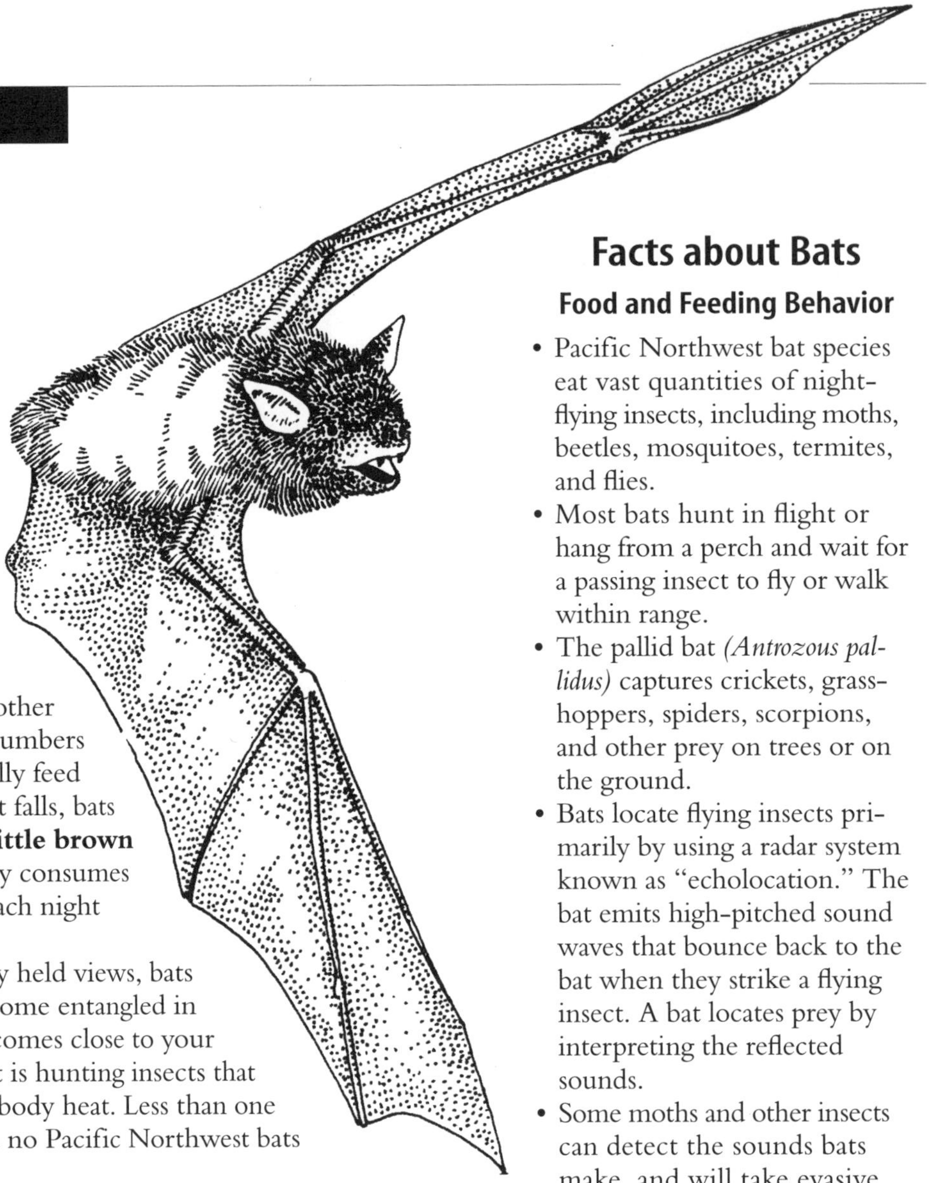

Bats are highly beneficial to people, and the advantages of having them around far outweigh any problems you might have with them. As predators of night-flying insects (including mosquitoes!), bats play a role in preserving the natural balance of your property or neighborhood.

Although swallows and other bird species consume large numbers of flying insects, they generally feed only in daylight. When night falls, bats take over: a nursing female **little brown bat** *(Myotis lucifugus)* probably consumes her body weight in insects each night during the summer.

Contrary to some widely held views, bats are not blind and do not become entangled in peoples' hair. If a flying bat comes close to your head, it's probably because it is hunting insects that have been attracted by your body heat. Less than one bat in 20,000 has rabies, and no Pacific Northwest bats feed on blood.

More than 15 species of bats live in the Pacific Northwest, from the common **little brown bat** *(Myotis lucifugus)* to the rare **Townsend's big-eared bat** *(Corynorhinus townsendii)*. Fur colors range from dark gray to hues of red or buff. Head to tail, bats range in length from the small 2.5-inch-long (6.5 cm) **Western pipistrelle** *(Pipistrellus hesperus),* to the large 6-inch (15 cm) long **hoary bat** *(Lasiurus cinereus).* The hoary bat has a body approximately the size of a house sparrow and a wingspan of 17 inches (43 cm).

The species most often seen flying around human habitat include the **little brown bat** *(Myotis lucifugus)*, **Yuma myotis** *(Myotis yumanensis)*, **big brown bat** *(Eptesicus fuscus)*, **pallid bat** *(Antrozous pallidus)*, **long-eared myotis** *(Myotis evotis)*, and **California myotis** *(Myotis californicus)*.

Figure 1. Bats are the only true flying mammals and they belong to the mammalian order Chiroptera, which means "hand-wing." The bones in a bat's wing work like those of the human arm and hand, but bat finger bones are greatly elongated and connected by a double membrane of skin to form the wing. (Oregon Department of Fish and Wildlife.)

Facts about Bats

Food and Feeding Behavior

- Pacific Northwest bat species eat vast quantities of night-flying insects, including moths, beetles, mosquitoes, termites, and flies.
- Most bats hunt in flight or hang from a perch and wait for a passing insect to fly or walk within range.
- The pallid bat *(Antrozous pallidus)* captures crickets, grasshoppers, spiders, scorpions, and other prey on trees or on the ground.
- Bats locate flying insects primarily by using a radar system known as "echolocation." The bat emits high-pitched sound waves that bounce back to the bat when they strike a flying insect. A bat locates prey by interpreting the reflected sounds.
- Some moths and other insects can detect the sounds bats make, and will take evasive action by staying out of the area, flying in zigzags, or dropping to the ground.
- Bats often capture insects when flying by scooping them into their tail or wing membranes, and then putting the insects into their mouth (Fig. 2). This results in the erratic flight most people are familiar with when they observe bats feeding in the evening.
- Most bats will fly half a mile to 6 miles (1–10 km) from their roost to a feeding site, using temporary roost sites there until returning to their main roost.

Figure 2. Bats often capture insects when flying by scooping them into their tail or wing membranes, and then putting the insects into their mouth. (Oregon Department of Fish and Wildlife.)

Roost Sites

- Bats use day roosts to sleep, keep safe from predators, raise young, and maintain their body temperature.
- Depending on the roost size, the species using it, and the time of year, bats roost singly, in groups of a few individuals, or in colonies of hundreds and occasionally thousands.
- Natural day roosts include rock crevices, caves, trees, cavities, and spaces under tree bark and in large birds' nests; bats also use attics, bat boxes, openings behind shutters, and space under shakes and shingles.
- In spring and summer, female bats collect in maternity colonies where they raise their young.
- As the surrounding air temperature changes, bats change their location to meet the comfort level they need. A good day roost prevents temperature extremes and allows bats to adjust their body temperatures by moving around the roost.
- Big brown bats are less tolerant of high temperatures than other species. Temperatures above 95°F (35°C) will often force them to move to a cooler area within the roost, or to change roosts altogether.

- Bats use night roosts to pause and eat large prey, digest food, rest, and socialize. Night roosts are more exposed than daytime roosts, and include tree branches and areas under bridges, piers, carports, and the eaves of buildings.

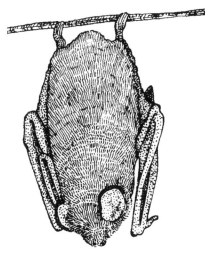

Figure 3. Some bats roost in the open, hanging from a branch or foliage, folding the wing membranes against their body to prevent injury. (Washington Department of Fish and Wildlife.)

Hibernation Sites

- To cope with winter conditions, most bats use a hibernation site, called a "hibernaculum." (See "Bats in the Winter.")
- Hibernation sites include cavities in large trees, caves, mine shafts, tunnels, old wells, and attics.

- The hibernaculum protects the bats from predators, light, noise, and other disturbances. Temperatures in the hibernaculum must be cool enough to allow bats to maintain a low body temperature but not freeze; humidity must be high and constant enough to prevent bats from dehydrating.
- Bats hibernate alone or in groups, and enter hibernation sites in late September or October.

Nursery Sites

- Most Pacific Northwest bats breed during late fall or winter at their hibernation site. Females store sperm until the following spring, when fertilization takes place after the females rouse from hibernation.
- The young, called "pups," are born and raised in nursery colonies occupied only by breeding females and their young.
- Males roost alone or in small groups during this time, leaving the warm roosts and optimal feeding grounds to females.
- Attics are often used as nursery sites because they maintain the warm temperatures, from 85 to 104°F (30–40°C) needed for raising young; the heat actually speeds the development of bat pups, both during fetal development and after birth.

Bats in the Winter

With few flying insects available to them during winter in the Pacific Northwest, bats survive by hibernating, migrating to regions where insects are available, or a combination of these strategies.

During hibernation, metabolic activities are greatly reduced—a bat's normal body temperature of 104°F (40°C) is reduced to just one or two degrees higher than that of the hibernaculum, and their heart rate slows to only one beat every four or five seconds. A hibernating bat can thus survive on only a few grams of stored fat during the five- to six-month hibernation period.

Banding studies indicate that little brown bats will migrate 120 miles (200 km) between hibernacula and summer roosts, and, if undisturbed, they occupy the same site year after year. They select areas in the hibernaculum where there is high humidity (70–95 percent), and the temperature is 34 to 41°F (1–5°C). Still, there are some species, such as the big brown bat, that can hibernate in relatively exposed situations in buildings where there is considerable fluctuation in temperature. Hibernation lasts until April or early May, but in coastal regions bats may arouse in late winter—little brown bats have been found feeding in the Puget Sound region of Washington and coastal Oregon in mid-March.

It is important not to disturb hibernating bats. If a bat rouses early from hibernation, it must use its fat reserves to increase its body temperature. A single disturbance probably costs a bat as much energy as it would normally expend in two to three weeks hibernating. Thus, if disturbed multiple times, hibernating bats may starve to death before spring.

It is important not to disturb roosting bats at any time of the year. In the spring, disturbing a maternity colony when flightless young are present may cause young bats to be dropped to their deaths, or abandoned, by panicked females. Because some bats hibernate in buildings during the winter months, batproof a building only when you are sure no bats are hibernating in it. If bats are found hibernating inside after October 15, they should be left alone until early spring (prior to the birthing period in May) after the weather has warmed enough for insects to be out regularly. Meanwhile, seal all potential entry points into human living spaces, and develop a plan so the exclusion process can be accomplished effectively in spring (see "Bats Roosting in Buildings").

- Most bat species produce one pup in May, June, or July; the specific dates depend on the species, locality, and weather.
- In the nursery, the young remain with their mothers and suckle frequently.
- Pups are unable to fly for about a month, and are left in the nursery when the mother hunts.
- Female bats may return to the exact location where they were born year after year.

Mortality and Longevity

- Bats have few predators. Owls, house cats, and other small mammals occasionally prey upon them.
- Heavy storms kill bats during migrations, and long winters can kill even more of them.
- Through habitat alteration, commercial pesticide use, and control practices, humans are a main source of mortality among bats. Exposure to pesticides kills bats either directly through exposure, or indirectly through ingestion of sprayed insects.
- For their size, bats are the world's longest-lived mammals. The big brown bat has been recorded to live 19 years in the wild. The record for a little brown bat is 33 years.

Viewing Bats

The safest way to view and enjoy bats is to watch them in action. Bats are fascinating flyers, zigging and zagging about as they chase and eat insects. Little brown bats and Yuma bats prefer to hunt over water. Big brown bats are often seen hunting along the margins of wooded areas, or silhouetted against the lighter sky as they twist and turn high above the tree canopy.

It's also fun to watch bats drink, which they usually do first thing after leaving their day roost. They scoop up mouthfuls of water with their lower jaws as they fly over lakes, streams, ponds, or water troughs. Most bats do not come out to eat or drink in heavy rain or when the air temperature remains below 50°F (28°C).

To view bats, follow these tips:

- Choose a warm summer evening and a place where you can sit and view the place where bats will emerge from a roost site or have been spotted in the air.
- When waiting for bats to emerge from a roost site, such as an attic or bat house, remain still and quiet, and listen for the squeaks or clicks that many species make before emerging.
- Some species of bats begin their night flights 20 to 30 minutes before dark—the common big brown bat may be out foraging earlier. The rare Western pipistrelle from the arid regions of the Pacific Northwest emerges before the sun goes down, and has been seen foraging with violet-green swallows. Other species don't emerge until after dark.
- The best places to see bats in flight are where night-flying insects abound, such as next to a stream, lake, or pond, over a meadow or large lawn, along a forest edge, or around bright streetlights or porch lights.
- With the aid of a bat detector, listen for the echolocation calls bats make when navigating and locating prey. (See "Eavesdropping on Bats.")

Attracting and Protecting Bats

Bats may already be hunting the night skies in your area without you knowing it. To improve their hunting grounds:

- Landscape with plants that attract night-flying moths— which then attract bats. Adult moths seek out nectar found in phlox, stock, fireweed, lilac, mock-orange, and flowering tobacco. Female moths lay eggs on the leaves of apple, cottonwood, maple, poplar, willow, and cherry trees; the developing caterpillars feed on the leaves, rarely causing significant damage. Avoid using insecticides on these and other plants.
- Install a light to attract insects. Outdoor lights that run on solar batteries are available commercially.
- Provide open water, such as a pond, for bats to drink. Water is also essential during the life cycle of many flying insects, and most bats eat flying insects.
- Retain large-diameter hollow trees and large trees with deep bark crevices or loosened bark as potential roost sites. Rather than remove any large tree completely, remove only those portions determined to be a safety hazard.
- Retain old barns and other structures as potential roost sites.
- Install and maintain bat houses (see "Bat Houses").

Also:
- Reduce disturbances around hibernation sites, nursery colonies, and other roosting areas.
- Report any observations of maternity or hibernating

Eavesdropping on Bats

The majority of Pacific Northwest bats produce sounds that are above the range of human hearing. The best human ear can detect sounds with frequencies up to 20 kilohertz. With a couple of exceptions, bat echolocation calls (a sonar analogous to the sonar used by the military) range from 20 to 120 kilohertz (kHz).

However, a small device (about the size of a cell phone), called a bat or ultrasound detector or bat locator, converts a bat's echolocation calls into signals audible to our ears and broadcasts them over a speaker.

When a flying insect is located, the bat needs additional information to zero in on its prey. So, it calls faster and faster until it sounds like a buzz. This "feeding buzz" is clearly audible with these detectors.

Bat locators are not likely to be helpful in determining if bats are occupying a building. The reason is that a locator is "searching" for ultrasonic sounds the bats create during feeding, and not the chirping or clicking noises they make while roosting.

To eavesdrop on roosting bats, try placing a stethoscope against the wall or ceiling of the roost.

(For additional information on bat detectors, see "Bat Houses and Bat Detectors" in Appendix F.)

colonies in wild areas to your local wildlife office.

- Protect forested areas, wetlands, and cave systems—particularly those within a mile of your property.
- Learn about bats and pass that knowledge on to others.
- Support bat research conducted by local universities, government agencies, and conservation groups.

Bat Houses

Some bat species prefer man-made structures to their natural roosts, whereas others are forced to roost in buildings when natural roosts, such as caves and hollow trees, are destroyed.

Two bat species that frequently use bat houses are the little brown bat and the big brown bat.

A well-designed, well-constructed, and properly located bat house may attract these and other bats if they live in or pass through your general area.

For the Pacific Northwest, especially west of the Cascades in British Columbia, Washington, and Oregon, bat houses should be painted with multiple coats of flat black exterior latex paint and placed where they will receive full sun. A house baking in the sun is what our bats need and seek—a nice warm place to raise their young, and that lets them decrease their metabolic needs during roosting.

Build or buy a bat house that is at least 2 feet (60 cm) tall and 14 or more inches (36 cm) wide. Bigger is better. A roughened or screen-covered landing platform measuring 3 to 6 inches (8–16 cm) should extend below the house.

The house can be single-chambered or multi-chambered, but chambers should be ¾ to 1 inch (2–2.5 cm) wide—a variety of sizes is good to provide for the needs of different species. This is particularly true for bat houses built for eastern Washington, eastern Oregon, and British Columbia.

The houses should be caulked during construction and preferably be screwed together. The idea is to create a tight microclimate inside the house capable of trapping both the heat captured during the day and the warmth generated by the bats.

Place the house in full sun, preferably on its own pole; the next-best location is on the southern side of a building in full sun. The optimal temperature range is between 85°F (30°C) and 104°F (40°C). Don't put it on a tree, as it will be in too much shade. Keep the area around the entrance clear of obstructions for 15 feet (4.5 meters).

Don't worry that adding a bat house to your property will encourage bats to move into your attic or wall space. If bats liked your attic or wall spaces, they would probably already be living there.

(See Appendix F for resources on bat house design, placement, and maintenance.)

Preventing Conflicts

For some people bats don't present a problem. For others, bats can be a worry, especially when they become unwanted guests in an attic, inside a wall of a home, or inside the home itself.

Unlike rodents, bats only have small teeth for eating insects, so they do not gnaw holes in walls,

shred material for nests, chew electrical wiring, or cause structural damage to buildings. Damage caused by bats is usually minimal, but they can be noisy and alarming, and the smell of bats and their droppings can be offensive. It is possible to learn to coexist with bats, and to benefit from their presence.

If a conflict arises, first make sure bats are the cause by following the tips listed under "Viewing Bats," or by observing the following:

Bat droppings: Bats defecate before entering buildings and places where they roost. In buildings where there is an attic roost or a roost in a wall, an accumultation of droppings may fall through cracks and stain ceilings and walls. Insects associated with bat droppings rarely bother humans.

Droppings are usually the size of a grain of rice, crumble easily between the fingers, and contain shiny, undigested bits of insects. The droppings of mice are much harder and more fibrous.

Bat sounds: Bats often squeak before leaving their roost at night, and may chatter on hot days when they move around seeking refuge from the heat. Baby bats separated from their mothers will squeak continuously. All these sounds are loud enough to be heard from a distance of up to 30 feet (9 m). Thus, an increase in such noises near dusk probably indicates bats.

Scrambling, scratching, and thumping sounds coming from attics and walls at night may be caused by rats, mice, flying squirrels, opossums, or raccoons. In rare cases, chirping and rustling sounds in a chimney may be

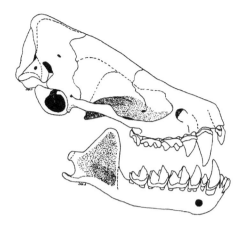

Figure 4. The lateral view of a big brown bat skull. Unlike rodents, bats do not gnaw holes in walls, shred material for nests, chew electrical wiring, or cause structural damage to buildings. (From Verts and Carraway, Land Mammals of Oregon.*)*

caused by swifts (birds similar to swallows; see Chapter 39).

Bat odors: Bats produce a musky solution from their scent glands and their roosts may take on a strong musky odor. Areas occupied by rats and mice also often have a musky smell.

Bats Roosting in Buildings

Like other wild animals, bats will seek shelter in an attic, wall, chimney, or other area of a structure. Bats are able to squeeze through surprisingly narrow slits and cracks; the smaller species need only a ½-inch (12 mm) opening.

When bats occupy a house, homeowners often feel they have a crisis they must deal with immediately. But in fact they may have been living with the bats for years. The following are suggestions for reducing conflicts:

Excluding Bats

The best way to get rid of bats is also the safest—both for the bats and the humans involved. This

is to humanly exclude them. However, because old buildings offer many points of entry it may be impossible to completely exclude bats from them, or from those with shake or cedar shingle roofs that have no underlayment.

A wildlife damage control company experienced in excluding bats can be hired, or you can do the exclusion work yourself (see Appendix C "Hiring a Wildlife Damage Control Company"). In attics and areas where large numbers of bats have been roosting for years, it is safer for you to hire a professional to do the work, including the cleanup of accumulated droppings.

Note: Never trap flightless young or adult bats inside a structure; this is needlessly cruel to the bats inside and can create a serious odor problem (see "Bats in the Winter" for important information about when not to exclude bats).

Trapping and relocating bats is not recommended. Traps can be fatal to bats if left unattended and can quickly become overcrowded. In addition, bats have excellent homing instincts and, when released, they may simply return to the capture area. Yuma myotis bats released 240 miles (400 km) from their roost have found their way back.

Prior to excluding bats, consider partitioning bats off from the area where they are in conflict with humans, and allowing them to roost elsewhere in the structure. An effective partition can be made from construction grade plastic sheeting and wooden battens. Another consideration is to provide an alternate roost site, such as a properly designed and installed bat house mounted close

to one of their exits. Install the bat house before excluding the bats as described below. (See "Bat Houses" and Appendix F for resources on bat house design and placement.)

The following will work to exclude bats from most structures:

Option A – Build bats out: From mid-October to mid-March, when bats should still be hibernating, or after you have made sure no bats are roosting in the attic or other area, seal all potential entry holes (see Fig. 5). Entering the attic during the day may reveal light shining through otherwise unnoticed cracks and holes. Insert pieces of fiberglass insulation or bits of stick in these holes to mark them for repair from the outside.

Large openings can be sealed off with aluminum flashing, wood, or ¼-inch (6 mm) mesh hardware cloth. Small holes around pipes, cracks, and gaps in shakes and tiles can be stuffed with balled-up galvanized window screening, pieces of fiberglass insulation, copper Stuff-it®, or copper or stainless-steel mesh scouring pads (steel wool will quickly corrode after becoming wet). Use weather-stripping, caulk, or expandable foam to seal spaces around doors, windows, and vents, and replace loose boards and roofing materials. Close the damper in the fireplace.

If caulk or expandable foam is used, **apply it early in the day** so that it is set up and no longer sticky if bats inspect the area in the evening. If there are large areas to be foamed, it may be worth purchasing a foam gun of the type used in building construction. Foam is very messy to use, so wear

Figure 5. Common entry points used by bats.
- *Down chimneys and where chimneys and other masonry meet the side of a house*
- *Joints between window frames and house siding*
- *Joints around large exterior beams*
- *At building corners*
- *Where pipes or wires penetrate the ceiling or walls in attics*
- *Between porches or other additions and the main house*
- *At roof edges, ridge caps, soffits, and fascia boards*
- *Where walls meet eaves at the gable ends of an attic*
- *In gaps under shingles. (Bat Conservation International.)*

gloves when applying it, and don't get it on your clothes or skin.

The advantage of caulk over foam is that it comes in a variety of colors and it is easier to apply. Before purchasing, check the label to make sure the caulk can be painted.

Insulation blown into wall spaces may be an effective barrier, but it must be done when bats are absent to avoid trapping them in the fill.

If bats are present, holes can also be blocked over a period of days early in the evening after the bats have left the structure to feed. Do this only from mid-August to mid-October (after the young bats have learned to fly and before cold weather arrives). Another window of opportunity occurs in early spring, before the birthing period in May.

For several days, bat counts should be made as holes are closed, leaving the main exit open. On the night of the final count after the bats have left, the main hole should be plugged to prevent their reentry. The following evening, the plugging should be removed to allow any remaining bats to leave before the exit is sealed.

Option B – Harassment: If bats are present and have to be excluded, persuade them to move to one of their alternate roost sites by creat-

ing an undesirable atmosphere. The time to do this is from mid-August to mid-October, after the young bats have learned to fly and before cold weather arrives. Another window of opportunity occurs in early spring, before the birthing period in May.

Bats don't like to roost under bright, windy, or noisy conditions. Therefore, locate the area where bats are roosting and light the area with a bright light, such as a mechanic's drop-light or trouble light, located away from burnable objects. (Use a fluorescent light to save on electricity and keep the heat level down.) In addition, aim a fan and a loud radio at the bats. Begin the harassment process shortly before dark and keep it in place day and night.

Because bats may move to a dark, protected area, you may need to move the lights and other equipment, or install them in various areas. Putting up sheets of plastic to separate the bats from the rest of the area can be effective, but make sure you don't block the bats' exit or exits.

Commercially available ultrasonic devices **may** be effective if they are placed in a small, confined area with the roosting bats. Since bats can hear high-frequency sounds, these devices, inaudible to humans, supposedly bombard the bat's range with jackhammer-like noise.

Naphthalene flakes or mothballs should not be used to exclude bats. These contain chemicals that can be toxic to humans and other life forms; poisoned bats may fall to the ground where they die slowly and are more likely to come into contact with children or pets.

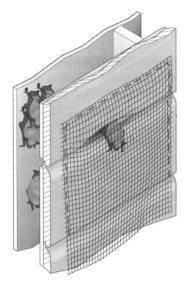

Figure 6. A one-way door allows bats to exit a structure, but prevents them from reentering. Hang a sheet of construction-grade plastic, screen-door material, or light-weight polypropylene netting (⅛ inch mesh) over the exit. Use staples or duct tape to attach the material to the building. The one-way door should extend 18 to 24 inches (46–61 cm) below the bottom edge of the opening. Leave the material loose enough to flop back after each bat exits. (Bat Conservation International.)

If the exclusion process was successful, immediately seal up the exits to prevent bats from reentering. If necessary, install a chimney cover, available from home improvement centers.

Option C – Install exclusion devices: Again, from mid–August to mid–October (after the young bats have learned to fly and before cold weather arrives), or in early spring (before the birthing period in May), identify the exit(s) bats are using. Have friends or family members stationed at the corners of the structure after sunset on a warm calm night. They need to be far enough away to see as much of the structure as possible without having to turn their heads; it takes only a second for a bat to exit and take flight. Note which side of the structure bats are seen from. On subsequent nights, focus your attention there to locate the exit hole. Remember this hole can be as small as ½-inch (12 mm).

Bats often defecate when exiting and reentering a building, so look closely for rice-sized black droppings clinging to the side of the structure. If droppings are observed, the exit hole will be directly above it. (To make sure droppings are new, remove the existing droppings or lay down newspaper over them to see if more droppings appear.) Bat body oils may also discolor a well-used opening.

Seal all entry holes but one using the methods described in Option A.

Exclude bats by covering the one existing entry hole with a device that allows bats to exit the structure, but prevents them from reentering (see Figs. 6–10). Install the exclusion devices during the day and leave them in place for five to seven days (longer during particularly cool or rainy weather).

When bats are using multiple openings to exit and enter, exclusion devices should be placed on each opening, unless you can be sure that all roosting areas used by the bats are connected. If all the roosting areas are connected, all but one or two exit holes can be sealed as described below. Place exclusion devices over the one or two remaining exit holes.

However, if the colony contains a hundred bats or more, which is common, leaving only one exit point can create a "bat log jam." In these cases, some bats might start looking for alternative ways out of the roost area, lead-

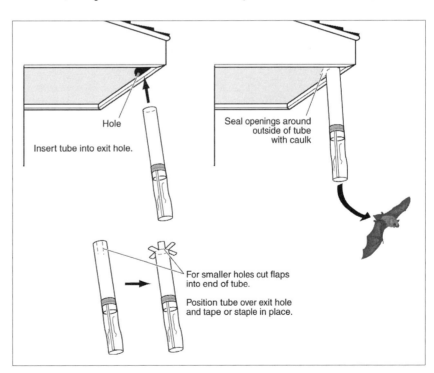

Figure 7. One-way tubes work where one-way doors won't, such as on horizontal surfaces. A flexible pipe or cardboard tube is easy to fit into a crevice or cut to create flaps that can be fit over an opening and be stapled, nailed, or taped to a building (Fig. 9). Do not let the tube project more than ¼-inch (6 mm) into the opening to make sure that bats can easily enter the tube. (Bat Conservation International.)

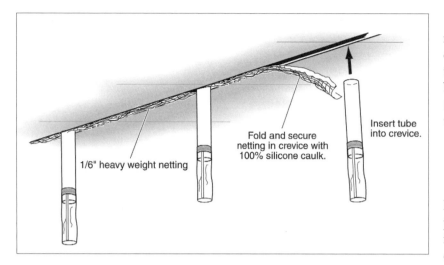

Figure 8. Some areas have lengthy crevices used by bats. Multiple exclusion tubes will need to be placed every few feet along the length of each crevice; spaces between the tubes should be closed with heavyweight netting or other material. The same procedure can be used in lengthy crevices created where flashing has pulled away from a wall. (Bat Conservation International.)

ing to bats finding their way into human-occupied areas. So, always watch to make sure bats are able to exit freely. If they do not appear to be exiting, or appear to be having trouble doing so, open additional exits.

After all bats have been excluded, remove the exclusion devices and immediately seal up the exits to prevent bats from reentering. If necessary, install a chimney cover, available from home improvement centers.

Bats Roosting above Porches and Other Areas

Bats temporarily roost above porches or under overhangs at night to eat large prey, digest, rest, and socialize. In such cases, they may frighten humans, or their droppings may accumulate. Nontoxic aerosol sprays, designed to repel dogs and cats, can prevent bats from night-roosting in these areas.

The spray is applied by day when bats are not present, and is reported to be effective for several months. However, aerosol repellents are not an adequate substitute for excluding bats that are using the area as a day roost, and should never be applied when bats are in a roost.

Mylar balloons or strips of aluminum foil hung from the porch ceiling and allowed to move in the breeze may also discourage bats from roosting in that area (see Chapters 29 and 40 for additional ideas).

Public Health Concerns

Large accumulations of bat droppings may harbor histoplasmosis fungi spores, which when inhaled can result in a lung infection referred to as "histo" (see "Public Health Concerns" in Chapter 29). No histo cases have been reported in the Pacific Northwest, but the recommended precautions should be followed when cleaning or removing large accumulations of bat droppings.

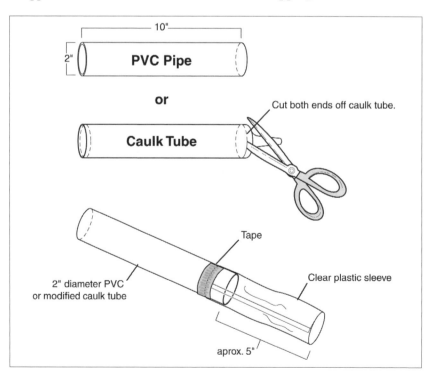

Figure 9. One-way tubes should be at least 2-inches (5 cm) in diameter, 10 inches (25 cm) in length, and have a smooth interior so bats are unable to cling to the inside. One-way tubes can be made from PVC pipe, flexible plastic tubing, empty caulking tubes, or dryer vent hose. To reduce the likelihood of bats reentering, a piece of plastic sheeting can be taped around the exit end of the tube. (Bat Conservation International)

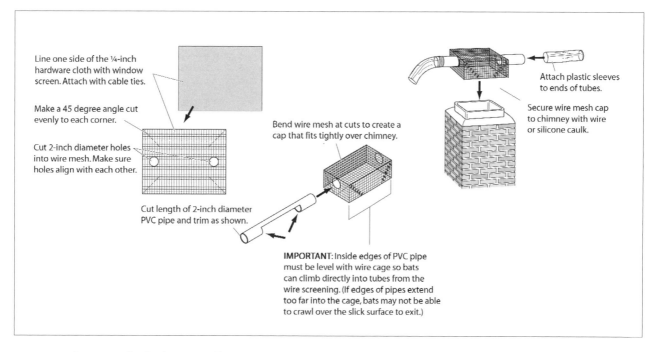

Line one side of the ¼-inch hardware cloth with window screen. Attach with cable ties.

Make a 45 degree angle cut evenly to each corner.

Cut 2-inch diameter holes into wire mesh. Make sure holes align with each other.

Cut length of 2-inch diameter PVC pipe and trim as shown.

Bend wire mesh at cuts to create a cap that fits tightly over chimney.

Attach plastic sleeves to ends of tubes.

Secure wire mesh cap to chimney with wire or silicone caulk.

IMPORTANT: Inside edges of PVC pipe must be level with wire cage so bats can climb directly into tubes from the wire screening. (If edges of pipes extend too far into the cage, bats may not be able to crawl over the slick surface to exit.)

Figure 10. One-way tubes for chimneys. If bats are roosting inside a chimney, construct a wire cage from ¼-inch (6 mm) mesh hardware cloth. Insert a modified section of 2 inch (5 cm) PVC pipe through holes cut in the sides of the wire cage. To further reduce the likelihood of bats reentering, a piece of plastic sheeting can be taped around the exit ends of the tube. (Bat Conservation International.)

Rabies

People are more often concerned about bats because of rabies, a virus that affects the nervous system of all mammals, including humans.

Rabies is spread when the saliva of an infected animal enters another body through a bite or scratch, or makes contact with their eyes, nose, mouth, or a break in the skin. There is little risk of contracting rabies from a bat as long as you exercise caution. People cannot get rabies from touching bat droppings, blood or urine, or fur.

Only 5 to 10 percent of sick, injured, or dead bats tested by the Washington State Department of Health (WDH) have rabies. Because rabid bats are more likely to be sick or injured, they have a higher potential to expose a person to rabies via a bite or some contact. WDH estimates that probably less than 1 percent of the native wild bat population has rabies.

If a bat does contract rabies, it is unlikely to be a threat to humans as long as simple precautions are followed. Most bats infected with rabies become paralyzed and fall to the ground. (**Note:** Young bats also fall to the ground when learning to fly. They may also have hit a window and been stunned, or simply be cold and unable to fly.) This means a person may contract rabies from a bat only if they pick up a sick bat, which then bites the person in self-defense. **Thus, if you do not handle bats, your odds of contracting rabies are extremely small.**

If you think you have been bitten, scratched, or exposed to rabies via a bat:

1. Wash any wound or other area that came into contact with the bat thoroughly with soap and water.
2. Capture or isolate the bat, if you can, without risking further contact (see "Bat Encounters Inside or Outside Your Home" for safe capture techniques). The captured bat will be sent to a laboratory for rabies testing.
3. Call your doctor or local health department. An evaluation of the potential of rabies exposure and the need for follow-up treatment will be done. Arrangements to have the bat tested for rabies, if necessary, will also be made.

People usually know when a bat has bitten them. However, because bats have small teeth and claws, the marks may be difficult to see. Contact your local health department or your doctor in the

What About Rabies and Your Pet?

Once dogs presented the major domestic risk of rabies, but now cats do. Routine rabies vaccination is not as widely practiced with our domestic cats as with dogs. Yet cats often play or hunt in wildlife areas.

All cats should be vaccinated for rabies, even indoor cats. The National Association of State Public Health Veterinarians publishes the Compendium of Animal Rabies Control yearly. These guidelines are clear: An unvaccinated animal that comes in contact with a potentially rabid animal (bats in the Pacific Northwest) that **cannot** be tested should (1) be euthanized; or (2) be held in strict quarantine for six months. Few people are willing to do either.

The message for everyone is, vaccinate your pets! Dogs require vaccination against several diseases. Puppies begin their series of vaccinations at six to eight weeks of age, and annual boosters are necessary to maintain immunity.

following situations, even in the absence of an obvious bite or scratch. In such cases, the bat should be captured for testing:

1. A bat is found in a room with a sleeping person.
2. A bat is found in a room with an unattended child.
3. A bat is found near a child outside.
4. A bat is found in a room with a person under the influence of alcohol or drugs, or who has another sensory or mental impairment.

Legal Status

Many Pacific Northwest bats are currently being studied and may be recommended for protection under the Endangered Species Act. For current legal status and other information, contact your local, state, or provincial wildlife office, or visit their Web site (see Appendix E for contact information).

Oregon: Protected bat species include: Townsend's big-eared bat, pallid bat, fringed myotis, silver-haired bat, western small-footed myotis, long-eared myotis, long-legged myotis, and

the Yuma myotis. All other bat species are not protected and can be removed without a permit.

Property owners, persons lawfully occupying land, or their agents may remove protected bats with a permit issued by Oregon Department of Fish and Wildlife if a bat is causing damage.

Washington: All species of bats are classified as protected wildlife and cannot be hunted, trapped, or killed (WAC 232-12-011). The Department of Fish and Wildlife makes exceptions for bats found in or immediately adjacent to a dwelling or other occupied building. In such cases, these animals may legally be removed and no permit is necessary (WAC 232-12-011).

British Columbia: All bat species are designated as wildlife under the Wildlife Act and cannot be hunted, trapped, or killed unless under permit. A permit signed by a regional manager is required to remove any bats that are considered a threat to public safety (B.C. Reg. 253/200, Section 2(c)(iii)). A person may kill on his or her own property, without a permit, any bat that is a menace to a domestic animal or bird (Wildlife Act, Section 26(2)).

Figure 11. Bats can be caught and released outdoors away from people and pets. (Bat Conservation International.)

Bat Encounters Inside or Outside Your Home

In spring and fall, migrating bats may temporarily roost outside on window screens, fence posts, piles of lumber, and other unlikely places. If a bat is seen roosting outside during daylight hours, leave it alone. It will probably be gone the following morning.

If a bat flies into your home it's probably a juvenile learning to fly, a solitary male following prey, or an adult that has been excluded from its roost. Bats often enter through an open door or window, or by coming down a chimney into an unused fireplace.

If a bat is found inside during the day, confine it to one room. Place a towel under doors to prevent the bat from moving into other parts of the house. Leave the area alone until nightfall.

At nightfall (if you are sure the bat has not been in contact with humans or pets), turn off any lights in the room where the bat is confined, open all doors and windows that lead outside, and stand in the corner. This allows you to watch the bat while staying out of its way. (If you must move around the room, stay as near to the wall as possible.) Be prepared to watch the bat for up to 20 minutes. Normally, the bat will fly around the room to orient itself, and then leave.

If the bat seems to have disappeared but you didn't see it leave, it may be perched somewhere, such as behind a curtain, in hanging clothes, or in a houseplant. The bat will generally choose a high place to roost. Moving these things around with a broomstick may arouse the bat.

If the bat doesn't leave, it can be caught and released outdoors away from people and pets.

Approach the bat slowly and place a container (small box, large glass, Tupperware container, coffee can) over it. Next, gently slide a piece of cereal-box paper or cardboard underneath the bat (be gentle—bats are fragile animals). Using the paper as a cover, take the bat outside. The ideal release procedure is to place the container against a tree, slowly slide the paper away, and then remove the container. Releasing the bat against a tree allows the bat to rest safe from potential predators—like the neighbor's cat.

You may also catch the bat using a pair of leather gloves and a pillowcase. **(Never handle a bat with your bare hands.)** Put your gloved hand inside the pillowcase and gently place it over the bat. Then fold the pillowcase over the bat so it is inside. Take the bat outdoors and safely release it on a rough tree trunk or lightly shake the pillowcase until the bat flies off. In the absence of a container or pillowcase and gloves, a thick towel can be used. Roll the bat up gently and release it outside.

Note: State or provincial wildlife offices do not provide bat removal services, but they can provide names of individuals or companies that do. To find such help yourself, look up "Animal Control," "Wildlife Control," or "Pest Control" in you phone directory.

See "Public Health Concerns" for what to do if a bat is in your home and you think it may have bitten or scratched a human or pet.

CHAPTER 3

Beavers

Beavers *(Castor canadensis)* are the largest living rodents in North America, with adults averaging 40 pounds (18 kg) in weight and measuring more than 3 feet (90 cm) in length, including the tail (Fig. 1). These semiaquatic mammals have webbed hind feet, large incisor teeth, and a broad, flat tail.

Once among the most widely distributed mammals in North America, beavers were eliminated from much of their range in the late 1800s because of unregulated trapping. With a decline in the demand for beaver pelts, and with proper management, they became reestablished in much of their former range and are now common in many areas. The beaver is the national mammal of Canada and the state mammal of Oregon.

Beavers are found where their preferred foods are in good supply—along rivers, and in small streams, lakes, marshes, and even roadside ditches containing adequate year-round water flow. In areas where deep, calm water is not available, beavers that have enough building material available will create ponds by building dams across creeks or other watercourses and impounding water.

Beavers dams create habitat for many other animals and plants. In winter, deer and elk frequent beaver ponds to forage on shrubby plants that grow where beavers cut down trees for food or use to make their dams and lodges. Weasels, raccoons, and herons hunt frogs and other prey along the marshy edges of beaver ponds. Migratory waterbirds use beaver ponds as nesting areas and resting stops during migration. Waterfowl often nest on top of beaver lodges since they offer warmth and protection, especially when lodges are formed in the middle of a pond. The trees that die as a result of rising water levels attract insects, which in turn feed woodpeckers, whose holes later provide homes for other wildlife.

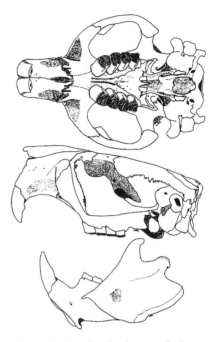

Figure 2. Dorsal and side view of a beaver skull. The beaver's incisors (front teeth) are harder on the front surface than on the back, and so the back wears faster. This creates a sharp edge that enables a beaver to easily cut through wood. (From Verts and Carraway, Land Mammals of Oregon.*)*

Figure 1. Beavers fell trees not only for building material, but also for the leaves, buds, and bark that compose their diet. (From Christensen and Larrison, Mammals of the Pacific Northwest: A Pictorial Introduction.*)*

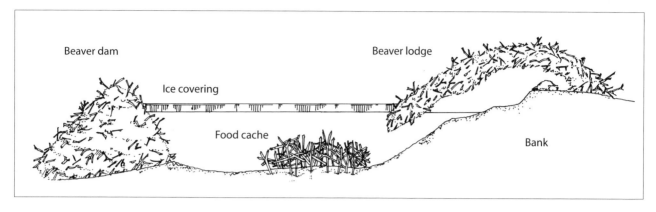

Beaver dam

Ice covering

Beaver lodge

Food cache

Bank

Figure 3. Like many rodents, beavers construct nesting dens for shelter and for protection against predators. These may be burrows in a riverbank or the more familiar lodges built in the water or on the shore (shown here). However, the basic interior design varies little and consists of one or more underwater entrances, a feeding area, a dry nest den, and a source of fresh air. (Drawing by Jenifer Rees.)

Facts about Beavers

Food and Feeding Habitats

- Beavers eat the leaves, inner bark, and twigs of aspen (a favorite food), alder, birch, cottonwood, willow, and other deciduous trees. Beavers also eat shrubs, ferns, aquatic plants, grasses, and crops, including corn and beans.
- Coniferous trees, such as fir and pine, are eaten occasionally; more often, beavers will girdle and kill these trees to encourage the growth of preferred food plants, or use them as dam-building material.
- Beavers have large, sharp, upper and lower incisors, which are used to cut trees and peel bark while eating. The incisors grow their entire lives, but are worn down by grinding them together, tree cutting, and feeding (Fig. 2).
- Fermentation by special intestinal microorganisms allows beavers to digest 30 percent of the cellulose they ingest.
- When the surface of the water is frozen, beavers eat bark and stems from a food "cache" (a safe storage place) they have anchored to the bottom of the

waterway for winter use (Fig. 3). They also swim out under the ice and retrieve the thick roots and stems of aquatic plants, such as pond lilies and cattails.
- Food caches are not found consistently where winters are comparatively mild, such as in the lowlands of western Oregon, Washington, and British Columbia.

Beaver Dams

- Beavers flood areas for protection from predators, for access to their food supply, and to provide underwater entrances to their den. Flooded areas also wet the soil and promote the growth of favored foods.
- Beavers living on water bodies that maintain a constant level (e.g., lakes, large rivers) do not build dams.
- Dams are constructed and maintained with whatever materials are available—wood, stones, mud, and plant parts (Fig. 3). They vary in size from a small accumulation of woody material to structures 10 feet (3 m) high and over 165 feet (50) m wide.

- The feel and sound of flowing water stimulate beavers to build dams; however, they routinely let a leak in a dam flow freely, especially during times of high waters.
- Beavers keep their dams in good repair and will constantly enlarge the dams as the water level increases in their pond. A family of beavers may build and maintain one or several dams in their territory.
- In cold areas, dam maintenance is critical. Dams must be able to hold enough water so the pond won't freeze to the bottom, which would eliminate access to the winter food supply.

Lodges and Bank Dens

- Depending on the type of water body they occupy, beavers construct freestanding lodges or bank dens.
- Lodges and bank dens are used for safety, and a place to rest, stay warm, give birth, and raise young.
- Freestanding lodges are built in areas where the bank or water levels aren't sufficient for a safe bank den.

A Beaver's Tail

The tail of a large beaver may be 15 inches (38 cm) long and 6 inches (15 cm) wide. It is covered with leathery scales and sparse, coarse hairs.

The beaver's tail has important uses both in the water and on land. In the water, the animal uses its flexible tail as a four-way rudder. When diving after being frightened, a beaver loudly slaps the water with its tail; the sound warns all beavers in the vicinity that danger is near, and perhaps serves to frighten potential predators.

On land, the tail acts as a prop when a beaver is sitting or standing upright. It also serves as a counterbalance and support when a beaver is walking on its hind legs while carrying building materials with its teeth, front legs, and paws. Contrary to common belief, beavers do not use their tails to plaster mud on their dams.

The tail stores fat, and because it is nearly hairless, releases body heat, helping the beaver to regulate its body temperature.

- Lodges consist of a mound of branches and logs, plastered with mud. One or more underwater openings lead to tunnels that meet at the center of the mound, where a single chamber is created.
- Bank dens are dug into the banks of streams and large ponds, and beavers may or may not build a lodge over them (Fig. 3). Bank dens may also be located under stumps, logs, or docks.
- One family can have several lodges or bank dens, but will typically use only one den during winter.

Reproduction and Family Structure

- A mated pair of beaver will live together for many years, sometimes for life.
- Beavers breed between January and March, and litters of one to eight kits (average four) are produced between April and June. The number of kits is related to the amount of food available (more food, more kits), and the female's age.
- The female nurses the kits until they are weaned at 10 to 12 weeks of age.
- Most kits remain with the adults until they are almost two years old. (Some leave at 11 months and a few females may stay until they are three years of age.) The kits then go off on their own in search of mates and suitable spots to begin colonies, which may be several miles away.
- Beavers live in colonies that may contain 2 to 12 individuals. The colony is usually made up of the adult breeding pair, the kits of the year, and kits of the previous year or years.

- Populations are limited by habitat availability, and the density will not exceed one colony per ½ mile (0.8 km) under the best of conditions.

Mortality and Longevity

- Because of their size, behavior, and habitat, beaver have few enemies.
- When foraging on shore or migrating overland, beavers are killed by bears, coyotes, bobcats, cougars, wolves, and dogs.
- Other identified causes of death are severe winter weather, winter starvation, disease, water fluctuations and floods, and falling trees.
- Humans remain the major predator of beavers. Historically, beavers have been one of the most commonly trapped furbearers. In Washington, from 1991 to 2000, an annual average of 5,289 beavers were trapped. However, the average for the past three years has dropped to just over 1,000.
- Beavers live 5 to 10 years in the wild.

Viewing Beavers

Beavers are nocturnal, but are occasionally active during the day. They do not hibernate, but are less active during winter, spending most of their time in the lodge or den.

Probably no animal leaves more obvious signs of its presence than the beaver. Freshly cut trees and shrubs, and prominent dams and lodges are sure indicators of their activity.

Look for signs of beavers during the day; look for the animals themselves before sunset or sunrise. Approach a beaver site slowly and downwind. (Beavers have poor eyesight but excellent hearing and sense of smell.) Look for a V-shaped series of ripples on the surface of calm water. A closer view with binoculars may reveal the nostrils, eyes, and ears of a beaver swimming.

If you startle a beaver and it goes underwater, wait quietly in a secluded spot and chances are that it will reemerge within one or two minutes. However, beavers are able to remain underwater for at least 15 minutes by slowing their heart rate.

When seen in the water, beavers are often mistaken for muskrats or nutria (see Chapter 16). Try to get a look at the tail: Beavers have a broad, flat tail that doesn't show behind them when swimming, whereas muskrats and nutria have a thin tail that is either held out of the water or sways back and forth on the water's surface as the animal swims. In addition, a muskrat is the size of a football, while a beaver and a nutria are four or more times that size.

Beavers may also be seen lying in the sun, combing through their thick fur with the naturally split grooming claws on each hind foot.

Beavers stand their ground and should not be closely approached when cornered on land. They face the aggressor, rear up on their hind legs, and hiss or growl loudly before lunging forward to deliver extremely damaging bites.

Harvest Sites

Beavers cut down trees, shrubs, and other available vegetation for food and building materials. Large stumps are pointed, 1 to 2½ feet (30–75 cm) high, and sometimes the tree trunk is still attached.

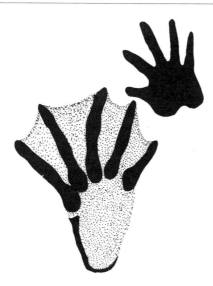

Figure 4. The beaver's front track looks like a small hand, about 3 inches (7.5 cm) long, with five long fingers. Sometimes only three or four fingers are visible. The hind track is 5 to 6½ inches (12.5–16.5 cm) long and 4 to 5 inches (10–12.5 cm) wide at the front, with long toes and webbing between the toes. (From Pandell and Stall, Animal Tracks of the Pacific Northwest.*)*

Tooth marks look like twin grooves, each groove measuring ⅛ inch (3 mm) or more. There will be a pile of wood chips on the ground around the base of recently felled trees. Limbs that are too large to be hauled off are typically stripped of bark over the course of several days. The cut on small wood usually involves a 45-degree cut typical of rodents, but at a larger scale. Branches and twigs under ¾ inch (2 cm) in diameter are generally eaten entirely.

Most harvesting is done within 165 feet (50 m) of the water's edge. In areas with few predators, but a lean food supply, toppled trees and other signs of feeding may be found twice that distance from the den site. Beavers transport woody material even farther through upstream and downstream sites.

By late fall (earlier in cold winter areas), all family members concentrate on repairing and building up dams and the family lodge in preparation for winter. Harvesting is at its most intense level at this time of year.

Slides

Slides are the paths beavers make where they enter and leave the water. They are 15 to 20 inches (38–50 cm) wide, at right angles to the shoreline, and have a slicked-down or muddy appearance.

Channels

Beavers construct channels or canal systems leading to their ponds, using them to float food—such as small, trimmed trees—from cutting sites. Canals are also safe travel ways for swimming instead of walking. With receding water levels during summer, beaver activity shifts toward building and maintaining channels to access new food supplies. Channels often look man-made, have soft, muddy bottoms, and are filled with 15 to 25 inches (38–62 cm) of water.

Food Storage Sites

Beavers that live in cold climates store branches of food trees and shrubs for winter use by shoving them into the mud at the bottom of ponds or streams near the entrance to their bank den or lodge.

Tracks

Beaver tracks can be obvious if the ground is suitable for leaving tracks (Fig. 4).

Droppings

Beaver droppings are seldom found on land; those that are will commonly be found in the early morning at the water's edge.

Individual beaver droppings are usually cylindrical, up to 2½

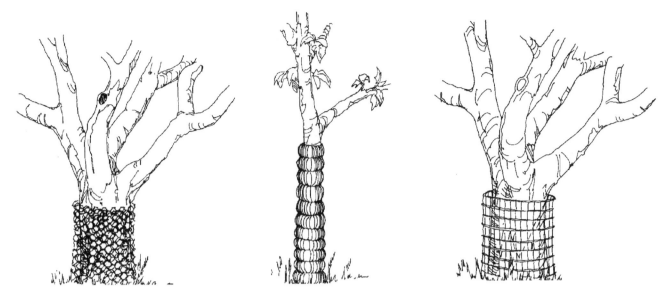

Figure 5. Various barriers can be used to protect plants from beaver damage. All plants should be protected to at least 3 feet (90 cm) above ground—or the snow line—and inspected regularly. (Drawing by Jenifer Rees.)

inches (6.4 cm) long (sometimes shorter), and look as if they were formed of compressed sawdust. The diameter is an indication of the animal's size, with 1 inch (2.5 cm) being average for adults. The color of fresh deposits is dark brown, with lighter-colored bits of undigested wood, all turning pale with age.

Sounds

In order to warn each other of danger, beavers slap their tails against the water, creating a loud splash. Sounds also include whining (noises made by kits), a breathy greeting noise, and loud blowing when upset.

Preventing Conflicts

Despite an appreciation for beavers and our best intentions to live with them, beavers can become a problem if their eating habits, and dam or den building activity, flood or damage property.

Before beginning any beaver-control action, assess the beaver problem fairly and objectively. Are beaver really causing damage or creating hardship requiring con-

trol action? The very presence of beavers is often seen as a problem when, in fact, the beavers are causing no harm. You should also determine the type of damage or problem the animals are causing, and then match the most appropriate and cost-effective controls to the situation.

Once you have decided to control beaver damage, you have three control options: prevention, beaver translocation, or lethal control.

To prevent conflicts or remedy existing problems:

Choose plants carefully. Plant areas with spruce, elderberry, cascara, osoberry (Indian plum), ninebark, and twinberry, because they are not the beavers' preferred food plants. Plant aspen, cottonwood, willow, spirea, and red-twig dogwood because, once their roots are well established, the upper parts of the plants often resprout after being eaten.

Note: Beavers do use plants as construction materials that they might not eat.

Install barriers. The trunks of individual large trees can be loosely

wrapped with 3 foot (90 cm) high, galvanized welded wire fencing, hardware cloth, or multiple layers of chicken wire (Fig. 5). The barriers can be painted to make them less noticeable. Welded wire fencing coated with green vinyl that helps the fencing blend in is also available.

Lengths of corrugated plastic drainpipe can be attached around the trunks of narrow-diameter trees (Fig. 5).

Note: Dark-colored pipe can burn trunks in full sun; wider-diameter pipe or pipe with holes in it may prevent overheating problems.

Painting tree trunks with a sand and paint mix (⅔ cup masonry grade sand per quart of latex paint) has proven somewhat effective at protecting trees from beaver damage.

Note: Preventing access to food sources may force beavers to eat other nearby plants, including roses and other ornamentals.

Surround groups of trees and shrubs with 3-foot (90 cm) high barriers made of galvanized, welded wire fencing or other sturdy material (Fig. 6). (A beaver's

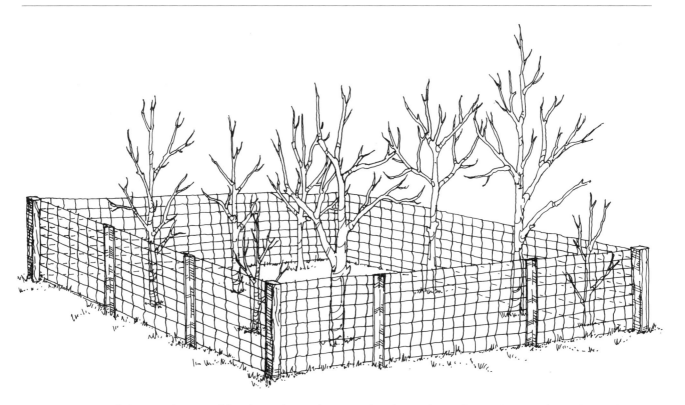

Figure 6. Groups of plants can be protected from beaver damage by surrounding them with wire fencing. (Drawing by Jenifer Rees.)

weight will pull down chicken wire and similar lightweight material.) Stake the barriers to prevent beavers from pushing them to the side or entering from underneath.

An electric fence with two hot wires suspended 8 and 12 inches (20 and 30 cm) off the ground is also effective at protecting groups of plants. See "Electric Fences" in Chapter 7 for information on electric fences.

Protect large areas that border beaver habitat by installing 4-foot (1.25 m) high field fencing. Keep the bottom of the fence flush to the ground to prevent beavers from entering underneath.

Apply repellents. Commercial taste and odor repellents have provided mixed results, perhaps because they need to be reapplied often, particularly in moist weather. Taste and odor repellents are most effective when applied at the first

sign of damage, when other food is available, and during the dry season. Two repellents that have had some success are Big Game Repellent® and Plant-skydd®. (See "Repellents" in Chapter 7 for information.)

Control the height of water behind a beaver dam to prevent flooding. It may be legal to make a small change in the depth of a beaver pond by installing a flow device at the intended depth, extending upstream and downstream of the dam. This keeps the rise in the water level at a minimum by using one or more plastic pipes to continually drain the pond area. For leveling systems to work properly, you will have to have at least 3 feet (90 cm) of water in the pond area for the beaver to stay.

The flow device can be constructed from plastic pipe measuring 4 to 12 inches (10–30 cm) in diameter, depending on the volume of water in the stream. The

end extending upstream from the dam must be baffled to prevent beavers from damming or blocking the pipe. (See Appendix H for construction details.)

Prevent beavers from plugging culverts. To a beaver, a culvert probably looks like a hole in an otherwise fine dam. When they plug the hole, a flooded road can result. However, V-shaped, semicircular, or trapezoidal fences of woven wire mesh can prevent culverts from being plugged. Large flow-control devices that include a solid framework can be covered and used as a deck or wildlife viewing spot. (See Appendix G for construction details.)

Note: Installation of flow-control devices is complicated and generally requires a permit (see "Legal Status"). Call your local Fish and Wildlife office for design assistance and permit information.

Dam removal. It is pointless to destroy a beaver dam because beavers (frequently many at one time) often begin rebuilding them immediately after they are removed. Furthermore, it may be illegal to remove a beaver dam without a permit (see "Legal Status").

Live Trapping

Live trapping and moving (translocating) beavers elsewhere is often cost-prohibitive. Also, research has shown that beavers seldom survive relocation, and those that do often move great distances from the release site. However, in urban areas where lethal trapping may be illegal or unsafe, live trapping and euthanising beavers, or moving them elsewhere, may be the only alternative. If you are moving beavers to a different site, particularly a site where other beavers are not already present, there must be cooperation between adjacent landowners and local wildlife officials. A cooperative evaluation of existing habitat quality and potential adverse beaver activity is also very important. (Expect beaver to cut and use a large number of trees for dam construction during the first year or two.)

To help ensure the survival of beavers, the gradient of the watercourse at the release site should be less than 3 percent, and the site should have adequate food supply. Move beavers during their principal dam-building period, August to October. This will allow them time to gather a food cache, but limit their time to explore before having to settle in for the coming of winter. It may be helpful to provide beavers with a pickup-truck load of aspen or other trees to use as building material at or near the release site. This may encourage the beavers to stay nearby.

See Appendix A for important information on trapping and moving beavers.

Lethal Control

Lethal control may become necessary when all efforts to dissuade problem beavers fail. Removing beavers is rarely a lasting solution since survivors have larger litters, and others will resettle good habitats.

Trapping

Lethal trapping has traditionally been the primary form of controlling beaver damage. If you feel you need to have a beaver trapped, private individuals who work directly with property owners on a fee basis to resolve problem beaver situations can be hired. Where the use of traditional beaver traps and the sale of beaver pelts is legal, a local trapper may trap beavers at no cost. Call your local Fish and Wildlife office for contact information.

Note: State or provincial wildlife offices do not provide animal removal services.

See Appendix A for additional information on trapping beavers.

Shooting

Shooting beaver requires skilled marksmanship, but can be an effective control technique where it is safe. For safety considerations, shooting is generally limited to rural situations and is considered too hazardous in more populated areas, even if legal.

Public Health Concerns

Wild animals, including rabbits, hares, voles, skunks, muskrats, and beavers, can be infected with the bacterial disease **tularemia.** Tularemia is fatal to animals and is transmitted to them by ticks, biting flies, and via contaminated water. Animals with this disease may be sluggish, unable to run when disturbed, or appear tame.

Tularemia may be transmitted to humans if they drink contaminated water, eat undercooked, infected meat, or allow an open cut to contact an infected animal. The most common source of tularemia for humans is to be cut or nicked by a knife when skinning or gutting an infected animal. Humans can also get this disease via a tick bite, a biting fly, ingestion of contaminated water, or by inhaling dust from soil contaminated with the bacteria.

A human who contracts tularemia commonly has a high temperature, headache, body ache, nausea, and sweats. A mild case may be confused with the flu and ignored. Humans can be easily treated with antibiotics.

Beavers are among the few animals that regularly defecate in water, and their droppings (like those of humans and other mammals) may cause a flu-like infection when contaminated water is ingested. The technical name for this illness is **"giardiasis."** It is more commonly referred to as "giardia"—derived from *giardia,* the single-cell protozoan that causes the disease. Another popular term, **"beaver fever,"** may be a misnomer. It has never been demonstrated that the type of giardia beavers carry causes giardiasis in humans. Giardia has been found in many animal species, including pets, wildlife, and livestock.

Legal Status

Because beavers' legal status, trapping restrictions, and other information change, contact your local, state, or provincial wildlife office, or visit their Web site for updates. (See Appendix E for contact information and where to access the state and provincial laws mentioned below.)

Oregon: The beaver is classified as a furbearer. It is protected, and trapping and hunting regulations apply. Property owners, persons lawfully occupying land, or their agents may trap beavers at any time with a permit issued by the Oregon Department of Fish and Wildlife if beavers are causing property damage.

There are no restrictions preventing private landowners from removing beaver dams on their property, unless their activities are under the jurisdiction of the State Forest Practices Act (OAR 629-660-050). Under the Forest Practices Act, beaver dams cannot be removed without prior approval of the State Forester, except for road maintenance if the beaver dam is

within 25 feet (7.62 m) of a culvert. For further information, contact the Oregon Department of Forestry.

Washington: The beaver is classified as a furbearer (WAC 232-12-007). A trapping license and open season are required to trap beaver.

The owner, the owner's immediate family, an employee, or a tenant of property may shoot or trap a beaver on that property if a threat to crops exists (RCW 77.36.030). In such cases, no special trapping permit is necessary for the use of live traps. However, a special trapping permit is required for the use of all traps other than live traps (RCW 77.15.192, 77.15.194; WAC 232-12-142). There are no exceptions for emergencies and no provisions for verbal approval. All special trapping permit applications must be in writing on a form available from the Department of Fish and Wildlife (WDFW).

It is unlawful to release a beaver anywhere within the state, other than on the property where it was legally trapped, without a permit to do so (RCW 77.15.250; WAC 232-12-271).

Any person, organization, or government agency wishing to remove or modify a beaver dam

must have a Hydraulic Project Approval (HPA)—a permit issued by WDFW for work that will use, obstruct, change, or divert the bed or flow of state waters (RCW 77.55). A permit application can be obtained from your local WDFW office or from the Department's Web site (www.wa.gov/wdfw).

In emergency situations (when an immediate threat to property or life exists), verbal approval from WDFW can be obtained for work necessary to solve the problem. A 24-hour hotline (360) 902-2537 is available for emergency calls during non-working hours. During normal hours, contact your nearest WDFW office.

British Columbia: Beavers are designated as furbearing animals. They may be legally trapped under license during open seasons as defined by regulation and there is no bag limit. Beavers that are involved in property damage may not be captured or killed, outside of general open seasons, unless persons first receive a permit from the regional Fish and Wildlife Office. It is also an offence for a person, other than a licensed trapper, to disturb, molest, or destroy a beaver house or den or beaver dam without lawful authority.

CHAPTER 4

Black Bears and Grizzly Bears

American black bears (*Ursus americanus,* Fig. 1) are the most common and widely distributed bears in North America. In the Pacific Northwest, black bears live in a diverse array of forested habitats, from coastal rainforests to the dry woodlands of the Cascades' eastern slopes. In general, black bears are strongly associated with forest cover, but they do occasionally use relatively open country, such as clearcuts and the fringes of other open habitat.

Grizzly bears (*Ursus arctos,* Fig. 2) once had a distribution similar to black bears throughout the Pacific Northwest. However, grizzly bears more commonly occupied open environments. Grizzlies were eradicated from Oregon and are rare in Washington, and occupy less than one percent of their former distribution in the contiguous 48 states. They still inhabit most of the mainland of British Columbia, except the coastal islands.

In the Pacific Northwest, bear populations have been impacted by habitat loss from development and other human uses and by direct removal of bears. In the past, unregulated hunting largely eliminated grizzlies from their historical ranges. Black bears have fared better, largely because they are more compatible with dense human populations. The statewide black bear population in Washington likely ranges between 25,000 and 30,000 animals.

As human populations encroach on bear habitat, people and bears have greater chances of encountering each other. Bears usually avoid people, but when they do come into close proximity of each other, the bear's strength and surprising speed make it potentially dangerous. Most confrontations with bears are the result of a surprise encounter at close range. All bears should be given plenty of respect and room to retreat without feeling threatened.

This chapter focuses on the black bear, which is more common. However, where noted, information is also applicable to grizzly bears.

Figure 1. The black bear (shown here) has a flatter facial profile than the dished or concave profile of the grizzly bear. (Oregon Department of Fish and Wildlife.)

Facts about Black Bears
Food and Feeding Behavior

- Black bears are omnivores. They eat both plants and animals; however, their diet consists mostly of vegetation.
- In the spring, black bear diets consist mostly of herbaceous plants, from emerging grasses and sedges to horsetail and various flowering plants.
- In summer, bears typically add ants, bees, grubs, and a host of later emerging plants to their diets.
- During late summer and fall, bears typically shift their diets toward berries, nuts, and acorns, but they still may consume a variety of plants.
- Fall is a critical season for black bears and they commonly acquire most of their annual fat accumulation at this time. Bears may forage up to 20 hours a day during fall, increasing their body weight by 35 percent in preparation for winter.
- Typically, a small proportion of the black bear's annual diet is made up of animal matter, including insects, mice, voles, ground squirrels, fawns and elk calves, eggs, carrion (animal carcasses), and fish, but their availability varies and is often unpredictable. An occasional bear may take livestock.
- Black bears have adequate senses of sight and hearing, but their keen sense of smell and innate curiosity make them skilled scavengers. They consume carrion when they can find it, and are notorious for taking advantage of human irresponsibility with food. Bears will eat anything that smells appealing and that could prepare them for

Figure 2. The muscles on a grizzly bear's back often appear as a distinct shoulder hump, which black bears lack. (From Larrison, Mammals of the Northwest.*)*

their long winter sleep. See suggestions in "Preventing Conflicts" to avoid placing either the bear or yourself at risk.

- Black bears move in response to the seasonal availability of food, roaming constantly throughout their home range.

Den Sites and Resting Sites

- Black bears den during the winter months (typically from mid-October into April) when food is scarce and the weather turns harsh.
- Denning black bears enter a state of torpor, a modified form of hibernation. This drowsy condition allows bears to defend themselves (and their cubs) more effectively should a predator visit the den.
- Bears do not urinate or defecate during denning—they recycle their waste into proteins and other nutrients. By not defe-

cating, bears keep their dens essentially scent-free, protecting them from potential predators like cougars.

- Black bears in coastal areas of the Pacific Northwest may remain active throughout the winter, except for pregnant females, which den to give birth to cubs.
- Black bears can take up residence in small dens, some scarcely bigger than a garbage can. Den sites include tree cavities, hollow logs, small caves, and areas beneath large roots, stumps, logs, and rural buildings. They'll occasionally excavate a den in the side of a hill near shrubs or other cover.
- Summer beds are merely concealed places scratched in the ground among dense vegetation, by a rock, or under the branches of a fallen tree. Young bears rest in trees (Fig. 3).

Reproduction and Family Structure

- Female black bears breed for the first time at 3½ to 5½ years of age. Mating takes place in June and July.
- Males compete for the right to breed, and breeding fights between males may be intense. Older males frequently have extensive scars on their heads and necks from fights in previous breeding seasons.
- Following a gestation period of about seven months, females normally give birth to one or two cubs in the winter den during January or February. Females have one litter every other year.
- Bears have a reproductive pattern known as delayed implantation. Following fertilization in early summer, a bear's embryo goes dormant, free-floating in the uterus. After the female dens in late fall, the embryo implants in the uterine wall and development of the fetus proceeds rapidly. Although the total gestation time is approximately seven months, the actual developmental period for the bear fetus is less than three months.
- At six months, cubs are able to locate food, but generally remain with their mother for over a year—usually denning with her during their second winter.
- Parental care is solely the responsibility of females; males sometimes kill and eat cubs.

Differences Between Black Bears and Grizzly Bears

The distinguishing characteristics of black bears and grizzly bears are their **shoulders, facial profiles, and claws:**

- Grizzly bears have well-developed shoulder muscles for digging and turning over rocks and boulders when foraging for food, and for bursts of speed needed to capture moose and caribou calves and deer fawns. These muscles often appear as a distinct shoulder hump, which black bears lack (Fig. 2).
- The facial profiles of grizzly bears are dished or concave, whereas black bears have a flatter facial profile from their forehead to the tip of their nose (Fig. 1).
- The front claws on grizzly bears, as long as a human finger, are relatively straight, and are adapted for digging. Black bear claws are much shorter—seldom more than 1½ inches (4 cm) long—are more curved than grizzly claws, and are adapted for climbing trees and tearing logs apart when foraging for insects (Figs. 5, 7).

Color cannot be used for reliable identification, as both species show considerable variation. Grizzly bears range from very blonde, through various shades of brown, to almost black. Although many grizzlies show the distinct silver-tipped hairs from which they derive their name, many do not. Black bears can be white (very rare), blonde, cinnamon, dark brown (most common), or black.

Size is also not a reliable indicator of species. Young grizzlies can be smaller than mature black bears. Most people overestimate the size of grizzlies. A typical adult female grizzly bear weighs 200 to 350 pounds (90–160 kg), and mature adult males weigh 400 to 700 pounds (180–315 kg). An adult male black bear can easily weigh as much as a female grizzly, and more than a young or subadult grizzly of either sex. At maturity, most female black bears will weigh 120 to 200 pounds (45–90 kg), whereas mature males will commonly weigh 300 to 500 pounds (135–225 kg). However, the largest wild male black bear on record weighed a whopping 880 pounds (360 kg).

Figure 3. Young black bears sometimes rest in trees (Oregon Department of Fish and Wildlife).

Mortality and Longevity

- Other than humans and grizzly bears, black bears have few predators—cougars, bobcats, and coyotes attack cubs if given the opportunity. Male bears may eat cubs.
- In the year 2000, hunters harvested 977 black bears in Oregon and 1,148 in Washington (see Table 1).
- Female black bears have the potential to live into their mid-20s. Male black bears do not typically live as long, rarely attaining 20 years of age.

Table 1. Washington black bear harvest, hunter effort, and median black bear age information, 1990–2001

Year	Sex of Bears		Total Harvest	Number of Hunters	Percent Success	Median Age of Bears	
	Male	Female				Males	Female
1990	NA	NA	NA	NA	NA	2.5	4.5
1991	876	503	1,379	10,839	13%	3.5	4.5
1992	921	521	1,442	13,642	11%	4.5	4.5
1993	986	521	1,507	12,179	12%	3.5	5.5
1994	654	419	1,073	11,530	9%	3.5	4.5
1995	850	368	1,218	11,985	10%	3.5	4.5
1996	951	359	1,310	12,868	10%	4.5	5.5
1997	546	298	844	11,060	8%	4.5	5.5
1998	1,157	645	1,802	20,891	9%	4.5	5.5
1999	757	349	1,106	37,033	3%	4.5	5.5
2000	777	371	1,148	37,401	3%	3.5	5.5
2001	919	512	1,431	25,141	6%	3.5	4.5

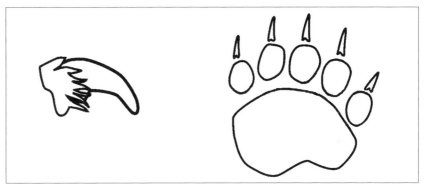

Figure 5. Black bear claws are much shorter, are more curved than grizzly claws (Fig. 7), and are adapted for climbing trees and tearing logs apart when foraging for insects. (Washington Department of Fish and Wildlife.)

Figure 4. The hind feet of an adult black bear average 7 to 9 inches (18–23 cm) long by 3 to 5 inches (8–13 cm) wide; the front feet are 4 to 5 inches (10–13 cm) long by nearly the same width. (Drawing by Kim A. Cabrera.)

Viewing Black Bears

Except for females with cubs, black bears are usually solitary animals. Depending on their food supply, they move about during the day or night. In late summer and fall, feeding keeps them active throughout the day so they can gain the weight needed for winter. When bears find a human food source, their schedule may change. If they are receiving handouts they can be most active at midday; if they are feeding at dumps or trash cans, they become active at night.

Both black bears and grizzly bears should be treated with respect and safely observed from a distance of at least 100 yards (90 m). This is especially important with females accompanied by offspring, as mother bears are very protective of their young.

Tracks

All black bear prints usually show five digits (Figs. 4, 5). The toes form a rough semicircle in front of each foot, with the middle toe being the longest. Front foot tracks have small footpads, whereas hind foot tracks characteristically show an extended footpad,

Figure 6. The hind feet of an adult grizzly bear average 9 to 10 inches (23–25 cm) long by 5 to 6 inches (13–15 cm) wide; the front feet are 5 to 6 inches (13–15 cm) long by nearly the same width. (Drawing by Kim A. Cabrera.)

resembling a human foot. The claw marks are about ½ inch (1.3 cm) in front of the toe pads, but often the claw marks do not show in a track.

Grizzly bear footprints are generally larger than black bear tracks and have claw marks up to 4 inches (10 cm) in front of the toe marks (Figs. 6, 7).

Droppings

When plants, insects, and animal carcasses make up most of a bear's diet, its droppings are cylindrical and typically deposited in a coiled form, sometimes in individual segments. Segments are 2 to 3 inches (5–7.6 cm) long and 1¼ to 1½ inches (3.2–3.8 cm) in diameter. Bits of hair, fur, bone, insect parts, and plant fibers distinguish these droppings from human feces, as does the large size of the deposit. Color ranges from dark brown to black, and when

grasses are being heavily eaten droppings are often green. When fruits and berries are in season, droppings assume a moist, "cow-pie" form and seeds are visible.

Bear Trees

Black bears commonly leave a variety of marks on trees. Because young bears often climb trees, trees in high bear density habitats will show the telltale claw marks and hairs indicating that a bear has previously climbed the tree.

On young conifers, such as spruce and fir trees, bears will rip strips of bark off with their teeth to reach insects or the sweet-tasting, nutrient-laden sap found inside (Fig. 8). The bear's teeth leave long vertical grooves in the sapwood and large strips of bark are found around the bases of trees they peel. These marks are typically made from April to July, but the results may be seen all year. This foraging activity is common in tree plantations where large stands of trees are similarly aged and of a single species.

A bear may also rub its back against a tree or other object. Rubbing is a favorite summer pastime among black bears, reliev-

ing the torment of parasites and loosening their thick, matted winter coat. Good scratching trees may be used repeatedly for several years, and are easily identified by the large amounts of long black or brown fur caught in the bark and sap. Rough-barked trees often serve as rubbing posts.

It has been debated whether bears mark trees to convey social information akin to territorial marking in other carnivores. Such marks are most easily seen on smooth-barked species of trees—alder, aspen, birch, and white pine—on which tooth and claw marks will contrast most visibly, but any live or dead standing trees may be heavily chewed. Human structures such as utility poles, footbridges, and even outbuildings may also be chewed.

Feeding Areas

Rotting logs and stumps are commonly turned over and torn apart to get at fat-rich grubs, ants, termites, worms, and spiders. A bear will also knock the top of an anthill or beehive off to get to the insects.

Black bears may break off entire limbs of fruiting trees, such as apple and chokecherry, to reach

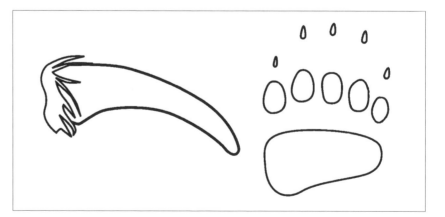

Figure 7. The front claws on grizzly bears, as long as a human finger, are relatively straight, and are adapted for digging. (Washington Department of Fish and Wildlife.)

Figure 8. Marks on trees made by black bears vary from claw marks left by climbing to peeling and biting left when larger bears (generally females) feed on insects and sap found under the bark. (Oregon Department of Fish and Wildlife.)

the fruit. Huckleberries and other fruiting shrubs may show signs of being crushed under a bear's feet. Bears may also dig for the starchy roots of some plants, to excavate seed caches of squirrels and mice, and to capture mice, voles, and ground squirrels. Evidence of digging ranges from well-defined holes to large areas that appear to have been rototilled.

Bear Encounters

Bears tend to avoid humans. However, *human-habituated bears* are bears that, because of prolonged exposure to people, have lost their natural fear or wariness around people. *Human-food-conditioned bears* are those that associate people with food. Such

bears can become aggressive in their pursuit of a meal. The most unpredictable type of bear is one that feels threatened.

Do everything you can to avoid an encounter with any bear. Prevention is the best advice.

If you are recreating in bear country, always remember: Never travel alone, keep small children near you at all times, and always make your presence known— simply talking will do the trick. Most experts recommend carrying pepper spray when recreating in areas of high bear density. An EPA-recommended pepper spray that has a pepper content between 1.3 and 2 percent can be an effective deterrent to an aggressive bear if it is sprayed directly into the bear's face within 6 to 10 feet (2–3 m).

Should you come in close contact with a bear, here are some tips:

- Stop, remain calm, and assess the situation. If the bear seems unaware of you, move away quietly when it's not looking in your direction. Continue to observe the animal as you retreat, watching for changes in its behavior.
- If a bear walks toward you, identify yourself as a human by standing up, waving your hands above your head, and talking to the bear in a low voice. (Don't use the word bear because a human-food-conditioned bear might associate "bear" with food . . . people feeding bears often say "here bear." Don't throw anything at the bear and avoid direct eye contact, which the bear could interpret as a threat or a challenge.
- If you cannot safely move away from the bear or the bear con-

tinues toward you, scare it away by clapping your hands, stomping your feet, yelling, and staring the animal in the eyes. If you are in a group, stand shoulder-to-shoulder and raise and wave your arms to appear intimidating. The more it persists the more aggressive your response should be. If you have pepper spray, use it.

- **Don't run from the bear,** unless, of course, safety is very near and you are absolutely certain you can reach it (knowing that bears can run about 35 mph).
- Climbing a tree is only a good evasive maneuver when dealing with a grizzly, but you must climb out of the bear's reach, which may mean you have to get 10 feet (3 m) or more off of the ground. Grizzlies rarely climb trees, but will attempt to knock one down. Climbing a tree is generally not recommended as an escape from an aggressive black bear, as black bears are adept climbers and may follow you up a tree.

Bear Attacks

In the unlikely event a **black bear attacks you** (where actual contact is made), **fight back** aggressively using your hands, feet, legs, and any object you can reach. Aim for the eyes or spray pepper spray into the bear's face.

If the attack continues, lie on your stomach with your legs spread slightly apart and your hands locked behind your head. This position protects your face and neck, which are bear targets. If the bear flips you over, continue rolling until you are back in that position. Don't struggle or cry out; that will cause the bear to continue. A bear will stop attacking when it thinks you are no longer a threat. Stay still and wait for the bear to leave before moving.

In the unlikely event a **grizzly bear attacks you** (where actual contact is made), **don't fight back.** Lie on your stomach with your legs spread slightly apart and your hands locked behind your head. This position protects your face and neck, which are primary bear targets. If the bear flips you over, continue rolling until you are back in that position. Don't struggle or cry out; that will cause the bear to continue. A grizzly bear will stop attacking when it thinks you are no longer a threat. Stay still and wait for the bear to leave before moving.

Preventing Conflicts

State and provincial wildlife offices throughout the Pacific Northwest receive thousands of black bear complaints each year regarding urban sightings, property damage, attacks on livestock, and bear/human confrontations.

Conflicts arise for several reasons: Because young bears (especially young males) are not tolerated by adult bears, they sometimes wander into areas occupied by humans. Females with cubs may also be forced to feed near human settlements because adult male bears may kill cubs. In addition, food may be scarce in some years—a late spring and poor forage conditions may be followed by a poor huckleberry crop, causing bears to seek food where they ordinarily would not.

If you live in areas where black bears are seen, use the following management strategies

around your property to prevent conflicts:

Don't feed bears. Often people leave food out for bears so they can take pictures of them or show them to visiting friends. Over 90 percent of bear/human conflicts result from bears being conditioned to associate food with humans. A wild bear can become permanently food-conditioned after only one handout experience. The sad reality is that these bears will likely die, being killed by someone protecting their property, or by a wildlife manager having to remove a potentially dangerous bear.

Manage your garbage. Bears will expend a great amount of time and energy digging under, breaking down, or crawling over barriers to get food, including garbage. If you have a pickup service, put garbage out shortly before the truck arrives—not the night before. If you're leaving several days before pickup, haul your garbage to a dump. If necessary, frequently haul your garbage to a dumpsite to avoid odors.

Keep garbage cans with tight-fitting lids in a shed, garage, or fenced area. Spray garbage cans and dumpsters regularly with disinfectants to reduce odors. Keep fish parts and meat waste in your freezer until they can be disposed of properly.

If bears are common in your area, consider investing in a commercially available bear-proof garbage container. Ask a local public park about availability or search the Internet for vendors.

Only grass and hedge clippings should be placed in outdoor compost bins. To reduce odors, cover compost with soil or lime.

Remove other attractants. Remove bird feeders (suet and seed feeders), which allow residue to build up on the ground below them, from early May through late October. Bring in hummingbird feeders at night. Harvest orchard fruit from trees regularly (rotting fruit left on the ground is a powerful bear attractant). If you have bear problems and do not use your fruit trees, consider removing them. If dogs and cats are fed outside, feed them in the morning or mid-afternoon, and clean up leftovers and spilled food before nightfall. Clean barbecue grills after each use. Wash the grill or burn off smells, food residue, and grease; store the equipment in a shed or garage and keep the door closed. If you can smell your barbecue then it is not clean enough. Avoid the use of outdoor refrigerators—they will attract bears.

Protect livestock and bees. Place livestock pens and beehives at least 150 feet (46 m) away from wooded areas and protective cover. Confine livestock in buildings and pens, especially during lambing or calving seasons. Livestock food also attracts bears and must be kept in a secure barn or shed behind closed doors. If bears are allowed access to livestock food, they may learn to feed on livestock. Immediately bury any carcasses or remove them from the site.

Install fences and other barriers. Electric fencing can be used where raids on orchards, livestock, beehives, and other areas are frequent (Fig. 9). Electric fencing only works, however, if it is operating **before** conflicts occur. Bears will go right through electric fencing once they are food-conditioned and know that food is available.

Bears can be lured into licking or sniffing the electrified wire by rubbing molasses, bacon grease, or peanut butter on the fence. (See "Preventing Conflicts" in Chapter 7 for additional information on electric fences.)

Traditional wire fencing can also be used as a barrier. Use heavy chain-link or woven-wire fencing at least 6 feet (1.85 m) high. Install 24-inch (60 cm) long wood or metal bar extensions at an outward angle to the top of the fence with two strands of barbed wire running on top. If necessary, a 2-foot (60 cm) wide underground apron of chain-link fencing or steel mesh can be staked down and attached to the fence to keep bears from digging under the fence (Fig. 9).

Bears can be dissuaded from climbing a tree by attaching 4-foot (1.25 m) long, 1 x 4 inch (2.5 x 15 cm) boards with 2-inch (5 cm) long wood screws screwed all the way through them every 6 inches (15 cm). (To prevent the board from splitting, drill pilot holes.) Attach at least four boards around the trunk of the tree using strong wire.

Use temporary scare tactics. Bears can be temporarily frightened from a building, livestock corral, orchard, and similar places by the use of a night light or strobe light hooked up to a motion detector on a tripod, loud music, or exploder cannons. The location of frightening devices should be changed every other day. Even so, over a period of time, bears will become accustomed to them. At this point, scare devices are ineffective and human safety can become a concern. (See "Preventing Conflicts" in Chap-

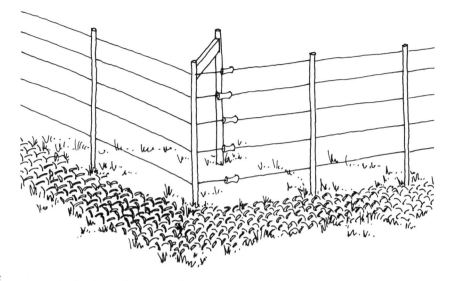

Figure 9. An electric fence designed to keep bears out of an area. A five-wire electric fence has been effective at keeping adult bears and their cubs out. If necessary, a 2-foot (60 cm) wide underground apron of chain-link fencing or steel mesh can be staked down and attached to the fence to keep bears from digging under the fence. If wood or other heavy-duty corner stakes are not used, the corner posts will need to be carefully braced. (Drawing by Jenifer Rees.)

ter 27 for additional information on scare tactics.)

Professional Assistance

Wildlife offices throughout the Pacific Northwest respond to bear sightings when there is a threat to public safety or property. A sighting or the presence of a bear does not constitute a threat to property or public safety. Typically, no attempt will be made by a wildlife agency staff to remove, relocate, or destroy the animal.

Problem bears can be live-trapped by specially trained wildlife professionals and moved to more remote areas; however, such removals are expensive, time consuming, and seldom effective. (Once a bear has tasted human food or garbage, it will remember the source and return again and again—bears have been known to return over 100 miles to a human food source after having been relocated.) Using tranquilizing drugs on bears to facilitate removal is not without risks to bears and humans.

When other methods have failed, lethal removal of problem animals may be the only alternative.

Contact your local wildlife office for additional information and, in the case of an immediate emergency, call 911 or any local law enforcement office, such as the state patrol.

Public Health Concerns

Bears are not considered a significant source of infectious diseases that can be transmitted to humans or domestic animals. However, humans can become infected with trichinosis by eating undercooked bear meat.

Trichinosis is an infection caused by the worm *Trichinella spiralis*, which is sometimes found in bear meat, other wild game, and pork. The disease, which may begin with flu-like symptoms, is not spread person-to-person, and most people with trichinosis recover fully with treatment. To prevent becoming infected with trichinosis, thoroughly cook game meat to 160°F (71°C).

Anyone who believes they may have contracted trichinosis should consult a physician as soon as possible, explaining to the doctor the possible sources of infection.

Legal Status

Because bears' legal status, hunting restrictions, and other information change, contact your local, state, or provincial wildlife office, or visit their Web site for updates. (See Appendix E for contact information and where to access the state and provincial laws mentioned below.)

Oregon: The black bear is classified as a game animal and hunting regulations apply (OAR 35-066). No permit is required to kill bears damaging land, livestock, or agricultural or forest crops. However, any person killing a bear in these situations must have in their possession written authority from the landowner or lawful occupant of the land. Such person must immediately report the taking to the Oregon Department of Fish and Wildlife and must dispose of the wildlife as the department directs (ORS 498.012, 498.012, 498.013).

Washington: The black bear is classified as a game animal

(WAC 232-12-007). A hunting license and open season are required to hunt black bears. A property owner or the owner's immediate family, employee, or tenant may kill a bear on that property if it is damaging crops or domestic animals. You must notify your local Department of Fish and Wildlife (WDFW) office immediately after taking a black bear in these situations (RCW 77.36.030).

The grizzly bear is classified as endangered under state law and threatened under federal law, and cannot be hunted or trapped in Washington (WAC 232-12-014).

The killing of a black or grizzly bear in self-defense, or defense of another, should be reasonable and justified. A person taking such action must have reasonable belief that the bear poses a threat of serious physical harm, that this harm is imminent, and the action is the only reasonable available means to prevent that harm.

Any bear that is killed, whether under the direct authority of RCW 77.36.030, or for the protection of a person, remains the property of the state and must be turned over to WDFW.

British Columbia: The black bear is considered both a big game animal and a furbearer, and may be hunted and trapped under license during an open season. It is illegal, however, to hunt and

kill any black bear less than two years old, or any black bear accompanying it (B.C. Reg. 190/84, Section 13.1), as well as any white (Kermodei) or blue (Glacier; see below) color phases of black bear (B.C. Reg. 190/84, Section 13.2).

The Glacier bear *(Ursus americanus emmonsii),* found in the Alsek River drainage in the extreme northwest corner of the province and characterized by its bluish tinged fur, has been put on the provincial Blue List (see "Legal Status" in Chapter 1).

The grizzly bear is defined as a big game animal under the Wildlife Act, and can be hunted under license during an open season. It is illegal to hunt and kill a grizzly bear less than two years old.

The British Columbia Conservation Data Centre has placed the grizzly bear on the Blue List and the COSEWIC has placed it in their "Special Concern" category.

A person may kill a black bear or grizzly bear on their property without a permit if it is a menace to a person, domestic animal, or bird (Wildlife Act, Section 26(2)). However, to capture or kill (on behalf of the government) a black bear or grizzly bear that is dangerous to public safety, a permit must be issued from a regional manager (B.C. Reg. 253/2000, Section 2(h), and 2(c)(iv)).

CHAPTER 5

Cougars (Mountain Lions) and Bobcats

S leek and graceful, cougars (*Puma concolor,* Fig. 1 and back cover) are solitary and secretive animals rarely seen in the wild. Also known as mountain lions or pumas, cougars are known for their strength, agility, and awesome ability to jump. Their exceptionally powerful legs enable them to leap 30 feet (9 m) from a standstill, or to jump 15 feet (4.5 m) straight up a cliff wall. A cougar's strength and powerful jaws allow it to take down and drag prey larger than itself (Fig. 2).

Cougars are the largest members of the cat family in North America. Adult males may weigh 180 pounds (80 kg) and measure 8 feet (2.5 m) long from nose to tip of tail. Adult males stand about 30 inches (76 cm) tall at the shoulder. Adult female cougars average about 25 percent smaller than males. Cougars vary in color from reddish-brown to tawny (deerlike) to gray, with a black tip on their long tail.

Cougars occur throughout Oregon and Washington where suitable cover and prey are found. In Washington the cougar population for the year 2002 was estimated to be 2,400 to 4,000 animals. In Oregon, the year 2002 cougar population was estimated to be 4,000 animals—a dramatic increase since the 1960s, when continued hunting pressure brought the population down to about 200 animals. Cougars primarily occupy the southern third of British Columbia; Vancouver Island has one of the highest densities of cougars in North America.

Wildlife offices throughout the Pacific Northwest receive hundreds of calls a year regarding urban sightings, attacks on livestock and pets, and cougar/human confrontations. Our increasing cougar and human populations and decreasing habitat create more opportunities for such encounters.

For information on bobcats, see "Notes on Bobcats."

Facts about Cougars

Habitat and Home Range

- Cougars use steep canyons, rock outcroppings and boulders, or vegetation, such as dense brush and forests, to remain hidden while hunting.
- Adult male cougars roam widely, covering a home range of 50 to 100 square miles (80–160 km), depending on the age of the cougar, the time of year, type of terrain, and availability of prey.
- Adult male cougars' home ranges will often overlap those of several females, which travel much less.

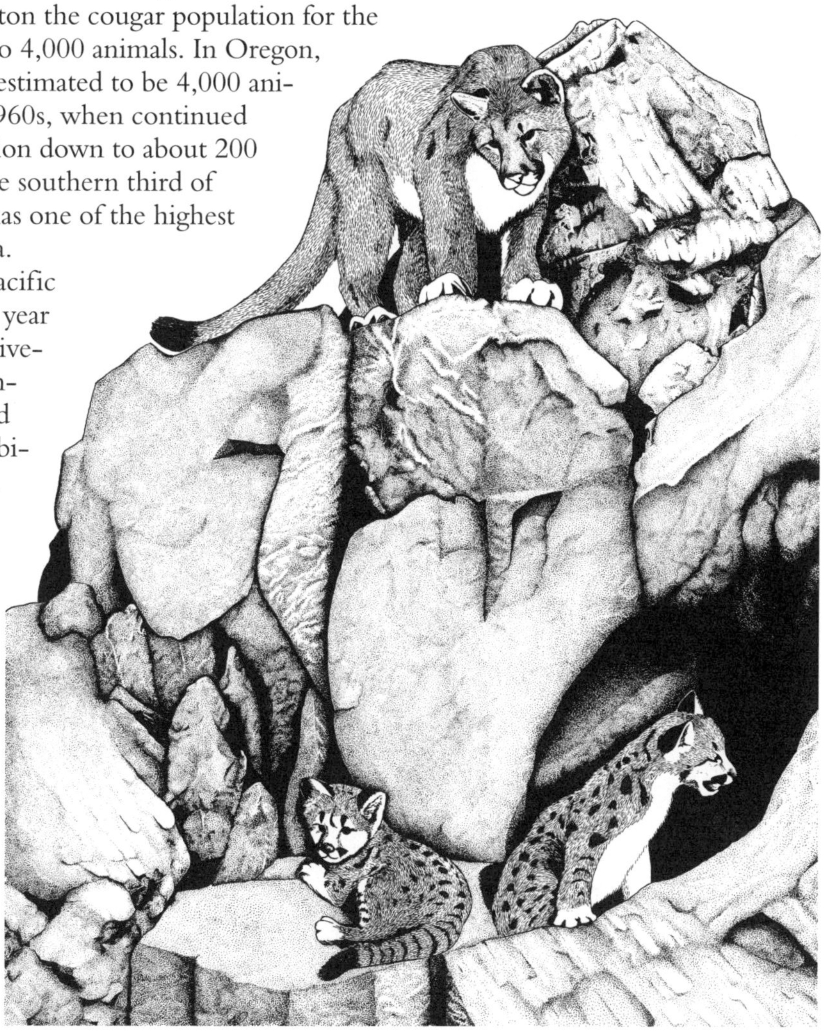

Figure 1. In rough terrain, cougar dens are usually in a cave or a shallow nook on a cliff face or rock outcrop. An average of two kittens are born each year. (From Christensen, Mammals of the Pacific Northwest: A Pictorial Introduction.*)*

Food and Feeding Habits

- Cougars are most active from dusk to dawn, although they sometimes travel and hunt during the day.
- Adult cougars typically prey on deer, elk, moose, mountain goats, and wild sheep.
- Other prey species, especially for younger cougars, include raccoons, coyotes, rabbits, hares, small rodents, and occasionally pets and lifestock.
- A large male cougar living in the Cascade Mountains kills a deer or elk every 7 to 12 days, eating up to 20 pounds (9 kg) at a time and burying the rest for later.
- Except for females with young, cougars are lone hunters that wander between places frequented by their prey, covering as much as 15 miles (25 km) in a single night.
- Cougars rely on short bursts of speed to ambush their prey. A cougar may stalk an animal for an hour or more (Fig. 3).

Den Sites

- A cougar's den is used for rest, protection from the weather, and to raise young.
- In rough terrain, dens are usually in a cave or a shallow nook on a cliff face or rock outcrop (Fig. 1). In less mountainous areas, dens are located in forested areas, thickets, or under large roots or fallen trees.
- Cougars prefer dens in secluded areas near water. No den preparation takes place.

Reproduction and Family Structure

- Cougars can breed year-round, but breeding is more common in winter and early spring. Sev-

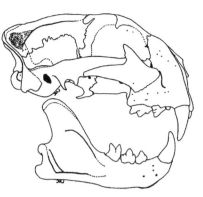

Figure 2. A cougar's strength and powerful jaws allow it to take down and drag prey larger than itself. (From Verts and Carraway, Land Mammals of Oregon.*)*

eral females may breed with a resident male whose home range overlaps theirs.
- After 91 to 97 days of pregnancy, one to four (but usually two) kittens are born.
- The bond between male and female is short-lived (about ten days), and the male cougar plays no role in raising the kittens.
- Kittens stay with their mothers for 12 to 16 months following their birth.
- Female cougars usually breed every other year.

Mortality and Longevity

- The two most common natural causes of death among cougars are being killed by other cougars, or by the prey during an attack.
- Humans, through hunting, depredation, and vehicle collisions, are probably the main source of mortality among cougars.
- Male cougars can live 10 to 12 years in the wild; females normally live longer.

Viewing Cougars and Bobcats

Because of their elusive nature and caution around humans, cougars and bobcats are seldom seen. In areas disturbed by humans, these cats typically limit their activity to night hours. (In dim light, cougars and bobcats see up to six times better than humans.) In undisturbed areas, they can be active at dawn or dusk if prey is active at that time.

Tracks

Cougar and bobcat tend to leave "soft" tracks, meaning the animals make very little impact on the ground, and their tracks may be virtually invisible on packed earth or crusted snow (Fig. 4). In addition, to preserve their sharpness for gripping prey, these ani-

Figure 3. While slowly stalking prey, cougars crouch and freeze, blending into the landscape. (Washington Department of Fish and Wildlife.)

Figure 4. Cougar tracks are about the size of a baseball, 3 to 3½ inches (7.5–9 cm) in diameter. Bobcat tracks are similar to cougar tracks, but only about 2 inches (5 cm) in diameter, which is about twice the size of house-cat tracks. Note the lack of claw marks, which are visible in tracks left by members of the dog family. (From Pandell, Animal Tracks of the Pacific Northwest.*)*

mals keep their claws retracted most of the time, and so claw marks are rarely visible in their tracks.

Because cougars carry their heavy tail in a wide U shape at a normal walk, in snow, the lowermost portion may leave drag marks between each print.

Droppings

Cougars and bobcats generally cover their droppings with loose soil. When visible, their droppings typically resemble those of most species in the dog and cat families. However, cougars have well-developed premolars that can slice through bone. Therefore, their droppings often show chunks and fragments of chewed bone. Members of the dog family gnaw on bones but usually don't chew them up into cut fragments.

Cougar droppings are generally cylindrical in shape, segmented, and blunt at one or both ends. An

Notes on Bobcats *(Lynx rufus)*

Although seldom seen, bobcats are found in suitable habitat from sea level to above timberline throughout the Pacific Northwest (Fig. 5). Rock cliffs, outcroppings, and ledges are important to bobcats for shelter, raising young, and resting sites. Large brush or log piles and hollow trees or logs are used in wooded areas.

Bobcats tend to avoid open areas, preferring to stay close to cover they can quickly disappear into. Evidence of a bobcat's presence may include droppings, tracks in snow or mud, and claw marks on tree trunks.

A bobcat's home range varies from 2½ to 6 square miles (4–10 km) for adult males, and about half that for adult females.

Adult male bobcats are slightly larger than females, weighing 20 to 30 pounds (9–14 kg) and averaging 3 feet (92 cm) in length.

Cougars and bobcats are easy to tell apart. A cougar's tail is nearly 3 feet (90 cm) long (roughly a third of the cougar's total length), while a bobcat's tail averages 3 inches (7.5 cm) long.

Bobcats hunt primarily by sight and sound, which means they spend much of their time sitting or crouching, watching, and listening. Once they've located prey, they stalk within range of a quick dash and then pounce. Bobcats are opportunistic and will prey upon a wide variety of animals. Food sources include various small mammals, from mice and rabbits to young beaver and deer fawns, as well as insects, reptiles, birds, and carrion.

Domestic animals occasionally taken by bobcats include house cats, poultry, small pigs, and lambs. To prevent conflicts, follow the suggestions under "Preventing Conflicts." (See Chapter 21 for information on fencing.)

Adult bobcats fall prey to cougars, coyotes, feral dogs, and humans. Young bobcats are killed by eagles, great horned owls, foxes, bears, and adult male bobcats.

Figure 5. Bobcats have a short, bobbed tail but their ears are long, with black tufts at the end. **Note:** *A small female bobcat may weigh less than a large house cat. (Washington Department of Fish and Wildlife.)*

average dropping measures 4 to 6 inches (10–15 cm) long by 1 to 1½ inches (2.5–4 cm) in diameter. The size of the dropping is a good indication of the size of the cougar. Bobcat droppings are similar to cougar droppings, but smaller.

Feeding Areas (caches)

Cougars usually carry or drag their kills to a secluded area under cover to feed, and drag marks are frequently found at fresh kill sites. After killing a large animal and having eaten its fill, a cougar often will cover the remains with debris such as snow, grass, leaves, sticks, or soil. Even where little debris is available, bits of soil, rock, grass or sticks may used to cover the carcass. The cougar will remain in the immediate vicinity of its kill, guarding it against scavengers and eating it over a period of one day to two weeks. **Do not linger around a recently killed or partially covered deer or elk.**

A bobcat will also eat the carcass of a large mammal. Like a cougar, it will cover the carcass remains and frequently return to feed on it. Being smaller than a cougar, a bobcat only reaches out 15 inches (40 cm) to rake up debris to cover the food cache. These marks, and the bobcat's much smaller tracks, help distinguish between bobcat and cougar caches.

Scratching Posts

Like house cats scratching furniture, cougars and bobcats mark their territory boundaries by leaving claw marks on trees, stumps, and occasionally fence posts. Claw marks left by an adult cougar will be 4 to 8 feet (1.25–2.5 m) above the ground and consist of long,

deep, parallel scratches running almost vertically down the trunk. These gashes rarely take off much bark; tree-clawing that removes much bark is probably the work of a bear (see Chapter 4). Bobcat claw marks are normally 2 to 3 feet (60–90 cm) above the ground; domestic cat scratching occurs at a height of about 1½ to 2 feet (45–60 cm).

Calls

Cougars and bobcats hiss, purr, mew, growl, yowl, chirp, and cry. The most sensational sounds they make are the eerie wailings and moans heard at night during mating season, especially when competing males have intentions toward the same receptive female. Such wails have been likened to a child crying, a woman's scream, and the screeching of someone in terrible pain.

Preventing Conflicts

The cougar's ability to travel long distances occasionally brings these cats into seemingly inappropriate areas, even places densely settled by humans. Such appearances are almost always brief, with the animal moving along quickly in its search of a suitable permanent home. However, where humans are encroaching on wildlife habitat and cougar numbers are increasing, the number of cougar sightings and attacks on livestock and pets is on the rise.

Cougar attacks on humans are extremely rare. In North America, fewer than 20 fatalities and 75 nonfatal attacks have been reported during the past 100 years. However, more cougar attacks have been reported in the western

United States and Canada over the past 20 years than in the previous 80. In Washington, of the one fatal and five nonfatal attacks reported here since 1924, four attacks occurred during the 1990s.

A high percentage of cougars attacking domestic animals or people are one- to two-year-old cougars that have become independent of their mothers. When these young animals, particularly males, leave home to search for territory of their own, and encounter territory already "owned" by an older male cougar, the older one will drive off the younger one, killing it if it resists. Some young cougars are driven across miles of countryside in search of an unoccupied territory.

If you are living in cougar country, prevent a conflict with them by using the following management strategies around your property, and, if possible, encourage your neighbors to do the same.

With bobcats, although far less extreme problems exist with them, the following information on pets and domestic animals will help prevent conflicts.

Don't leave small children unattended. When children are playing outdoors, closely supervise them and be sure they are indoors by dusk. (See "Cougars and Kids.")

Modify the habitat around your home. Light all walkways after dark and avoid landscaping with plants that deer prefer to eat. Where a deer goes, a cougar may follow. Shrubs and trees around kids' play areas should be pruned up to prevent cougars from hiding behind them.

A chain-link or heavy woven-wire fence that is 10 feet (3 m) high with 3-foot (90 cm) extensions installed at a 65-degree angle on each post may keep cougars out of an enclosed area. To increase effectiveness, string barbed wire or four electric wires between the extensions, alternating positive and negative wires. (See "Preventing Conflicts" in Chapter 7 for information on electric fences.)

Don't feed wildlife and feral cats (domestic cats gone wild). This includes deer, raccoons, and other small mammals. Remember predators follow prey.

Close off open spaces under structures. Areas beneath porches and decks can provide shelter for prey animals. (See "Preventing Conflicts" in Chapter 21 for information).

Feed dogs and cats indoors. If you must feed outside, do so in the morning or midday, and pick up food and water bowls, as well as leftovers and spilled food, well before dark. Pet food and water attract small mammals that, in turn, attract cougars.

Keep dogs and cats indoors, especially from dusk to dawn. Left outside at night, small dogs and cats may become prey for cougars or bobcats (which have attacked cocker-spaniel-size dogs).

Use garbage cans with tight-fitting lids. Garbage attracts small mammals that, in turn, attract cougars. See "Preventing Conflicts" in Chapter 21 for information on garbage management.

Keep outdoor livestock and small animals confined in secure pens. For a large property with livestock, consider using a guard ani-mal. There are specialty breeds of dogs that can defend livestock. Donkeys and llamas have also successfully been used as guard animals. As with any guard animal, pros and cons exist. Purchase a guard animal from a reputable breeder who knows the animal he or she sells. Some breeders offer various guarantees on their guard animals, including a replacement if an animal fails to perform as expected.

See "Preventing Conflicts" in Chapter 6 for additional information on livestock management.

Encountering a Cougar

Relatively few people will ever catch a glimpse of a cougar much less confront one. If you come face to face with a cougar, your actions can either help or hinder a quick retreat by the animal.

Here are some things to remember:

- Stop, pick up small children immediately, and don't run. Running and rapid movements may trigger an attack. Remember, at close range, a cougar's instinct is to chase.
- Face the cougar. Talk to it firmly while slowly backing away. Always leave the animal an escape route.
- Try to appear larger than the cougar. Get above it (e.g., step up onto a rock or stump). If wearing a jacket, hold it open to further increase your apparent size. If you are in a group, stand shoulder-to-shoulder to appear intimidating.
- Do not take your eyes off the cougar or turn your back. Do not crouch down or try to hide.
- Never approach the cougar, especially if it is near a kill or with kittens, and never offer it food.
- If the cougar does not flee, be more assertive. If it shows signs of aggression (crouches with ears back, teeth bared, hissing, tail twitching, and hind feet pumping in preparation to jump), shout, wave your arms and throw anything you have available (water bottle, book, backpack). The idea is to convince the cougar that you are not prey, but a potential danger.
- If the cougar attacks, fight back. Be aggressive and try to stay on your feet. Cougars have been driven away by people who have fought back using anything within reach, including sticks, rocks, shovels, backpacks, and clothing—even bare hands. If you are aggressive enough, a cougar will flee, realizing it has made a mistake. Pepper spray in the cougar's face is also effective in the extreme unlikelihood of a close encounter with a cougar.

Professional Assistance

Wildlife offices throughout the Pacific Northwest respond to cougar sightings when there is a threat to public safety or property. Problem cougars may be live-trapped by trained fish and wildlife personnel and moved to more remote areas; however, such removals are expensive, time consuming, and seldom effective. Using tranquilizing drugs on cougars to facilitate removal is difficult and dangerous for cougars and humans. When other methods have failed, lethal removal of problem animals may be the only alternative.

Cougars and Kids

Cougars seem to be attracted to children, possibly because their high-pitched voices, small size, and erratic movements make it difficult for cougars to identify them as human and not prey. To prevent a problem from occurring:

- Talk to children and teach them what to do if they encounter a cougar.
- Encourage children to play outdoors in groups, and supervise children playing outdoors.
- Consider getting a dog for your children as an early-warning system. A dog can see, smell, and hear a cougar sooner than we can. Although dogs offer little value as a deterrent to cougars, they may distract a cougar from attacking a human.
- Consider erecting a fence around play areas. (See "Modify the habitat around your home.")
- Keep a radio playing when children are outside, as noise usually deters cougars.
- Make sure children are home before dusk and stay inside until after dawn.
- If there have been cougar sightings, escort children to the bus stop in the early morning. Clear shrubs away around the bus stop, making an area with a 30-foot (9 m) radius. Have a light installed as a general safety precaution.

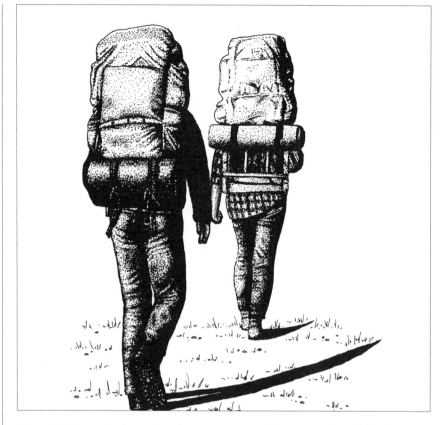

Figure 6. To avoid a close and unpleasant encounter with a cougar, do not hike alone in cougar country. (Washington Department of Fish and Wildlife.)

Contact your local wildlife office for additional information, and in the case of an immediate emergency, call 911 or any local law enforcement office, such as the state patrol.

Public Health Concerns

Cougars and bobcats rarely carry any communicable diseases that are regarded as threats to humans in the Pacific Northwest.

Feline distemper *(Feline panleukopenia)* antibodies have been documented in Pacific Northwest cougar populations, but the degree that the *Feline panleukopenia* virus causes cougar mortality, or is transferred to domestic cats, is unknown.

Legal Status

Because the legal status, hunting restrictions, and other information relating to cougars and bobcats change, contact your local, state, or provincial wildlife office, or visit their Web site for updates. (See Appendix E for contact information and where to access the state and provincial laws mentioned below.)

Oregon: The cougar is classified as a game animal and hunting regulations apply (OAR 635-067). The bobcat is classified as a furbearer and a license is required to trap or hunt bobcats during established open seasons.

No permit is required to kill cougars harming livestock and no permit is required to kill bobcats damaging land, livestock, or agricultural or forest crops. However, any person killing a cougar or bobcat in these situations must have in their possession written authority from the landowner or lawful occupant of the land, must immediately report the taking to the Oregon Department of Fish and Wildlife, and must dispose of the wildlife as directed by the department (ORS 498.012 (Sections 2 and3)).

Washington: Cougars and bobcats are classified as game animals and an open season and a hunting license are required to hunt them (WAC 232-12-007). A property owner or the owner's immediate family, employee, or tenant may kill a cougar or bobcat on that property if it is damaging domestic animals (RCW 77.36.030). No permit is required.

The killing of a cougar or bobcat in self-defense, or defense of another, should be reasonable and justified. A person taking such action must have reasonable belief that the cougar or bobcat poses a threat of serious physical harm, that this harm is imminent, and the action is the only reasonable available means to prevent that

harm.

The body of any cougar or bobcat, whether taken under the direct authority of RCW 77.36.030, or for the protection of a person, remains the property of the state and must be turned over to the Department of Fish and Wildlife immediately.

British Columbia: Cougars and bobcats are considered big game under the Wildlife Act and may be legally hunted under license during open seasons as defined by regulation. All harvested cougars must be compulsory inspected by a ministry officer within 15 days of the date of kill. It is compulsory to report all harvested bobcats within 30 days after the date of kill.

A permit must be issued from a regional manager to capture or kill cougars or bobcats that pose a threat to public safety (B.C. Reg. 253/2000, Section 2(h) and 2(c)(iii)). A person may trap or kill a cougar or bobcat on their own property that is a menace to a domestic animal or bird. No permit is required in such cases (Wildlife Act Section 26(2)).

Precautions for Hikers and Campers

While recreating in a cougar's territory, you can avoid close encounters by taking the following precautions:

- Hike in groups and make enough noise to prevent surprising a cougar.
- Avoid hiking after dark.
- Keep small children close to the group, preferably in plain sight ahead of you.
- Do not approach dead animals, especially recently killed or partially covered deer and elk.
- Be aware of your surroundings, particularly when hiking in dense cover or when sitting, crouching, or lying down. Look for tracks, scratch posts, and partially covered droppings.
- Keep a clean camp. Reduce odors that might attract mammals such as raccoons, which in turn could attract cougars. Store meat, other foods, pet food, and garbage in double plastic bags.

CHAPTER 6

Coyotes and Wolves

In pioneer days, coyotes *(Canis latrans)* were restricted primarily to the sagebrush lands, brushy mountains, and open prairies of the American West. Wolves occupied the forests. Coyotes have since taken advantage of human activities (including the reduction of gray wolf populations) to expand their range throughout North and Central America.

In the Pacific Northwest, these intelligent and adaptable animals manage to occupy almost every conceivable habitat type, from open ranch country to densely forested areas to downtown waterfront. Despite ever-increasing human encroachment and past efforts to eliminate coyotes, the species maintains its numbers and is increasing in some areas. The coyote's tenacity tries some people's patience and inspires others' admiration.

At first glance, the coyote resembles a small German shepherd dog, yet its color can vary from animal to animal (Fig. 1 and back cover). Shades include black, brown, gray, yellow, rust, and tan. Coyotes also have shorter, bushier tails that are carried low, almost dragging the ground, and longer, narrower muzzles than their dog cousins. Adult coyotes weigh 20 to 35 pounds (9–16 kg), with males being slightly larger than females. At the shoulder, an adult male coyote is about 25 inches (65 cm) tall.

For information on wolves, see "Notes on Gray Wolves."

Figure 1. Coyotes in the Pacific Northwest are usually solitary. The so-called "bands" or "packs" of coyotes are almost always family groups composed of a mother and her young. (During the mating period, a group may include a female and several males.) (From Christensen and Larrison, Mammals of the Pacific Northwest: A Pictorial Introduction.)

Facts about Coyotes
Food and Feeding Habits

- Coyotes are opportunists, both as hunters and as scavengers. They eat any small animal they can capture, including mice, rats, voles, gophers, mountain beavers, rabbits, and squirrels, also snakes, lizards, frogs, fish, birds, and carrion (animal carcasses). Grass, fruits, and berries are eaten during summer and fall.

- Grasshoppers and other insects are important to juvenile coyotes learning the stalk-and-pounce method of hunting.

- Pairs of coyotes or family groups using the relay method pursue deer and antelope. These large mammals are important food in winter; fawns may be eaten in spring.
- Coyotes eat wild species, but they are known to eat pet food, garbage, garden crops, livestock, poultry, and pets (mostly cats).
- Coyotes occasionally kill domestic dogs (and foxes) that they consider territorial intruders. Coyotes are also very protective of their young and will attack dogs that get too close to their den and pups. *Note:* The list of killers of domestic cats and dogs includes other dogs and cats, vehicles, bears, cougars, bobcats, foxes, disease, and furious neighbors!
- Most hunting activity takes place at night. Undisturbed and hungry coyotes will hunt during daylight hours, and may be seen following farm machinery, catching voles and other small prey.

Den Sites

- The female coyote digs her own den under an uprooted tree, log, or thicket; may use a cave, hollow log, or storm drain; or take over and enlarge another mammal's burrow.
- The den will have an entrance 1 to 2 feet (30–60 cm) across, be dug 5 to 15 feet (1.5–4.5 m) long, and terminate in an enlarged nesting chamber.
- Coyotes usually have several dens and move from one to the other, minimizing the risk that a den containing young will be detected. These moves also help to prevent an accumulation of fleas and other parasites, as well as urine, droppings, and food refuse.

- Coyotes use the same dens yearly or make new dens in the same area.

Reproduction and Family Structure

- Occasionally, a mated pair of coyotes will live, hunt, and raise pups together for many years, sometimes for life.
- Breeding occurs in late winter. After a gestation (pregnancy) of 63 days, an average of four pups are born from early April to late May. (Litter size can be affected by population density and food availability.)
- The young are principally cared for by the female; occasionally a nonbreeding sibling will assist with raising the litter. The male provides some food for the mother and the young.
- Pups emerge from the den in two to three weeks and begin to eat regurgitated food. Because food requirements increase dramatically during pup rearing, this is a period when conflicts between humans and coyotes are common.
- Juvenile coyotes usually disperse alone or sometimes in groups at six to eight months of age. A few may stay nearby, while others seek new territory up to 50 miles (80 km) away. The greater the amount of food available in a given area, the closer the juveniles will stay to their den.
- Coyotes can interbreed with domestic dogs and wolves; however, such crosses are rare.

Mortality and Longevity

- Coyote numbers are controlled by social stress, diseases, parasites, competition for food, and predators.

- Predators include humans, cougars, wolves, black bears, and other coyotes. Eagles, dogs, and adult coyotes kill some coyote pups.
- Where coyotes are hunted, trapped, and poisoned, females produce more pups per litter than in areas where they are protected.
- Coyotes in captivity live as long as 18 years. In the wild, few coyotes live more than four years; the majority of pups die during their first year.

Viewing Coyotes

Coyotes are extremely wary. Their sense of smell is remarkable, and their senses of sight and hearing are exceptionally well developed.

Sightings of coyotes are most likely during the hours just after sunset and before sunrise. To view a coyote, locate a well-used trail and wait patiently from an area overlooking a canyon, ravine, or other area. A coyote will often come down the trail the same time every morning or evening. Also, you could watch a coyote's feeding area, such as a livestock or big game carcass.

By six months of age, pups have permanent teeth and are nearly fully grown. At about this time, female coyotes train their offspring to search for food, so it is not unusual to observe a family group.

Never approach an occupied coyote den. A mother's protective instincts can make her dangerous if she has young in or nearby the den. Den sites, and coyote activity, should be observed with binoculars or a spotting scope from a distance that does not visibly disturb the animals. Unfamiliar or

Notes on Gray Wolves (Canis lupus)

When Europeans settled Oregon and Washington, wolves were common from the Cascade foothills to the Pacific Coast. However, by the 1940s wolves had been effectively eliminated by control programs that involved shooting, trapping, and poisoning. Although a few widely scattered individuals or pairs of wolves may exist in Oregon and Washington, no documented wolf breeding pairs or packs currently are known in these states. Wolves that have been seen in Oregon and Washington are believed to have been lone wolves that strayed into the state from British Columbia or Idaho or pet wolf/dog hybrids that have escaped or been released by owners. (See Appendix D for information.)

Idaho wolf populations have increased significantly during the past several years. The fall 1998 population was estimated to be 114 animals. The fall 2003 population was estimated to be 346 wolves. As a result of this continued growth in numbers, biologists expect wolves to eventually begin dispersing from Idaho into neighboring states.

- **Distribution:** In British Columbia, wolves are distributed from sea level to above the timberline throughout most of the province, except the Queen Charlotte Islands. They are most abundant on northern Vancouver Island and in the northern half of British Columbia.

- **Description:** The wolf is nearly twice as large as the coyote and has much smaller ears in proportion to its large, broad head (Fig. 2). When running, a wolf's tail is held straight out behind or raised high over its back, whereas the coy-ote's tail is usually held lower than its body. In general, a wolf's coloration is grayish and resembles that of a coyote, but some wolves are nearly white or black.

- **Family structure:** Wolves are highly social and almost always live in family groups of two to nine individuals consisting of a set of parents (alpha pair), their offspring, and other (non-breeding) adults.

- **Food and feeding habits:** Wolves frequently hunt in groups, or packs, and feed on elk, deer, moose, caribou, mountain goat, and bighorn sheep. In summer, wolves eat many small mammals, including ground squirrels, rabbits, hares, and beaver. Grouse, waterfowl, spawning fish, carrion, and garbage may be eaten year-round.

  Although wolves can sometimes cause significant damage to livestock, they often live near livestock without causing more than, at most, occasional harm. Some livestock losses blamed on wolves are actually caused by domestic dogs, coyotes, bears, bobcats, and cougars.

  Wolves are able to live on little or no food for long periods of time.

When prey is plentiful, wolves may eat only part of what they kill, leaving uneaten portions that provide food for crows, ravens, eagles, small birds, foxes, and other scavengers.

Wolf packs hunt within a specific territory, which may be 50 to 500 square miles (80–800 km), depending on the availability of food. Although wolves usually lope along at 5 miles (8 km) per hour, they can attain speeds as high as 45 miles (72 km) per hour.

Wolves will howl before and after a hunt, to sound an alarm, or to locate other members of the pack; they may also howl at other times, for reasons not understood. They howl more frequently in the evening and early morning, especially during winter breeding and pup-rearing.

- **Reproduction and activity:** Breeding occurs in late February to mid-March with an average of six pups being born in May. Dens are usually a hole dug in the ground but may also be in rock crevices, hollow logs, or overturned stumps. Dens may be used for two or

Figure 2. The wolf is nearly twice as large as the coyote and has much smaller ears in proportion to its large, broad head. (Washington Department of Fish and Wildlife.)

more years, or several dens may be used during one season.

Wolf pups are weaned gradually during midsummer. In mid or late summer, pups are usually moved some distance away from the den and by early winter are capable of traveling and hunting with adult pack members. Sometime after reaching one or two years of age, a young wolf will leave to form its own pack.

During the summer, wolves are primarily nocturnal, since activity is focused around dens and sites where the pack gathers to hunt. During winter, movements are most extensive and may occur at any time of night or day.

Wolves travel as far as 30 miles (48 km) a day, using rivers or timbered areas as travel corridors in the winter, and dirt roads and game trails in the summer.

• **Mortality:** In spite of a generally high birth rate, wolves rarely become abundant because mortality is high. In much of British Columbia, hunting and trapping are the major sources of mortality, although diseases, malnutrition, accidents, and particularly interspecies strife act to regulate wolf numbers.

Fears that wolves might attack people or threaten outdoor activities are largely unfounded. Wolves are reclusive by nature and avoid contact with humans.

Note: There has never been a documented human fatality involving a healthy, wild wolf in British Columbia or the United States.

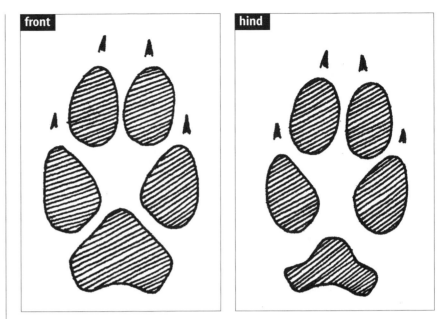

Figure 3. Coyote tracks are more oblong-shaped than dog and wolf tracks. The normal track is about 2 inches (5 cm) wide and 2½ inches (6.5 cm) long, with the hind track slightly smaller than the front. The toenails nearly always leave imprints. (Drawing by Kim A. Cabrera.)

new human activity close to the den, especially within one-quarter mile (0.4 km), will often cause coyotes to move, particularly if the pups are older, if the adults see you, or if the den is in an open area with little protective cover.

Tracks, Trails, and other Signs

Look for coyote tracks in mud, sand, dust, or snow (Fig. 3). Their trails are often found along draws, fence lines, game and livestock trails, next to roads, in the middle of dirt roads, and on ridge tops.

When a tree falls across a trail, coyotes have to either go over or under it, depending on their size. Those that go over tend to rub the bark off the top of the log; those that go under sometimes leave their hairs on the underside. Also look for coyote hairs on a wire fence where a trail runs next to or under the fence.

Droppings

Coyote droppings are found in conspicuous places and on or near their trails. The droppings are extremely variable in size, shape, and composition. Individual droppings average 3 to 4 inches (7.5–10 cm) long with a diameter of 1 inch (2.5 cm). Droppings consisting of a lot of hair may be larger. The residue from pure meat is likely to be semiliquid and black. Those resulting from a diet of cherries, apples, blackberries, huckleberries, elderberries, or other fruits tend to crumble.

Feeding and Hunting Sites

When small mammals such as rabbits are eaten, the head, feet, and hide will have been eaten, leaving a scattering of fur at the feeding site. Bones, feathers, and fur can be seen next to den entries. Signs of digging occur where coyotes follow promising scents and excavate prey, including moles, voles, and gophers.

Figure 4. Construct a clapper by hinging together two 24-inch (60 cm) 2 x 4s. Smack the two sides together. (Drawing by Jenifer Rees.)

Coyotes Too Close for Comfort

Coyotes are curious but timid animals and will generally run away if challenged. However, remember that any wild animal will protect itself or its young. Never instigate a close encounter.

If a coyote ever approaches too closely, pick up small children immediately and act aggressively toward the animal. Wave your arms, throw stones, and shout at the coyote. If necessary, make yourself appear larger by standing up (if sitting) or stepping up onto a rock, stump, or stair. The idea is to convince the coyote that you are not prey, but a potential danger.

Where coyote encounters occur regularly, keep noisemaking and other scare devices nearby. A starter pistol can be effective; so can a vinegar-filled super soaker or a powerful spray of water from a hose. Where pyrotechnics are out of the question, construct a "clapper" (Fig. 4). A solid walking stick and/or pepper spray are powerful deterrents at close range.

If a coyote continues to act in an aggressive or unusual way, call your local wildlife office (see Appendix E) or state patrol.

Figure 5. Juvenile coyotes are often heard in summer, trying out their voices. (Washington Department of Fish and Wildlife.)

Scent Stations

Coyotes mark their territory with urine and droppings. They use their feet to scratch near piles of droppings, spreading the odor from their droppings and foot glands to identify their scent stations. A prominent tree, rock, or even a hiking-trail sign along a coyote trail can also serve as a scent station. If it is active, you'll find scratch marks, loose grass, sod, and other material as far as 6 feet (1.8 m) away. The area may also draw the attention of your pet dog.

Calls

Coyotes create a variety of vocalizations. Woofs and growls are short-distance threat and alarm calls; barks and bark-howls are long-distance threat and alarm calls; whines are used in greetings; lone and group howls are given between separated group members when food has been found; and a yip-howl is often done after a group reunites. Juvenile coyotes are often heard in summer, trying out their voices (Fig. 5).

Preventing Conflicts

To date, there have been relatively few documented coyote attacks on humans in the Pacific Northwest. Prior to 1981, coyote attacks on humans throughout North America were also rare. However, in the 1990s more than 50 coyote-related human health and safety incidents were reported nationally to the U.S. Department of Agriculture. From 1988 to 1997 in the West's most densely populated area—southern California—53 coyote attacks on humans, resulting in 21 injuries, were documented by a University of California Wildlife Extension Specialist. A study of those incidents indicates that attacks on pets may precede and even predict more serious coyote/human conflicts, and that human behavior is contributing to the problem.

Research suggests that humans help to create the conditions for conflicts with coyotes. People do this by deliberately or inadvertently providing the animals with food, such as handouts or carcasses of farm animals. When humans provide food, young coyotes quickly learn not to fear humans; this makes them bold. They will also become dependent on the easy food source humans have come to represent. Once a coyote stops hunting on its own and loses its fear of people, it becomes dangerous and could attack without warning.

Outwitting coyotes begins with prevention. Once a coyote causes damage for the first time, it gets easier for the animal to do it again. To prevent conflicts with coyotes, use the following management strategies around your property and, if possible, encourage your neighbors to do the same.

Don't leave small children unattended where coyotes are frequently seen or heard. If there are coyote sightings in your area, prepare your children for a possible encounter. Explain the reasons why coyotes live there (habitat/food source/species adaptability) and what they should do if one approaches them **(don't run, be as big, mean, and**

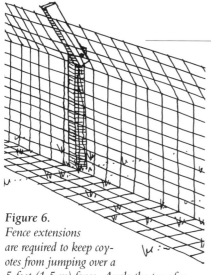

Figure 6.
Fence extensions are required to keep coyotes from jumping over a 5-foot (1.5 m) fence. Angle the top of a woven-wire fence out about 15 inches (40 cm) and completely around the fence. An effective fence extends below the surface, or has a wire apron in front of it to prevent digging. (Drawing by Jenifer Rees.)

loud as possible). By shouting a set phrase such as "go away coyote" when they encounter one, children will inform nearby adults of the coyote's presence as opposed to a general scream. Demonstrate and rehearse encounter behavior with the children.

Never feed coyotes. Coyotes that are fed by people often lose their fear of humans and develop a territorial attitude that may lead to aggressive behavior. Try to educate your friends and neighbors about the problems associated with feeding coyotes. If you belong to a homeowner's association or neighborhood watch, bring up the subject during one of the meetings.

Don't give coyotes access to garbage. Keep garbage can lids on tight by securing them with rope, chain, bungee cords, or weights. Better yet, buy quality garbage cans with clamps or other mechanisms that hold lids on. To prevent tipping, secure the side handles to metal or wooden stakes driven into the ground. Or keep your cans in tight-fitting bins, a shed, or a garage.

Prevent access to fruit and compost. Keep fruit trees fenced, or pick up fruit that falls to the ground. Keep compost piles within a fenced area or securely covered. Cover new compost material with soil or lime to prevent it from smelling. Never include animal matter in your compost; it attracts coyotes. If burying food scraps, cover them with at least 8 inches (20 cm) of soil, and don't leave any garbage above ground in the area—including the stinky shovel.

Feed dogs and cats indoors. If you must feed your pets outside, do so in the morning or at midday, and pick up food, water bowls, leftovers, and spilled food well before dark every day.

Don't feed feral cats (domestic cats gone wild). Coyotes prey on these cats as well as any feed you leave out for the feral cats.

Prevent the buildup of feeder foods under bird feeders. Coyotes will eat bird food and are attracted to the many birds and rodents that come to feeders. (See "Preventing Conflicts" in Chapter 37 for information on feeder management.)

Keep dogs and cats indoors, especially from dusk to dawn. If left outside at night in an unprotected area, cats and small to mid-size dogs may be killed by coyotes. Pets can be easy prey for coyotes. Being raised by humans leaves them unsuspecting once they leave the safety of your home. If you suspect losing a dog or cat to a coyote, notify your neighbors. Once a coyote finds easy prey it will continually hunt in the area.

Modify the landscape around children's play areas. Shrubs and trees should be pruned several feet above ground level so coyotes can't hide in them. Keep deterrents nearby in times of increased sightings. An old hockey stick, broom, or a pile of stones near the play area can help prepare children for an encounter and will remind them of effective encounter behavior.

Build a coyote-proof fence. Coyotes don't leap fences in a single bound but, like gray foxes and domestic dogs, they grip the top with their front paws and kick themselves upward and over with the back legs. Their tendency to climb will depend on the individual animal and its motivation. A 5-foot (1.5 m) woven-wire fence with extenders facing outward at the top of each post should prevent coyotes from climbing over (Fig. 6).

However, all coyotes are excellent diggers, and an effective fence needs to extend at least 8 inches (20 cm) below the surface, or have a galvanized-wire apron that extends out from the fence at least 15 inches (38 cm).

Electric fences can also keep coyotes out of an enclosed area (Figs. 7 and 8). Such a fence doesn't need to be as high as a woven-wire fence because a coyote's first instinct will be to pass through the wires instead of jumping over them. Digging under electric fences usually doesn't occur if the bottom wire is electrified. (For information on electric fences, see "Preventing Conflicts" in Chapter 7.)

Alternatively, install a commercial device, such as the Coyote Roller™, to prevent coyotes from being able to get the foothold necessary to hoist them over a fence (Fig. 9). (See Appendix D for similar barriers.)

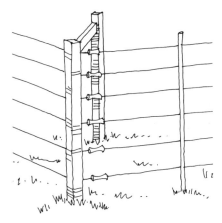

Figure 7. A six-wire electric fence can keep coyotes out of an enclosed area. (Drawing by Jenifer Rees.)

Figure 8. Two electrified wires, 8 and 15 inches (20 and 38 cm) above ground and offset from an existing wood fence by 12 inches (30 cm) will prevent coyotes from accessing the fence. A single strand may be sufficient, but two electrified wires will provide added insurance. (Drawing by Jenifer Rees.)

Enclose ducks or chickens at night in a secure coop. Like all predators, coyotes seek the easiest available food, and ducks and chickens are often vulnerable. See "Preventing Conflicts" in Chapter 21 for information.

Keep livestock and small animals that live outdoors confined in secure pens during periods of vulnerability. All animals should be confined from dusk to dawn. (Temporary or portable fencing keeps livestock together so that they can be guarded more effectively.) During birthing season, keep young and vulnerable animals confined at

all times. Do not use remote pastures or holding areas, especially when there has been a recent coyote attack. Remove any sick and injured animals immediately. Ensure that young animals have a healthy diet so that they are strong and less vulnerable to predators.

Livestock producers have discovered that scare devices, such as motion detectors, radios, and other noice makers, will deter coyotes—until they realize that they aren't dangerous.

Note: Many ranchers now attempt to kill coyotes only when damage has occurred. If your property is the home territory of coyotes that don't harm livestock, they will keep away other coyotes that are potential livestock killers. Coyotes also benefit ranchers and other property owners by helping control populations of mice, rats, voles, moles, gophers, rabbits, and hares.

Remove or bury dead livestock.
Coyotes, with their keen sense of smell, quickly find dead animals. Cover the carcass with a minimum of 2 feet (60 cm) of soil.

For a large property with livestock, consider using a guard animal.
There are specialty breeds of dogs that can defend livestock. Donkeys and llamas have also successfully been used as guard animals. As with any guard animal, pros and cons exist. Purchase a guard animal from a reputable breeder who knows the animal he or she sells. Some breeders offer various guarantees on their guard animals, including a replacement if an animal fails to perform as expected.

Lethal Control

If all efforts to dissuade a problem coyote fail and it continues to be

a threat to humans, or animals in their care, the animal may have to be killed.

In suburban areas of southern California, trapping and euthanizing coyotes has been shown not only to remove the individual problem animal, but also to modify the behavior of the local coyote population. When humans remove a few coyotes, the local population may regain its fear of humans in areas where large numbers of humans are found. It's neither necessary nor possible to eliminate the entire population of coyotes in a given area. Contact your local wildlife office for additional information (see Appendix E for contact information).

See Appendix A for additional information, including live-trapping coyotes.

Public Health Concerns

Coyote diseases or parasites are rarely a risk to humans, but could be a risk to domestic dogs in the Pacific Northwest. Anyone handling a coyote should wear rubber gloves, and wash their hands well when finished.

Canine distemper, a disease that affects domestic dogs, is found in our coyote populations. Have your dogs vaccinated for canine distemper to prevent them from contracting the disease. (For more information on canine distemper, see "Public Heath Concerns" in Chapter 21.)

Canine parvovirus, or **"parvo"** is another disease that affects domestic dogs and is found in our coyote populations. Parvo vaccinations have helped to control the spread of this disease. Despite being vaccinated, some

Figure 9. The Coyote Roller™ prevents coyotes from being able to get the foothold necessary to hoist them over a fence. (See "Wildlife/ Human Conflicts" in Appendix F for contact information.)

dogs—especially puppies and older domestic dogs—still contract and die from parvo.

Parvo is usually spread to coyotes and domestic dogs by direct or indirect contact with infected droppings. Exposure to domestic dogs occurs where dogs assemble, such as parks, dog shows, kennels, pet shops, and where they have contact with coyotes. Contact your veterinarian for vaccination information if your dog is ill.

Mange occurs in coyote and red fox populations in the Pacific Northwest. Mange is caused by a parasitic mite that causes extreme irritation when it burrows into the outer layer of the animal's skin. The mite causing mange is fairly species-specific, and thus it would be difficult for a human to contract mange from an infected wild animal.

If a person is bitten or scratched by a coyote, immediately scrub the wound with soap and water. Flush the wound liberally with tap water. In other parts of North America coyotes can carry **rabies**. Contact your physician and the local health department immediately. If your pet is bitten, follow the same cleansing procedure and contact your veterinarian.

Legal Status

Because legal status, trapping restrictions, and other information on coyotes and wolves changes, contact your local, state, or provincial wildlife office, or visit their Web site for updates. (See Appendix E for contact information and where to access the state and provincial laws mentioned below.)

Oregon: The coyote is classified as an unprotected mammal, but to hunt or trap a coyote a license is required. The coyote is also classified as a "predatory animal" by the Oregon Department of Agriculture, which allows a landowner, a person lawfully occupying the land, or their agent to control coyotes without a permit (ORS 610.105). It is illegal to capture coyotes from the wild or to hold coyotes in captivity for any reason except in publicly owned parks or zoos (ORS 497.298).

Washington: Coyotes are unclassified wildlife; however, a license and an open season are required to hunt or trap them (RCW 77.32.010). A property owner or the owner's immediate family, employee, or tenant may kill or trap a coyote on that property if it is damaging crops or domestic animals (RCW 77.36.030). A license is not required in such cases.

Except for bona fide public or private zoological parks, persons and entities are prohibited from importing a coyote into Washington State without a permit from the U.S. Department of Agriculture and written permission from the Department of Health. Persons and entities are also prohibited from acquiring, selling, bartering, exchanging, giving, purchasing, or trapping a live coyote for a pet or export (WAC 246-100-191).

The **gray wolf** is a federal and state Threatened species that cannot be hunted or trapped (WAC 232-12-014).

British Columbia: Coyotes are designated as both furbearing animals and small game. They may be legally hunted or trapped only with a license during open seasons as defined by regulation; there is no limit on the number harvested. Coyotes may be hunted or trapped outside of general open seasons by a person on their own property if the coyote is a menace to a domestic animal or bird (Wildlife Act, 26(2)). No permit is required in such cases.

Wolves are considered big game under the Wildlife Act and may be legally hunted under license during open seasons as defined by regulation. A permit must be issued from a regional manager to capture or kill wolves that pose a threat to public safety (B.C. Reg. 253/2000, Section 2(h) and 2(c)(iii)). A person may trap and kill on their own property without a permit, however, any wolf that is a menace to a domestic animal or bird (Wildlife Act, 26(2)).

For information on importing coyotes or wolves, contact the B.C. Director of Wildlife (see Appendix E for contact information).

CHAPTER 7

Deer

Deer are among the most familiar animals of the Pacific Northwest, and in many places they are the largest wildlife that people encounter. Their aesthetic beauty is appreciated and admired, although their fondness for garden and landscape plants tries some peoples' patience.

Two species and four subspecies of deer occur in the Pacific Northwest (see Table 1).

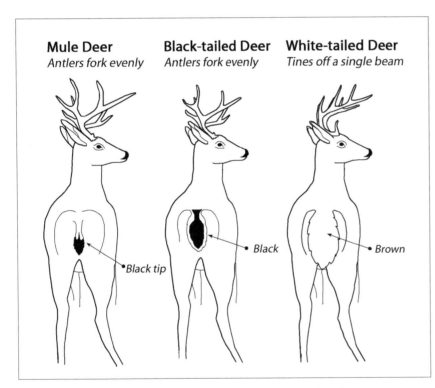

Figure 1. *Characteristics to help identify Pacific Northwest deer. (Washington Department of Fish and Wildlife.)*

Facts about Deer
Food and Feeding Habits

- Deer eat a wide variety of plants, but their main food item is browse—the growing tips of trees and shrubs. In late winter and early spring, deer eat grass, clover, and other herbaceous plants (Table 2).
- Deer also eat fruit, nuts, acorns, fungi, lichens, and farm and garden crops if available.
- For their first few weeks of life, fawns thrive on milk, which is more than twice as rich in total solids as the best cow milk.
- Deer eat rapidly and, being ruminants, initially chew their food only enough to swallow it. This food is stored in a stomach called the "rumen." From there it is regurgitated, then re-chewed before being swallowed again, entering a second stomach where digestion begins. From there it is passed into a third and then a fourth stomach, finally entering the intestine.

Shelter and Range Needs

- Deer are sometimes referred to as "edge" species, meaning they thrive at the interface of openings and cover patches. This allows deer to feed in productive openings while being close to escape cover.
- Many wooded suburban environments, such as parks, greenbelts, golf courses, and roadsides, meet the needs of deer.
- Mule deer can move long distances during spring and fall migrations to avoid mountain snow. Mule deer summering in the Cascade mountains migrate as far as 80 miles (128 km) to reach adequate winter range.

Table 1. Pacific Northwest Deer

Rocky Mountain mule deer (*Odocoileus hemionus hemionus,* Figs. 1 and 2), or mule deer, inhabit areas east of the Cascade mountains in Oregon and Washington, and interior British Columbia, preferring open forests and sagebrush meadows.

During summer, mule deer are tan to light brown; during winter, they are a salt-and-pepper gray. They have large, dark-edged ears, from whence they derive their name. The 7 to 8 inch (18–20 cm) tail of a mule deer is white, except for a black tip. Mule deer are the largest deer in the Pacific Northwest. Adult bucks weigh up to 250 pounds (112 kg); adult does weigh 120 to 170 pounds (54–76 kg).

Columbian black-tailed deer (*Odocoileus hemionus columbianus,* Fig. 1), also called coast deer, are the most common deer subspecies in the Pacific Northwest. They occur from the crest of the Cascade mountains in Oregon and Washington west to the ocean, preferring brushy, logged lands and coniferous forests. In British Columbia they can be found on nearly all the islands, and west of the summit of the Coast and Cascade ranges.

Many of the physical characteristics of black-tailed deer are similar to those of the larger mule deer. The tail is broader and the backside of the tail is covered with dark brown hair that grades to black near the tip. When alarmed or fleeing from danger, the tail may be raised, displaying the broad, white underside. Adult black-tailed deer bucks weigh 140 to 200 pounds (63–90 kg) and adult does weigh 90 to 130 pounds (40–58 kg).

White-tailed deer (*Odocoileus virginianus idahoensis,* Fig. 1 and back cover) occur in the Peace River region of northeastern British Columbia, and in southeastern British Columbia, eastern Washington, and irregularly throughout Oregon. They are typically seen on farmlands, in low-elevation stream and river corridors, and near populated areas.

White-tailed deer are usually reddish tan in summer and brownish gray in winter. They derive their name from their broad, 10 to 11 inch (25–28 cm) long tail. When alarmed, white-tailed deer raise their flaglike tail, displaying the white underside. White-tailed bucks weigh 150 to 200 pounds (68–90 kg) and adult does tip the scales at 110 to 140 pounds (50–63 kg).

Columbian white-tailed deer (*Odocoileus virginianus leucurus),* a subspecies of white-tailed deer, are found in limited areas along the lower Columbia River in Oregon and Washington, and in the Umpqua River Basin near Roseburg, Oregon. Once common in these and other areas, this species was federally listed as an Endangered species in Washington and Oregon in 1967.

Hybrids of mule deer/black-tailed deer and mule deer/white-tailed deer are known to occur. Mule deer/black-tailed deer hybrids are common where their ranges overlap. Mule deer/white-tailed deer hybrids are not common, but are occasionally seen where both species occur in close proximity.

Figure 2. Mule deer are the largest deer in the Pacific Northwest. (Washington Department of Fish and Wildlife.)

All About Antlers

One unique characteristic of the deer family is their antlers. While horns remain permanently affixed to the skull, antlers are shed every year. The exception to this rule is the American pronghorn antelope; they shed their horn sheaths once each year (the pronghorn's bony horn cores are not shed).

Pacific Northwest species that grow antlers each year include deer, elk, moose, mountain caribou, and bighorn sheep.

Antlers form beneath a covering of living skin. This skin, often called velvet, is complete with hair, a very sensitive nerve network, and blood vessels. When antler growth is complete the velvet dries, shreds, and peels off, leaving the hard, mineralized antler ready for the breeding season.

Male fawns develop buttons (small bumps on top of the head) at six to eight months of age. These buttons are the rudimentary beginnings of the young buck's first antler set. Just before the fawn's first birthday, these velvet covered buttons begin to elongate, growing from bony extensions of the skull known as pedicels. By September these first antlers are fully grown spikes, or small, forked antlers with two points (Fig. 3).

Each year, the antlers tend to grow in mass and diameter. Older bucks tend to have more antler points than younger bucks, but the number of points is not a reliable indicator of actual age. Antler size and conformation also respond to nutrition, and thus serve to advertise the physical condition of the buck. Rich feeding in captivity has produced five-point antlers on yearlings, while a meager food supply can limit even dominant bucks to forks. Bucks generally attain adult-size antlers when they are four to five years of age, but the size and weight of the antlers may continue to increase each year until age ten.

Antlers serve to establish dominance hierarchies among bucks. Big antlers, like bright feathers on male songbirds, are an example of fitness evolved through sexual selection. Because large antlers mean a buck has either survived many years, has superior genetics, or uses high-quality areas, bucks with large antlers make good sires for a doe's fawn. Does tend to select dominant bucks with large antlers for their mates, and this selection enhances the success of bucks with large antlers even more.

Bucks carry their antlers through the fall, dropping them between late December and early March. Hormonal changes cause a weakening of the bone at the tip of the pedicel, where the antler-growing center is located, and the pedicel/antler connection eventually becomes so weak that the antler separates and falls from the pedicel.

Most antlers that have been on the ground for more than a few weeks will show considerable signs of gnawing by smaller animals, and after a year most of the antler points will have been considerably shortened by these mineral craving critters. Dropped antlers are chewed by mice, rats, squirrels, hares, and porcupines, helping them to sharpen their front teeth while supplying them with calcium, phosphorus, and other minerals. (For additional information, see "Elk Rubs" and "Collecting Antlers" in Chapter 9.)

Figure 3. A deer's first antlers are fully grown spikes or small, forked antlers with two points. (Washington Department of Fish and Wildlife.)

- Black-tailed and white-tailed deer normally reside within a ½ to 3 square-mile (0.8–4.8 km) area; in mountainous locations, they move to lower elevations for the winter.

Reproduction and Family Structure

- Deer breed during a rutting season that normally occurs in November and December. Bucks compete for the right to breed using ritualized posturing and movements, and occasionally through intense fighting.
- Unlike elk, deer bucks do not herd groups of females; however, a single mature buck may breed with several females.
- Pregnancy lasts 180 to 200 days. Younger does give birth to one

fawn, while does three to nine years of age and in good condition often have twins. White-tailed deer will occasionally have triplets.

- Newborn fawns nurse soon after birth and can walk on spindly legs almost immediately.
- Adult bucks take no part in raising fawns, and generally remain solitary or form bachelor groups throughout the summer.
- Family groups usually consist of a doe and her fawns, and sometimes her fawns from the previous year. Occasionally, groups of several does may be seen together.
- In winter, deer may be observed in larger groups of 15 to 30, usually grouping because they are concentrated in limited winter habitat.

Mortality and Longevity

- Cougars, bears, coyotes, and domestic dogs prey on adult deer; young fawns fall victim to these species as well as to eagles and bobcats.
- Hunting, vehicles, and diseases all take their toll on deer. In many deer populations, hunting dampens the effects of other mortality factors; as hunting mortality decreases, other forms of mortality tend to increase, and vice versa.
- Probably few deer live longer than ten years, and most live for no more than five.

Viewing Deer

Deer often become very habitual in their activities. They show up at the same time and follow the same trails, taking paths of least resistance. Although deer may be active at any time of day, they

Figure 4. Deer tracks are 1½ to 3¼ inches (4–8.5 cm) long. The smallest prints belong to fawns and the largest to mature bucks. The small dewclaws shown here on a mule deer track may not register. (From Pandell and Stall, Animal Tracks of the Pacific Northwest.*)*

are most active near dawn and dusk (a pattern of activity called "crepuscular"). Typically, deer feed in open habitats such as meadows and clearcuts, retreating to more secure areas, such as thickets and closed canopy forests, to rest and chew their cud.

To observe deer, position yourself at dawn or dusk near cover in a good deer-feeding area. Remain absolutely still, because deer are alert for any movement. They also have a good sense of smell; stay downwind of the feeding area to prevent deer from detecting your scent. Deer in wild hunted areas will probably not stay around long if they notice you; deer in areas where hunting is not permitted are more likely to tolerate your presence.

Conditions in the fall can make for good deer viewing. Bucks are battling each other for females and are not as concerned about staying hidden. Leafless trees afford you greater visibility, and when it is raining there is less

chance of you being heard crunching through an area. However, be aware of open hunting seasons during this time of year and wear bright orange clothing for your safety.

Winter can be a good time to view deer because they are often concentrated at lower elevations. This is a critical time for deer and, if harassed, deer will expend vital energy to flee the harassment. However, during winter deer can often be observed without harassment by using binoculars and spotting scopes to scan open, sunny areas, especially those with significant shrub cover.

Finally, never approach a deer closely; if threatened it can cause serious injury. Doe deer especially will go to great lengths to defend their young.

Tracks and Trails

Deer tracks are easy to identify. In a normal hoof print, the two roughly teardrop-shaped halves print side by side to form a split heart (Fig. 4). When a deer is walking on a slippery surface, such as mud or snow, its hooves are likely to be spread into a V, which helps keep the deer from sliding forward.

Deer have regular routes through their home range; these become well-worn trails that look a little like narrow human footpaths. The trails are clear of low vegetation, but are not bare unless they are in shade or are heavily used by deer and other mammals.

Droppings

Deer droppings vary greatly in size and shape, but are easy to identify. Most of the year they are deposited in a group of 20 to 30 dark cylindrical pellets with one flat or concave end and one pointed end. Individual pellets are ½ to ¾ inch (1.3–2 cm) long; individual piles are 4 to 6 inches (10–15 cm) in diameter. When deer are feeding on moist vegetation, the pellets stick together and form clumps. New droppings have a shiny, wet appearance for a few days and then lighten in color as they age.

For comparison of deer and elk droppings, see "Droppings" in Chapter 9.

Feeding Areas

In areas where many deer live, a noticeable "browse line" appears on trees where the deer have repeatedly reached up to eat low-hanging twigs and branches. Similarly, the tops of shrubs may be browsed, leaving only a few inside branches extending upward. Browsing seldom occurs more than 4 feet (1.25 m) above the ground, except in areas with deep snow.

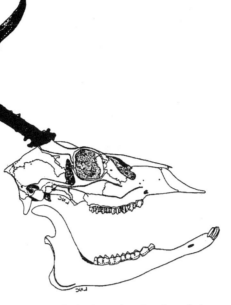

Figure 6. As shown here in a lateral view of a mule deer scull, deer lack upper incisors and canine teeth, and must press foods between their hard upper palates and their bottom teeth. (From Verts and Carraway, Land Mammals of Oregon.)

Browsing by deer can usually be identified since twig ends deer have browsed have a ragged appearance, while those browsed by rabbits, porcupines, and other rodents have a neat, clipped appearance. This is because deer lack upper incisors and canine teeth, and cannot nip off twigs. Instead they must press foods between their hard upper palates and their bottom teeth, and jerk their heads up to tear it free (Figs. 5 and 6).

However, when deer browse new growth they leave a clean, blunt stem-end, where the tender shoots break off. The height of the clipped plant will then be the indicator of what species ate it.

When browse and other green foods are no longer available, deer strip bark from young trees.

Rubbing Sites

Bucks scrape off the velvet covering their antlers by rubbing them against young trees and shrubs

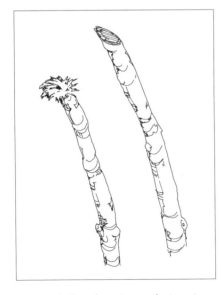

Figure 5. Deer browsing results in twig ends having a ragged appearance; plants browsed by rabbits, porcupines, and other rodents appear neatly clipped. (Drawing by Jenifer Rees.)

(Fig. 7). These rubbing sites also communicate their presence and breeding readiness to other deer. This communication has several facets: the visual sign left by the buck's rubbing, chemical signals left from glands on the buck's face, and the sound of the buck thrashing branches of the tree on which it is rubbing.

Although antlers are bone-white when the velvet is first removed, they become stained various shades of brown as plant compounds accumulate through constantly being rubbed by brush and trees. The color results from a chemical reaction of the plant compounds with the air, in a process known as oxidation.

Bedding Areas

An area of flattened vegetation 3 to 4 feet (90 cm–1.2 m) long and 2 to 3 feet (60–90 cm) wide indicates where a deer has bedded down. Deer sleep in dense cover or tall grasses and may return to the same spot over many days. Since deer often travel in small groups, there may be several "deer beds" in the same vicinity. During winter, similarly sized depressions in the snow, often littered with old hairs, characterize bed-sites.

Calls

The best-known vocalization is an alarm whistle, "snort," or "blow," made when deer exhale forcefully through their nostrils. The sound resembles a sudden release of high-pressure air. The snort is a danger call that alerts all deer in the area of a potential predator.

Older fawns commonly bleat when frightened, and older deer sometimes do as well. Doe deer call to their hidden fawns with a

Figure 7. Bucks rub their antlers against trees and shrubs. This leaves the bark in a shredded condition, often with long frayed strips of bark hanging at the top of the barked area. Rubbing is generally restricted to an area between 1.5 to 3.5 feet (45 cm–1 m) above ground. (Drawing by Jenifer Rees.)

soft, gentle mewing sound, and fawns respond quickly to this call by seeking the side of their mother. Bucks make a grunt during the rut.

Preventing Conflicts

In most places deer are valued as watchable wildlife or as game animals. However, where hunting is limited or no longer permitted and natural predators are few, deer populations can increase to a point where human/deer conflicts become a concern.

Problems associated with high deer populations include damage to crops, ornamental plants, restoration and reforestation projects, and deer/vehicle collisions.

Problem areas are often where new development has appeared in traditional deer habitat. Those who live on the edges of new develop-

Deer Fawns and What to Do If You Find One

To reduce the risks of a predator locating her fawn, a doe seeks seclusion just prior to birth, trying to be less conspicuous by avoiding other deer. For the first few weeks of the fawn's life, the doe keeps the fawn hidden except for suckling bouts. The doe may also feed and bed a considerable distance from the fawn's bed site. This way, even if a predator detects the doe, the fawn may still have a chance of avoiding detection.

To further keep her fawn safe from predators, the doe consumes the fawn's urine and droppings to help keep the fawn as scent-free as possible. The droppings provide the mother with further nutrition at a time when it is much needed.

When not nursing, the fawn curls up in a bed site and remains motionless, its white spots blending in well with the sun-flecked ground. Fawns lose their spots at 90 to 120 days of age, when they begin growing their winter coats.

Every year, wildlife departments and wildlife rehabilitators receive calls about "orphaned" fawns. Callers are told that in the spring it is a perfectly natural occurrence to come across a fawn that is seemingly by itself in the woods. The fawn is probably not alone; its mother is nearby, aware, and attentive.

The advice to anyone encountering a fawn lying quietly alone in the woods is to leave it alone. Mother will be nearby and will be taking care of it once you move away.

If you have handled the fawn, rub an old towel in the grass and wipe the fawn to remove human scent. Using gloves, return the fawn to where it was found. Fawns can often be returned to their mothers if taken back to where they were found within eight hours.

If a fawn appears cold, weak, thin, or injured, and its mother does not return in approximately eight hours, it may be orphaned. In such a case, you can call a local rehabilitator (look under "Animal" or "Wildlife" in your phone directory) or your local Department of Fish and Wildlife office for the name and phone number of a rehabilitator in your area. (See Chapter 20 for additional information on wildlife rehabilitators.)

Figure 8. In the spring it is a perfectly natural occurrence to come across a fawn that is seemingly by itself. (Washington Department of Fish and Wildlife.)

Tips for Attracting Deer

Although property owners with large acreage can provide significant deer habitat, those with small acreage can also contribute. Keep in mind that deer may damage ornamental plants and gardens, and might also attract animals that prey on deer, such as cougar and bear.

The best way to attract deer to your property is to protect and maintain deer habitat:

- Provide hiding cover. Deer use hiding cover year-round during resting periods throughout the day, but it is especially important during hunting season and the first few months of a fawn's life. Hiding cover can consist of stands of trees or dense shrubs. Where cover is limited, providing large, disturbance-free areas may still encourage use by deer.
- Conserve areas with forage plants that deer prefer (Table 2).
- Prevent infestations of noxious weeds that degrade areas containing preferred food plants (contact your county extension office for information).
- Conserve vegetation along streams and other freshwater areas and avoid placing roads near these areas, which are among the most favored habitats of deer. Consider developing artificial watering areas if water is scarce on your property.
- If a contractor is clearing vegetation, make sure the contract states that the contractor will be held responsible for plant restoration or alternate improvements if areas set aside for deer are inadvertently cleared. Temporarily fence important areas and supervise the work to keep disturbances to a minimum.
- Consult with local resource management agencies for advice on specific habitat-management activities that may be highly effective in your local area. These activities might include prescribed burning to rejuvenate shrub species, fertilizing of old fields to increase palatability of forage to deer, and not mowing areas that provide food and shelter.
- Do not let dogs run loose and chase deer. This fatigues and weakens deer, especially when they are forced to run on crusted snow. Deer can also be killed by dogs, either directly or indirectly, because fleeing burns up energy needed to combat cold and starvation. Dogs running in packs are even more threatening to deer. Some county laws provide for shooting dogs and/or fining their owners if dogs are observed harassing wildlife. (See Appendix D for additional information.)
- Inform guests, visitors, and contractors coming to your property that you do not allow dogs to roam free.
- Property fences and wire fences constructed on ranges used by deer should have a 17-inch (45 cm) gap at the bottom to let fawns and adult deer pass beneath them, and be no more than 4 feet (1.2 m) high to let adults jump safely over them.
- For additional information, see "Tips for Attracting Elk" in Chapter 9.

ments, or adjacent to undeveloped areas, may experience higher deer damage than others whose homes are within developments or otherwise buffered by urbanization. However, deer readily adapt to human activity and are seen in unlikely places at times.

If deer damage is occurring on commercial property, a wildlife professional from your local wildlife department can assist you in evaluating damage-control options. Your local wildlife office may also have cost-share or other programs available to help you manage deer on your property.

Deer Fences

When deer browsing is moderate to severe, or a landowner isn't willing to tolerate even a limited amount of damage, fencing to exclude deer is the only option. However, traditional deer fences are not always practical because of appearance, zoning restrictions, cost, or rugged terrain. In such cases, another type of barrier described below may be appropriate.

Questions to ask: Before installing a deer fence, ask these questions:

- Must my entire property be protected, or only certain parts or plants?
- Is this need temporary, such as to protect young trees for a few years?
- Are there visual constraints, including aesthetics, or your neighbor's or a passerby's view?
- Are there any community or local government regulations or restrictions?
- Is building a fence time- and cost-effective, or should other methods be considered, even though they are not as effective?

Table 2. Food Plants used by Pacific Northwest Deer

Mule Deer	Black-Tailed Deer	White-Tailed Deer
Trees and Shrubs		
Serviceberry, *Amelanchier alnifolia* Sagebrush, *Artemisia tridentata* Deer brush, *Ceanothus integerrimus* Snowbush, *Ceanothus velutinus* Rabbitbrush, *Chrysothamnus* spp. Red-twig dogwood, *Cornus sericea* Winterfat, *Eurotia lanata* Juniper, *Juniperus* spp. Mock-orange, *Philadelphus lewisii* Ninebark, *Physocarpus capitatus* Ponderosa pine, *Pinus ponderosa* Bitter cherry, *Prunus emarginata* Douglas-fir, *Pseudotsuga menziesii* Bitterbrush, *Purshia tridentata* Golden current, *Ribes aureum* Wild rose, *Rosa* spp. Thimbleberry, *Rubus parviflorus* Willow, *Salix* spp. Snowberry, *Symphoricarpos albus*	Vine maple, *Acer circinatum* Red alder, *Alnus rubra* Serviceberry, *Amelanchier alnifolia* Snowbush, *Ceanothus* spp. Deer brush, *Ceanothus integerrimus* Hazelnut, *Corylus cornuta* Hawthorn, *Crataegus columbiana* Salal, *Gaultheria shallon* Douglas-fir, *Pseudotsuga menziesii* Oak, *Quercus* spp. Cascara, *Rhamnus purshiana* Blackberry, *Rubus* spp. Thimbleberry, *Rubus parviflorus* Salmonberry, *Rubus spectabilis* Willow, *Salix* spp. Elderberry, *Sambucus* spp. Western red-cedar, *Thuja plicata* Red huckleberry, *Vaccinium parvifolium*	Serviceberry, *Amelanchier alnifolia* Sagebrush, *Artemisia tridentata* Deer brush, *Ceanothus integerrimus* Crabapple, *Malus* spp. Bitter cherry, *Prunus emarginata* Douglas-fir, *Pseudotsuga menziesii* Bitterbrush, *Purshia tridentata* Willow, *Salix* spp. Western red-cedar, *Thuja plicata*
Forbs and Legumes		
Balsamroot, *Balsamorhiza* spp. Prickly lettuce, *Lactuca serriola* Twinflower, *Linnaea borealis* Alfalfa, *Medicago sativa* Burnet, *Sanguisorba* spp. Dandelion, *Taraxacum* spp. Clover, *Trifolium* spp. Trefoil, *Trifolium* spp.	Pearly everlasting, *Anaphalis margaritacea* Balsamroot, *Balsamorhiza* spp. Fireweed, *Epilobium angustifolium* Cat's ear, *Hypochaeris* spp. Alfalfa, *Medicago sativa* Clover, *Trifolium* spp. Vetch, *Vicia* spp.	Alfalfa, *Medicago sativa* Burnet, *Sanguisorba* spp. Dandelion, *Taraxacum* spp. Clover, *Trifolium* spp.
Grasses and Others		
Wheatgrass, *Agropyron* spp. Oats, *Avena fatua* Cheatgrass, *Bromus tectorum* Bluegrass, *Poa* spp. Wheat, *Triticum aestivum* Lichen Mushrooms and other fungi	Oats, *Avena fatua* Deer fern, *Blechnum spicant* Bluegrass, *Poa* spp. Sword fern, *Polystichum munitum* Wheat, *Triticum aestivum* Lichen Mushrooms and other fungi Seaweed	Wheatgrass, *Agropyron* spp. Orchard grass, *Dactylis glomerata* Fescue, *Festuca* spp. Lichen Mushrooms and other fungi

Figure 9. A 6 to 8 foot (1.8–2.4 m) woven-wire fence presents a formidable deer (and elk) barrier when properly constructed and maintained. The 20-year life span of a well-built fence can justify its cost.

Major materials include sturdy, rot-resistant wooden corner posts set in concrete (optional), wooden or studded steel T line posts, woven-wire fencing, and gates.

If needed, extensions can be attached to the top of the fence to prevent deer or elk from jumping over. A 2-foot (60 cm) high band of chicken wire can be added to the bottom to exclude rabbits and hares. (Drawing by Jenifer Rees.)

Before you build: If you decide to build or have a fence built, construct it properly. A poorly constructed deer fence is dangerous to the deer, and will not protect your valuable plants. If a deer fence exists nearby, ask the property owner about its effectiveness, its construction, and who built it. To locate a fence builder, look under "Fence Contractors" in your phone directory. Request references and follow up on them before hiring any contractor.

If you build a deer fence yourself, carefully measure the area to insure the efficient use of fence rolls. (You don't want to end up having to cut a small length of fence from a new and potentially expensive roll). In addition, make sure you know where your property line is—existing fences may not be on your property. Never fence across an easement without notifying the necessary authority.

Fencing facts:
- It is easier to build a fence while the land is vacant; when possible, fence an area before you plant an orchard or a garden.

- Enclose the entire area needing protection (including driveways). Deer will wander the perimeter of the fenced area until they find an opening.
- Keep fencing material flush to the ground (including under gates). Fill dips with gravel, rocks, logs, or other suitable material. Incredibly, deer will try to either crawl under or squeeze through a fence before jumping over it.
- Deer can be excluded from areas with a properly constructed and maintained 6- to 8-foot (1.8–2.4 m) high fence (Figs. 9 and 10). The higher fence will be needed in an area with many deer and a low supply of wild food.
- A board fence or hedge that prevents deer from seeing a safe landing zone on the other side need be only 5½ feet (1.7 m) high.
- The larger the area being enclosed, the more travel patterns will be disrupted, and the

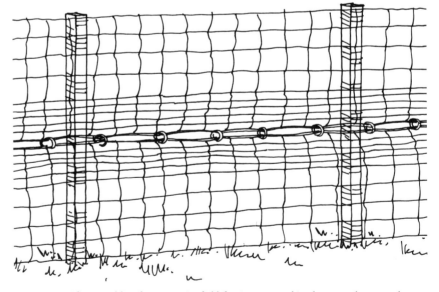

Figure 10. If two widths of woven-wire field fencing are combined, secure them together with hog rings at approximately 18-inch (45 cm) intervals. To allow small mammals access through the fence, invert the lower fence run so the larger openings are at the bottom. (Drawing by Jenifer Rees.)

Deer 61

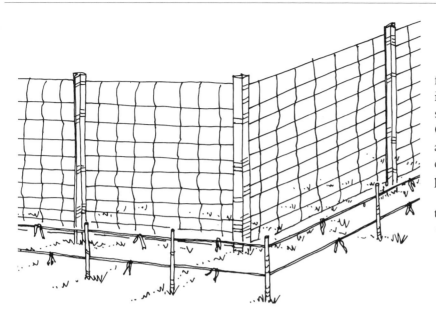

Figure 11. When a deer jumps over a fence, it will start its jump just inches from the barrier. In places with low deer pressure, people have had success with stringing two wires 3 feet (90 cm) out from the existing fence. Electric wires with flags placed on them will make the barrier more effective. (Drawing by Jenifer Rees.)

more pressure the fence will receive.

- Large areas with high deer pressure should be fenced with woven-wire deer fencing or a high-tensile electric wire. Heavy-duty black polypropylene deer fencing is commercially available and useful where other fencing is visually obtrusive. See "Wildlife/Human Conflicts" in Appendix F for resources.
- Information on corner bracing, stretching wire, and other fence-building details can be obtained from fencing material suppliers.

Electric fences: A properly designed and maintained electric fence can be very effective at preventing deer from entering an enclosed area as small as a vegetable garden or an area as large as a commercial orchard. One or two hot wires can also be strategically placed to keep other animals out of chicken coops, ponds, and other areas.

Electric fences work by delivering a high-voltage but low-amperage jolt that won't set fire to plants or injure animals or humans. Some fences are built with alternating positive and negative wires so an animal receives a shock when it touches both wires simultaneously. More commonly, the animal completes the circuit when it touches a hot wire while standing on the ground. The advantage of this design is that the animal only has to touch one wire to receive a shock. One disadvantage is that plants must be kept from contacting the wire or the fence will short circuit. (Newer low-impedance chargers make this less of a problem.)

A fence with eight wires evenly spaced to 80 inches (2 m) is believed to be adequate to keep deer out of an enclosed area (Fig. 12). Due to the variables affecting your selection of a power source, and fence design and operation, it is best to consult a reputable dealer for the specifics regarding its use (look under "Fence Contractors" in your phone directory). Information is also available from farm supply centers. Most home improvement centers carry units suitable for protecting gardens. Consult your local zoning office and neighborhood covenants to determine if electric fences are permitted where you live.

Here is some general information on electric fences:
- An electric fence is not a physical barrier to entry. It acts as a psychological barrier that some deer will continually test. Because of this, it is important that your fence be properly designed, installed, powered, and maintained.
- Maintenance visits are required at least every two weeks to check the voltage and to rectify any problems, such as fixing sagging wires. However, inspections every two days will be necessary during the first three weeks after installation. This is to make certain that animals encountering the fence for the first time don't damage the fence or get injured.

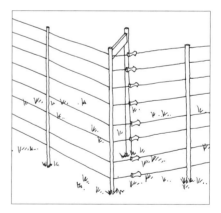

Figure 12. A fence with eight wires evenly spaced to 80 inches (2 m) is believed to be adequate to keep deer (and elk) out of an enclosed area. The fence should alternate positive and negative wires. Twisting aluminum foil dipped in peanut oil onto electric fence wires helps "initiate" animals to the shocking properties of the fence. The animals will lick the treated foil and receive a full-charge shock. (Drawing by Jenifer Rees.)

- An electric fence can be less expensive to build than woven-wire fences, but requires more effort to maintain and will not last as long.
- Recently available electric poly-wire/tape fencing is much wider and is meant to also work as a visual repellent. It is less expensive, relatively portable, and easy to install. Also, it can easily be removed for mowing or farming and moved to other locations as the need arises.
- Poor conductivity occurs in rocky or dry ground. Most projects fail because of poor grounding.

Mini-Barriers

Barriers to protect small areas, individual plants, or vulnerable parts of plants can be purchased or made at home. These have the advantage of being less expensive and obtrusive than full fences, allowing deer access to surrounding food plants while protecting others. They can also protect plants from bucks rubbing their antlers, which breaks branches and strips bark off trunks.

To prevent deer from pushing over or moving a mini fence surrounding a tree or shrub, the fence should be 5 feet (1.5 m) high and staked to the ground (Figs. 13 and 14).

Netting—normally sold to protect berries and fruit from birds—can be draped over individual plants or used as a temporary fence. However, deer can easily break lightweight netting with their hooves to get to desirable plants and songbirds can get entangled in excess netting. Stronger netting material is commercially available from bird-control outlets and companies selling

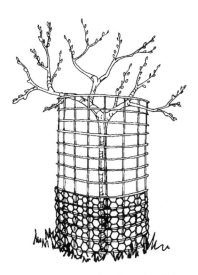

Figure 13. A mini deer fence should be at least 5 feet (1.5 m) high, placed far enough out from the plant to prevent deer from accessing the plant and causing damage, and be firmly staked to the ground. Prior to installation, remove all grass and weeds within the barrier. Add mulch to reduce maintenance needs. A 2-foot (60 cm) high band of chicken wire can be added to the bottom to exclude rabbits. Such barriers designed for elk should be 6 feet (1.8 m) high. (Drawing by Jenifer Rees.)

polypropylene deer fencing. (See "Wildlife–Human Conflicts" in Appendix F for resources.) When draped over plants, any netting will need continual rearranging to protect new growth.

An inexpensive and subtle deer barrier can be made from 100-pound test monofilament fishing line tied to sturdy, 5-foot (1.5 m) tall stakes, or attached to a structure. Nearly invisible at a 12-inch (30 cm) spacing, fishing line works best to protect small enclosures, such as surrounding several rose bushes.

Commercially available tree guards protect trees from damage done to the bark from deer antlers and gnawing from other wildlife (Fig. 7). They can be wrapped around nearly any size tree, cut to different heights, and expand as the tree grows (Fig. 15). (See "Wildlife/Human Conflicts" in Appendix F for resources.)

Plastic or nylon tubes, netting, and bud caps have all been used successfully to protect small transplants and growing tree tips (Fig. 16). For small plants, use tubes that match the plant's height and allow room for growth. Be sure to hold the tube upright with a wood or metal stake.

Cattle Guards

Some people see cattle guards as eyesores, but they provide the most effective protection for ungated driveways on properties that are otherwise fenced to keep deer out.

Before installing a cattle guard, it's important to determine how many deer use the driveway (assuming a deer fence encloses the perimeter of your property). In areas where deer concentrations are heavy, deer will test the cattle guard and are known to jump over one that is undersized.

Most cattle guards are about 14 feet (4.5 m) wide and 10 to 14 feet (3–4.5 m) long. You will need

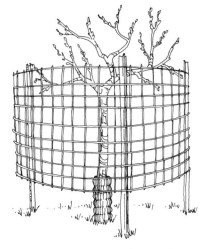

Figure 14. A variation of a mini-deer fence that allows for mowing. Note the plastic guard to prevent damage from mowing and gnawing animals. (Drawing by Jenifer Rees.)

Figure 15. Loosely wrapping vulnerable areas with commercially available tree-guards or chicken wire can protect tree bark from a buck's antlers. Placing branches or another solid material, such as rebar, on four sides of trees can also prevent bucks from accessing tree trunks. (Drawing by Jenifer Rees.)

a backhoe to excavate a hole and to lower the sections into place; each section weighs over 1,000 pounds (450 kg).

Repellents

Deer repellents use a disagreeable odor or taste, or a combination of both, to dissuade deer from eating the treated plant. They are easy to apply and homemade solutions are inexpensive (see "An All-in-One Homemade Deer Repellent").

Numerous odor and taste repellents have been developed to reduce deer damage, and new products are continually becoming available. There have been numerous studies to test the effectiveness of these repellents, often producing conflicting results. No repellent eliminates deer damage entirely.

Before you apply: Most repellents function by reducing the palatability of the treated plant to a level below other available plants. Hence, repellent effectiveness depends upon the availability of wild deer food. Repellents are more appropriate for short-term rather than long-term problems and are the most practical for non-

commercial users experiencing low to moderate deer damage.

Repellents work best if applied **before** the deer develop a routine feeding pattern. This means applying repellents before leaves or flower buds emerge and as new growth appears. It's easier and more effective to prevent a feeding habit from forming than to try to break an established one.

Repellent facts:

- Spray-on repellents need to be applied frequently to protect the new plant growth, and will need to be reapplied after rain and long exposure to hot, dry, or windy weather.
- Deer may become accustomed to the same repellent over time, and eventually ignore it. Alternating repellents may help keep deer confused and more wary of eating your plants.
- Repellents that are applied to plant surfaces are generally more effective than capsules containing garlic oil, bags of hair, or other devices that pro-

duce an odor intended to protect a specific area.

Finally, before putting complete faith in a repellent, first try it on a small area. Always use commercial repellents according to the manufacturer's directions.

Scare Tactics

Like most animals, deer are neophobic (fearful of novel objects), and many scare tactics take advantage of this behavior. However, deer soon get accustomed to new things and damage resumes after they realize no actual harm will come to them. As with repellents, a given tactic will work on some deer, but no single one seems to work on all of them. If the animals are already used to feeding in the area, scare tactics will last an even shorter length of time.

Scare tactics can be visual (scarecrows, bright lights, spare blankets), auditory (noisemaking devices such as exploders, whistles, etc.), or olfactory (predator urine or droppings).

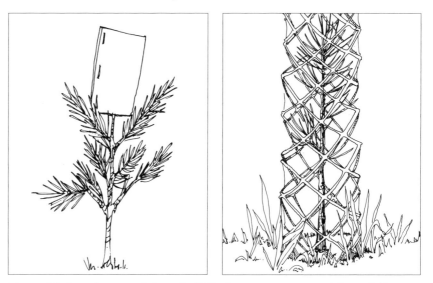

Figure 16. Several types and sizes of individual protectors are commercially available to protect entire plants or their growing tips. Studies indicate that a rolled or folded sheet of standard size paper provides inexpensive, effective control. Bud covers are recommended for protection of small, newly planted conifers. (Drawings by Jenifer Rees.)

An All-in-One Homemade Deer Repellent

Mix the following in a 1-gallon tank sprayer:

– **2 beaten and strained eggs—strain them to remove the white strings surrounding the yolk, which otherwise will plug up your sprayer).** ***Note:*** **Studies indicate that repellents with active ingredients that emit sulfurous odors, such as egg solids, generally provide the best results.**

– **1 cup milk, yogurt, buttermilk, or sour milk**

– **2 tsp. Tabasco sauce or cayenne pepper**

– **20 drops essential oil of clove, cinnamon, or eucalyptus, found in small bottles at health food stores**

– **1 tsp. cooking oil or dormant oil**

– **1 tsp. liquid dish soap**

Top off the tank with water and pump it up. Shake the sprayer occasionally and mist onto dry foliage. One application will last for 2 to 4 weeks in dry weather.

One recent innovation is a motion sensor combined with a sprinkler that attaches to a hose. When a deer comes into its adjustable, motion-detecting range, a sharp burst of water is sprayed at the animal. This device appears to be effective by combining a physical sensation with a startling stimulus. Similar in approach but less effective are radios and lights hooked up to a motion detector.

A dog can help keep deer away, especially if it is large and awake. To keep the dog at home while simultaneously repelling deer from your property, use a "dog trolley" or an invisible (buried electric) fence, where paractical. Avoid tethering a dog near stairways and fences, and provide at least 15 feet (4.5 m) of cleared space for it to move around in. Do not use a choke chain, and remove all debris that could tangle or injure your dog. Provide shade, water, and shelter for the dog at all times. (See Appendix D.)

Landscaping with Deer-Resistant Plants

Although a deer fence or other barrier is the best insurance against damage, landscaping with deer-resistant plants is a more aesthetic alternative. In addition, there may be areas where a deer fence isn't practical. A walk or drive through the neighborhood or a visit to the neighbors can give you an idea of what plants are less palatable to deer.

Whether or not a particular plant will be eaten depends upon several factors: the deer's nutritional needs, its previous feeding experience, plant palatability, time of year, and availability of wild foods. When preferred foods are scarce, there are few plants that deer will not eat. A large deer population can create competition for food, causing deer to eat many plants that they normally would avoid.

Deer develop predictable travel patterns, and prior damage is often a good indicator of potential future problems. Any new plantings added to an existing landscape or garden already suffering from severe deer damage will likely also be browsed.

The list of deer-resistant (or close to it) plants is a general guide. Deer sometimes will browse the plants listed and sometimes will avoid plants not listed. ***Note:*** A few vigorous native plant species are included in both Tables 2 and 3.

Public Health Concerns

Deer are not considered a significant source of infectious diseases that can be transmitted to humans or domestic animals. However, as you would when dealing with any wild animal, it is recommended that you wear rubber gloves if you need to handle a sick or dead deer, and wash your hands afterwards. Fully cook all deer meat to 160°F (71°C).

Anyone who believes they may have contracted any disease from a deer should consult a physician as soon as possible, explaining to the doctor the possible sources of infection.

Legal Status

Because legal status, hunting restrictions, and other information about deer change, contact your local, state, or provincial wildlife

Table 3. Deer-proof (or close to it) Plants for Pacific Northwest Landscapes

Plants are listed alphabetically by their botanical name. (N) Indicates that the plant is native to Oregon, Washington, or British Columbia. *Note:* **All small, newly planted plants are subject to being pulled out of the ground by browsing deer. All trees are vulnerable to damage until they are at least 4 feet (1.25 m) tall, at which time they can generally handle being browsed by deer.**

Deciduous Trees
Birch, *Betula* spp. (N)
Fig, *Ficus carica*
Oregon ash, *Fraxinus latifolia* (N)
Sumac, *Rhus* spp. (N)
Willow, *Salix* spp. (N)
Little-leaf linden, *Tilia cordata*

Evergreen Trees
Fir, *Abies* spp. (N)
False cypress, *Chamaecyparis* spp.
Juniper, *Juniperus* spp. (N)
Tan oak, *Lithocarpus densiflorus* (N)
Spruce, *Picea* spp. (N)
Pine, *Pinus* spp. (N)
Douglas-fir, *Pseudotsuga menziesii* (N)
Western red-cedar, *Thuja plicata* (N)
Hemlock, *Tsuga* spp. (N)
Bay (Oregon-myrtle), *Umbellularia californica* (N)

Deciduous Shrubs
Barberry, *Berberis* spp.
Red-twig dogwood, *Cornus sericea (stolonifera)* (N)
Hazelnut (filbert), *Corylus* spp. (N)
Winter jasmine, *Jasminum nudiflorum*
Potentilla, *Potentilla fruticosa*
Chokecherry, *Prunus virginiana* (N)
Golden currant, *Ribes aureum* (N)
Red-flowered currant, *Ribes sanguineum* (N)
Wild gooseberry, *Ribes* spp. (N)
Wild rose, *Rosa* spp. (N)
Elderberry, *Sambucus* spp. (N)
Spirea, *Spiraea* spp. (N)
Snowberry, *Symphoricarpos* spp. (N)
Lilac, *Syringa* spp.

Evergreen Shrubs
Manzanita, *Arctostaphylos* spp. (N)
Sagebrush, *Artemisia tridentata* (N)
Evergreen barberry, *Berberis* spp.
Rabbitbrush, *Chrysothamnus* spp. (N)
Mexican-orange, *Choisya* spp. (N)
Silverberry, *Elaeagnus pungens*
Silk-tassel bush, *Garrya elliptica* (N)
Salal, *Gaultheria shallon* (N)
Juniper, *Juniperus* spp. (N)
Mountain-laurel, *Kalmia latifolia* (N)
Oregon-grape, *Mahonia aquifolium* (N)
Wax-myrtle, *Myrica californica* (N)

Oregon-boxwood, *Pachystima myrsinites* (N)
Dwarf mugho pine, *Pinus mugo mugo*
Coffeeberry, *Rhamnus californica* (N)
Rhododendron, *Rhododendron* spp.
Evergreen huckleberry, *Vaccinium ovatum* (N)

Perennial Flowers
Yarrow, *Achillea* spp. (N)
Rockcress, *Arabis* spp.
Seathrift, *Armeria maritima* (N)
Snow-in-summer, *Cerastium tomentosum*
Daisy, *Chrysanthemum maximum*
Coreopsis, *Coreopsis* spp.
Bleeding heart, *Dicentra* spp. (N)
Coneflower, *Echinacea purpurea*
Globe thistle, *Echinops exaltus*
Wild buckwheat, *Eriogonum* spp. (N)
Sea-holly, *Eryngium amethystinum*
Wallflower, *Erysimum* spp.
Blanket flower, *Gaillardia aristata* (N)
Baby's breath, *Gypsophila paniculata*
Hellebore, *Helleborus* spp.
Daylily, *Hemerocallis* spp.
Hosta, *Hosta* spp.
Iris, *Iris* spp. (N)
Poker plant, *Kniphofia* spp.
Gayfeather, *Liatris spicata*
Lobelia, *Lobelia cardinalis*
Lupine, *Lupinus* spp. (N)
Bee balm, *Monarda didyma*
Catmint, *Nepeta* spp.
Poppy, *Papaver* spp.
Russian sage, *Perouskia atriplicifolia*
Solomon's seal, *Polygonatum* spp. (N)
Lungwort, *Pulmonaria* spp.
Black-eyed Susan, *Rudbeckia* spp.
Fall sedum, *Sedum spectabile*
Blue-eyed grass, *Sisyrinchium* spp. (N)
California fuchsia, *Zauschneria* spp.

Annual Flowers
Ageratum, *Ageratum houstonianum*
Calendula, *Calendula officinalis*
Bachelor buttons, *Centaurea cyanus*
Clarkia, *Clarkia* spp. (N)
Larkspur, *Consolida ambigua*
Cosmos, *Cosmos bipinnatus*
California poppy, *Eschscholtzia californica* (N)

Geranium, *Pelargonium* spp.
Sunflower, *Helianthus annuus*
Sweet alyssum, *Lobularia maritima*
Zinnia, *Zinnia* spp.

Bulbs, Corms, and Tubers
Crocosmia, *Crocosmia crocosmiflora*
Crocus, *Crocus* spp.
Fritillary, *Fritillaria* spp.
Garden corn-lily, *Ixia* spp.
Trillium, *Trillium* spp. (N)

Ground Covers and Low Shrubs
Kinnikinnik, *Arctostaphylos uva-ursi* (N)
Dwarf coyote brush, *Baccharis pilularis* (N)
Bunchberry, *Cornus unalaschkensis (canadensis)* (N)
Cotoneaster, *Cotoneaster* spp.
Heather, *Erica* spp.
Wild strawberry, *Fragaria* spp. (N)
Wintergreen, *Gaultheria procumbens*
Salal, *Gaultheria shallon* (N)
Sunrose, *Helianthemum* spp.
Juniper, *Juniperus* spp. (N)
Lithodora, *Lithodora diffusa*
Oregon-grape, *Mahonia* spp. (N)
Oxalis (wood sorrel), *Oxalis oregona* (N)
Trailing rosemary, *Rosmarinus officinalis*
Trailing raspberry, *Rubus pedatus* (N)
Trailing blackberry, *Rubus ursinus* (N)

Garden Herbs
Garden chive, *Allium schoenoprasum*
Garlic chive, *Allium tuberosum*
Hyssop, *Hyssopsis officinalis*
Garden mint, *Mentha* spp.
Rosemary, *Rosmarinus officinalis*
Lavender, *Lavandula* spp.
Thyme, *Thymus* spp.
Sweet marjoram, *Origanum majorana*
Oregano, *Origanum vulgare*
Rue, *Ruta graveolens*
Santolina, *Santolina* spp.

Vines
Clematis, *Clematis* spp. (N)
Honeysuckle, *Lonicera* spp. (N)
Wisteria, *Wisteria* spp.

From: Link, *Landscaping for Wildlife in the Pacific Northwest.*

Tips for Driving in Deer Country

Vehicles kill thousands of deer each year in the Pacific Northwest. Deer will cross roads at any time of day or night, creating a hazard for the vehicles, passengers, and deer.

More than half of all deer/vehicle collisions occur in October and November. The rut (mating season) and peak days for hunting may account for this.

Here are driving tips to help prevent collisions:

- Deer are most active at dawn and dusk. Be especially watchful during these times.
- One deer crossing the road may be a sign that more deer are about to cross. Watch for other deer—they will move fast to catch up with leaders, mothers, or mates and may not pay attention to traffic.
- When you see brake lights, it could be because the driver ahead of you has spotted a deer. Stay alert as you drive by the spot, as more deer could try to cross.
- Wonder why the person ahead is driving so slowly? The driver may know where to slow down and be extra alert for deer. Don't be too quick to pass, and watch out.
- Take note of deer-crossing signs and drive accordingly. They were put there for a reason.
- Try to drive more slowly at night, giving yourself time to see a deer with your headlights. Lowering the brightness of your dashboard lights slightly will make it easier to see deer.
- Be especially watchful when traveling near steep roadside banks. Deer will pop onto the roadway with little or no warning.
- Be aware that headlights confuse deer and may cause them to move erratically or stop. Young animals in particular do not recognize that vehicles are a threat.
- Deer hooves slip on pavement and a deer may fall in front of your vehicle just when you think it is jumping away.
- Deer whistles, small devices that can be mounted on your vehicle, emit a shrill sound that supposedly alerts deer nearby. (Humans cannot hear the sound.) How well the devices work is not scientifically known.

If a collision with a deer seems imminent, take your foot off the accelerator and brake lightly. But—and this is critical—keep a firm hold on the steering wheel while keeping the vehicle straight. **Do not swerve in an attempt to miss the deer.** Insurance adjusters claim that more car damage and personal injury is caused when drivers attempt to avoid collision with a deer and instead collide with guardrails or roll down grades.

If you accidentally hit and kill a deer, try to move the animal off the road—providing you can do so in complete safety. Otherwise, report the location of the deer's body to the city, county, or state highway department with jurisdiction for the road. If no action is taken, contact the non-emergency number of the local police department, and the agency will arrange for the body to be removed. This will prevent scavengers from being attracted onto the road, and eliminate a potential traffic hazard.

If the deer is wounded, call the non-emergency number of the local police department and describe the animal's location. Emphasize that the injured deer is a traffic hazard to help ensure that someone will come quickly.

office or visit their Web site for updates. (See Appendix E for contact information and where to access the state and provincial laws mentioned below.)

Oregon: Mule deer, black-tailed deer, and white-tailed deer are classified as game animals and hunting regulations apply (OAR 635-068 and OAR 625-069). Columbian white-tailed deer are protected as a "State Sensitive" species (OAR 635-100-125).

If deer are causing damage, property owners, persons lawfully occupying land, or their agents may kill the offending deer at any time with a permit issued by Oregon Department of Fish and Wildlife (ODFW).

Washington: Mule deer, black-tailed deer, and white-tailed deer are classified as game animals (WAC 232-12-007). A hunting license and open season are required to hunt them.

After obtaining a permit issued by the Department of Fish and Wildlife (WDFW), a property owner or the owner's immediate family, employee, or tenant may kill a deer on that property if it is damaging crops (RCW 77.36.030). You must notify WDFW immediately after taking a deer in these situations.

The Columbian white-tailed deer is classified as a state and federal Endangered animal and may not be hunted or killed (WAC 232-12-014).

British Columbia: All species of deer are classified as game animals and hunting regulations apply. Deer that are involved in property damage may not be captured or killed, outside of general open seasons, unless you first receive a permit from the regional Fish and Wildlife Office.

CHAPTER 8

Deer Mice and House Mice

The Pacific Northwest is home to both native and non-native mice. The **house mouse** *(Mus musculus)* is found in and around human dwellings, barns, sheds, and in nearby wild areas where it has a reliable source of food and shelter. Originally from Asia, the house mouse arrived in the Pacific Northwest in the mid-1800s on board ships, trains, and covered wagons. After humans, it is now thought to be the most common mammal in cities and towns. The white mice used in research laboratories are albinos bred from this species. (See Table 1 in Chapter 18 for information on other introduced mammal species.)

Because of its wide distribution, the **deer mouse** *(Peromyscus maniculatus)* is considered the most abundant mammal in North America (Map 1). This native mouse is equally common around human habitation and wild areas, including Pacific Northwest forests, beaches, and arid grasslands.

Both the house mouse and the deer mouse have slightly pointed noses, small, black, somewhat protruding eyes, and sparsely haired ears. They are about 3 inches (7.5 cm) long, with another 3 inches (7.5 cm) of tail (Fig. 1).

Voles *(Microtus* spp.), also called "meadow mice" or "field mice," have larger, chunky bodies and rarely enter structures. For information on voles, see Chapter 25.

Figure 1. The color of deer mice varies greatly with habitat and geographic area. They are often grayish to reddish brown above and white below. The house mouse ranges in color from pale gray to brown, usually with lighter sides and underparts. Deer mice differ from house mice by having a sharp line of demarcation between their dark upper parts and white under parts. (From Christensen and Larrison, Mammals of the Pacific Northwest: A Pictorial Introduction.)

Facts about Deer Mice and House Mice

Food and Feeding Habits

- Deer mice are opportunistic feeders, eating seeds, green vegetation, insects, berries, and fungi.
- House mice are known to eat anything caloric that has the possibility of being digested.
- Unlike Old World rats, mice can live without access to water if their diet contains other liquids in adequate amounts.

Map 1.

Because of its wide distribution (shown here in black), the deer mouse is considered the most abundant mammal in North America. (From Verts and Carraway, *Land Mammals of Oregon*.)

Nest Sites

- Deer and house mice will nest in any dark, sheltered location in or near human habitation.
- Deer mice, being excellent climbers, have been found nesting in abandoned squirrel, woodrat, and bird nests 80 feet (24 m) up in trees.
- Female mice construct 4- to 6-inch (10–15 cm) cylindrical, loosely woven nests from available materials—grass, moss, thistledown, paper, fabric, and insulation.

Reproduction and Family Structure

- Within heated structures, deer and house mice breed year-round; outside structures, breeding slows from October to February.

- The young are born after a three-week gestation, with an average of four to five young per litter.
- Deer and house mice young are fully furred after ten days, weaned at about three weeks, and capable of breeding at two months of age.
- The social unit of house mice is made up of a mature male, two to five mature females, and several young. The social unit of deer mice is less structured, perhaps because of the tendency of male deer mice to kill unrelated young.

Mortality and Longevity

- Practically all predators of suitable size prey on deer and house mice, and since they are so common, they serve as a diet mainstay of many animals.
- Humans kill many mice with traps and poisons.
- The average life span of deer and house mice is less than one year.

Signs of Deer Mice and House Mice

Deer and house mice are active all year, mostly at night. Daytime activity doesn't indicate a locally high population, as it normally does with rats.

Mouse activity centers around food and a nest. House mice seldom travel more than 30 feet (9 m) from their nests, and when food is nearby this distance may only be a few feet. Deer mice travel up to 500 feet (150 m) in search of food.

Nests

Nests consist of finely shredded fibrous materials and are found as people clean garages, closets,

desks, basements, old cars, and other places where mice are present. (See "Public Health Concerns.")

Trails and Tracks

Because deer and house mice are so light, their trails or "runways" are seen only on rare occasions when surface conditions are perfect, as in mud, dust, or snow. In structures, frequently used runways may be seen in the dry dirt along walls and foundations.

To help make footprints show more clearly, place a tracking patch in suspected mouse areas. A tracking patch is a light dusting of an inert material such as clay, talc (unscented baby powder), or powdered limestone. Don't use flour, which may be eaten by mice and other animals. A good patch size is 12 x 4 inches (30 x 10 cm). When inspecting tracking patches, shine a flashlight at an angle that causes the tracks to cast a distinct shadow.

Rub Marks

Body oil and dirt rub off of mice coats and can become noticeable along frequently used runways. Look along wall/floor junctions, where mice move around obstacles, and at regularly used openings in walls, floors, and ceilings. Mouse rub marks are smaller than those created by rats and are not as easy to detect.

Droppings

Mice produce more than 50 droppings per day; these are left near food, shelter, and in other places mice frequent. Mice droppings are about half the size of rat droppings—¼ inch (6 mm) long with pointed ends.

Bats produce similar droppings; however, if you tap mouse

Figure 2. The deer mouse is a member of the rodent family, which is the largest family of mammals. The word rodent means "gnawing animal." This figure shows the two long pairs of incisors that are used like chisels to gnaw on hard foods and other objects. These incisors must grow continuously since they are worn down by gnawing. (From Verts and Carraway, Land Mammals of Oregon.*)*

droppings with a spoon, they will be hard. Bat droppings will crumble into dust.

Gnawing

Fresh accumulations of wood shavings (the consistency of coarse sawdust) and similar-size gnawed materials are a sign of mouse activity (Fig. 2).

Odors

Mice have an easily detectable musky odor. The smell of their urine is intensified in a heated vehicle if a mouse has spent time inside the heater unit. The same smell can come from a heated stove or other appliance where a mouse has been.

Sounds

Sounds are common at night where mice are present. Listen for squeaks, scrambling, and gnawing.

Preventing Conflicts

In the process of seeking food and shelter, mice may contaminate human or domestic animal food with their droppings, destroy insulation, and chew on electrical

wiring. Mice also create disturbing noise in attics, walls, and crawl spaces. Deer mice may spread a virus that can cause serious respiratory illness and death in humans (see "Public Health Concerns"). Mouse problems tend to become especially severe in winter, as the onset of cold weather causes them to move into buildings, garages, and other structures in search of shelter and food.

Successful long-term mouse control is often not possible in barns and old buildings. Mice populations may also be a consequence of community-wide activities over which you have little control, such as improper garbage disposal and poorly maintained bird-feeding stations.

If you're considering addressing a mouse problem with a house cat, remember that cats allowed to roam outdoors may also kill songbirds, chipmunks, squirrels, rabbits, snakes, and other wildlife. (For additional information, see Appendix D.)

Preventing a mice problem within a building begins by eliminating their food source and access to the inside. The next step is to remove any mice that remain inside, and then follow up to keep others from gaining entry inside.

The following recommendations ideally should be followed *before* mice enter areas where they are unwanted.

Eliminate food sources. Store human and animal food in mouse-proof buildings, rooms, or containers. Good sanitation procedures include keeping stovetops clean, frequently cleaning under and behind appliances, and cleaning under bottom drawers in built-in kitchen cabinets.

Most buildings in which food is stored, handled, or used will support mice if the mice are not excluded, no matter how good the sanitation. While good sanitation will seldom completely control mice, poor sanitation is sure to attract them and will permit them to thrive in greater numbers.

Detailed information on this topic is provided in the chapter on Old World rats—see "Preventing Conflicts" in Chapter 16.

Prevent access to shelter. In the long term, the most successful form of mouse control is to build them out. Also called mouse-proofing, this approach makes it impossible for mice to get inside or under a building where they could do damage. But mouse-proofing a structure can be a challenge. Mice have collapsible ribs and can squeeze through any hole they can get their heads through. They will also enlarge an opening that can be gnawed around. Examples of mouse signs are given under "Signs of Deer Mice and House Mice."

Seal openings around drain-pipes, vents, cables, and other utilities that enter structures. Seal cracks and holes in and under building foundations and exterior walls (including warped siding).

It can be difficult to find mouse entry points from the exterior of a structure. The best inspection method is to enter the building's crawl space during daylight hours. Potential entry points will be visible at locations where light enters from the outside. If you see holes, insert a plastic straw or other item through the hole to mark the location, then return to the area from the outside to make the needed repairs.

Of Mice in Vehicles

Mice will climb up into a vehicle to get warm, reproduce, or eat. In the process, they may gnaw on the small tubes, hoses, and wires that can be found under the hood of any car or truck. This can become both an aggravating and a costly problem.

To prevent problems, keep the interior of vehicles clean to avoid attracting mice. Locate holes that lead into the interior and fill them with balled-up galvanized window screen or copper scouring pads.

To fix an existing mouse problem, don't use poison bait. Mice may die inside the vehicle and create an odor problem that may be intensified when the heater is turned on. Instead, use snap traps to capture mice living in a vehicle. **Note:** Leave a note on the inside of the window reminding you traps are set inside! If the problem is occurring under the hood, set traps there or outside around the vehicle. Use trapping boxes to prevent problems with kids, pets, or other wildlife (see "Trapping" in Chapter 16 for information on trapping boxes).

To prevent gnawing problems under the hood, treat exposed hoses, tubes, and wires with a liquid repellent such as Ropell®. This product tastes terrible and will stop them from gnawing treated surfaces. You will need to apply a repellent about twice a month on average during their active season.

See "Preventing Conflicts" in Chapter 16 for detailed information.

Remove the existing mouse population. If mouse problems persist after you've eliminated all known sources of food and shelter, some form of population reduction, such as trapping or baiting, is almost always necessary.

There is no scientific evidence that commercially available ultrasonic devices will drive mice from buildings. Because sound from the devices does not penetrate objects and quickly loses its intensity with distance, mice move to a noise-free area. Mice naturally avoid some odors and tastes, however no repellents have been found to solve a mouse problem.

Although cats, dogs, and other predators may kill mice, they do not provide effective mouse control in most circumstances. Mice often live in very close association with dogs and cats. Mouse problems around homes often are encouraged by the presence of food, water, and shelter provided for pets.

Lethal Control

When food, water, and shelter are available, mice can reproduce quickly and populations grow fast. While the most permanent form of control is to limit food, water, shelter, and access to buildings, lethal control is often necessary.

Lethal control includes trapping and poisoning; however, they are seldom long-term solutions around buildings if the preventive measures mentioned above and in Chapter 17 are not carried out.

Many companies and individuals make a portion of their

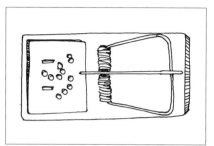

Figure 3. Common snap mousetraps, or one of their variations, kill mice instantly. Traps made with an expanded plastic trigger (shown here) are much more effective than those with a small metal trigger. Mice will easily clean the bait off the metal-trigger trap, but doing so is harder with the expanded version. (Drawing by Jenifer Rees.)

living from handling mice-related problems through trapping and baiting programs. Before hiring someone for the job, read "Hiring a Wildlife Damage Control Company" in Appendix C.

Trapping

Trapping is an effective method for controlling small numbers of mice. Although time-consuming, it is the preferred method in homes, vehicles, garages, and other structures where only a few mice are present. Trapping has several advantages: It does not rely on potentially hazardous poisons; it permits users to verify their success; and it allows for disposal of trapped mice, thereby eliminating dead-mouse odors that may occur when poison is used. Mice are generally much easier to trap than rats.

Common snap mousetraps, or one of their variations, kill mice instantly (Fig. 3). Most supermarkets, hardware stores, and farm supply stores carry mousetraps. Professional models are available from pest-control companies and from sources on the Internet.

Locate traps in areas where there are large numbers of droppings, along walls, behind objects, in dark corners, and where runways narrow, causing mice to be funneled into a limited area. Mice travel along walls and not in the middle of a floor, unless they feel very secure. (See "Trapping" in Chapter 16 for an example of trap placement.)

Peanut butter is effective bait. However, sometimes mice steal it by licking it off the small metal triggers found on older model traps. Any bait must be fresh to be effective. Odors of humans and dead mice on traps are not known to reduce trapping success.

Use several traps to make the trapping period short and decisive. A dozen or more traps may be necessary for a heavily infested home. Keep small children and pets out of the area where traps are set.

Don't set unprotected snap traps outside (see "Trapping" in Chapter 16 for information on trapping boxes).

Check all traps daily to remove mice, and reset any sprung traps. Because mice may carry diseases, do not handle them without gloves; you can use a plastic bag slipped over your hand and arm as a glove. Once the mouse is removed from the trap, hold it with your bagged hand and turn the bag inside out while slipping it off your arm and hand.

For information on glue boards, electronic traps, and live traps, see "Trapping" in Chapter 16.

Poisoning

Using poison bait to kill mice around homes (and in vehicles) is the least preferred method of control due to the possibility of poisoning non-target species, and because one dead mouse can cre-

ate an odor problem. Contrary to common myth, mice don't always go outside to die.

If odor does become a problem, there are odor-eliminating products that mask smells or remove them at the molecular level. These are available from farm supply centers, pet stores, and from the Internet using the keywords "Pest Control Supplies."

Only use baits labeled for mouse control and carefully follow label instructions and understand all precautions.

Poison baits can be purchased packaged in a variety of different formulations. Baits formed into paraffin or wax blocks are useful in damp locations where other baits spoil quickly. Unfortunately, mice may not accept these blocks as readily as they do other baits.

Bait boxes protect bait from weather and provide a safeguard to people, pets, and other animals. Bait boxes should have at least two openings about 2 inch (5 cm) in diameter and should be large enough to accommodate several mice at one time. Place bait boxes next to walls (with the openings close to the wall) or in other places where mice are active. (See "Poisoning" and the sidebar "Pets, Wildlife, and Rodent Control" in Chapter 16 for detailed information on poison baits and bait boxes.)

Remove and properly dispose of all uneaten bait at the end of a control program. Also, collect and properly dispose of any dead mice found while baiting. Pick them up, using a sturdy plastic bag inverted on your hand, and seal them in the bag for disposal with household garbage, or bury them in a location where pets or scavengers will not easily dig them up.

Follow-up: It is difficult to permanently mouse-proof a structure. As buildings settle new opportunities open up for reentry, and mice can chew a new hole into a structure at any time. Therefore, watch for any sign of mice (droppings, food damage, gnawing damage, holes, etc.) and take steps to control a mouse problem as soon as you notice it.

Mice are here to stay, and extermination is impractical, especially if your property borders wildlands or another area that historically has served as mouse habitat. While you may be able to remove the existing population, or drive them elsewhere temporarily, other mice will quickly move into vacated areas if conditions for them are present and nearby mice populations exist.

Public Health Concerns

Mice can carry a wide variety of diseases transmissible to humans. **Salmonella** (food poisoning) can be transmitted by mice and is an important concern in food storage and preparation areas. To prevent contracting salmonella, follow the recommendations in this chapter and Chapter 16.

Hantavirus pulmonary syndrome (HPS) is a human disease caused by a virus carried by deer mice in the western United States, including the Pacific Northwest. Human cases have occurred rarely in the Pacific Northwest, despite the thousands of persons who are exposed to these rodents every day.

Rodents, particularly deer mice, shed the virus in their urine, saliva, and droppings. A person may be exposed to HPS by breathing dust after cleaning rodent droppings or disturbing nests, or by living or working in rodent-infested settings. Other less common means of transmission are by touching contaminated rodent urine, droppings, or saliva and then touching your nose or mouth; eating food contaminated by a HPS-carrying rodent; or being bitten by an HPS-carrying rodent.

Symptoms of HPS usually begin one to three weeks after exposure to an infected rodent. Symptoms include fever, chills, and muscle aches, followed by the abrupt onset of respiratory distress and shortness of breath. The muscle aches are severe, involving the thighs, hips, back, and sometimes the shoulders. Other symptoms include nausea, vomiting, and abdominal pain.

You can't tell whether a mouse is infected by looking at it or at its droppings, so avoid any contact with all mice, use control measures to keep them out of your living space, and clean up after them appropriately.

Before working or cleaning up in suspected mice areas, put on latex rubber gloves and a tightly fitting dust mask. Don't stir up dust by sweeping up or vacuuming up droppings, urine, or nest materials. Instead, thoroughly wet contaminated areas with detergent in liquid to deactivate the virus. Most general-purpose disinfectants—

which state they kill viruses on the label—are effective. A mix of 1½ cups of household bleach in 1 gallon of water may be substituted in place of a commercial disinfectant. When using bleach, avoid spilling the mixture on clothing or other items that might be damaged.

Once everything is wet, take up contaminated materials with a damp towel, sponge, or mop, and then mop or sponge the area with disinfectant.

Double-bag dead rodents along with all cleaning materials and dispose of them by burying or burning—or throwing them out in an appropriate waste-disposal system.

Finally, disinfect gloves **before taking them off** with a disinfectant or soap and water. After taking off the clean gloves, thoroughly wash hands with soap and warm water.

When going into cabins, outbuildings, or work areas that have been closed for a while, open them up and air them out for at least an hour before cleaning.

For updates on the spread of HPS and recommendations for prevention and risk reduction, contact your public health office (see Appendix E for contact information).

If a person is bitten or scratched by a wild mouse, immediately scrub the wound with soap and water. Flush the wound

liberally with tap water. Contact your physician or the local health department immediately.

Legal Status

Because legal status and other information about deer mice and house mice change, contact your local, state, or provincial wildlife office, or visit their Web site for updates. (See Appendix E for contact information and where to access the state and provincial laws mentioned below.)

Oregon: Deer mice are classified as unprotected wildlife; the killing of deer mice is unrestricted. House mice are not under the jurisdiction of the Oregon Department of Fish and Wildlife, and are not protected.

Washington: All species of mice are unclassified and may be trapped or killed without a permit.

British Columbia: House mice are not included under the definition of wildlife and do not fall under the regulatory authority of the Ministry of Water, Land, and Air Protection.

Deer mice are listed under Schedule "B" of the Wildlife Act, which means they may be captured or killed for the purposes of protection of property by the person that owns or occupies the property or his/her spouse, parent, guardian, or child (Wildlife Act, Designation and Exemption Regulation 2 and 3).

CHAPTER 9

Elk

Elk are members of the deer family and share many physical traits with deer, moose, and caribou. They are much larger than deer, but not as large as the moose. Adult bull (male) elk weigh 600 to 800 pounds (270–369 kg), and adult cows (female elk) typically weigh 400 to 500 pounds (180–225 kg). With thick bodies, short tails, and long legs, adult elk stand 4½ to 5 feet (1.3–1.5 m) high at the shoulder (Fig. 1).

Elk range in color from light brown in winter to reddish tan in summer, and have characteristic buff-colored rumps. In winter, a dark brown, shaggy mane hangs from the neck to the chest.

Bull elk have large, spreading antlers. Like other members of the deer family, the antlers of bull elk grow during spring and summer beneath a hairy skin covering known as velvet. In late summer the velvet dries and falls off to reveal the bonelike structure of the fully grown antlers. Elk shed their antlers beginning in late February for the largest males, extending to late April and even early May for younger ones. New antler growth begins soon after shedding. (See "All About Antlers" in Chapter 7 for addition information on antlers.)

Two subspecies of elk are found in the Pacific Northwest (Table 1).

Facts about Elk
Food and Feeding Behavior

- Elk require large amounts of food because of their body size and herding tendencies.
- In spring and summer, when food is plentiful, elk are mainly grazers—eating grasses, sedges, and a variety of flowering plants.

Figure 1. Elk have been an intrinsic part of Northwest tribal culture for thousands of years (a bull Roosevelt elk is shown here). They have helped Indian people survive throughout the centuries by providing a continual source of meat and marrow for sustenance and vitamins. Elk also have been used for religious purposes, clothing, and drum making. To this day, the elk is part of traditional ceremonies and is essential for maintaining tribal culture. (Washington Department of Fish and Wildlife.)

Table 1. Pacific Northwest Elk

Roosevelt elk (*Cervus elaphus roosevelti,* Fig. 1), named after U.S. President Theodore Roosevelt, occur in the Coast Range and west slopes of the Cascade mountains in Oregon and Washington, and in Washington's Olympic Range. In British Columbia they are found only on Vancouver Island, having disappeared from the lower mainland at least 100 years ago. The Roosevelt elk is the state mammal of Washington. Olympic National Park in northwest Washington holds the largest number of elk living anywhere (about 5,000).

Rocky Mountain elk *(Cervus elaphus nelsoni)* occur primarily in the mountain ranges and shrublands east of the Cascades crest in Oregon and Washington, and in the East Kootenays and Muskwa areas of British Columbia. Small herds have been established, or reestablished, throughout the Pacific Northwest. Many Rocky Mountain elk populations currently in Washington stem from elk transplanted from Yellowstone National Park in the early 1900s.

Rocky Mountain elk are slightly lighter in color than Roosevelt elk, and some experts believe they are slightly smaller in size. The antlers of Rocky Mountain elk are typically more slender, have longer tines, and are less palmated than Roosevelt elk antlers.

"Wapiti" is the name for Rocky Mountain elk in the Shawnee language and means "white rump."

Hybrids, or genetically mixed populations of Roosevelt elk and Rocky Mountain elk, are common in the Cascade Ranges of Washington and Oregon.

- In fall, elk increasingly become browsers, feeding on sprouts and branches of shrubs and trees, including conifers as a last resort when snow covers other plants.
- During fall and winter, elk continue to eat grasses when these are available and not covered by deep snow.
- Like deer and moose, elk are ruminants. They initially chew their food just enough to swallow it. This food is stored in a stomach called the "rumen." From there, the food is regurgitated, then re-chewed before being swallowed again, entering a second stomach where digestion begins. Then it passes into third and fourth stomachs before finally entering the intestine.

Cover and Range Needs

- Elk are hardy animals that have few physiological needs for cover. They do, however, use cover during extreme weather, to avoid hunters, or when they are harassed. Cover also conceals newborn calves from predators.
- Ideal elk habitat includes productive grasslands, meadows, or clearcuts, interspersed with closed-canopy forests.
- Year-round ranges for Rocky Mountain elk vary from 2,500 to 10,000 acres (1000–4000 h), and usually include distinct summering and wintering areas.
- Year-round ranges for Roosevelt elk are smaller, usually 1,500 to 4,000 acres (600–1600 h), because they are generally found where the climate is less severe and where food and cover are more readily available.

Social Structure

- Elk are social animals, living in herds for much of the year. During spring, summer, and winter, elk tend to split into cow–calf herds and bull herds.
- Cow–calf herds are usually led by older, experienced cows and may include adolescent bulls.
- During the mating season (rut) in early fall, adult and subadult bulls find and temporarily join cow herds. The larger, more aggressive bulls try to gather harems of cows, which they defend against competing bulls (Fig. 2).
- Harems range in size from 3 to 4 cows to as many as 20 to 25 cows. Bulls socially dominate the cows within their harems, but the movements of these breeding groups are still determined by older, lead cows.

Figure 2. *During the mating season (called the rut) in early fall, the larger, more aggressive bull elk gather harems of cows, which they defend against competing bulls. (Oregon Department of Fish and Wildlife.)*

- Adolescent males form small bachelor groups or patrol the edge of breeding harems.
- Breeding activities cease by mid-October; bulls usually leave the cow–calf groups then and the herds disperse into wintering areas.

Reproduction

- Mating occurs during the fall rut, and successful bulls breed with numerous females each year.
- Once the rut begins, mature bulls challenge each other vocally, emitting high-pitched calliope-like whistles, or "bugles."

- Cows have an eight- to nine-month pregnancy, which results in the birth of a single spotted calf in late May or early June.
- The timing of birth seems to optimize calf survival by being late enough that the risk of cold, inclement weather has passed, but early enough so that there is considerable time for calves to grow before the onset of next winter.
- Just before giving birth, a cow elk will leave the herd and select a birthing place. Because predators would easily detect large groups of elk, cow elk appear to avoid grouping with other

elk until their calves are large enough (usually about two weeks of age) to run effectively to escape predators.
- Other cows sometimes tend calves when mothers are feeding; a mother may nurse her calf for up to nine months.
- Calves grow quickly and lose their spots by summer's end. By the onset of winter, a calf that entered the world weighing 35 pounds (16 kg) may tip the scales at 225 to 250 pounds (101–112 kg).

Mortality and Longevity

- With a superb sense of smell, excellent hearing, and a top running speed of 35 mph (56 kph), elk are well equipped to avoid the few predators capable of bringing them down.
- Cougars and wolves prey upon adult elk; calves may also fall victim to bears, bobcats, domestic dogs, and coyotes.
- Hunting, automobiles, predation, and habitat loss all take their toll on elk populations.
- Most elk are physically declining by age 16, and a 20-year-old wild cow elk is very old. Bulls generally do not live as long as cows, rarely surpassing 12 years.

Viewing Elk

Elk are primarily crepuscular (active mostly at dawn and dusk), so early morning and late evening are the best times to observe them. But when temperatures soar or when they are harassed, elk may become more active at night.

When disturbance levels are low and temperatures mild, elk may be observed feeding in short bouts throughout the day. When not hunted, elk adapt well to humans and find lawns and golf courses excellent places to graze.

A good time of year to observe elk is in fall. In late September and October, bulls are battling each other over females and are not as concerned about being seen. This is a fascinating time to observe elk because the shrill bugles of the bulls can often be heard near dawn and dusk.

Leafless trees allow greater visibility, and when it is raining there is less chance of being heard crunching through an area. However, be aware of open hunt-ing seasons during this time of year and wear bright orange clothing for your safety. Also, care needs to be taken when around adult male elk during the mating season, particularly in areas where they are accustomed to people, such as national parks.

The best way to view wild elk is to find a meadow, clearcut, or other open grassland elk have been using and to wait quietly nearby. Because elk have a keen sense of smell, it is best to be downwind of where you expect them to come from. (Contact your local fish and wildlife office for information on where to view elk in your area.)

Feeding Areas

In winter, look for pits dug in snow where elk have been pawing for food, or for the well-worn trails or crisscrossing tracks in the snow typical of foraging elk.

Gnawed aspen and other deciduous tree trunks are also common in elk country during winter. The bottom-teeth-only scrape marks of elk and moose are virtually identical. Gnawings may also be found on downed trees and branches and are easily distinguished from the chisel-like cuttings of beaver.

Aspen trunks that have been gnawed year after year eventually develop a rough, blackened trunk as far up as the animal can reach. A grove of black-trunked aspen is a sign that winter range has been heavily used by elk or moose.

Tracks and Trails

Elk, much easier to track than most animals due to their weight, leave marks in or on almost anything they walk over. Tracks, often found in large numbers indi-

Figure 3. A 4-inch (10 cm) long and 3-inch (7.5 cm) wide elk print is larger and rounder than that of a deer, and somewhat rounder and smaller than that of a moose. (From Pandell, Animal Tracks of the Pacific Northwest.*)*

cating a passing herd, are easy to identify and follow (Fig. 3).

Like all members of the deer family, elk have cloven hooves that normally resemble a split-heart shape on soft earth. The dewclaws on all four feet may register in several inches of mud or snow. Hoof prints may be splayed wide on slippery surfaces, or when the animals were running.

Elk trails are often several animals wide and quite noticeable at the transition from grassland into brush or woodlands.

Droppings

Given a steady, consistent diet, pellets deposited by deer and elk may be the same general shape and texture. Individual pellets are usually dimpled at one end and have a small projection at the other, giving them an almost acorn-like shape. However, elk droppings are slightly larger, and whereas an adult deer may leave 20 to 30 pellets at a time, elk may deposit twice that many. This difference in volume becomes especially apparent when a rich diet causes the animals' droppings to become a soft mass, similar to a domestic cowpie, but smaller. "Elkpies" average 4 to 6 inches (10–15 cm) in diameter, while those of deer run about 2 inches (5 cm) across. Even when elk are eating mostly grass, elkpies will still show more distinct edges among the individual pellets than cowpies, which may be an amorphous mass.

Elk Rubs

In late summer, as antler growth ceases, it finishes mineralizing and the blood supply to the velvet begins to deteriorate. This causes the velvet covering of the antlers to dry up and shred. As it dies, bulls begin to vigorously rub their antlers on shrubs and trees, to help rid them of the velvet. This rubbing behavior may

also be the first ritualized use of the bull's newly hardened antlers—it is quite noisy and attracts the attention of other elk.

It has been theorized that this "horning" of shrubs frequently causes shrub branches to be broken off and intertwined with the bull's antlers, effectively making them look larger and more threatening to rivals and more impressive to potential mates. The rubbing also covers the bone-white antler with plant compounds that subsequently oxidize and stain the antlers to their characteristic dark brown color.

Regardless of the cause of this behavior, the result is obvious: small saplings and shrubs are

Collecting Antlers

Finding the simple treasures of shed deer or elk antlers can brighten up a casual winter hike in the woods (Fig. 4). Although antler hunting season gets under way in late winter, because most bucks and some bulls have shed them then, biologists recommend delaying serious antler hunting to late spring to avoid inadvertently harassing animals on winter range.

Collecting naturally shed deer and elk antlers is legal, but there are some ethical considerations to keep in mind and a few places that are restricted or off-limits. The easiest antler hunting is, of course, where deer or elk concentrate in the winter. But if many antler hunters descend on that area before wintering animals have left, the disturbance can threaten their survival at the harshest time of year.

Public lands across the Pacific Northwest may have rules, so antler hunters should do their homework before going afield. By long-standing policy, shed antlers or anything else naturally found in national parks cannot be removed.

Figure 4. The number of tines an elk antler has is a poor indication of the animal's age. Older bulls usually have more tines than younger bulls, but poor food or mineral supply may retard antler development. A mature bull elk with six points (shown here) on each side is known as a "royal bull." A bull with seven points on each side is called an "imperial bull," and one with eight points per antler is called a "monarch bull." (Oregon Department of Fish and Wildlife.)

left looking like someone with a hedge trimmer went on an angry rampage. In areas where elk are abundant, mangled shrubs and small trees are extremely obvious signs of the presence of bulls and their preparation for breeding.

Wallow Sites

Probably the most easily identified elk sign is the mud wallow scented with urine and droppings. Bull elk roll in wallows to cover their bodies with scent, creating bathtub-size depressions with low walls of displaced mud ringing their perimeters. Receptive cow elk, drawn by the odor, will also roll and urinate in the wallow, indicating their willingness to mate. Elk that use these wallows may become so foul smelling that, when downwind, humans can easily detect their presence.

Elk also roll in mud wallows to loosen their dead winter coats and help dislodge annoying parasites. A coating of mud also provides some degree of protection from bloodsucking insects.

Mud wallows are musky smelling and have light-colored hairs lining their bottoms as well as fresh hoofprints all around. Wallows are found where the ground is wet and muddy, usually near water and almost always in a secluded area where elk feel relaxed enough to drop their guard a bit. Abandoned wallows will likely be filled with water, have grasses growing around them, and may develop into breeding sites for frogs and salamanders.

Calls

Elk are the noisiest member of the deer family in North America. Males are known for their eerie bugles during the rutting season. The bugle starts with a guttural groaning that quickly yields to a high-pitched whistle, and often ends with a few repetitive low-toned grunts.

Calves often bleat to locate their mothers, and adult females commonly bark loudly to alert other elk to danger.

Tips for Attracting Elk

The large ranges required by elk, especially elk that migrate between summer and winter ranges, mean that most property owners are unable to manage or provide year-round elk habitat. Thus, habitat management for elk requires considerable coordination among landowners.

Contact the agency that owns large areas of elk habitat near you. Ask them to carefully manage such things as the timing and distribution of firewood cutting, logging, and the density of roads in order to minimize elk disturbances, especially in high-use summer areas.

Because lands traditionally used by elk are severely impacted by land subdivisions, changes in agricultural practices, and invading noxious weeds, policies that control these activities can substantially benefit elk.

Key winter range areas should be identified and given high priority in future land acquisitions, leases, easements, and incentives to create or preserve elk habitat. (Contact your local wildlife office for cost-share or other programs that may help you manage elk on your property.)

Mineral blocks are attractive to elk and deer, especially during the spring. However, mineral blocks, if placed on the ground, will leach minerals into the soil. Even after removing the block, these leached minerals will continue to attract use and in almost no time at all you will have a pit in the ground, excavated by elk and/or deer. These can become huge areas that could be an injury hazard to an animal or person, and they may continue to be excavated several seasons after removing the mineral block. So, once you commit to putting a mineral block on the ground, it's no simple matter to change your mind and remove it. Once there, its presence, even if only temporary, will continue to affect the local environment.

Leaching may be prevented by putting the block in a covered area or in a plastic tub with high-enough walls to prevent an animal from rolling the block out onto the ground.

If a landowner intended to hunt deer or elk in the area where a mineral block was located, they should check with enforcement personnel to make sure the use of a mineral block would not constitute "baiting" or be restricted for any reason.

Plants that elk prefer are listed in Table 2.

For additional information, see "Tips for Attracting Deer" in Chapter 7.

Preventing Conflicts

In most areas, elk summer ranges are on public lands, whereas winter ranges largely are on private lands. Herein lies the source of most complaints of damage to crops and property.

Problems associated with elk include damage to tree farms and conifer plantations, hay and alfalfa fields, orchards, and other agricul-

Table 2. Food Plants Used by Pacific Northwest Elk

Roosevelt Elk	Rocky Mountain Elk
Trees	
Aspen, *Populus tremuloides* Cottonwood, *Populus trichocarpa* Red alder, *Alnus rubra* Vine maple, *Acer circinatum* Willow, *Salix* spp.	Aspen, *Populus tremuloides* Chokecherry, *Prunus virginiana* Cottonwood, *Populus trichocarpa* Rocky mountain maple, *Acer glabrum* Willow, *Salix* spp.
Shrubs and Groundcovers	
Blackberry, *Rubus* spp. Huckleberry, *Vaccinium* spp. Oregon-grape, *Mahonia* spp. Salal, *Gaultheria shallon* Salmonberry, *Rubus spectabilis* Thimbleberry, *Rubus parviflorus* Wild rose, *Rosa* spp.	Bitterbrush, *Purshia tridentata* Currant, *Ribes* spp. Deer brush, *Ceanothus integerrimus* Elderberry, *Sambucus* spp Huckleberry, *Vaccinium* spp. Oceanspray, *Holodiscus* spp. Red-twig dogwood, *Cornus sericea* Serviceberry, *Amelanchier alnifolia* Snowberry, *Symphoricarpos albus* Sumac, *Rhus* spp. Wild rose, *Rosa* spp.
Forbs, Ferns, and Legumes	
Bear grass, *Xerophyllum tenax* Cat's ear, *Hypochaeris* spp. Clover, *Trifolium* spp. Cow-parsnip, *Heracleum lanatum* Fireweed, *Epilobium angustifolium* Foamflower, *Tiarella trifoliata* Oregon oxalis, *Oxalis oregana* Pearly everlasting, *Anaphalis* spp. Sword fern, *Polystichum munitum*	Alfalfa, *Medicago sativa* Clover, *Trifolium* spp. Dandelion, *Taraxacum* spp. Fireweed, *Epilobium angustifolium* Sweet clover, *Melilotus* spp. Yellow salsify, *Tragopogon* spp.

tural crops. When frightened, elk damage wire fences by running through them rather than jumping them. Finally, many dangerous vehicle–elk collisions occur each year in the Pacific Northwest.

The likelihood of human/elk conflicts is influenced by the number of elk in the area, the availability of alternative food sources and hiding cover, and winter weather conditions. If elk are damaging your property, personnel from your local wildlife department can help you evaluate damage-control options. Typical nonlethal damage-control techniques include but are not necessarily limited to herding, hazing, scare devices, fencing and fence repair, land purchases, purchasing or leasing crops, crop-damage payments, and winter feeding.

Depredation permits may be issued to individual landowners, and special elk permits may be issued to hunters to reduce local populations. Removing elk by any means is probably a short-term solution; other elk are likely to move in if attractive habitat remains available.

For information on fencing, see below. For information on repellents, scare tactics, and other ways to reduce human/elk conflicts, see "Preventing Conflicts" in Chapter 7. For tips on how to prevent a vehicle collision with an elk, see "Tips for Driving in Deer Country" in Chapter 7.

Elk fences and other barriers: Fencing can provide relief from elk damage in situations where plants cannot be protected individually. A well built, 6- to 8-foot (1.8–2.4 m) high woven-wire fence will keep elk out of enclosed areas. The higher fence will be needed in an area with many elk and a low supply of wild food.

Recently, electric fences have proven to be a cheaper alternative. These fences feature eight to ten strands of high tensile steel wire supported by conventional fence post systems.

For any fence to be effective, it must be seen by elk. A group of elk led by the dominant cow will go through any type of fence, except perhaps a cyclone fence, if the fence is in their path and they don't see it before the group is upon it.

Placing branches along the top of fences and draping survey or similar tape from electric wires makes fences more visible to elk.

Polytape electric fencing is much wider and more visible than traditional electric wire fencing and is meant to work as a visual repellent as well as a shocking device. Once an elk has been shocked by a polytape fence, it is likely to both remember and recognize the wide, brightly colored tape and avoid going near it again. The tape is also more visible to people. There is less maintenance required for a polytape than for single-strand fences simply because the greater visibility of the tape prevents it from being knocked down as often.

For small orchards and other areas needing protection, individual trees can be protected by 6-foot (1.8 m) tall cylinders made of welded wire.

Individual protectors for conifer seedlings are effective until the leader (growing tip) or lateral branches grow out of the protectors and are once again subject to elk browsing. Plastic or nylon tubes, netting, and bud caps have all been used successfully. Studies indicate that paper bud covers provide effective control, and are cheaper. They are recommended for protection of conifer seedlings.

See "Preventing Conflicts" in Chapter 7 for more information and examples of all the above described fences and barriers.

Public Health Concerns

Elk are not considered a significant source of infectious disease that can be transmitted to humans or domestic animals. However, as when working with any wild animal, it is recommended that you wear rubber gloves if you need to handle a sick or dead elk; wash your hands afterwards, and fully cook all deer meat to 160°F (71°C).

Legal Status

Because legal status, hunting restrictions, and other information about elk change, contact your local, state, or provincial wildlife office, or visit their Web site for updates. (See Appendix E for contact information and where to access the state and provincial laws mentioned below.)

Oregon: Elk are classified as game mammals and hunting regulations apply (OAR 635-070 and OAR 635-071). Property owners, persons lawfully occupying land, or their agents may kill elk at any time with a permit issued by Oregon Department of Fish and Wildlife (ODFW) if the elk are causing damage.

Washington: Roosevelt elk and Rocky Mountain elk are classified as game animals (WAC 232-12-007). A hunting license and open season are required to hunt them. After obtaining a permit issued by the Department of Fish and Wildlife (WDFW), a property owner or the owner's immediate family, employee, or tenant may kill an elk on that property if it is damaging crops (RCW 77.36.030). You must notify WDFW immediately after taking an elk in these situations.

British Columbia: Elk are defined as big game under the Wildlife Act, and can be hunted with a license during an open season defined by regulation. Permission of the landowner is required to hunt on private land. Animals that damage property may not be captured or killed, outside of general open seasons, without first receiving a permit from the regional Fish and Wildlife Office.

CHAPTER 10

Foxes

Foxes belong to the dog family, *Canidae.* They resemble small dogs, with erect, relatively large pointed ears, long pointed muzzles, and long bushy tails. Three species of foxes occur in the Pacific Northwest.

In the early and mid-1900s, Eastern red foxes (*Vulpes vulpes,* Fig. 1) were introduced throughout the Pacific Northwest by hunting clubs and farmers who raised foxes for their valuable pelts. Released foxes and escapees established populations throughout most of Oregon, Washington, and south-central British Columbia.

Native forms of red fox are still found in the Pacific Northwest. The **Cascade red fox** *(Vulpes vulpes cascadensis)* is found throughout many low-, mid-, and high-elevation areas in the Cascade mountains of Oregon and Washington; the **Rocky Mountain red fox** *(Vulpes vulpes macroura)* occupies the mountains of northeastern Oregon and southeastern Washington. There is no means of visually distinguishing native from non-native red foxes.

Adult red foxes stand about 14 inches (35 cm) tall at the shoulder and average 3 feet (90 cm) in length, including a 12 to 18 inch (30–45 cm) long bushy tail. Adults weigh 10 to 15 pounds (4.5–7 kg). Males, or "dogs," are usually heavier than females, or "vixens." Both have reddish coats, bushy, white-tipped tails, and black "stockings," although there are color phases that include silvery tones to nearly black.

The **gray fox** (*Urocyon cinereoargenteus,* Fig. 2) occurs in Oregon, mostly in interior valleys throughout the state's western half, and along the Cascade mountains' eastern side. Gray foxes are slightly smaller than red foxes. The upper parts of their coats are blackish gray, and the under parts are reddish brown or buff colored. They have a white throat and a black stripe on top of their tails. Their tails end in a black tip.

The **kit fox** *(Vulpes macrotis),* also known as the **swift fox,** occurs sparsely in the open deserts of southeast Oregon. It averages 28 inches (70 cm) in length—its bushy, dark-tipped tail is 40 percent of its total length. Its coat is a buff color and the chin, throat, and belly are white.

Because of the great variety of color types among foxes, a red-colored fox is not necessarily a "red fox" and a gray-colored fox is not necessarily a "gray fox." When a close examination can be made, a white tip at the tail's end is indicative of a red fox.

This chapter focuses on the more commonly seen gray and red foxes. However, where noted, information is also applicable to kit foxes.

Facts about Foxes

Habitat and Home Range

- Gray foxes are more associated with woodlands than are red foxes, which are seen in more open areas, such as grasslands interspersed with brush and timber.
- Both gray and red foxes hunt and/or live in urban and suburban parks, golf courses, and other developed areas. Red foxes adapt readily to human-made environments.
- Home ranges are 1 to 3 square miles (1.5–5 km), depending on habitat type and prey availability.

Food and Feeding Behavior

- Foxes are opportunists, eating a wide variety of food, including rabbits,

Figure 1. Red foxes are intelligent, cautious, cunning, swift, and alert. Their sight, hearing, and smell are acute, and they use these senses to stalk prey with patience and stealth. (Washington Department of Fish and Wildlife.)

The Tree-Climbing Gray Fox

The gray fox is the only member of the dog family in North America to climb trees. It does so to rest, get access to berries and fruit, and to escape enemies. Gray foxes have even been found curled up in abandoned nests of hawks in treetops. They aren't agile enough to pursue prey through a tree's branches, nor do they pounce on victims from trees.

When tree limbs are close to the ground, a fox hops from branch to branch as it ascends the tree. When necessary, a gray fox can also use its strongly curved claws to climb catlike up the limbless trunk of a tree—grasping the trunk with its forefeet and pushing with its hind feet. Gray foxes run down sloping trunks head first, but descend vertical trunks tail first (Fig. 2).

The gray fox's bushy tail helps it maintain balance and the animal uses it to insulate its face and toes from extreme cold when the animal is curled up.

Figure 2. When necessary, a gray fox can use its strongly curved claws to climb catlike up the limbless trunk of a tree. (From Christensen and Larrison, Mammals of the Pacific Northwest: A Pictorial Introduction.*)*

hares, ground and tree squirrels, chipmunks, voles, and other small rodents. Fruits and berries are eaten during summer and fall. Foxes also eat grass, insects, snakes, lizards, frogs, animal carcasses, birds, and bird eggs.
- In coastal areas, red foxes eat crabs, stranded fish, and dead seabirds.
- Gray foxes depend on fruits and other plant foods more than other species of fox.
- A fox diet is composed mostly of wild species, but they also eat pet food, garbage, handouts, garden crops, poultry, and small pets.

Den Sites
- Foxes use their dens for raising young and for protective cover, but often sleep out in the open.
- Foxes dig their own burrows in the ground or use abandoned burrows dug by other mammals.
- Dens are also located in rock outcroppings, culverts, hollow trees and logs, and under buildings and abandoned automobiles.
- Dens usually have multiple entrances, with escape exits that may be disguised in thick, brushy vegetation.
- Foxes usually have several dens and move from one to the other, minimizing the risk that a den containing young will be detected by potential predators. These moves also help to prevent accumulation of fleas and other parasites, as well as urine, droppings, and food waste.
- Foxes use the same dens in successive years or make new dens in the same area.

Reproduction and Family Structure
- The breeding season occurs from early February to April; young are born between mid-March and mid-May.
- After a gestation (pregnancy) period of 51 to 53 days (red fox) or 53 to 55 days (gray fox), an average litter of five kits is born.
- After giving birth, the female remains with the kits while the male hunts and provides food.
- At three months of age pups begin accompanying adults and learn to hunt.
- In September and October most kits disperse to establish their own territories.
- Adult males and females come together during the breeding season, but tend to be solitary at other times.

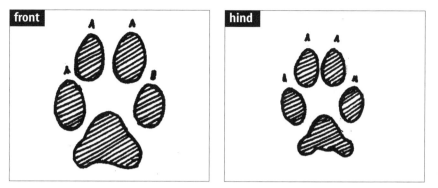

Figure 3. The front paw track of the red fox measures about 2½ inches (6.5 cm) in diameter, and the hind paw track measures about 2 inches (5 cm) in diameter. The tracks of the gray fox measure about a half-inch (1.3 cm) smaller. (Drawings by Kim A. Cabrera.)

Mortality and Longevity

- Predators of adult and young foxes include humans, eagles, wolves, coyotes, domestic dogs, bobcats, cougars, and bears. Great horned owls may also prey on kits.
- Where legal, humans kill foxes to control predation on small farm animals, and trap foxes for their fur.
- Over 50 percent of foxes die during their first year; most of the remainder die before they are four years of age.

Viewing Foxes

Foxes are most likely to be seen during the hours just after sunset and before sunrise. In spring they are occasionally seen during the day near their dens or when pursuing prey that is active by day, such as squirrels. In winter, when prey is scarce, foxes may continue their search for food into daylight hours.

In areas where you suspect foxes are present, watch (with binoculars) for movement along hedges, fences, bushes lining streams, and other places where a fox could use nearby cover to feel safe. A fox will often travel a particular trail the same time every morning and evening.

Den Sites

A fan of loose soil spreading from an 8 to 10 inch (20–25 cm) wide entrance indicates a recently created den site. Fur, bones, feathers, and well-worn areas next to an entry indicate an occupied den site. There are usually two or more entrances to each den, with openings in different directions for safe getaways. Dens are often on or near the tops of south-facing slopes to take advantage of the warming spring sun.

Note: Den sites should be observed from a distance with binoculars. Approaching the den, especially one containing kits, may cause the fox to abandon that site.

Tracks

Look for fox tracks in mud, sand, dust, or snow (Fig. 3). Fox tracks can be difficult to distinguish from coyote or domestic dog tracks.

Droppings

Since foxes, coyotes, and bobcats have similar diets, their droppings are nearly identical. In fall, winter, and spring, droppings are 2 to 4 inches (5–10 cm) long and a half-inch (1.3 cm) in diameter. They will usually contain hair, bits of insects, fruit, and will have pointed ends. In summer the droppings are shorter, contain more insects and fruit, and may have blunt ends.

Droppings are usually left on a prominent object, such as a rock or a tuft of grass along a trail, or at the edge of a well-traveled area. Droppings are often repeatedly left in the same spot, so that droppings of varying ages accumulate.

Calls

Foxes don't howl the way dogs, coyotes, or wolves do, but they do communicate with a variety of high-pitched yelps and barks.

Preventing Conflicts

Problems associated with foxes are few. However, a fox may prey on a house cat or a farm animal, such as a rabbit, chicken, duck, or goose, if it can gain access to the animal. This is generally less typical of gray foxes. Loss of domestic animals to foxes most often occurs in spring when adults need to feed kits. Damage can be difficult to detect because the prey is usually carried off to the den and uneaten parts are buried.

The same information on how to prevent coyotes from killing domestic animals applies to foxes (see "Preventing Conflicts" in Chapter 6).

Occasionally a fox will create a den where it is unwanted, such as under a porch, house, shed, or other structure. If you can be patient, the parents and young will abandon the den by the time the young are three months old. If you cannot wait that long, mild harassment will encourage a move. See Appendix B, "Evicting Animals from Buildings," for more information.

Trapping or lethal control should be a last resort and can never be justified without first applying the preventative measures listed under "Preventing Conflicts" in Chapter 6. If all efforts to dissuade a problem fox fail, the animal may have to be live-trapped or killed. See Appendix A for pertinent information.

Public Health Concerns

Diseases or parasites associated with foxes are rarely a risk to humans but could be a risk to domestic dogs. Anyone handling a fox should wear rubber gloves, and wash their hands well when finished.

Canine distemper, a disease that affects domestic dogs, is found in Pacific Northwest fox populations (see "Public Heath Concerns" in Chapter 21).

Canine parvovirus, or **"parvo,"** is another disease that affects domestic dogs and is found in Pacific Northwest fox populations (see "Public Heath Concerns" in Chapter 6).

Mange occurs in fox populations in the Pacific Northwest (see "Public Heath Concerns" in Chapter 6).

If a person is bitten or scratched by a fox, immediately scrub the wound with soap and water. Flush the wound liberally

with tap water. In other parts of the United States foxes can carry **rabies.** Contact your physician and the local health department immediately. If your pet is bitten, follow the same cleansing procedure and contact your veterinarian.

Legal Status

Because legal status, trapping restrictions, and other information about foxes change, contact your local, state, or provincial wildlife office, or visit their Web site for updates. (See Appendix E for contact information and where to access the state and provincial laws mentioned below.)

Oregon: The red fox and gray fox are classified as furbearers, so trapping and hunting regulations apply. A license is required to trap or hunt foxes during established open seasons.

No permit is required to kill a red fox damaging land, livestock, or agricultural or forest crops. However, any person killing a red fox in these situations must have in their possession written authority from the landowner or lawful occupant of the land, must immediately report the taking to the Oregon Department of Fish and Wildlife (ODFW), and must dispose of the wildlife as ODFW directs (ORS 498.012 (2 & 3)).

Property owners, persons lawfully occupying land, or their agents may kill gray fox with a permit issued by ODFW if a fox is causing damage.

The kit fox is protected as a state Threatened species. The

hunting, trapping, pursuing, killing, taking, catching, or having possession, either dead or alive, whole or in part of kit fox is prohibited (ORS 635-044-0130(1)).

Washington: All species of fox are classified as game animals, so trapping and hunting restrictions apply (WAC 232-12-007).

A property owner or the owner's immediate family, employee, or tenant may kill or trap a fox on that property if it is damaging crops or domestic animals (RCW 77.36.030). No permit is necessary, but there are restrictions on the type of trap that can be used, and the local Washington Department of Fish and Wildlife office must be notified immediately after taking a fox in these situations.

It is unlawful to release a fox anywhere within the state, other than on the property where it was legally trapped, without a permit to do so (RCW 77.15.250 and WAC 232-12-271).

British Columbia: The red fox is designated as small game under the Designation and Exemption Regulation (B.C. Reg. 168/90), and so may be hunted under license if legal anywhere in the province.

The red fox, kit fox, and gray fox are all designated as furbearing animals, and may be legally trapped under license during open seasons as defined by regulation A person may capture or kill any fox that is on the person's property and is a menace to a domestic animal or bird (Wildlife Act, Section 26(2)).

CHAPTER 11

Ground Squirrels and Marmots

Ground squirrels range in size from the chipmunk to the tree squirrel. All ground squirrels make their nests in the ground and, when frightened, will retreat to a burrow, whereas tree squirrels will climb a tree, fence, or other structure.

Nine species of ground squirrels occur in the Pacific Northwest, with most species occurring in the hot, arid regions east of the Cascade mountains in Oregon and Washington (Table 1). Due to agriculture conversion, commercial building, residential housing developments, and roads, many species no longer occur in areas where they were once common.

Ground squirrels are brown, yellow-gray, or gray, and have plain or speckled backs. Two distinctly striped species are the 11-inch (28 cm) long **golden-mantled ground squirrel** and **North Cascades ground squirrel,** which are abundant in campgrounds and picnic areas and are often confused with chipmunks. However, chipmunks are smaller and lighter, and have longer tails and stripes on the face (see Chapter 24).

Ground squirrels live in a variety of habitats but usually avoid thick brush, dense woods, and wet areas. The **Belding's ground squirrel, Columbian ground squirrel,** and **California ground squirrel** are the species most often seen in areas disturbed by humans, such as road or ditch banks, fence rows, croplands, irrigated pastures, and areas around buildings.

For information on marmots, see "Notes on Marmots."

Facts about Ground Squirrels

Food and Feeding Habits

- After emerging from hibernation, ground squirrels eat grass, clover, and other newly emerging vegetation.
- In spring and summer, flowers, fruits, bulbs, tree bark, roots, nuts, seeds, and grains are consumed in large quantities in preparation for hibernation.
- The Belding's and Columbian ground squirrels include some insects and other animal matter in their diets.
- Ground squirrels use inside-cheek pouches to transport food to safe places for consumption.

Figure 1. The Columbian ground squirrel is one of the species seen in areas disturbed by humans, such as road or ditch banks, fence rows, croplands, irrigated pastures, and areas around buildings. (From Christensen and Larrison, Mammals of the Pacific Northwest: A Pictorial Introduction.*)*

Table 1. Distribution of Pacific Northwest Ground Squirrels

Columbian ground squirrel *(Spermophilus columbianus, Fig. 1)*, southeastern and south-central British Columbia, eastern Washington, northeastern Oregon.

California ground squirrel *(Spermophilus beecheyi)*, south-central Washington, western Oregon (the only ground squirrel found in the lowlands west of the Cascade mountains in Oregon and Washington).

Belding's ground squirrel *(Spermophilus beldingi)*, eastern Oregon.

Merriam ground squirrel *(Spermophilus canus)*, central and eastern Oregon.

Golden-mantled ground squirrel *(Spermophilus lateralis,* Fig. 5 and back cover*)*, central and eastern Oregon.

Townsend's ground squirrel *(Spermophilus townsendii)*, south-central Washington.

White-tailed antelope ground squirrel *(Ammospermophilus leucurus)*, southeastern Oregon.

Piute ground squirrel *(Spermophilus mollis)*, southeastern Oregon.

North Cascades ground squirrel *(Spermophilus saturatus)*, south-central British Columbia, eastern Cascade mountains in Washington.

Washington ground squirrel *(Spermophilus washingtoni)*, southeastern Washington, northeastern Oregon.

Burrow Systems and Home Range

- Ground squirrels use their burrow system as a place to sleep, rest, rear young, and avoid predators.
- Ground squirrels dig their own burrows and rarely, if ever, use burrows of other mammals.
- Burrows are excavated 5 to 30 feet (1.5–9 m) horizontally in hillsides and banks, or dug in flat ground, 2 to 4 feet (60 cm–1.25 m) below the soil surface.
- Ground squirrels spend most of their lives within a 75-yard (67 m) radius of their burrow system. The young, in search of new territories, often disperse much farther away.

Reproduction

- Ground squirrels breed after emerging from hibernation, and a single litter of two to ten young is produced three to five weeks later.
- The young grow rapidly and by six weeks of age they resemble adults. This rapid development permits young of the year to attain the body mass and fat reserves necessary for overwinter survival.
- Ground squirrels are moderately social animals. Depending on the species, they live in a small family group, or in aggregations of several hundred animals.

Mortality and Longevity

- Ground squirrels are eaten by snakes, long-tailed weasels, badgers, foxes, coyotes, and bobcats, as well as hawks, owls, and golden eagles.
- Many ground squirrels are poisoned, trapped, shot, or run over by motor vehicles.
- Ground squirrels may live four to eight years, but most newborn never reach the age of one due to a variety of natural factors.

Figure 2. The yellow-bellied marmot is the most widely distributed species of marmot in the Pacific Northwest. These animals have distinct yellow speckles on the sides of their neck, white between their eyes, and a yellow to red-yellow belly. (From Larrison, Mammals of the Northwest: Washington, Oregon, Idaho, and British Columbia.*)*

Notes on Marmots

Marmots are the largest members of the squirrel family in the Pacific Northwest (Fig. 2). They range in total length from 21 to 30 inches (53–75 cm), with their bushy tails being 25 to 50 percent of their total length. Marmots are yellowish brown or gray, with a compact, chunky body supported by short strong legs. Their forefeet have long, curved claws well adapted for digging burrows.

Although marmots are wary and difficult to approach closely, the fascinating sight of a marmot community rewards persistent observers. See "Viewing Ground Squirrels and Marmots."

Five marmot species occur in the Pacific Northwest:

The **hoary marmot** *(Marmota caligata)* is found in the mountains near the timberline on talus slopes and in alpine meadows in Washington and mainland British Columbia.

The **yellow-bellied marmot** *(Marmota flaviventris,* Fig. 2 and back cover*),* also called the **rock chuck,** is the most widespread species, and is found throughout eastern and central Oregon, eastern Washington, and south-central British Columbia.

The **Olympic marmot** *(Marmota olympus)* is closely related to the hoary marmot and is found in the alpine meadows of the Olympic Mountains in northwestern Washington.

The **woodchuck** *(Marmota monax)* occurs in eastern, central, and northern British Columbia. Many people know the woodchuck by the name "groundhog."

The **Vancouver Island marmot** *(Marmota vancouverensis)* is restricted to mountains in southeastern Vancouver Island. At the time of writing, only 30 individuals remained in the wild.

Marmots are slow and physically defenseless, needing solid protective areas for mating, raising young, hibernating, and protection when threatened. Such sites include rockslides, rock outcroppings, and rockwork along roads and highways. Marmots also use areas underneath log piles and old buildings, and when a food source is nearby they'll burrow in meadows and forest edges.

In spring, marmots eat anything that is green and tender; as the season changes their diet includes flowers and seeds (either green or ripe), tree bark, and some insects and other animal matter. Since marmots hibernate and don't store food for winter, they must accumulate body fat, which seems to be their primary mission during their aboveground activities. Once they are in hibernation, which may begin in midsummer when the heat dries up spring's herbs and grasses, marmots don't reappear until February, March, or April, depending on weather conditions and elevation.

Marmot burrows are more than 3 feet (90 cm) in depth, but hibernation burrows may be 15 feet (4.5 m) deep.

Marmots breed soon after they emerge from hibernation and give birth a month later. Litters range from four to six young.

Marmots benefit other animals and plants. Their abandoned holes can become homes for other mammals, and their digging and droppings loosen, aerate, and improve the soil. Animals that prey on marmots, especially the young, include golden eagles, hawks, bobcats, cougars, and coyotes; badgers and black bear dig them out of their burrows.

Most marmot populations occur in sparsely populated areas and they have little negative effect on humans. In rare instances, when they live near agricultural lands, yellow-bellied marmots and woodchucks can be a problem because of their raids on crops, and the hazards posed by their burrows to farm machinery and livestock. See "Preventing Conflicts" for information on how to prevent conflicts or remedy existing problems.

The Ecological Value of Ground Squirrels

Ground squirrels are keystone species in their ecosystems and their loss may have serious consequences. Of course, they are important food sources for many mammals, birds, and reptiles that prey on them. Burrowing owls and other wildlife also depend on their abandoned burrows for homes.

More importantly, however, ground squirrels loosen, aerate, move, and mix soils. For example, a Columbian ground squirrel population in the Canadian Rocky Mountains brought 1.1 to 1.4 metric tons/per hectare/per year to the surface. In the arid and semi-arid regions of the Pacific Northwest where earthworms are rare, ground squirrels are the primary mechanism for bringing nutrients from deep soil layers to the surface.

Burrowing also encourages water infiltration, and alters the soil's water-holding capacity and mix of organic matter. In semiarid southeastern Idaho, infiltration of water into the soil was 21 percent higher in the vicinity of ground-squirrel burrows than in nearby control areas without burrows. Water also penetrated deeper into the soil in the vicinity of burrows. The additional water enhances decomposition of organic matter as well as plant survival and seed germination rates.

Although ground squirrels can sometimes cause problems in agricultural situations, conflicts can be dealt with through focused control measures. The presence of ground squirrels brings ecological benefits that can far outweigh short-term economic losses, especially on arid lands.

Viewing Ground Squirrels and Marmots

Ground squirrels and marmots are active during the day and spend their time feeding, sunning, dust-bathing, and grooming.

Ground squirrels are often seen sitting straight up, motionless, with their arms held across their chests and their paws resting one upon the other (Fig. 3). From this position they survey the surroundings, using eyesight that is thought to be equivalent to that of humans. When sunbathing, ground squirrels lay with their belly and elbows on the ground and their head raised.

Some ground squirrels are climbers. California ground squirrels and the two golden-mantled ground squirrels can be seen on fence posts, boulders, stumps, and occasionally as high as 30 feet (9 m) up in trees. Yellow-bellied marmots are also able to climb into shrubs and trees.

Burrow Sites

The burrows they make are an obvious sign of ground squirrel and marmot presence. Burrow openings are 2 to 4 inches (5–10 cm) across, depending on the size of the squirrel occupying the burrow. The marmot's burrow opening is 5 to 6 inches (13–15 cm) across. Each burrow normally has one to several trails radiating from its mouth. These trails will lead to another burrow or to a favorite feeding site or lookout.

Marmots will have a main winter den and several summer dens. The summer dens are used as islands of safety while they are out foraging away from the main site.

Newly dug ground squirrel and marmot dens have piles of fresh, loose dirt near the entrances and older dens may blend into the landscape.

Note: Coyotes, badgers, and foxes are known to convert ground squirrel and marmot dens to their own use, sometimes after eating the original occupants. The entrance to a coyote den will be enlarged to 1 to 2 feet (30–60 cm) across. Badger holes are elliptical, with a width of 8 to 12 inches (20–30 cm). Fox den entrances are also 8 to 12 inches (20–30 cm) in diameter, but are more rounded and open onto an enlarged chamber just below the surface.

Tracks

The track shape and print pattern are similar for all ground squirrels, but both vary in size, depending on the species. The hind print of the California ground squirrel is 2 inches (5 cm) long, with five toe pads; the fore print is 1.5 inches (3.8 cm) long and shows only four toe pads. Long claws may show, especially on the hind print. The hind print of the yellow-bellied marmot is 2½ inches (6.5 cm) long, with five toe pads; the fore print shows only four toe pads (Fig. 4).

Droppings

Ground squirrel droppings are small, usually unconnected ovals, ¼ inch (6 mm) long.

Figure 3. The Washington ground squirrel is often seen sitting nearly straight up, surveying

Figure 4. The hind print of the yellow-bellied marmot is 2½ inches (6.5 cm) long, with five toe pads; the fore print shows only four toe pads. (From Pandell and Stall, Animal Tracks of the Pacific Northwest.*)*

Marmot droppings resemble those of an adult house cat in size and general configuration. Plant fibers are evident. Marmots are prolific when it comes to their droppings and they can foul public areas such as picnic tables, pathways and cemeteries, creating an unwanted mess.

Calls

Ground squirrels give a "chirp" in response to a hawk or other predator. Such alarm calls are costly, because callers are twice as likely to be caught by a predator than non-callers. A "chur" is made when encountering a strange object.

The alarm shriek (not a whistle because it is made with the vocal cords) of a marmot is an abrupt, high-pitched "tweeeet" that sounds like a person whistling

through their teeth. The hoary marmot's version of this whistle is particularly loud and prolonged. Sounds of agitation or aggression include a fast chattering of teeth, hissing, growling, and squealing.

Preventing Conflicts

Ground squirrels and marmots will feed in gardens, orchards, and croplands, and create burrows in golf courses, ball fields, agricultural fields, and pastures. Burrows may pose a danger to livestock and farm machinery and have been known to weaken ditch banks and canals, and to undermine foundations and collapse gravesites in cemeteries.

Because ground squirrels and marmots hibernate, damage will not occur during their periods of inactivity; plant-related damage then is more likely caused by deer, rabbits, hares, or pocket gophers. Nearby tracks and droppings can help determine the species causing the problem.

Many species of ground squirrel in the Pacific Northwest are in decline and some are, or soon will be, of conservation concern (see "Legal Status"). The presence of an endangered or threatened species of ground squirrel, marmot, or other animal in the area where you plan to take action—chemical, nonchemical, mechanical, or otherwise—could preclude use of this action. If you are aware of an endangered or threatened species in an area where you plan one of these controls, you must contact the U.S. Fish and Wildlife

Hibernation and Aestivation

Ground squirrels spend much of the year beneath the ground in a state of dormancy known as hibernation and aestivation. Hibernation and aestivation are physiologically equivalent, but aestivation happens in the summer and hibernation in the winter. The length of this dormant period is dependent on many factors: species, location, elevation, weather, age, sex, and physical condition of the ground squirrel.

Unlike bears, which merely enjoy long periods of sleep, ground squirrels go into what might be thought of as a state of near suspended animation. The heart rate slows down to as little as a tenth of normal, and respiration slows down to the point where the animal takes a breath only once every couple of minutes. Even in this state, ground squirrels do wake up for short periods.

Ground squirrels have one of the longest hibernation periods of any North American mammal. By late spring, some adult males have entered aestivation. Females typically follow later, and young of the year, needing more time to accumulate fat, go last of all.

Since their food is not stored in the burrow, ground squirrels forage voraciously on all parts of plants, and on grasshoppers, crickets, caterpillars, and other insects.

In just four months above ground, some species of ground squirrels must reproduce and then store enough fat to survive the remaining eight months underground.

the surroundings. (From Verts and Carraway, Land Mammals of Oregon.*)*

Figure 5. Two distinctly striped ground squirrel species are the golden-mantled ground squirrel (shown here) and the North Cascades ground squirrel, which are abundant in campgrounds and picnic areas and are often confused with chipmunks. (From Christensen and Larrison, Mammals of the Pacific Northwest: A Pictorial Introduction.*)*

Service in the United States or the Canadian Wildlife Service in British Columbia for further information before moving forward (see Appendix E for contact information).

The following are suggestions for preventing conflicts or remedying existing problems around homes and small properties. For control on commercial sites, contact your County Extension Office, the U.S. Department of Agriculture, or the Ministry of the Environment (see Appendix E for contact information).

Don't feed ground squirrels or marmots: Alternative feeding has, on occasion, been effective in drawing ground squirrels away from feeding on a crop until it's harvested; however, such feeding also has some negative consequences. It is rarely good for the health of ground squirrels or marmots and will generally, over time, attract more animals to the site. Animals that have been repeatedly fed often lose their fear of humans and may, in turn, become aggressive when begging for food.

Harassment: Becoming a "bad neighbor" may cause ground squirrels or marmots to leave an area, especially if they haven't lived there long. Harassment techniques include partially digging out burrow entrances, clearing vegetation away from around entrances, and packing entrances with hay, dirt, or rocks, and then repacking them when reopened.

A repellent can also be used. Repellents include used kitty litter or commercially available predator urine, such as coyote or fox urine, sold at hunting stores and from the Internet (see "Wildlife–Human Conflicts" in Appendix F). Pour the kitty litter down the hole or sprinkle the repellent on a rag and stuff it down the hole. Harassment must be continuous and concentrated to be effective.

A tethered dog or cat in the area where ground squirrels or marmots are digging can be an effective form of harassment. (See Appendix D for information on how to safely tether the dog or cat.)

For humane reasons, when harassing the animals in their burrows, do so only after young are mobile. In some areas, snakes can be found in the den sites. As a safety measure, do not put your hands or face near the entrance until it has been determined it is safe to do so.

Frequently, a ground squirrel or marmot burrow is abandoned or unoccupied for weeks and even months, but then is reopened. A highly developed sense of smell allows the animals to locate places where other animals have been living months after the previous occupants left, even when the entrance is partly filled and full of vegetation.

Because of this, ground squirrel and marmot burrows will need to be permanently removed or closed after the animals have left. Before any burrow system is disturbed, make sure no animals are occupying it. Stuff a wad of newspaper in each hole and inspect the holes over a 48-hour period for signs of movement during the time you know the ground squirrels or marmots are not hibernating.

To permanently seal the tunnels, use ½-inch (12 mm) mesh hardware cloth, available from building supply stores and home improvement centers. Cut the wire in 3-square-foot (90 cm) sections and bury it at least 1 foot (30 cm) deep after excavating a suitable area around the tunnel entrances.

Monitor all the entrances frequently to make sure a new resident is not trying to get established.

Install barriers: Ground squirrels and marmots are visually oriented animals that have a strong need to keep their burrow entrances in sight at all times. Thus, a solid 3-foot-high (90 cm) fence placed along the perimeter of a colony, obstructing the line of vision, may be effective in limiting the spread of the colony, possibly even forcing existing colonies to relocate (Fig. 6). Silt fencing and construction grade black plastic have been used to create visual barriers.

Note: Ground squirrels that live in urban and suburban environments become highly adaptable and lose much of their need for unobstructed vision. Hence, the creation of additional visual barriers is less likely to move an exist-

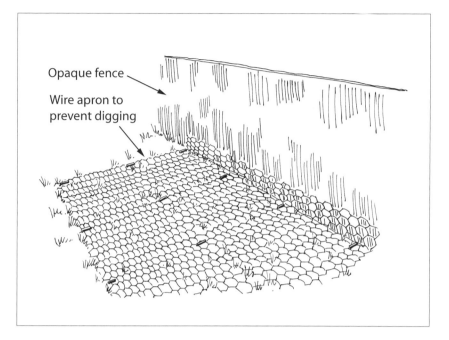

Figure 6. A visual barrier may be effective in limiting the spread of a ground squirrel colony, possibly even forcing an existing colony to relocate. On the animal side of the barrier, secure a 6-foot (1.8 m) wide horizontal strip of 1-inch (2.5 cm) mesh poultry wire. Secure the perimeter of the poultry wire into the ground using landscape staples every 3 feet (60 cm). Extend the poultry wire 6 inches (15 cm) up onto the animal side of the barrier. (Drawing by Jenifer Rees.)

ing established population of ground squirrels.

To prevent animals from tunneling under the barrier, place the bottom of the fence 12 inches (30 cm) underground, or install an apron on top of the ground on the animal side of the fence. Be sure to secure the apron firmly with stakes.

The use of ½-inch (1.25 cm) mesh hardware cloth to close openings will prevent ground squirrels and marmots from entering buildings. The same wire can be used to prevent the animals from digging beneath concrete slabs and foundations. (See both Chapter 23 and Appendix B for examples.)

Metal flashing can be installed around tree trunks to prevent damage to the base of the trees and to keep ground squirrels and

marmots from climbing trees to eat fruit or nut crops. See "Preventing Conflicts" in Chapter 21 for diagrams showing construction details.

Seeds and seedlings can be covered with a temporary wire cage or netting. Where bulbs are being dug up, poultry wire can be laid down and securely staked.

Habitat management: Historically, ground squirrels roamed vast areas of Pacific Northwest grasslands. Deer and elk once kept the taller grasses clipped, which allowed ground squirrels to migrate onto the short-grass areas.

Today, refraining from mowing tall grasses can restrict ground squirrel movement from unwanted areas. Fast-growing tall perennial grasses can also be planted to keep colonies from expanding. Plant these in close rows that will produce a dense, self-maintaining barrier. Shrubs should be used in conjunction with the barrier to create a more permanent vegetative barrier that is aesthetically pleasing.

Because ground squirrels and marmots also find protection beneath rock and lumber piles, as well as under old agricultural implements and in similar areas, removing these can reduce the areas' attractiveness to the animals.

Apply repellents: Commercially produced repellents available from nurseries and farm supply centers can help to keep ground squirrels and marmots from chewing on vegetation. Spray or paint the repellent on valuable shrubs and other plants per label directions. Apply often for best results. (See "Repellents" in Chapter 7 for more information.)

Encourage natural control: Sometimes conflicts occur because people move into habitat occupied by ground squirrels and marmots and force them to live in and around human settlements. If such areas are maintained with as much of the native plant communities intact as possible, conflicts with these animals will be minimized.

Encouraging the natural predators of ground squirrels and marmots—or at least not interfering with them—aids in reducing damage. One suggestion is to provide perches for owls and hawks (see "Maintaining Hawk Habitat" in Chapter 33).

Lethal Control

A few ground squirrels or an occasional marmot are generally tolerated, but when the conflicts they create are severe enough, lethal control is often considered. Such action is taken only when and where the advantages significantly outweigh the assessed negatives, such as hazards to pets or non-target wildlife.

The persecution of ground squirrels and marmots is a big factor in the range-wide decline of burrowing owls, and perhaps other species.

Trapping

Trapping ground squirrels and marmots can be effective for removal of small populations where other control methods are unsatisfactory or undesirable. However, trapping is rarely a permanent solution near large populations of ground squirrels and marmots, since other animals are likely to move in if food, water, or shelter remains available. For information on trapping, see Appendix A.

Toxic Bait and Fumigants

Bait and commercially made bait stations are available at most farm supply, hardware, or horticultural retailers.

Read and follow all label instructions on any poisonous bait, and pay particular attention to safety information and possible effects of bait on non-target species.

Fumigants are available for the home gardener and can be effective if used when ground squirrels are active. They work best when the soil is moist and dense, which helps keep the gas in the burrows.

Look closely for signs that non-target species are inhabiting inactive ground squirrel or marmot burrows. Foxes, coyotes, badgers, and burrowing owls use vacant marmot burrows; weasels and burrowing owls use vacant ground squirrel burrows.

Shooting

Ground squirrels and marmots that are repeatedly shot at become hunter- and gun-shy quickly. Rarely can one get close enough to use a pellet gun effectively, and the noise of a shotgun scares the animals sufficiently that after the first shot, the remaining ground squirrels and marmots will be very hesitant to emerge from their burrows.

If local laws permit, shooting with a .22 rifle may provide some control where ground squirrel and marmot numbers are low. For safety considerations, shooting is generally limited to rural situations and is considered too hazardous in more populated areas, even if legal.

Public Health Concerns

Ground squirrels can harbor diseases harmful to humans, particularly when populations are dense. A major concern in the Southwest region of the United States is **bubonic plague,** which can be transmitted to humans by fleas carried on the squirrels. Plague rarely occurs in humans and animals in the Pacific Northwest. Ground squirrels are susceptible to plague, which has wiped out entire colonies in the Pacific Northwest in the past. If you find unusual numbers of ground squirrels dead for no apparent reason, notify public health officials.

Marmots are not considered to be a significant source of any infectious disease that can be transmitted to humans.

Do not handle ground squirrel or marmot carcasses without protective gear. Use plastic bags slipped over each hand and arm as gloves. To dispose of the dead animal, turn the bag inside out while slipping it off your arm and hand. Report any bites to your doctor or your local health department.

Legal Status

Because legal status, trapping restrictions, and other information about ground squirrels and marmots change, contact your local, state, or provincial wildlife office, or visit their Web site for updates. (See Appendix E for contact information and where to access the state and provincial laws mentioned below.)

Oregon: The California ground squirrel, Belding's ground squirrel, Columbian ground squirrel, Townsend's ground squirrel (including *S. mollis* and *S. canus*) are classified as unprotected wildlife; the killing of these squirrels is unrestricted.

The golden-mantled ground squirrel is a protected species and may not be hunted; however, these squirrels may be captured and held as pets, but only with a holding permit issued by Oregon Department of Fish and Wildlife (ODFW) prior to capture. Prop-

erty owners, persons lawfully occupying land, or their agents may capture and/or kill a golden-mantled ground squirrel with a permit issued by ODFW if the squirrel is causing damage.

The white-tailed antelope ground squirrel is a state sensitive species of undetermined status and may not be hunted or held live in possession. The Washington ground squirrel is a state endangered species (OAR 635-100-0125) and may not be hunted or held live in possession. The yellow-bellied marmot is classified as unprotected wildlife.

Washington: The golden-mantled ground squirrel, North Cascades ground squirrel, Washington ground squirrel, hoary marmot, and Olympic marmot are protected wildlife. Two subspecies of ground squirrels are state candidates: the Washington ground squirrel and the Townsend's ground squirrel *(S. t. townsendii)*, [ie. the population south of the Yakima River; not the *nancyae* subspecies north of the Yakima River]; the Washington ground squirrel is also a Federal Candidate for ESA listing. Under extreme circumstances, verbal permission followed by a written permit may be obtained to trap or kill these species (RCW 77.36.030).

All other ground squirrels and marmots are unclassified and may be trapped or killed year-round without a permit. Although trapping these species is not illegal,

there are restrictions on what types of traps may be used.

It is unlawful to release a ground squirrel or marmot anywhere within the state, other than on the property where it was legally trapped, without a permit to do so (RCW 77.15.250 and WAC 232-12-064).

British Columbia: All species of ground squirrels are designated as wildlife under the Wildlife Act and, with the exception of Columbian ground squirrels, cannot be hunted, trapped, or killed unless under permit. Columbian ground squirrels may be legally hunted during open seasons as defined by regulation and there is no bag limit.

Hoary marmots, yellow-bellied marmots, and woodchucks are not considered game animals under the Wildlife Act, and there are no legal trapping or hunting seasons for these species.

Yellow-bellied marmots, woodchucks, and Columbian ground squirrels are listed under Schedule "B," which means they may be captured or killed for the purposes of protection of property by the person that owns or occupies the property or his/her spouse, parent, guardian, or child (Wildlife Act, Designation and Exemption Regulation 2 and 3). No permit is required in such cases.

The Vancouver Island marmot is listed as "Endangered," and as such is fully protected under the provincial Wildlife Act.

CHAPTER 12

Moles and Shrews

Though moles are the bane of many lawn owners, they make a significant positive contribution to the health of the landscape. Their extensive tunneling and mound building mixes soil nutrients and improves soil aeration and drainage. Moles also eat many lawn and garden pests, including cranefly larvae and slugs.

Moles spend almost their entire lives underground and have much in common with pocket gophers—small weak eyes, small hips for turning around in tight places, and velvety fur that is reversible to make backing up easy. Moles also have broad front feet, the toes of which terminate in stout claws faced outward for digging (Fig. 1). (The Chehalis Indian word for mole translates into "hands turned backward.")

However, moles are predators of worms and grubs, while gophers are herbivores. (See Chapter 18 for comparison of their scull shapes.)

Four species of moles occur in the Pacific Northwest. Curiously, none are known to inhabit any British Columbian island, and moles occur on only a few of the islands in Puget Sound.

At a total length of 8 to 9 inches (20–23 cm), the slate black **Townsend mole** *(Scapanus townsendii,* back cover*),* is the largest mole species in North America. It occurs in meadows, fields, pastures, lawns, and golf courses west of the Cascade mountains.

The **Pacific mole** *(Scapanus orarius,* Fig. 1), also known as the coast mole, is similar in appearance to the Townsend mole, and ranges from 6 to 7 inches (15–18 cm) in total length. It inhabits drier, brushier, and more wooded habitats than the Townsend mole, including interior sagebrush areas.

The dark gray or coppery brown **broad-footed mole** *(Scapanus latimanus),* also called the California mole, is 6 to 7 inches (15–18 cm) long and found in valleys and mountain meadows in south-central Oregon.

At a total length of 4 to 5 inches (10–13 cm), the gray to black **shrew-mole** *(Neurotrichus gibbsii)* is the smallest mole in North America and is unique to the Pacific Northwest. These tiny moles have many shrewlike features—they lack the mole's developed forelegs for digging—and are found in shady ravines and along stream banks. Unlike other moles, which create characteristic molehills, shrew-moles do not create mole-hills and are commonly active above ground.

Facts about Moles
Territory

- Except during the breeding season and for a female mole with her young, moles are solitary animals that live in established burrow systems with limited overlap into other moles' territories.
- Typically, the ranges of neighboring male moles do not overlap, but each male's range may overlap those of a number of females, and female ranges may overlap with those of other females.
- Population densities vary from one mole in 6 acres (2.4 h) to more than five moles per acre (0.4 h).
- The average city lot will rarely harbor more than one male or two female moles.

Figure 1. Moles have broad front feet, the toes of which terminate in stout claws faced outward for digging. (From Christensen and Larrison, Mammals of the Pacific Northwest: A Pictorial Introduction.*)*

Figure 2. Unlike other moles, which create characteristic molehills, shrew-moles (shown above) do not create molehills and are commonly active above ground. The shrew-mole is rarely ever considered a pest. (From Christensen and Larrison, Mammals of the Pacific Northwest: A Pictorial Introduction.)

Food and Feeding Habits

- Although most North American moles are insectivorous, Pacific Northwest species are somewhat omnivorous, eating both plants and animals.
- A mole's diet is mostly insects and other invertebrates, including earthworms, centipedes, millipedes, snails, slugs, grubs, ants, sowbugs, termites, beetles, and crickets.
- Stomach samples in Oregon revealed that 70 to 90 percent of the Townsend mole's diet is earthworms; however it also ate grass roots, vegetable crop roots and bulbs, and seeds.
- Moles patrol their complex arrangement of tunnels in search of prey that use or fall into the tunnels.
- The shrew-mole is unique among moles in being able to climb low bushes in search of insects.

Nest Sites and Reproduction

- Depending on the species and elevation, mating can take place from winter to early summer, producing a single litter of three to five young.
- The nest is constructed from soft vegetation that is matted down and interwoven to form a compact, protective shell about 2 inches (5 cm) thick. The nest is located within a cavity 5 to 18 inches (12–45 cm) underground, usually in elevated areas to prevent the nest from being flooded by late-winter or early-spring rains.
- Green vegetation may be added to the nest after the young are born. As the wet plant material decays, it generates heat that is retained in the nest cavity, keeping the young warm when their mother is absent.
- Young moles disperse above ground at night during the month after weaning, forming their own territories within about 30 yards (27 m) of their birth site.
- Moles reach sexual maturity at about ten months of age and breed in their first winter.

Mortality and Longevity

- Moles may come to the surface at night to search for food, water, nesting material, and new territories. Moles move slowly on

Notes on Shrews

Shrews (*Sorex* spp.) are the Northwest's smallest mammals; the pygmy shrew is no bigger than your entire thumb. Shrews are also one of our most common mammals, inhabiting areas from sea level to high mountain meadows. Even so, they are surprisingly one of our least well-known mammals.

Shrews are mouselike in proportion, but differ from mice in having long, pointed muzzles and minute eyes. Most are less than half the size of adult mice, and blackish or brownish in color, with a pale-colored belly (Fig. 3).

Because of the similarities between shrew species, even specialists find it difficult to differentiate between them. Ten species of shrews are found in the Pacific Northwest:

At 6 inches (15 cm) in total length, the **Pacific water shrew** (*Sorex bendirii*) is the largest Pacific Northwest species; at 3½ inches (9 cm) and with a total weight about the same as that of a dime, the **Pygmy shrew** (*Sorex hoyi*) is the smallest.

The 4-inch (10 cm) long **vagrant shrew** (*Sorex vagrans*) is the most widespread species and is found in marshes, wet meadows, forests, streamsides, and gardens throughout the Pacific Northwest.

The **Trowbridge's shrew** (*Sorex trowbridgii*) is by far the most abundant species in western Oregon and Washington forests.

The semiaquatic **Northern water shrew** (*Sorex palustris*) and Pacific water shrew are able to run on top of water for several seconds, gaining buoyancy from air bubbles trapped in their partially fringed toes. They are also able to spend up to 3½ minutes underwater in search of prey. See Appendix F for reference information on these and other Pacific Northwest shrew species.

Water shrews eat stonefly nymphs, caddisflies, mayflies, leeches, and small fish. The terrestrial shrews eat a variety of insects and other invertebrates, including crickets, spiders, earthworms, slugs, and snails. A hunting shrew pounces on and subdues its prey with a series of rapid bites; it may have to pin a larger invertebrate to the ground. Small snails are cracked open and eaten whole. Slugs and snails are a problem because their heavy mucus covering often coats the whiskers and fur, which necessitates a good cleaning after a meal. Nevertheless, slugs have been identified in the stomachs of larger shrews.

Adult shrews lead solitary lives and are active throughout the year. Various shrew species are active at periods throughout the day and night, others are strictly nocturnal. Shrews curtail their activity in winter, and feeding periods tend to be brief, as they spend much time in their well-insulated nests.

Shrews prefer moist environments because their high metabolic rates create high moisture requirements and they can easily become dehydrated. Moist environments also tend to have a diverse and abundant food supply.

To discover shrews, go into the nearest woods and carefully lift up fallen leaves to look for their tunnels, which are common in leaf litter just above the soil surface. If you don't find any in one spot, choose a new spot about 30 feet (9 m) away, and try again.

Shrew tunnels are 1 inch (2.5 cm) or less in diameter, with vegetation pressed to the side. They also use vole and mole tunnels.

Shrews tunnel through the snow in the same manner as voles (see Chapter 25). The only way to distinguish between shrew and vole tunnels is by the smaller size of the shrew tunnels, which are about an inch (2.5 cm) or less in diameter unless melting has occurred. When the snow begins to melt, shrew tunnels, like vole tunnels, show up as meandering trails.

Shrews produce one to several litters a year. Their nests are similar to those of voles and mice, being composed of shredded grass and leaves. Nests are located along tunnels or under discarded boards, logs, tree roots, or rocks.

Shrews are preyed on by owls, snakes, and Pacific giant salamanders. House cats, opossums, foxes, and similar-size mammalian predators kill but may not eat shrews, presumably because, when frightened or agitated, shrews produce a musky odor from their anal glands.

A shrew's life span rarely exceeds 16 months, and most probably die before they are 12 months old. Extremely sensitive, shrews have died of fright from loud noises, even thunder. Unfortunately, shrews also commonly enter and become entrapped in discarded soft-drink and beer bottles, where they die of starvation.

Shrews are rarely considered pests. They will occasionally enter homes but seldom cause any trouble other than perhaps startling a resident.

Techniques used to mouse-proof structures will also exclude shrews. Place ¼-inch (6 mm) mesh hardware cloth over potential entrances. Copper or stainless-steel mesh scouring pads (steel wool quickly corrodes after becoming wet) placed in small openings can also exclude shrews. See "Preventing Conflicts" in Chapter 8 for additional information on excluding shrews.

Figure 3. At 6 inches (15 cm) in total length, the Pacific water shrew is the largest Pacific Northwest shrew species. (From Christensen and Larrison, Mammals of the Pacific Northwest: A Pictorial Introduction.*)*

the surface and are preyed upon by owls, hawks, and snakes.

- Because of their secluded life underground and their strong, musky odor, moles are unpalatable to most mammalian predators; however, raccoons and coyotes do dig them out, presumably to eat them.
- Moles are also killed by domestic dogs and cats, but are rarely eaten.
- Spring floods kill many moles, especially young ones; humans also kill many moles.
- The maximum life span of moles ranges from four to six years.

Signs of Moles

Moles are active all year round at any time of day, but are rarely seen due to their underground existence. They are best recognized by their molehills, which they push up along their tunnel systems. Both moles and pocket gophers construct tunnels and mounds, but there are distinct differences (see Table 2 in Chapter 18 for comparisons).

Landscaped areas, which provide a perfect food source for moles, are often where you see their telltale signs. The soil is kept rich with heavy applications of organic materials such as mulch, compost, and manure. The soil is kept moist through watering or irrigation, and the presence of man-made borders, edgings, and timber or stone walls. All of these provide a wonderful environment for worms and soil insects.

Mole Tunnels

A mole's territory is a mazelike system of connecting, intertwining underground tunnels located at various depths (Fig. 4). It's a perfect fortress in which to survive threats, either natural or man-made—drought, freezes, predators, toxic gases, and other poisons. They routinely scent-mark their tunnels while patrolling for insects and other invertebrates that travel or fall into their tunnel systems.

Moles are fast diggers and can tunnel at a rate of 15 feet (4.5 m) per hour. In favorable areas, shallow tunnels can be built at a rate of 12 inches (30 cm) per minute.

Moles construct two kinds of tunnels: surface tunnels and deep tunnels, or runways. Surface tunnels are located 1 to 4 inches

(2.5–10 cm) below the surface. These appear as 3-inch (7.5 cm) wide ridges or rips in the lawn or in soil, or as puffed-up areas in mulch. In lawns, surface tunnels are often held together only by the surrounding grass roots, and you may see the ridges—or feel them as you step on them. Surface tunnels wind around with no apparent direction or plan; they are used once or revisited several times for feeding purposes, and possibly for locating mates in the breeding season.

Surface tunnels connect with deeper runways that are located 3 to 12 inches (7.5–30 cm) below the surface, but may be as deep as 40 inches (1 m) under roads or buildings. Deep runways are main passageways that are used daily as the mole travels to and from surface tunnels and its nest.

Digging is most pronounced in fall and winter when the soil is moist and easy for moles to work. In periods of dry weather or drought, moles tunnel deeper, near moist, cool areas where insects and worms congregate—along sewer drainfields and under sidewalks, rocks, and fencerows.

Molehills

To create tunnels, the mole muscles its way through the soil with swimming motions, pushing the soil aside with alternating left and right paw strokes and compressing it against the tunnel walls. The large, thick, clawed forepaws do the digging, while the small hind feet provide leverage against the tunnel sides.

The soil excavated or cleaned out from the deep tunnels is pushed to the surface through vertical tunnels and forms the surface mounds, or molehills.

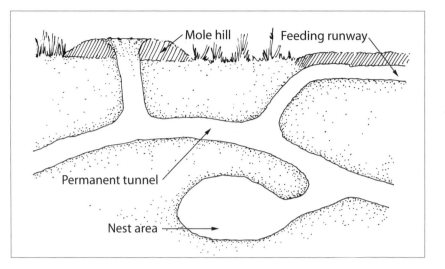

Figure 4. Cross-section of a mole's runway system. A single mole can construct 200 mounds over the course of a winter. (Drawing by Jenifer Rees.)

Molehills occur in the moist, loose earth found at the edges of woods and in fields, lawns, and other grasslands where food is available. Excavated materials are piled in roughly circular mounds that are 6 to 24 inches (15–60 cm) in diameter and 2 to 8 inches (5–20 cm) high. The opening to the burrow is near the center of the mound and is always left plugged, but the plug often lacks definition.

Preventing Conflicts

Because of the surface tunnels and mounds they create, moles may be considered pests in yards, ball fields, golf courses, horse pastures, and other locations. Moles may also inadvertently heave small plants out of the ground as they tunnel, or damage plants when their mounds cover small seedlings. The easiest way to prevent minor mole damage is to regularly visit problem areas, taking a few moments to reposition or uncover plants as needed.

Moles, gophers, and voles (which use mole tunnels) can be found in the same location, and determining which creature is causing damage can be confusing. Positive identification is needed, as control methods differ for each species. See the chapters on gophers and voles for information.

Moles are here to stay and extermination is impractical, especially if your property borders an area that has historically served as a source of moles. While you may be able to remove an existing mole population or drive moles elsewhere, if suitable conditions exist and moles occur nearby, other moles will eventually move into vacated areas. In addition, it is important to understand that mole problems rarely can be resolved by a quick fix method, but that a continuing commitment to whatever solutions are adopted is required.

To prevent conflicts or remedy existing problems, consider the following:

Repellents: No repellents currently available will reliably protect lawns or other plantings from moles. However, people mistakenly think they have successfully repelled a mole because they don't see new molehills for long periods following use.

The reasons for this are simple: moles are relatively solitary animals except for when breeding and rearing young, and they have large, complex tunnel systems that may extend for several hundred lineal feet. Moles may work one portion of their tunnel system for a few days and then move on some distance away to another portion of the system, which may be in the neighbor's yard. Hence, the application of some obnoxious substance just prior to or immediately following the mole's shift in its feeding location will be credited to the effect of the repellent. When the mole returns a week or two later, the gardener is convinced it is a new mole.

Mothballs, garlic, or spearmint leaves placed in the tunnels, and a perimeter of mole plants (*Euphorbia* spp.) planted around gardens, have all produced mixed results. Similarly, ground or broken pieces of glass, used razor blades, sections of barbed wire, or thorned rose bush canes have all been placed in mole tunnels. Some of these are actually more hazardous to the gardeners themselves than to the moles.

When moles run into the unfamiliar foreign object in their tunnels, they may simply circumvent the object by blocking those tunnels off with soil and then proceed to dig new tunnels, just as they do with a poorly set trap. There is no convincing evidence that these sharp, potentially harmful items cause any mortality or that they have resulted in the mole leaving the immediate area.

Another Homemade Mole Repellent?

Commercially available castor oil–based repellents have been scientifically tested on moles in the Eastern United States with some success. In theory, the repellent coats earthworms and other prey with castor oil. This renders the prey distasteful and, if eaten, gives the moles diarrhea. The moles supposedly then leave the treated area in search of a new food source.

The following home-prepared formula has been around for many years and some gardeners swear by it; others claim it does not work. Because moles move around within their territorial burrow systems, repellents are very difficult to evaluate.

The formula for the castor-oil repellent can be made by using a blender to combine ¼ cup of unrefined castor oil (can be purchased at most pharmacies) and 2 tablespoons of a dishwashing liquid. Blend the two together, add 6 tablespoons water, and blend again. Combine the concentrated mixture with water at a rate of 2 tablespoons of solution to 1 gallon of water. Use a watering can or sprayer to liberally apply the solution to areas where moles are active. The above mixture will cover approximately 300 square feet.

The repellent will be most effective where it can be watered into the moist soil surrounding surface tunnels made by moles. Areas that receive extensive irrigation will quickly loose the repellent to leaching. For best results, spray the entire area needing protection; moles will burrow under a perimeter treatment.

The repellent may need to be reapplied before moles depart. Once moles move elsewhere, the solution usually remains effective for 30 to 60 days.

Practicing Tolerance

Before trying to control moles, be sure that they are truly a problem. If the individual mole is not really a problem, consider it an asset.

To remove the visible presence of the mole's little-understood lifestyle, try the following:

Molehills: Remove them as they appear or before mowing by shoveling up the earth, scooping up the earth with your hands, or spreading it in place with a rake. Grass seed can be spread over large bare areas throughout the rainy months.

Surface ridges: Flatten these ridges with your foot.

Run depressions: Bring in sand or screened dirt to fill the depressions, and then reseed.

Passive acceptance: The subterranean life your wild neighbor leads beneath your feet is there for your understanding and enjoyment if you so choose.

To render mole activity less obvious, try adopting a more naturalistic landscape style and let the lawn grow up to hide mole activity. Cut grass with a weed-whacker to the desired height as required to keep a semi-tidy look. You can take advantage of the soil preparation done by moles by planting shrubs and other plant material directly into mole mounds. This process eventually transforms the lawn area into a wildlife-friendly landscape setting where mole activity goes unnoticed.

Scare tactics: Although numerous devices are commercially available to use to frighten moles (vibrating stakes, ultrasonic devices, pinwheels, etc.), moles do not frighten easily. This is probably because of their repeated exposure to noise and vibrations from sprinklers, people moving about, and lawnmowers and other power equipment. Consequently, frightening devices have not proven to be effective in the Pacific Northwest. *Note:* Be skeptical of commercial products and claims, and make sure the manufacturer offers a money-back guarantee if the product proves ineffective.

Food reduction: It is often suggested that if you eliminate grubs from an area you will get rid of moles. Grubs, however, make up only a portion of the mole's diet. During dry periods, moles are known to frequent well-irrigated lawns just for moisture. Thus, moles often are present even in grub-free yards. If all the earthworms, grubs, and other soil animals in a lawn are eliminated by repeated insecticide application, moles may be forced to seek other areas.

The use of soil insecticides is an expensive approach with no immediate reduction of damage and little likelihood of long-term control. In the process, soil insecticides may poison the groundwater, kill beneficial soil invertebrates, and damage songbirds and other desirable wildlife.

Barriers: Constructing an underground barrier to keep moles from tunneling into an area can be labor-intensive and costly; however, it is recommended for exceptional situations. See "Preventing Conflicts" in Chapter 18 for information and designs.

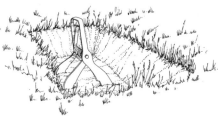

Figure 5. The Out O' Sight® scissor-jawed trap is set so that the jaws straddle the runway. (Drawing by Jenifer Rees.)

Natural control: A long-term way to help prevent conflicts with moles is natural control. Predators—including snakes, dogs, and coyotes—kill moles. In addition, attracting barn owls and other raptors, which prey on juvenile moles when they disperse, may help control a mole population, particularly in rural areas (see "Maintaining Hawk Habitat" in Chapter 33 for information).

Predators alone won't always keep mole populations below the levels that cause conflicts in gardens and landscaped areas. However, when combined with other control techniques, including practicing tolerance, natural control can contribute to overall control.

Lethal Control

Before using a lethal control, make sure it's moles, and not pocket gophers (especially a listed species—see "Legal Status" in Chapter 18) you are trying to control.

Because moles are territorial, removing them from an area may appear to solve the problem. However, other moles will eventually enter the area if attractive habitat is available. Long-term control is possible by first reducing or eliminating the mole population by trapping, and then continuing with a maintenance-trapping program to remove invading moles as they become evident.

Moles can be controlled any time, but it is best to concentrate the effort in late winter and early spring, before they give birth.

If you decide to hire someone to do the trapping, consider signing a long-term contract with the trapper, which states that they will trap all moles for a year. See Appendix C, "Hiring a Wildlife Damage Control Company" for important information.

Trapping

Though time consuming and often giving only temporary results, trapping is the most effective method of mole control. An understanding of mole behavior will help improve trapping success. When a mole's sensitive snout encounters something strange in the tunnel, the mole is likely to plug off that section and dig around or under the object. For this reason, traps are generally set straddling or encircling the runway, or are suspended above it.

Several types of mole traps are available from hardware stores and farm supply stores, and work well if used according to the manufacturer's instructions. Pacific Northwest trappers generally rank the Out O' Sight® (Figs. 5 and 6), the Victor® harpoon-type, and the Nash® mole traps as the best, but some have good success using the Cinch® and NoMol® traps as well. Before choosing a trap to purchase, consider your ability to set the trap. Good hand strength is needed to set many mole traps. See "Legal Status" for other information.

Avoid newly marketed mole traps until they have been proven effective in the field, as most new traps will not measure up to the best mole traps currently marketed. *Note:* Traps advertised for both gophers and moles are generally

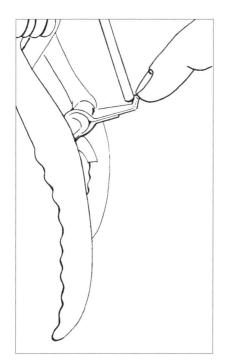

Figure 6. Regardless of the type of mole trap used, set the trigger so it will spring easily. A hair trigger setting on the scissor-jawed trap is shown here. (Drawing by Jenifer Rees. Adapted from Hygnstrom et al., Prevention and Control of Wildlife Damage.*)*

ineffective for moles, although they may be quite good for gophers.

For related information on trapping, see "Preventing Conflicts" in Chapter 18.

Poisoning and Stunning

Since moles feed on insects and worms, poisoned baits have proven to be ineffective on moles. A new gel-type bait has been registered for mole control, however, it has not been on the market long enough to determine its control value for Pacific Northwest moles. If toxic mole bait is used, follow all label directions to prevent the possibility of poisoning non-target wildlife species, domestic animals, or humans.

Gas cartridges and smoke bombs are unreliable. Their effec-

tiveness is probably compromised by the extensive nature of mole tunnel systems and because gas diffuses in soil. Moles will seal off their burrows in seconds when they detect smoke or gas.

If using gas cartridges, use them only on moles that have just invaded an area, as their burrow systems will be less extensive. Apply a cartridge into the main tunnel and not into the shallow feeding tunnels. A cartridge should be placed in two or more locations of what is believed to be the burrow system of one mole. Turning on the sprinkler to wet the soil surface of the garden or turf prior to the application will aid in retaining the toxic gas in the burrow system.

Because moles are sensitive to concussion, smacking a shovel on the ground above a mole in its surface tunnel often will quickly kill it.

Flooding

Moles can easily withstand normal garden or home landscape irrigation, but flooding can sometimes be used to force them from their burrows where they can be dispatched with a shovel or caught by a dog. The entire tunnel system will need to be quickly and completely flooded to evict its tenets. Five-gallon buckets of water poured in the hole will flood the area more quickly than a running hose.

Flooding has the greatest chance of succeeding if moles are invading the property for the first time. Where they are already well established, their systems are too extensive.

For best results and for humane reasons, concentrate the effort in late winter and early spring, before moles give birth. Be careful when

attempting to flood out a mole near a building; doing so could damage the foundation or flood the basement or crawl space.

Shooting

Since moles spend most of their time underground, shooting is impractical unless you have the time and patience to wait for one to be active at or near the surface. For safety considerations, shooting is generally limited to rural situations and is considered too hazardous in more populated areas, even if legal.

Follow-up

Once moles have been controlled, monitor the area on a regular basis for signs of their return. If resident moles are trapped out, nearby moles often migrate into, and use, established tunnels. (Moles always take the path of least resistance, so taking over established mole tunnels fits that pattern.) To help monitor the area, level all existing mounds so fresh mounds can be easily seen.

Public Health Concerns

Diseases or parasites associated with moles and shrews are rarely a risk to humans or domestic animals.

Cats that are allowed to hunt outside may bring dead, uneaten moles or shrews inside the home. Dispose of these by placing a plastic bag over your hand, picking up the dead mole or shrew, turning the bag inside out while holding the animal, sealing the bag, and discarding it with the garbage. Using a plastic bag in this manner reduces the potential for flea, tick, or disease transmission.

Legal Status

Because legal status, trapping restrictions, and other information about moles and shrews change, contact your local, state, or provincial wildlife office, or visit their Web site for updates. (See Appendix E for contact information and where to access the state and provincial laws mentioned below.)

Oregon: Moles and shrews are classified as unprotected wildlife; the killing of moles and shrews is unrestricted.

Washington: Moles and shrews are unclassified and people may trap or kill moles and shrews on their own property when they are causing damage to crops, domestic animals, or their property (RCW 77.36.030).

Although trapping moles is not illegal, there may be restrictions on what devices may be used.

The Merriam's shrew is a state candidate species being considered for listing as a State Sensitive, State Threatened, or State Endangered species. It cannot be trapped without a permit from the state.

British Columbia: All species of moles and shrews are designated as wildlife under the Wildlife Act, and cannot be hunted, trapped or killed unless under permit. The Townsend mole and coast mole, however, are also listed under Schedule "B," which means they may be captured or killed for the purposes of protection of property (Wildlife Act, Designation and Exemption Regulation 2 and 3).

The Townsend mole and Pacific water shrew are Red-listed and designated as "Threatened." The water shrew (S. palustris brooksi), and Tundra shrew (S. tundrensis) are Red-listed. The Trowbridge's shrew *(S. trowbridgii)* is Blue-listed. See "Legal Status" in Chapter 1 for more information on listing of Canadian animals.

Moose

Moose (*Alces alces*, Fig. 1) are the largest land mammals in the Pacific Northwest. A mature bull (male) moose can measure 6 feet (1.8 m) at the shoulder and weigh up to half a ton—nearly three times the weight of a large black bear.

The great size and long legs of the moose allow it to browse high on trees, wade into deep water to feed on aquatic plants, and stride through dense vegetation and deep snow.

Moose are uniformly dark brown, turning grayer in winter, with paler legs. They have prominent muzzles with an overhanging upper lip. The large flap of hair-covered skin that hangs beneath their throat is called a "bell" or "dewlap." Some theorize the dewlap helps to drain water off of the moose's face. The male's dewlap has a scent.

Moose are rarely found in Oregon, but are occasionally reported in eastern Oregon when they wander across the Idaho border. Approximately 1,000 moose inhabit northeast Washington. Although they are occasionally spotted elsewhere, the most significant population is in the Selkirk Mountains of Washington, Idaho, and British Columbia. In British Columbia, moose are found throughout the mainland, excluding coastal areas and the arid region centered in the Okanagan Valley.

Facts about Moose

Food and Feeding Behavior

- In winter, moose feed primarily on the twigs of shrubs and trees, including willows, red-twig dogwood, birch, aspen, and cottonwood.
- As winter progresses and snow continues to accumulate, moose

Figure 1. Bull (male) moose weigh 800 to 1,000 pounds (360–450 kg). Cows (females) are smaller than bulls, weighing 600 to 800 pounds (270–360 kg). Adult bulls have broad, flat, palmated antlers tipped with a number of points, depending on the age and health of the animal. The antlers of a large bull moose may spread 6½ feet (2 m) and weigh 80 pounds (36 kg). Antlers are shed as early as November, but usually in December and January. (From Christensen and Larrison, Mammals of the Pacific Northwest: A Pictorial Introduction.*)*

feed increasingly on coniferous trees.
- In the warm months, moose add aquatic plants to their diets—sedges, pondweeds, and water lilies.
- In summer, a moose can consume as much as 24 pounds (11 kg) of forage per day; in winter, when food is more difficult to find, consumption falls to half that per day. Stored fat helps sustain moose through the winter.
- Moose will seek out needed minerals on roadside vegetation that is coated in road salt.

Range Needs and Territory

- Optimum moose habitat consists of patches of shrubby growth with nearby mature timber, lakes, marshes, and other wetlands. Areas with early stages of regrowth following disturbances such as fires and logging are part of their home range.
- Mature forests help keep moose warm in winter and cool in summer; they also provide foraging areas in the shallow snow under the canopy.
- With its great size and forage demands, the average moose in any given season requires a home range of 3 to 6 square miles (4.8–9.6 km).
- In mountainous areas, moose concentrate at elevations below 3,500 feet (1067 m) during winter, moving to higher elevations during summer.

Reproduction and Family Structure

- The mating season, or rut, takes place in early fall and lasts for a month.

- After a gestation period of eight months, single or twin calves are born in May or June.
- New calves are awkward and kept hidden for two to three days. Cows are protective of their calves and will not let other moose or animals, including humans, approach them.
- A calf stays with its mother throughout its first year and is fully weaned by the time the mother gives birth the next spring.
- Adult moose are solitary in summer. Yearling calves sometimes remain together after they leave their mother.
- In winter, moose tend to form small groups of 20 or so. These groups include bulls, cows with their calves, and adolescent moose.

Mortality and Longevity

- Black bears and cougars prey on moose calves; grizzly bears and wolves prey on calves and adults.
- Hunting, vehicles, and habitat loss all take their toll on moose populations. Collisions with trains also can be a major cause of winter mortality. This occurs when moose become trapped in the railroad right-of-way by high snowbanks piled alongside the tracks.
- Bull moose reach their prime when they are 6 to 8 years old; in the wild they live to be 15 years old.

Viewing Moose

Moose may be active at any time during the day or night, alternating a few hours of feeding with a few hours of resting and digesting. Like deer and elk, moose

are ruminants and spend hours "chewing their cud." In areas with substantial human activity, moose are more secretive during the day, and they feed in more exposed places only between dusk and dawn.

Moose don't spook or shy away from people as readily as deer and elk; at times they appear docile or even curious. But, like other wild animals, moose can be dangerous if not treated with respect.

Be careful when observing moose, because being approached too closely by humans may provoke them. Although they could outrun us, moose—unlike deer and elk—will often choose "fight" over "flight" when threatened. A charging moose kicks forward with its front feet, knocking down the threat, then stomps and kicks with all four feet. Antlered bulls can use their racks just as aggressively.

See "Preventing Conflicts" for precautions to use when living in or visiting moose country.

Tracks and Trails

Moose are easier to track than most animals due to their weight and large hoofprints, which leave marks in or on almost anything they walk over (Fig. 2). They are adapted to marshy areas and their hooves spread broadly when they step, helping to displace weight on mud or pond bottoms. Adult moose tracks are larger and more pointed than those made by elk; they are similar in shape to deer tracks, but twice as large.

Droppings

During the summer, when moose feed extensively on aquatic vegetation, droppings are normally

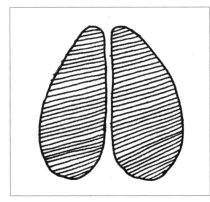

Figure 2. Moose tracks show two toes that range in length from 2½ inches (6.5 cm) in a young calf to 6½ inches (16.5 cm) in a mature bull. The tracks of a moose calf may look like those of a large deer or elk, but the moose calf would be accompanied by its mother, and her tracks would be unmistakable. (Drawing by Kim A. Cabrera.)

massed or even liquid, and some resemble domestic cowpies. In winter or at other times when moose feed on trees and shrubs, their droppings take the form of pellets that resemble compressed sawdust. Individual pellets are about 1½ inches (4 cm) long and ¾ inch (2 cm) in diameter.

Feeding Sites

In areas where moose consume quantities of twigs from trees and shrubs, they can actually establish a "hedge" or browse line 6 to 8 feet (1.8–2.4 m) above the ground by clipping most of the terminal shoots of their favored food species.

The trunks of aspen and other deciduous trees that have been gnawed on by moose year after year eventually become rough and blackened as far up as the moose can reach. A grove of black-trunked aspen is a sign of winter range that has been heavily used by moose. Gnawings may also be found on felled trees and branches and are easily distinguished from the chisel-like cuttings of the beaver.

Wallows

Like elk, moose of both sexes and all ages wallow in mud. Because moose feed mainly on aquatic plants during the summer months, they're exposed to high concentrations of flies and mosquitoes. Rolling in mud applies a temporary bug-proof coating that offers relief.

Moose wallows typically measure 4 feet (1.25 m) square and 3 to 4 inches (7.5–10 cm) deep, with displaced mud around the perimeters. The bottoms contain dark-brown or gray hairs, as opposed to the blond or tan hairs found in elk wallows. Lakeshores and riverbanks are the best places to look for mud wallows.

Beginning about September, bulls scent the wallows with their musky urine, a territorial scent-post and an advertisement of availability to breeding females. A bull may repeatedly lie down and roll about in the wallow. A cow that has joined the bull will also lie down and roll in the wallow, sometimes with her calf.

Calls

During the breeding season, moose utter a variety of sounds including moans, croaks, and barks. Cows, calves, and yearlings maintain a dialogue of whimpers, grunts, and low mooing sounds (like those of domestic cattle) as they travel and forage.

Preventing Conflicts

Increasing human population and encroachment into moose habitat creates new management challenges. Wildlife offices in northeast Washington and British Columbia receive moose complaints that include sightings on urban streets, moose eating crops and young trees, moose standing on airfields, and moose colliding with automobiles and trains.

Excluding moose with fences or other barriers and using other tactics to prevent moose from gaining access to areas is not effective; moose are large enough to roam wherever they choose.

Close encounters: Taking the following precautions will help prevent unpleasant encounters with moose when visiting or living in moose country:

If you see a moose, give it lots of space. If you're hiking in the woods, yield the trail in whatever way provides the most space between yourself and the moose—back off or change directions, and enjoy the animal from a distance.

In the winter, exercise caution when snowmobiling, skiing, or snowshoeing. When there is deep snow on the ground, moose will try to conserve energy by using the same packed trails used by humans.

If you see a moose calf and not a cow, be extremely careful as you move out of the area; you may have walked between mother and baby. This situation is the most dangerous one with moose.

Never approach a moose, even if it seems quiet and gentle. Even a resting moose, if approached, may become stressed and aggressive. In the fall, breeding bulls in rut become highly competitive and agitated.

Use binoculars, a spotting scope, or telephoto camera lenses to view moose from a distance.

Never feed moose. Moose that are given food can become aggressive when they are not fed as expected. They may attack another

In Case of an Attack

Many moose charges are bluffs or warnings, but you should take them seriously. Even a calf, which weighs 300 to 400 pounds (135–180 kg) by its first winter, can injure you.

A moose that sees you and walks slowly toward you is not trying to be your friend. It is probably warning you to keep away (or looking for a handout if it has been fed in the past). It may signal an attack by laying its ears back, raising the long hairs on its shoulder hump, stomping the ground, or swinging its head in your direction. If this happens, you are too close!

If this does happen, look for the nearest tree, fence, building, or other obstruction to hide behind. Unlike cougars, bears, or even dogs, moose probably will not chase you far if you attempt to run away. So it's a good idea to run away from a moose. Use trees or other obstacles to run around—a human can usually do this faster than a moose can.

If a moose knocks you down, it may run off or start stomping and kicking. If the attack continues, curl up in a ball, protect your head with your arms and hands, and hold still. Don't move or try to get up until the moose moves a safe distance away or it may renew its attack.

person who has no food to offer. A moose with a history of unprovoked attacks on people may have to be killed to protect public safety.

Keep all dogs confined when in moose country. Dogs are related to wolves, the mortal enemies of moose. Moose have been known to go out of their way to kick at a dog, even one on a leash or in a fenced yard. Being chased by dogs or even just barked at may provoke a moose to aggression. If you see a moose in the vicinity of where you live, bring dogs inside. If you're walking with your dog and see a moose, keep your dog quiet and take an alternate route out of the area.

Moose in town: If a moose wanders into an area where it will have trouble wandering out because of traffic and other human congestion, contact your local wildlife office or police department for assistance. Meanwhile, confine dogs and other pets and keep children inside and quiet. Give the moose ample room to move out of your yard and don't block its escape routes. Draw curtains on large glass doors and windows so the moose doesn't mistake them for an escape route.

Moose and vehicles: During periods of restrictive snow cover, moose utilize railways and highways for travel routes, resulting in many moose/vehicle collisions. However, in fall, the threat of moose–vehicle collisions peaks, because bull moose are actively pursuing females. Once bulls locate a cow moose in heat, they'll pursue her until they can breed, or until they're driven away by a larger bull. If there's a road separating a bull from a receptive cow, the bull will inevitably cross the road.

Cars are no match for an adult moose, and because the animal's legs are long, a collision usually sends the moose crashing down on the roof and onto the occupants.

To compound the problem, a coat of dark hair makes moose difficult to see at night. Additionally, they often stand their ground, even when faced with an oncoming vehicle.

If you are driving and come upon a moose standing or walking in the road, yield to the moose. It may be trying to rest or save energy. If you try to scare it off your vehicle could be attacked. If you are driving at night in an area that is frequented by moose, slow down and be extra cautious—a collision with a moose could be fatal for both of you. (For additional information, see "Tips for Driving in Deer Country" in Chapter 7.)

Public Health Concerns

Moose are not considered a significant source of infectious disease that can be transmitted to humans or domestic animals. However, as with any wild animal, it is recommended that you wear rubber gloves if you need to handle a sick or dead moose. Wash your hands afterward, and fully cook all moose meat to 160°F (71°C).

Legal Status

Because legal status, hunting restrictions, and other information about moose change, contact your local, state, or provincial wildlife office, or visit their Web site for updates. (See Appendix E for contact information and where to access the state and provincial laws mentioned below.)

Oregon: Moose are classified as game mammals, but they are rarely found in Oregon and there is no open hunting season.

Washington: The moose is classified as a game animal (WAC 232-12-007) and hunting regulations apply. After obtaining a permit issued by the Department of Fish and Wildlife, a property owner or the owner's immediate family, employee, or tenant may kill a moose on that property if it is damaging crops. You must notify the Washington Department of Fish and Wildlife immediately after taking an elk in these situations.

British Columbia: Moose are classified as game animals and hunting regulations apply. Animals that are involved in property damage may not be captured or killed, outside of general open seasons, without first receiving a permit from the regional Fish and Wildlife Office.

CHAPTER 14

Mountain Beavers

M ountain beavers (*Aplodontia rufa*, Fig. 1) are considered by many taxonomists to be the world's most primitive living rodent species. They are not really beavers, but were so named because they gnaw bark and cut off limbs in a manner similar to true beavers.

Mountain beavers live in moist forests, on ferny slopes, and are occasionally found in damp ravines in urban areas. Their worldwide range is the coastal lowlands and coastal mountains of southern British Columbia (from the Fraser Valley to the Cascade mountains), western Washington, western Oregon, and south into California.

Mountain beavers are not considered game or fur animals today although, in the past, Native American Indians wore robes made of mountain beavers and valued their meat.

Most people don't know mountain beavers exist and some still continue to question that fact even after they've heard about the animals.

Facts about Mountain Beavers
Food and Feeding Habits

- Mountain beavers are herbivores and eat a wide variety of plants.
- Food items include all above- and below-ground parts of ferns, salal, nettles, fireweed, bleeding heart, salmonberry, brambles, dogwoods, vine maples, willows, alders, and conifers. Mountain beavers also eat rhododendrons and other ornamental perennials, shrubs, and trees.

Figure 1. Mountain beavers, also called boomers, are 12 to 14 inches (30–35 cm) long and resemble large, overgrown hamsters or tailless muskrats. They have small ears and eyes, short, rudimentary tails, and large curved front claws that are used for digging, grasping, and climbing. (From Christensen and Larrison, Mammals of the Pacific Northwest: A Pictorial Introduction.*)*

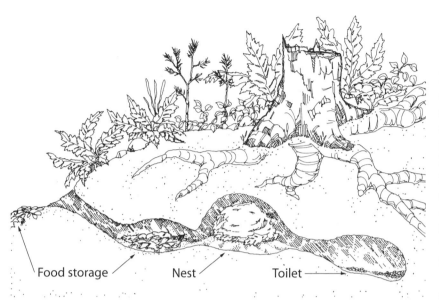

Food storage   Nest   Toilet

Figure 2. A cross section of a portion of a mountain beaver's burrow. Over time, their old nests, partially filled food pantries, and toilets are buried well below the surface, where the vegetation and droppings become fertilizer. (Drawing by Jenifer Rees.)

- Food items are eaten on site, temporarily stored outside burrow entrances, or placed in caches inside burrow systems.
- Mountain beavers will climb into trees to lop off living branches that are up to 1 inch (2.5 cm) in diameter.

Burrow System
- Mountain beavers dig tunnels 6 to 8 inches (15–20 cm) in diameter throughout their territories, which may be 2 acres (0.8 h) or more, depending on food and cover availability, and population density.
- Tunnel systems, or burrow systems, are located in or near thick vegetation, and tend to radiate out from a nest site (Fig. 2). Mountain beavers have been found using tunnels that are 10 feet (3 m) underground.
- Burrow systems may contain ten or more exits and special chambers for nesting, feeding, storing food, and storing droppings.

- Unoccupied mountain-beaver tunnels and chambers are used by mice, moles, voles, rats, cottontail rabbits, weasels, mink, spotted skunks, and salamanders.

Reproduction
- Mountain beavers are solitary except during the breeding season.
- In the Oregon Coast Range, breeding takes place from late February to mid-April.
- Two to four young are born after a 28- to 30-day gestation period.

Mortality
- Mountain beavers are eaten by bobcats, coyotes, large owls, and occasionally cougars and bears. Weasels and mink (primarily large males) eat young mountain beavers.
- Large numbers of mountain beavers are often trapped to prevent damage to newly seeded or planted commercial forests.

Viewing Mountain Beavers

Mountain beavers are abundant and active year-round, yet they are seldom observed due to their subterranean existence. Although active on and off throughout a 24-hour period, they are only occasionally seen wandering around on the ground or climbing in trees during daylight hours. They find the majority of their food and water within 150 feet (45 m) of their burrows.

Mountain beavers have various calls; the most frequent is a chattering produced by gnashing the tips of the lower and upper front teeth. This indicates irritation and at close range is best heeded, because mountain beavers have sharp teeth and can be swift, vicious biters if cornered (Fig. 3).

Active Burrow Systems
Active systems are most evident during the late spring and summer months when most of the digging and repairing is done. Look for newly excavated soil (sometimes called a "kick out") or freshly cut vegetation next to or within the entrance of a 6- to 8-inch (15–20 cm) diameter hole (Fig. 4). The presence of a mountain beaver

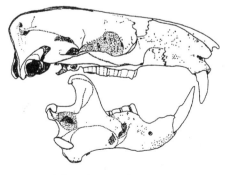

Figure 3. The side view of a mountain beaver skull. (From Verts and Carraway, Land Mammals of Oregon.)

Figure 4. A sure sign of mountain beaver is freshly cut vegetation next to or within the entrance of a 6- to 8-inch (15–20 cm) diameter hole. (Drawing by Jenifer Rees.)

(or other mammals using its tunnel system) can be recognized by the worn appearance of the tunnel floor and a lack or scarcity of spiderwebs at tunnel entrances.

In addition, after foraging above ground, mountain beavers carry or drag cut vegetation, which may vary in length from a few inches to several feet, to the burrow. There the material is cut into short sections at the burrow entrance and carried into the burrow to be eaten, stored, or used as nesting material. "Haystacks" of drying vegetation may be found near their burrows.

To check for occupants, you can install a temporary obstruction in the tunnel entrance. Cut three or four small-diameter (¼-inch) (6 mm) wide, 18-inch (45 cm) long woody stems and insert them vertically at the exits of several burrows. If mountain beavers are present, the inserted stems will be pushed aside or

clipped within a few days. Their musky odor may also be noticeable.

Feeding Sites

Look for signs of clipped twigs and branches and stripped bark on shrubs and trees.

Seedlings less than 1½ inches (38 mm) in diameter are most often eaten. These are usually clipped off at or close to ground level, making signs of activity difficult to locate and invisible when covered by soil, vegetation, or debris.

Multiple bites on the clipped plant can create a serrated edge, but more often a clean, slanted cut similar to those made by rabbits, hares, voles, and other rodents is evident. (See Chapter 7 for comparisons to deer.)

On small trees and large seedlings, the side branches are frequently clipped off high in the tree, leaving 1- to 3-inch (2.5–7.5 cm) stubs. (To distinguish moun-

tain beaver activity from that of porcupines, note that mountain beavers eat from the bottom up and porcupines eat from the top down.)

Feeding activity on the roots of trees may cause trees to lean at odd angles or develop a curved trunk. Eating roots and/or bark may also kill the trees. Often, the foliage of injured conifers remains green during the first year, but the needles turn reddish brown during a relatively brief defoliation period the following summer. The defoliated skeletons of these trees may remain standing and visible for many years, becoming excellent habitat for the birds and mammals that use such trees.

Tracks

Mountain beavers are generally slow-moving animals so they leave a trail of closely spaced tracks (Fig. 5). Look for tracks near active burrows. The imprints, in mud, show distinctively long and slender toes that are not apt to be confused with those of any other animals.

Droppings

Mountain beaver droppings are seldom seen because they are normally deposited inside the burrow system. If you find droppings, they are probably from another animal using the burrow.

Mountain beaver re-ingest their soft droppings, much as rabbits do, and store their hard droppings in underground chambers.

Preventing Conflicts

Mountain beavers serve an important function in nature owing to the amount of soil they move and the number of vacant burrows

Figure 5. A mountain beaver's hind tracks are 1¾ inches (4–5 cm) long and ¾ inch (2 cm) wide. The front tracks are slightly shorter. (From Pandell and Stall, Animal Tracks of the Pacific Northwest.*)*

they leave behind for other wildlife. Over time, their old nests, partially filled food pantries and toilets, are buried well below the surface, where the vegetation and droppings become fertilizer.

Most people would not tolerate problematic mountain beavers on their commercial property because of the logical concern that leaving them alone would lead to more damage. When they are feeding in Christmas tree farms, commercial timber farms, and other commercial operations, this may be true. For the homeowner, however, mountain beavers are more of an occasional nuisance in the landscape or garden, not a long-term problem or threat. There are of course, exceptions.

If the burrowing activity of mountain beavers is causing problems for livestock in pastures or undermining roadbeds, irrigation ditches, and earthen dams, see "Preventing Conflicts" in Chapter 15 for prevention strategies. Mountain beavers occasionally will get caught in window wells. See "Removing Skunks from Window Wells and Similar Areas" in Chapter 23 for information.

While you may be able to remove an existing mountain beaver population or force them elsewhere, if suitable conditions exist and mountain beavers occur nearby, others will eventually move into vacated areas. In addition, it is important to understand that mountain beaver problems rarely can be resolved by a quick fix method, but that a continuing commitment to whatever solutions are adopted is required.

To prevent conflicts or remedy problems:

Harass mountain beavers in their burrows. Becoming a "bad neighbor" may cause a mountain beaver to leave an area, especially if it hasn't lived there long. Fill all existing and new tunnel entrances with dirt, rocks, or wadded up newspaper. Some people have had success using freshly used cat litter in this way.

In addition, you can roll rags into tight balls the size of tennis balls and tie them with twine. Sprinkle predator urine (mink, coyote, or bobcat—available from trapper supply outlets and over the Internet) or ammonia on these. Using a piece of stiff wire, such as an opened clothes hanger, put the rag balls into the burrow as far as you can and cover the hole lightly with dirt or wadded newspaper.

Harass the mountain beaver *daily* for as long as necessary, and don't be surprised if it takes a couple of weeks for the animal to leave.

Flooding the burrow system with a garden hose is rarely effective unless the entire tunnel system can be quickly and completely flooded. Where mountain beavers are well established, their systems are extensive and flooding is unlikely to produce results.

When attempting to flood out a mountain beaver, concentrate the effort in late winter and early spring, before mountain beavers give birth. Be careful when attempting to flood out a mountain beaver near a building; doing so could damage the foundation or flood the basement or crawl space.

Install fences and other barriers. In areas where individual small trees or shrubs are being damaged, surround the plants with 24-inch (60 cm) tall smooth metal cylinders (Fig. 6). To prevent mountain beavers from climbing larger plants to access upper branches, install barriers as described under "Preventing Conflicts" in Chapter 21.

Multistemmed trees, large shrubs, and groups of plants can be enclosed in a mini floppy fence made from wire mesh (Fig. 7), silt fencing, plastic weed mats, or a similar smooth material (see "Preventing Conflicts" in Chapter 11 for an example). Mountain beavers have well-developed senses of smell, touch, and taste. However, they have poor eyesight, and this barrier should prevent them from finding the trees and shrubs. If they attempt to climb the fence, its tendency to flop will keep the animals from reaching the top.

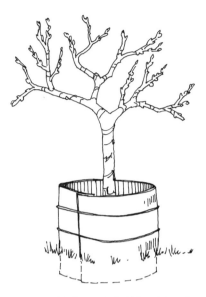

Figure 6. Eighteen-inch (45 cm) lengths of stovepipe or aluminum flashing can be placed around tree trunks to keep mountain beavers from accessing the bark and branches. The smooth-sided barriers can be held together with a top and bottom wire and painted to blend into the landscape. (Drawing by Jenifer Rees.)

A floppy fence can also be constructed as a barrier between an active mountain beaver colony and a large area needing protection. To prevent mountain beavers from walking around the fence, connect each end to an existing, impenetrable solid fence or structure.

To prevent the mountain beaver from digging under the fence, keep a 2-foot (60 cm) wide wire apron on top of the ground on the mountain beaver's side of the fence. Keep the apron flush to the ground with rocks and/or stakes, or the mountain beaver will shimmy under it.

An alternative to the floppy fence is a fence made of electrified netting, of the type used to exclude rabbits. Electrified netting is available from some farm supply centers and off the Internet (see "Wildlife/Human Conflicts" in Appendix F).

Commercially available solid plastic or plastic-mesh seedling protectors that are anchored with a heavy wire or wooden stake will protect most tree seedlings until they grow taller than the tube height (see "Preventing Conflicts" in Chapter 7 for information).

Small areas that need protection from burrowing mountain beavers can be covered with a 6-inch (15 cm) layer of gravel or 1-inch (2.5 cm) wire mesh laid over the area and anchored to the ground.

Repellents and Fumigants

Repellents applied to plants have not proven consistently effective. All repellents require special application procedures to assure that the plant stems are treated near the base of small plants. See "Repellents" in Chapter 7 for additional information.

Fumigants of all types have been tried and are generally ineffective, probably because of mountain beavers' maze of tunnels, and their ability to quickly close off entrances.

Trapping and Lethal Control

Because mountain beavers are territorial, removing them from an area may appear to solve the problem. However, other mountain beavers will eventually enter the area if attractive habitat is available.

Long-term control is possible by first reducing or eliminating the mountain beaver population by trapping, and then continuing with a maintenance-trapping program to remove invading animals as they become evident.

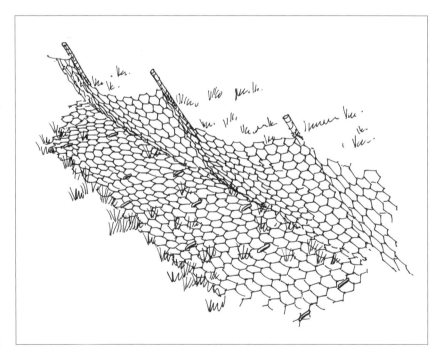

Figure 7. A mini floppy fence constructed of 1-inch (2.5 cm) mesh wire or heavy plastic needs to be at least 2 feet (60 cm) high and staked so that it's wobbly. The fence should not be pulled tight between the stakes, but rather there should be some "give" so that when the mountain beaver tries to climb the fence, it will wobble, discouraging further climbing. Constructing the fence so that it leans slightly toward the mountain beaver's side will increase its effectiveness. (Drawing by Jenifer Rees.)

Mountain beavers can be trapped anytime, but for best results and for humane reasons it is best to concentrate the effort in late winter, before they give birth.

A wildlife damage control company can be hired to do the trapping, or you can do it yourself (see Appendix C, "Hiring a Wildlife Damage Control Company"). Never attempt to handle trapped or wild mountain beavers. They are capable of producing a very bad bite and have very sharp claws.

For information on moving, trapping, and euthanizing mountain beavers, see Appendix A.

Public Health Concerns

Mountain beavers are not considered a significant source of any infectious disease that can be transmitted to humans or domestic animals. Anyone handling a living or dead mountain beaver should wear rubber gloves, and wash his or her hands well when finished.

Although the largest flea *(Hystrichopsylla schefferi)* in the world—it is up to ¼ inch (6 mm) long—is found on mountain beavers and in their burrows, it does not bother humans.

Legal Status

Because legal status, trapping restrictions, and other information about mountain beavers change, contact your local, state, or provincial wildlife office, or visit their Web site for updates. (See Appendix E for contact information and where to access the state and provincial laws mentioned below.)

Oregon: The mountain beaver is classified as an unprotected animal and the general furbearer regulations do not apply (OAR 635-050-0045 (10)). Mountain beavers may be captured and held in captivity, but they must be maintained in a humane manner (OAR 635-044-0132). Any persons may sell, purchase, or exchange the pelt, carcasses, or any part of a mountain beaver (OAR 635-200-0020 (1 & 5)); live mountain beavers can not be sold in Oregon.

Washington: The mountain beaver is unclassified and may be trapped or killed year-round without a permit. Although trapping mountain beaver is not illegal, there may be restrictions on what devices may be used.

It is unlawful to release a mountain beaver anywhere within the state, other than on the property where it was legally trapped, without a permit to do so (RCW 77.15.250 and WAC 232-12-271).

British Columbia: The mountain beaver is considered a species at risk. It is Blue-listed (vulnerable), and any animals that are involved in property damage may not be captured or killed unless a permit is first received from the regional Fish and Wildlife Office.

CHAPTER 15

Muskrats and Nutrias

Muskrats (*Ondatra zibethicus,* Fig. 1) get their common name from their resemblance to stocky rats and from the musky odor produced by their scent glands.

Muskrats weigh 2 to 4 pounds (0.9–1.8 kg) and reach lengths of 18 to 25 inches (45–65 cm), including their 8- to 11-inch (20–28 cm), sparsely haired tails. Their coat color is generally dark brown, but individuals can range from black to almost white. Muskrats have partially webbed hind feet that function as paddles and much smaller front feet used primarily for digging.

Muskrats are found throughout still or slow-moving waterways, including marshes, beaver ponds, reservoirs, irrigation canals and ditches, and marshy borders of lakes and rivers. Muskrats don't live in mountainous areas where cold weather makes their food unobtainable.

Muskrats make a valuable contribution to aquatic communities. By harvesting plants for food and den sites, they create open water for ducks, geese, shorebirds, and other wildlife. In addition, a variety of animals—including snakes, turtles, frogs, ducks, and geese—use muskrat lodges and platforms to rest and nest in. However, muskrats are considered pests when their burrowing activity damages dams and dikes, and when their feeding activity damages new plantings and crops.

For information on nutrias, see "Notes on Nutrias."

Figure 1. *The muskrat has a stocky appearance, with small eyes and very short, rounded ears. Its front feet, which are much smaller than its hind feet, are adapted primarily for digging and feeding. (From Christensen and Larrison,* Mammals of the Pacific Northwest: A Pictorial Introduction.*)*

Facts about Muskrats

Food and Feeding Habits

- Muskrats eat a wide variety of plants, including cattails, sedges, bulrush, arrowhead, water lilies, pondweed, and ferns. They also eat alfalfa, clover, corn, and other crops if muskrats find them in their territories.
- Although muskrats will eat shellfish, snails, fish, frogs, and salamanders, such animal foods are a small part of their diet, and are generally consumed when plant foods are scarce.
- Muskrats normally feed within 150 feet (45 m) of their main dwellings; however, they will travel much farther in search of food.
- When muskrats become too numerous, an "eat-out" can occur where nearly all the available food is eaten. The eat-out area becomes virtually uninhabitable for muskrats, and only a few animals may be found where dozens or more once were.

Den Sites

- Depending on site conditions, muskrat dens are located in banks or lodges.
- In dams, dikes, and banks, muskrats tunnel upward from below the water surface into the soil to make dens that remain dry (Fig. 4).
- Bank dens range from a short tunnel leading to an enlarged nest chamber, to a long, complex system of chambers, air ducts, and entrances.
- In marshes and other areas lacking steep banks, muskrats build dome-shaped lodges from leaves, stems, roots, and mud.

- Lodges are constructed in open water that is 2 to 4 feet (60 cm–1.25 m) deep, and are built high enough to keep the den above high-water levels.

Reproduction and Family Structure

- Muskrats are prolific breeders and under favorable conditions may raise 20 young per season.
- The first litter is born in early spring; one or two litters may follow.
- An average of six kits are born after a 30-day gestation period. Kits are dependent on their mother for approximately 30 days, after which time they can swim, dive, and eat green vegetation.
- At about six weeks of age, kits leave the den or live in a separate chamber.
- Adult females are thought to overwinter with surviving offspring from the last litter and one or more adult males. In spring, the young seek out their own territories, generally within 300 feet (90 m) of the maternal female's home range.

Mortality and Longevity

- Muskrats have many predators, including mink (a major predator), otters, bobcats, house cats, domestic dogs, coyotes, foxes, large hawks and owls, and large-mouth bass.
- Muskrats are fierce fighters and fights among males are common when densities are high and food supplies are low.
- Spring flooding can drown early litters and inundate burrows and lodges, exposing muskrats to predators.
- Historically, muskrats have been one of the most commonly trapped animals in the Pacific Northwest. In the early 1970s and through the 1980s, 25,000 to 60,000 muskrats were taken per year in Oregon, with prices for pelts ranging from 90 cents to $5.75 apiece. In Washington, from 1991 to 2000, an annual average of 6,189 muskrats were trapped.
- Most muskrats don't live more than one year.

Viewing Muskrats and Nutrias

Muskrats and nutrias are active throughout the year. Although they may be seen at any time, they are most active at twilight and throughout the night. Both species may be seen feeding during the day when food is scarce, or basking in the sun when temperatures are low.

Rarely will muskrats and nutrias be seen very far from water, and they are usually seen swimming. Muskrats and nutrias tend to swim with their narrow, pointed tails snaking in the water behind them, or arched out of the water; you never see a beaver's rounded tail as it swims.

When startled, muskrats and nutrias enter the water with a loud splash, and, being strong swimmers, they may swim long distances underwater before surfacing. (Nutrias can remain submerged for as long as 10 minutes while muskrats can stay under for 20 minutes.) Both species can also remain motionless under sparse vegetation, with only their noses and eyes above water.

When cornered or captured, muskrats and nutrias are aggressive biters and scratchers and can seriously injure pets and humans.

Living Areas

In marshes, ponds, and other water areas east of the Cascade mountains, prominent muskrat lodges are sure indicators of a present muskrat population.

Look for entrances into their bank dens along dams, dikes, and stream banks, particularly west of the mountains. Entry holes are particularly evident where muskrats are living in tidewater areas near the mouths of rivers. When the tide recedes, the entrances are exposed until the tide comes back in.

Similarly, in dry years the water in ponds and reservoirs can drop and expose den entrances. Muskrats will then usually dig new dens farther out in the pond.

Entry holes are 5 to 8 inches (13–20 cm) in diameter and are located 3 to 36 inches (7–90 cm) below the surface of the water.

Feeding Areas

Evidence of muskrat and nutria feeding includes plants gnawed to a stubble, floating cattail roots or other vegetation that has been clipped, and piles of clipped vegetation under overhanging vegetation or in a well-concealed spot at the water's edge.

Muskrats sometimes use feeding huts or eating platforms that they create from mud and compacted vegetation. Feeding huts look like small lodges about 1 foot (30 cm) above the water level and are hollow inside; feeding platforms look like small piles of cut vegetation.

Feeding platforms built and used by nutrias can measure 5 to 6 feet (1.5–1.8 m) across. Both feeding huts and platforms are built near dens or lodges, and there may be travel channels through the mud leading to them.

Notes on Nutrias

Nutrias (*Myocastor coypus,* Fig. 2) are semi-aquatic rodents native to southern parts of South America. In the 1930s, they were sold throughout North America as a "weed cutter," or a means of controlling unwanted aquatic vegetation. Various associations, magazine and newspaper articles, and demonstrations at county fairs promoted the sale of nutrias in Oregon and Washington.

More than 600 nutria farms existed in Oregon and Washington from the 1930s to the 1950s. Flooding and storms damaged holding structures, allowing nutrias to escape. Farmers also released their stock when nutria farming became uneconomical. By the 1940s, nutrias had been captured by trappers on both sides of the Cascade mountains in Oregon and Washington.

Figure 2. *A nutria is three times the size of a muskrat and its tail is round, not flattened vertically, as is the muskrat's. Nutria and beaver are similar in size, but the beaver has a large tail, which is flattened horizontally. Unlike beavers, neither nutrias nor muskrats build dams, cut down trees, or store food for the winter. (From Christensen and Larrison,* Mammals of the Pacific Northwest: A Pictorial Introduction.)

- **Distribution:** Nutrias are currently found in wetlands, drainage ditches, and irrigation canals in the Puget Sound area and the Willamette River valley, along the Columbia River, and along Oregon's coastal rivers. Only the Yakima River drainage in south-central Washington supports substantial numbers east of the Cascade Mountains. In British Columbia, nutrias have been trapped in the lower mainland and on Salt Spring and Vancouver Islands, but no large populations exist.

  Cold temperatures seem to reduce the distribution of nutrias, as they don't live in areas where water surfaces freeze for long periods.

- **Description:** A nutria's total length is 28 to 56 inches (70–142 cm), including 12 to 17 inches (30–43 cm) of tail. Depending on the nutria's ancestry and current habitat, its fur will vary from light yellowish brown to dark reddish brown, and black. Their hind legs are much larger than the forelegs. When moving on land, a

nutria may drag its chest and appear to hunch its back.

- **Foods and feeding habitats:** Nutrias are herbivores, with the succulent, lower portions of plants being preferred food. Roots, rhizomes, tubers, and tree bark are important during winter when the green parts of plants aren't available. Nutrias also eat crops and lawn grasses found next to water.

  Because their fore-paws are small and dexterous, nutrias can excavate soil and handle small food items.

- **Reproduction and family structure:** Nutrias reproduce from spring through fall, having one to three litters annually; litter size averages five young. The young are born in a grass-lined burrow or a floating platform nest made of vegetation. Nutrias will dig their own burrow, or use an abandoned burrow or lodge of a beaver or muskrat.

  Young nutrias grow rapidly. The young are capable of swimming within 24 hours after birth, and at

one week of age they can search for vegetation on their own. Weaning occurs when young are five to seven weeks old.

A rather unusual characteristic of the female nutria is that her teats are so high on her sides the babies can nurse even while their mother is lying on her stomach.

Two to 13 individual nutrias usually form a group, with one dominant male and female. Adult males are sometimes solitary.

- **Mortality:** Predators of adult nutrias include coyotes, domestic dogs, and humans; great horned owls, foxes, great blue herons, hawks, eagles, and raccoons prey on the young. In the early 1990s, 5,300 to 7,700 nutrias were taken per year in Oregon, with prices for pelts ranging from $2.00 to $4.00 apiece.

  In the wild, most nutrias live less than three years.

See the following sections for additional information.

Figure 3. Muskrat tracks are small, hand-like prints, with long, fingerlike toes. The rear print is 2 to 3 inches (5–7.5 cm) long and may look like a smaller version of a raccoon track. The front print may appear four-toed, as the inner toe is extremely small and barely shows in the track. (Washington Department of Fish and Wildlife.)

Tracks

Muskrat and nutria tracks can be found in mud or sand along shorelines (Fig. 3). The mark of a dragging tail is sometimes apparent.

Nutria tracks are similar to but larger than muskrat tracks, with the hind feet showing webbing between the inner four toes. Nutria have five clawed toes on each foot; the front feet are not webbed. Nutria tracks are easily confused with beaver tracks when the beaver's fifth toe webbing does not print.

Droppings

Muskrat and nutria droppings can be found floating in the water, along shorelines, on objects pro-

truding out of the water, and at feeding sites. The animals may repeatedly use these spots, and more than one muskrat or nutria may use the same spot.

Muskrat and nutria droppings are dark green, brown, or almost black. Muskrat droppings are slightly curved, cylindrical, and about ½ inch (12 mm) long and ⅜ inch (9 mm) in diameter.

Nutria droppings are 2 inches (5 cm) long and ½ inch (12 mm) in diameter. Nutria droppings are unique in that they have distinct parallel grooves along their entire length.

Slides

Slides are the narrow trails muskrats and nutrias make where they enter and leave the water. Muskrat slides are about the width of a hand, while nutria slides are twice that wide. (Beaver slides can be up to 20 inches or 50 cm wide.) Slides look like muddy trails and may be slicked down from the animals' sliding down them on their bellies.

Calls

Where large numbers of nutrias are present at dusk, a chorus of pig-like grunts may be heard.

Preventing Conflicts

Although muskrats are important contributors to natural aquatic systems, their burrowing may threaten the safety of dams, dikes, and other human-created embankments. Muskrats may also undermine retaining walls that shore up homes, bridges, and other structures. Muskrats occasionally eat new wetland plantings and agricultural crops growing in their territories.

Nutria damage is also related to burrowing and feeding. Their large size makes it possible for them to girdle orchard trees, landscape trees, and ornamental shrubs. In parts of Oregon, introduced nutrias are out-competing muskrats for food and places to live.

Muskrat and nutria numbers may increase to the point where an area is denuded of aquatic plants. After foraging on entire plants, including the roots, they leave the area pitted with digging sites and deep swimming canals. This feeding behavior can destroy existing root mats that bind and secure a wetland together, and the area can be quickly eroded by wind and wave action.

The following suggestions will help to reduce conflicts. You can do the work yourself or hire a company to do all or part of the work (see Appendix C, "Hiring a Wildlife Damage Control Company"). In cases where these methods are not practical, contact your local County Extension Agent, Department of Agriculture, or Ministry of Agriculture (B.C.) for further information (see Appendix E for contact information).

Water-level management:
Muskrats and nutrias (and occasionally voles and Old World rats) dig into dams, dikes, and other embankments to make dens (Fig. 4). Typically these dens have 2 feet (30–60 cm) or more of earth above them. However, when fluctuating water levels flood their initial den, muskrats and nutrias burrow farther into the bank or dig new, higher den chambers close to the surface. In such cases this can weaken the bank, or livestock and other large animals can

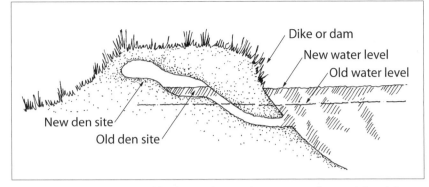

Figure 4. In dams, dikes, and banks, muskrats and nutrias tunnel upward from below the water surface into the soil to make dens that remain dry. When fluctuating water levels flood their initial den, they burrow farther into the bank or dig new, higher den chambers close to the surface. (Drawing by Jenifer Rees. Adapted from Hygnstrom et al., Prevention and Control of Wildlife Damage.*)*

pierce holes in the bank, starting the erosion process.

To prevent muskrats and nutrias from tunneling higher in an embankment, keep fluctuations in water levels to a minimum. This can require frequently monitoring the spillway to ensure an unobstructed flow, or widening the spillway to carry off surplus water so that it never rises more than 6 inches (15 cm) on the dam.

Water-level manipulation can also be used to force muskrats and nutrias to other suitable habitat. Raising the water level in the winter to a near-flood level, and keeping it there, will force the animals out of their dens. Similarly, dropping water levels during the summer will expose muskrat and nutria dens to predators, forcing them to seek a more secure area.

Slope management: Muskrats and nutrias prefer to burrow on steep slopes covered with vegetation. Hence, they can be discouraged by keeping side slopes to a 3:1 or less ratio, and by controlling vegetation growth. Managing vegetation by hand can be difficult in large areas, but routine mowing or cutting with a weed whacker

can be effective. Only herbicides registered for use next to water should be used, and then only per the manufacture's recommendations.

If possible, keep livestock off embankments to avoid the chance that an animal will put a hoof through a den chamber. If a roof is pierced, immediately fill in the cavity with soil, rocks, or a mudpack (see below).

Embankment barriers: A barrier installed 1 foot (30 cm) above to 3 feet (90 cm) below normal water level can prevent muskrats and nutrias from burrowing into an earth embankment.

A barrier can be made from 1-inch (2.5 cm) mesh hardware cloth (aluminum and stainless steel are also available), or heavy-duty plastic or fiberglass netting. The barrier should be placed flat against the bank and anchored every few feet along all edges. To extend the life of galvanized hardware cloth, spray it with automobile undercoat paint or other rust-proof paint before installation. Since the wire will eventually corrode, do not use this material where people are likely to swim.

Riprapping areas with stone creates an effective barrier and protects slopes from wave action. Stone should be at least 6 inches (15 cm) thick. Do not use rocks larger than 6 inches (15 cm) in diameter because when piled they tend to form cavities, providing hiding places for muskrats and Old World rats.

In situations where muskrats are burrowing into existing rock walls, place gravel or concrete in between the rocks to block up the holes. A permit may be needed when working near state waters (see "Legal Status").

Where a burrowing problem is extreme, use a gas-powered trenching machine (available at rental stores) to dig a narrow trench along the length of the embankment. Hand digging will be required to dig to the recommended depth—3 feet (90 cm) below the high-water level. Next, fill the trench with a mudpack. A mudpack is made by adding water to a 90 percent earth and 10 percent cement mixture until it becomes a thick slurry. The resulting solid core will prevent muskrats and nutrias from digging through the embankment.

Floating dock barriers: Muskrats will burrow into floating docks, generally those floating on Styrofoam, scattering the broken white foam along the shoreline. This becomes an environmental danger, due to birds and other small animal eating this foam. To solve this problem, the dock needs to be pulled up on shore and 1-inch (2.5 cm) mesh hardware cloth (aluminum and stainless steel are also available) needs to be used to cover the Styrofoam.

Fences and other barriers: Muskrats and nutria are not climbers. A properly designed and maintained 2-foot (60 cm) tall wire fence will prevent muskrats from entering an area; a 4-foot (1.25 m) tall fence will exclude nutrias. The fence must be taller if snow or other materials are likely to build up near it.

Because muskrats and nutrias are diggers, the fence will need to extend at least 12 inches (30 cm) below ground. Alternately, a tight fit to the ground and an L extension that runs 24 inches (24 cm) out on the soil surface toward the animal will also prevent entering from underneath. (For an example, see Chapter 14.)

A single strand of electric wire placed 4 to 5 inches (10–13 cm) above the ground and the same distance outside the fence will also help prevent nutrias from burrowing and climbing. Vegetation near any electric fence should be removed regularly to prevent the system from shorting out. (See Chapter 7 for additional information on electric fences.)

Welded-wire cylinders around individual plants are often used where only a few plants need to be protected. For examples of such barriers, see "Preventing Conflicts" in Chapter 20. *Note:* Lightweight plastic seedling protectors do not work because muskrats and nutrias can chew through them.

Harassment and repellents: Muskrats and nutrias are wary animals and will try to escape when threatened. When new burrows are discovered early on, the entry holes can be stuffed with rocks, balled-up window screen, and/or rags sprinkled with preda-

tor urine (mink, coyote, or bobcat—available from trapper supply outlets and over the Internet) or ammonia. Some people have had success applying used cat litter in this way. Exposing their tunnels from above may also work. The success of this type of control depends on persistence from the harasser and thus is often short-lived.

Loud noises, high-pressure water sprays, and other types of harassment have been used to scare nutrias from lawns and golf courses. However, the success of this type of control is usually short-lived and problem animals soon return.

Large dogs that are awake during the night can be effective at keeping muskrats and nutrias out of areas. Bold nutrias often intimidate small dogs. (See Appendix D for information on how to safely manage dogs.)

Commercially available taste repellents may be effective at preventing damage to crops and other plants. (For detailed information, see "Repellents" in Chapter 7.)

Crop location: Unfenced crops and gardens located close to water will be more attractive to muskrats and nutrias than those further from water. If you have a choice of where to locate your garden, consider muskrat and nutria damage. Natural vegetation buffers next to water bodies can provide feeding areas and reduce the attractiveness of vegetation further from the water.

Lethal Control

Since muskrats and nutrias are usually found in waterways, there is often an unlimited supply of replacement animals upstream

and downstream from where the damage is occurring. Rapid immigration coupled with a high reproductive rate makes ongoing lethal control a "high-effort" method of damage control that is often ineffective. (Lethal control can be effective in areas where the local population of muskrats and nutrias is still small.) The exclusion methods described and referenced above are often the best long-term solution.

Trapping
Lethal trapping has traditionally been the primary form of control. See Appendix A for information on trapping and relocating muskrats and nutrias. *Note:* State or provincial wildlife offices do not provide trapping services, but they can provide names of individuals or companies that do.

Shooting
Shooting has been an effective in eliminating small isolated groups of muskrats and nutrias. Shooting muskrats requires good marksmanship. For safety considerations, shooting is generally limited to rural situations and is considered too hazardous in more populated areas, even if legal.

Fumigants
No fumigants are currently registered for muskrat or nutria control.

Public Health Concerns

Rabbits, hares, voles, muskrats, nutrias, and beavers are some of the species that can be infected with the bacterial disease **tularemia** (see "Public Health Concerns" in Chapter 3).

Muskrats and nutrias are

among the few animals that regularly defecate in water, and their droppings (like those of humans and other mammals) can cause a flulike infection, which old-time trappers referred to as **"beaver fever."**

Anyone handling a dead or live muskrat or nutria should wear rubber gloves, and wash his or her hands well when finished.

Legal Status

Because legal status, trapping restrictions, and other information about muskrats and nutria change, contact your local state or provincial wildlife office, or visit their Web site for updates. (See Appendix E for contact information and where to access the state and provincial laws mentioned below.)

Oregon: The muskrat is classified as a furbearer and trapping regulations apply (OAR 35-066). Property owners, persons lawfully occupying land, and their agents may only trap muskrats with a permit issued by Oregon Department of Fish and Wildlife if the animals are causing damage.

Nutria are classified as unprotected nongame wildlife (OAR 635-044-0132). As unprotected wildlife, nutria may be trapped or shot (but cannot be relocated). No license is needed for a landowner to control nutria on his or her own property.

Washington: Muskrats are classified as furbearers (WAC 232-12-007). A trapping license is required and they can only be trapped during seasons set by the state.

A property owner or the owner's immediate family, employee, or tenant may kill or trap muskrats on that property if they are damaging crops or domestic animals (RCW 77.36.030). A permit is not necessary.

Nutria are unclassified and may be trapped or killed year-round without a permit.

It is unlawful to release a muskrat or nutria anywhere within the state, other than on the property where it was legally trapped, without a permit to do so (RCW 77.15.250 and WAC 232-12-271).

A permit is required by any person, organization, or government agency doing work that will use, obstruct, change, or divert the bed or flow of state waters (see "Legal Status" in Chapter 3).

British Columbia: Muskrats are designated as furbearers. They may be legally trapped under license during open seasons as defined by regulation and there is no bag limit. Muskrats that are involved in property damage may not be captured or killed—outside of general open seasons—unless a permit is first received from the regional Fish and Wildlife Office.

The nutria is listed under Schedule "C" and may be captured or killed anywhere or at any time in the province. No permit is required, but a person must have land owner/occupier permission on private property.

CHAPTER 16

Old World Rats and Woodrats (Packrats)

The Pacific Northwest is home to both native and non-native rats, the latter sometimes being referred to as Old World rats. Native to the Orient, Asia Minor, and Siberia, Old World rats were introduced to North America on the ships of the early voyagers. These highly adaptable rats continue to enter new regions on board ships, trains, and trucks carrying freight and other goods around the world. In the Pacific Northwest, they are now found wherever humans have established permanent residence or industry. (For information on other introduced mammal species, see Table 1 in Chapter 17.)

Old World rats include the **Norway rat** *(Rattus norvegicus,* Fig. 1), also known as the brown rat, sewer rat, or wharf rat, and the **black rat** *(Rattus rattus),* also called the roof rat.

Norway rats average 16 inches (41 cm) in length, which includes the animal's long, tapering, scaly tail that is slightly shorter than the combined length of the rat's head and body. Norway rats are grayish-brown in color from top to bottom; white, black, or mottled individuals are occasionally found. The rats used in laboratories and sold as pets are specially bred strains of Norway rats.

While early scientific descriptions of this species came from Norway, and it was once believed to have arrived in England aboard Norwegian ships, the "Norway rat" is neither a native of Norway nor more common there.

Black rats are similar in length to Norway rats, but are more slender and darker. The tail is longer than the combined length of the head and body. As one of their names implies, roof rats are agile climbers and are found on roofs and in the upper levels of buildings. Preferring saltwater-influenced warmer climates, and being more likely to board ships than Norway rats, black rats are often seen in port cities and coastal towns.

For information on our native **woodrats** *(Neotoma* spp.), known as packrats, see "Notes on Woodrats (Packrats)."

Near bodies of water it's possible to encounter the native **muskrat** *(Ondatra zibethicus),* which isn't a rat at all, but a member of the vole family that is the size of a bulky, half-grown house cat (see Chapter 15, "Muskrats and Nutrias").

Facts about Old World Rats

Food and Feeding Behavior

- Old World rats will eat anything humans or livestock will eat, plus many less palatable items including animal droppings, garbage, and other rats.
- Rats living apart from human habitation are known to eat seeds, nuts, and insects, as well as young birds and bird eggs.
- When given a choice, rats select a nutritionally balanced diet, choosing fresh, wholesome items over stale or contaminated foods.
- Rats begin foraging soon after dark and most of their food gathering occurs before midnight. They often store or hoard food in hidden areas.
- Food items in household garbage offer a fairly balanced diet and also satisfy their moisture needs.

Figure 1. The Norway rat has a long, tapering, nearly naked tail. (From Larrison, Mammals of the Northwest.*)*

Figure 2. Bushy-tailed woodrats are 12 to 15 inches (30–38 cm) long, including a bushy, squirrel-like tail that is as long as the combined length of the head and body. (From Christensen and Larrison, Mammals of the Pacific Northwest: A Pictorial Introduction.*)*

Territory and Family Structure

- Old World rats travel 50 to 300 feet (15–90 m) from their nests to look for food and water and patrol their territory. However, they can travel much farther when necessary.
- Rats generally live together in a group dominated by a large male that guards a harem of females and aggressively prevents other males from mating.
- Rats seen during the day are generally socially low-ranked individuals who have been denied access to food by dominant rats during the night.

Nest Sites

- Roof rats build nests in attics, trees, and overgrown shrubbery or vines. Roof rats rarely dig burrows for living quarters if off-the-ground sites exist.
- Norway rats prefer to nest at or under ground level, and in the lower floors of buildings.
- Rats can have several nest sites. They may spend a week in their primary nest site, and then move for a day or two into an alternate nest site.

- Rats prefer to nest where food and water are easily available.

Reproduction

- Old World rats breed year-round, but reproduction is concentrated in the warmer months.
- A litter of six to ten young is born after a gestation period of about three weeks.
- Young rats are weaned at around 20 days of age and can breed at three to four months of age.
- Younger rats will mate in the same location in which they were born or will migrate to a new, unoccupied territory.

Mortality and Longevity

- Old World rats are killed by vehicles, traps, poisons, or other rats. Some domestic cats and dogs capture rats, usually small ones.
- Owls, hawks, foxes, coyotes, and weasels prey upon rats; snakes eat immature rats.
- The average life span of a rat in the wild is less than one year, with females living longer than males.

Signs of Old World Rats

Old World rats are active all year, mostly at night. However, when disturbed (weather change, construction, etc.), hungry, or when living in crowded conditions they are seen at any hour.

It is not easy to tell how many rats are using an area in and around a structure. However, you can use their signs as a rough guide to whether the population is low, medium, or high. Use a powerful flashlight to search for their signs

Notes on Woodrats (Packrats)

Woodrats are also known by the vernacular name "packrats" because of their habit of moving almost any object small enough for them to carry—bones, sticks, jewelry, eating utensils, tin-can lids, keys, and other items. The reason for this behavior has never been established.

In appearance, woodrats differ from the Old World rats by having large ears and eyes, and tails that are not scaly but haired.

Three species of woodrats occur in the Pacific Northwest:

• **Bushy-tailed woodrats** (*Neotoma cinerea,* Fig. 2) are 12 to 15 inches (30–38 cm) long, including a bushy, squirrel-like tail that is as long as the head and body. They are found throughout the Pacific Northwest, but are absent from most islands, including Vancouver Island and the Queen Charlotte Islands. They occupy cliffs, rocky outcroppings, and human-made rocky areas along the banks of roads and railroad tracks. Where suitable rocky habitats are scarce, they live in hollow standing and fallen trees, deserted buildings, and abandoned mineshafts.

• **Dusky-footed woodrats** (*Neotoma fuscipes*) range from 15 to 18 inches (38–46 cm) long, including their 8-inch (20 cm), sparsely haired tail. They are found in dense thickets and tangles of small trees, shrubs, vines, and brambles of western Oregon (except for the Coast Range).

• **Desert woodrats** (*Neotoma lepida*) are the smallest of the woodrat species, measuring a total length of 10 to 12 inches (25–30 cm). They occur in the sagebrush areas of southeast Oregon, especially those associated with rocky areas.

Woodrats are nocturnal, but are occasionally seen during daylight hours. They have sharp claws and are good climbers. They eat a wide variety of plant materials, including the leaves, needles, inner bark, flowers, and fruit of wild and domestic plants, as well as fungi and invertebrates.

Dusky-footed woodrats live in colonies and construct large lodges of sticks and a wide range of other "treasures" such as bones, pieces of metal, and pine cones. The lodges are built at the base of trees, in dense brush or blackberry brambles, or up in trees. Their nests are more visible in inland areas than in coastal areas, where vegetation keeps the lodges hidden.

Dusky-footed woodrat lodges have several chambers connected to one another as well as several exits. The various chambers can be used as nurseries, living rooms, or storage rooms, with one as a toilet. Except for the toilet, all chambers are kept clean. The lodge grows over time as new generations add to it.

The more solitary bushy-tailed woodrats and desert woodrats don't build lodges. Instead, they loosely construct cuplike nests in hollow trees or other sheltered places. In some areas, they also build lodgelike structures for food storage and as barricades to guard nests. When nests are constructed in trees, the ground underneath is often littered with cut twigs and the rat's dung.

In the Northwest the bushy-tailed woodrat is the species most likely to utilize human-made structures. In an attic its nest may have the appearance of a large pile of sticks. The desert and the dusky-footed woodrats less commonly take up residence in structures.

Woodrats are born in spring. There is one litter per year, sometimes two, and two to four offspring per litter is the norm.

Woodrat predators include hawks, owls, coyotes, bobcats, martens, and domestic cats and dogs. Humans sometimes kill them in and around buildings with traps and poisons.

Conflicts with bushy-tailed woodrats occur when they shred upholstered furniture and mattresses to get nest material, take up residence in a parked vehicle, or gnaw on wires and other mechanical components. Woodrats can damage trees by gnawing on their bark, and clip plant parts from ornamentals. Damage is not usually extensive.

When problems occur in and around buildings, woodrats can be excluded using the methods described for Old World rats. To prevent damage to trees and ornamental plants, see "Preventing Conflicts" in Chapters 14 and 20.

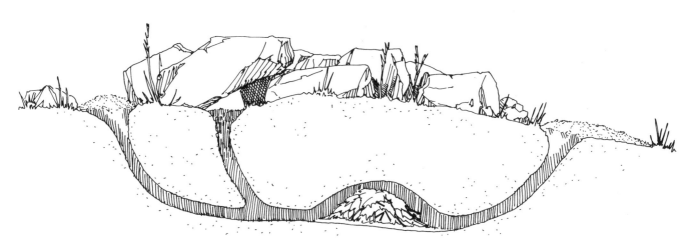

Figure 3. Norway rats often nest in burrows dug into the ground. Burrows are typically less than 18 inches (46 cm) deep and 3 feet (90 cm) long, and have a central nest. Extra "bolt holes" are used for emergency escapes. (Drawing by Jenifer Rees.)

in dark places and to spot the animals after dark.

Where rat numbers are low, no sign other than a burrow entrance 2 to 4 inches (5–10 cm) in diameter may be seen.

In a medium population area, droppings and gnawings can be found, and rats will be heard in or under a structure at night. In addition, cats and dogs may excitedly probe an area where rats are present, especially if rats have only recently entered.

When a high rat population is present, fresh droppings and signs of gnawing will be common. Also, rats may be seen and heard both day and night.

Burrows

Norway rat burrows are found singly or in groups along building foundation walls, under slabs, in overgrown weedy areas, beneath debris and buildings, and in moist areas in and around gardens and fields (Fig. 3). Active burrow entries are 2 to 4 inches (5–10 cm) in diameter, free of dirt, leaves and spiderwebs, and are surrounded with smooth, hard-packed soil. A fresh entry will have soil pushed out in a fan-shaped pattern.

To determine if an existing burrow is occupied, fill the entry with some wadded-up newspaper, leaves, or dry soil. If rats are using the burrow, they will reopen and

clear the hole in a couple of days during good weather.

Trails

Rats tend to travel the same route—along walls, along fences, under bushes—nightly. Trails appear as narrow (wide enough for only one rat), worn paths. Roof rats can often be seen at night running along overhead utility lines or fence tops.

Tracks

Rat tracks appear in dust or soft, moist soil. A rat's footprint is about ½ inch (12 mm) long and shows four or five toes. Norway rats may also leave a tail dragline in the middle of their tracks.

Rat (and mouse) tracking patches can be placed in suspected rat areas to reveal footprints. A tracking patch is a light dusting of an inert material such as clay, talc (unscented baby powder), or powdered limestone. Don't use flour, which can attract rats and other animals. A suitable patch size is 12 x 4 inches (30 x 10 cm). When inspecting tracking patches, shine a flashlight at a low angle, causing the tracks to cast distinct shadows.

Figure 4. Old World rat droppings are ½ inch (12 mm) long and somewhat blunted, with black rat droppings being slightly smaller and more pointed than Norway rat droppings. Deer mice droppings are shown here for comparison. (From Ingles, Mammals of the Pacific States.*)*

Figure 5. A side view of the Norway rat showing the upper and lower incisors that curve slightly inward. This inward curve makes it difficult for rats to gnaw into a flat, hard surface. When given a rough surface or an edge, however, they can quickly gnaw into most materials. (From Verts and Carraway, Land Mammals of Oregon.*)*

Droppings

A single rat can produce 50 droppings daily (Fig. 4). Most droppings are found where rats rest or feed. Fresh droppings are black or nearly black, look wet, and have the consistency of putty. After a few days, droppings become dry, hard, and have a dull appearance. After a few weeks, droppings become gray and crumble easily.

Note that old droppings moistened by rain look like new droppings; however, if crushed, they will crumble.

Gnawing

A rat's incisor teeth grow at the amazing rate of 5 inches (13 cm) per year (Fig. 5). Rats keep their extremely hard teeth worn down by continuously working them against each other and by gnawing on hard surfaces. Look for signs of gnawing on floor or ceiling joists, door corners, siding, and around pipes in floors and walls.

Rub Marks

Body oil and dirt rub off of rats' coats and can become noticeable along frequently used trails. Look along wall–floor junctions, where rats move around obstacles, and at regularly used openings in walls, floors, and ceilings.

Sounds and Smells

Old World rats make squeaks and fighting noises, as well as clawing, scrambling, and gnawing sounds in the walls and other parts of buildings. These sounds may be so loud it seems a larger animal is present. Areas occupied by rats often have a musky smell. One dead rat can cause considerable odor. (See "Poisoning" for information on odor control.)

Preventing Conflicts

In the process of seeking food and shelter, Old World rats can contaminate human or domestic animal food with their droppings, destroy insulation, and create noise in attics, walls, and crawl spaces. Rats also chew on electrical wiring and structural supports in buildings. Norway rats occasionally burrow into dikes and dams (see "Preventing Conflicts" in Chapter 15 for management options). Both species can damage garden crops and ornamental plantings.

Successful long-term rat control is not simple; a continuing commitment to whatever solutions are adopted is required. Within a population, some rats will be easy to control, some difficult. Complete control is often not possible in old barns and similar structures. Rat populations may also be a consequence of community-wide activities over which you have little control—improper garbage disposal, building demolition, and poorly maintained bird-feeding stations.

If you think you can solve a rat problem with a house cat, remember that cats allowed to roam outdoors may also kill songbirds, chipmunks, snakes, lizards, and young rabbits and squirrels. In addition, although a cat may kill a young rat, after one experience with an adult rat, cats often prefer to seek easier prey. (See Appendix D for additional information.)

The following recommendations ideally should be followed *before* rats enter areas where they are unwanted and before the numbers of rats becomes extreme.

Prevent Access to Food and Water

This includes managing areas both inside buildings and their surroundings.

Store human and animal food in rat-proof buildings, rooms, or containers. Old World rats are very capable of chewing through heavy-duty plastic garbage cans when they can get started chewing on a corner or the lid or a handle. Use metal garbage cans where this is a problem. (See "Preventing Conflicts" in Chapter 21 for detailed information on garbage management.)

Prevent access to fruit and compost. Don't put food of any kind in open compost piles; instead use a rat-proof composter or a covered worm box. If burying food scraps, cover them with at least 8 inches (20 cm) of soil and don't leave any garbage above ground in the area—including a stinky shovel. Keep all open or lightly covered compost piles the consistency of a wrung out sponge to make them unfavorable to nesting rats.

Pick up fruit that falls to the ground. Don't allow garden produce to rot on the vine. Compost it, or rototill or dig it into the soil.

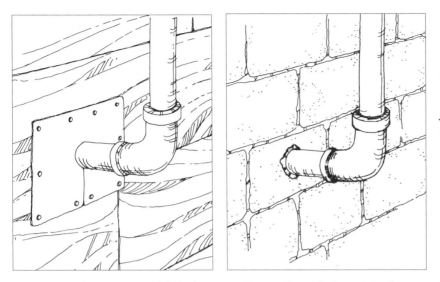

Figure 6. Seal openings around drainpipes, power lines, cables, and other utilities that enter structures to prevent rats from gnawing their way inside. (Drawings by Jenifer Rees.)

Feed dogs or cats inside and clean up droppings. One of the most common attractants around homes is pet food. The strong smell attracts rats from a distance. Once they get a taste of these nutritious foods they will try to feed there daily. If you must feed pets outside, pick up food and water bowls, as well as leftovers and spilled food, before dark. Also, clean up pet droppings—rats can subsist on a diet of droppings.

Prevent access to bird feed and feeders. Birdseed is another item that attracts rats around buildings. Once rats get a taste of this nutritious food they will try to feed there daily.

Place baffles above and below feeders to prevent rats from gaining access to feeder foods.

Rats are attracted to the smell of seed hulls, so rake up the shells or offer birds hulled sunflower seeds (also known as sunflower hearts or chips). For detailed information on feeder management, see "Tree Squirrels and Bird Feeders" in Chapter 24.

Eliminate access to water. Fix leaky outdoor faucets and, where prac-

tical, eliminate access to other sources of ground water.

Prevent Access to Shelter

In the long term, the most successful form of rat control is to build them out. Also called rat-proofing, this approach makes it impossible for rats to get inside or under a building where they could do damage.

Both Norway and roof rats may gain entry to structures by gnawing, climbing, jumping, or swimming through sewers and entering through the toilet or broken drains. While Norway rats are more powerful swimmers, roof rats are more agile and are better climbers.

It is always easier to keep rats out of buildings. Once they are inside, controlling them requires more work, cost, and aggravation.

Eliminate access into buildings. To enter a building, rats only require a hole ¼-inch (6 mm) in diameter to chew larger and squeeze through. Seal openings around drainpipes, vents, power lines, cables, and other utilities that enter structures (Fig. 6).

Seal cracks and holes in building foundations and exterior walls (including warped siding) and roof joints. Seal all the above-mentioned areas with ¼-inch (6 mm) mesh hardware cloth, metal flashing, copper Stuff-it®, mortar, or concrete patch. It's possible to temporarily stuff balled-up galvanized window screening, copper or stainless-steel mesh scouring pads (steel wool quickly corrodes after becoming wet) into cracks, holes, and other openings.

Install commercially available vent guards to prevent rats from entering bathroom exhaust vents and dryer vents. Basement drains should be screened. If the drain is no longer used, seal it.

Rodent-proofing against roof rats usually requires more time to find entry points than for Norway rats because of roof rats' greater climbing ability. Roof rats often enter buildings at the roofline area so be sure that all access points in the roof are carefully inspected and sealed.

Prevent rats from climbing buildings. Both roof rats and Norway rats are excellent climbers. Place metal or heavy plastic barriers around trees and over pipes and other places rats climb and gain access into buildings (Fig. 7). (See "Preventing Conflicts" in Chapter 21 for additional examples of barriers).

If rats are traveling on overhead utility wires, trim tree branches at least 4 feet (1.2 m) away from utility lines that lead into structures. Contact the utility company for information on measures that can be taken to prevent rats from using these lines.

Remove vines, such as English ivy, which provide rats a way

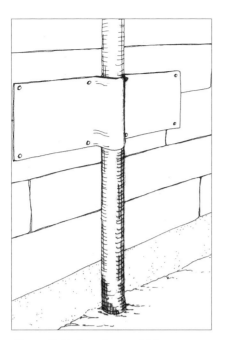

Figure 7. Both roof rats and Norway rats are excellent climbers. Place metal or heavy plastic barriers over pipes that rats climb to gain access into buildings. (Drawing by Jenifer Rees.)

to climb structures and hide their access points. (Ivy also facilitates rats' entry into the tree canopy, where they can prey on nestling birds.)

Keep doors closed, especially at night. Rats can find easy access into buildings under doors (including garage doors) and through open or poorly fitting doors. Once in a garage, rats gain entry into the main structure via holes around pipes, furnace ducts, and drains. Cover door bottoms that are subject to gnawing with metal flashing or hardware cloth, and keep the openings no larger than ¼ inch (6 mm).

Prevent rats from tunneling under buildings. To prevent rats from digging under a foundation, slab, or other area, create a barrier as you would for skunks (see "Preventing Conflicts" in Chapter 23).

Keep ground areas around structures open. Keep a 4-foot (1.2 m) wide

space next to buildings mowed and free of thick vegetation, wood piles, and debris to allow for easy inspection. Materials stored in stacks or piles should be on pallets at least 12 inches (30 cm) off the ground—higher will allow for easier inspection. Ornamental shrubs next to the house should be pruned up at least 18 inches (45 cm) from the ground.

Remove the Existing Rat Population

If rat problems persist after you've eliminated all known sources of food and shelter, some form of population reduction, such as trapping or baiting, is almost always necessary.

Rats are wary animals, easily frightened by unfamiliar or strange noises. However, they quickly become accustomed to repeated sounds, making the use of frightening sounds, including high-frequency and ultrasonic sounds, ineffective for controlling rats in home and garden situations.

Rats have an initial aversion to some odors and tastes, but no repellents have been found to solve a rat problem for more than a very short time.

Encouraging—or at least not discouraging—the natural predators of rats may aid in controlling their population in rural areas. Recently there has been interest in attracting barn owls and other raptors to an area for rodent control by installing nest boxes and perching poles (see "Maintaining Hawk Habitat" in Chapter 33). Such efforts are futile in urban areas where the number of natural predators is low.

Lethal Control

Trapping and poisoning are common methods to lethally control Old World rats. Neither will provide long-term control if attractions are not removed and areas where rats are entering buildings are not repaired.

Many companies and individuals make a portion of their living from handling rat-related problems. Before hiring any work to be done, read the guidelines under "Hiring a Wildlife Damage Control Company" in Appendix C.

Because rats are neophobic (wary of changes in their environment), a trap or bait station may be avoided until the rats become familiar with the new objects along their travel routes. Even then, they will approach cautiously. Rats may become trap shy if they set off a trap but manage to escape alive; these rats may subsequently be much more difficult to trap.

In addition, if the mother rat has become wary of poison baits or traps, her young may learn to avoid them. This learning experience can make trapping and poisoning difficult in sites where control programs have occurred recently.

Trapping

Trapping has several other advantages: It does not rely on potentially hazardous poisons; it permits users to verify their success; and it allows for disposal of trapped rats, thereby eliminating dead-rat odors that may occur when poison is used. Many styles of rat traps are available:

Snap traps: Snap traps and other lethal traps are thought to be more humane than the use of poison because traps generally kill the rats instantly.

The traditional snap trap, or one of its variations, is an effective tool for killing rats, especially when there are only a few rats in a limited area. Traps made with expanded triggers are much more effective than those with small metal triggers (see the example under "Trapping" in Chapter 8). Rats will easily clean the bait off the old metal-trigger traps, but doing so is harder with the expanded version. Most supermarkets, hardware stores, or farm supply stores carry snap rat traps. Professional models are available from pest control companies and Internet vendors (see "Wildlife/Human Conflicts" in Appendix F).

Don't set unprotected snap traps outside of a structure. Otherwise a chipmunk, raccoon, ground-feeding songbird, pet, or other animal may spring the trap—killing or injuring the animals.

If trapping outside is necessary, use a trapping box. Commercial trapping boxes hold two traps, and have one or two small entrance holes. A trapping box also can be made from a wooden box (Fig. 8). Another option is to use a large, plastic toolbox with a 2-inch (5 cm) hole in each end. Place two snap traps facing each way inside, making sure the traps can operate freely with the lid closed. To minimize the possibility of trapping non-targeted wildlife, especially in rural areas, set the trapping box out at night and retrieve it—or cover the holes—at daybreak. Not retrieving the box early enough could mean the death of a curious chipmunk or other small animal.

Snap traps, within suitable boxes and placed off the ground in fruit or nut trees, on fences, and even on roof tops, are effective for trapping roof rats.

Rats have a highly developed sense of touch due to sensitive body hairs and whiskers which they use to explore their environment. They prefer a stationary object on at least one side of them as they travel, thus they commonly move along walls. Such knowledge is helpful when placing traps (Fig. 9).

Check all traps daily to reset any sprung traps and remove dead rats as quickly as possible.

Because rats may carry diseases, do not handle them without gloves; you can use a plastic bag slipped over your hand and arm as a glove. Once the rat is removed from the trap, hold it with your bagged hand and turn the bag inside out while slipping it off your arm and hand.

Old World rats are wary animals and careful attention to detail is necessary to trap them. Here are some tips:

- Set traps out as soon as rats are detected.
- Set traps where evidence of activity is found—along walls, behind objects, in dark corners, or where the rat is forced through a narrow opening, such as in the tunnel where the rat enters and exits a building.
- Move boxes and objects around to create narrow runways leading to the traps.

- Use as many traps as are practical so trapping time will be short and decisive. A dozen traps may be necessary for a heavily infested home.
- Place the trigger side of the traps next to the wall (Fig. 9).
- Bait traps with peanut butter (which is difficult to lick off traps with expanded triggers), hot-dog slices, or bacon. If necessary, hold the bait on with a thread or a twist tie.
- If rats are traveling on rafters or pipes, fasten traps to them using screws, wire, or strong rubber bands. Secure the traps before setting them.

Glue boards: One of the alternatives to a snap trap is a glue board. Glue boards work on the same principle as flypaper: when a rat attempts to cross the glue board, the rodent gets stuck. One of the major drawbacks with glue boards is that the trapped animal may not die quickly, and for humane reasons, you will need to kill it. For this reason, glue boards are not a good alternative for many people and their use is not recommended. Glue boards also lose their effec-

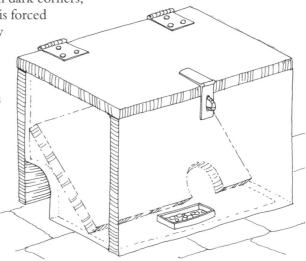

Figure 8. A trap box can be made from wood or a large tackle box. (Drawing by Jenifer Rees.)

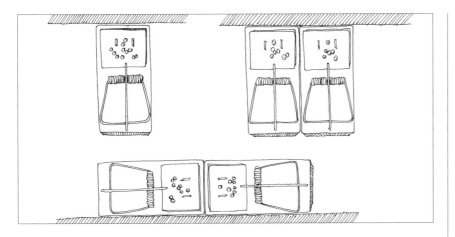

Figure 9. Various ways to place traps along a wall. Note that the baited end is against the wall. (Drawing by Jenifer Rees.)

tiveness in dusty areas unless covered, and extreme temperatures also affect their tackiness.

Electronic traps: Traps that kill rats by electrocution are available on the market. These traps are considerably more expensive than the common snap trap and can be used safely and effectively only in limited situations. Like the snap traps, these traps catch only one rat at a time and then must be emptied. Twenty or more snap traps can be purchased and put to use for the price of one of the electrocution units. When compared with snap traps, electrocution traps have not shown sufficient advantages to promote their use.

Live traps: Live traps are not recommended because trapped rats must either be killed or released elsewhere. Releasing rats outdoors is not recommended because of health concerns to people and the damage they may cause elsewhere. Because neither the roof nor Norway rat is native to this country, their presence in the wild is detrimental to native ecosystems. Elsewhere, they have been known to decimate some bird populations.

Poisoning

Using poison bait is the least preferred method of controlling rats. In addition to the possibility of poisoning children, pets, and non-target species, one dead rat can cause a major odor problem. Also, fleas and mites often leave rat carcasses and may enter the house if the carcass is not removed promptly.

If poison bait is used outdoors it must be registered for such use, applied according to label directions, and placed in a bait box. Bait boxes are designed so children, pets, and non-target wildlife cannot access the bait, but a rat can. (Bait trays and flimsy plastic or cardboard boxes are not tamper-resistant.)

Bait boxes vary in type and quality of construction, but they are usually metal or heavy plastic; those designed for rats are larger than those used for mice. Small children may be able to slide their hands inside, but the bait is tucked away in chambers, which will be out of their small hands' reach. Bait boxes can be purchased from a farm supply store, hardware store, or pest control business over the Internet.

Dealing with a Dead Rat in a Wall

Contrary to common myths, rats don't eat poison bait, then run outside to drink and die, and bait does not make rats mummify, or dry up into dust, so that they won't stink.

The stench of one dead rat can be unbearable, especially when the weather warms up.

In extreme cases, it may be necessary to drill a hole in a wall to remove the carcass. If you can locate the odor source, a small hole can be drilled in the wall, 6 inches (15 cm) inches above the floor between the two adjacent studs. An odor-eliminating product that masks smells or removes them at the molecular level can then be sprayed or poured in. Using an unscented product that removes the odor, then spraying with a scented product can be effective. A common mistake people make when deodorizing is not using enough deodorizing chemicals, and then applying too much scented material.

Commercial odor-eliminating products are available through hospital supply houses, drugstores, pet stores, and the Internet—use the keywords "Pest Control Supplies."

If left untreated, the odor will usually disappear on its own in about a month. If the odor cannot be precisely located, apply the deodorant in the general area according to the directions on the product's label.

Pets, Wildlife, and Rodent Control

Many of the methods and materials used to control rats and mice can affect pets and wildlife. All rodent baits are toxic to dogs, cats, and wildlife, so be cautious in their use. Because the anticoagulants are cumulative and slow acting, dead rats or mice may contain several lethal doses of toxicant, and secondary poisoning of pets and wildlife is possible if several carcasses are consumed over a few days.

While this secondary poisoning is possible, it is not common. Most fatalities in pets involve dogs and are due to the animal consuming the bait directly (primary poisoning) or to a combination of direct bait consumption and secondary poisoning.

Anticoagulant baits used to be thought of as relatively safe baits to use around the house and garden because they required multiple feedings to be effective. With a reaction time of three to four days, and vitamin K as an antidote, the risk of accidental poisoning of humans and pets was considered low. Newer anticoagulant baits, however, have been developed that only require a single feeding to be effective and are, therefore, more hazardous to pets, children, and wildlife than the older type of anticoagulant bait.

Baits that require multiple feedings over a period of several days contain warfarin, chlorophacinone, or diphacinone as their active ingredient, whereas the single-feeding anticoagulants contain brodifacoum, bromadiolone, or difethialone.

Use extra caution with the single-feeding anticoagulant baits; exposure to even a single dead rat or mouse killed by these might be enough to cause poisoning in a pet or other animal. The great advantage of multiple-feeding anticoagulants is that a good antidote, vitamin K, as well as whole-blood transfusions are available if medical attention is received early enough. Symptoms of anticoagulant poisoning in mammals include lethargy, loss of color in soft tissues such as the lips and gums, and bleeding from the mouth, nose, or intestinal tract.

Put bait in a tamper-resistant bait box in locations out of the reach of children, pets, domestic animals, and wildlife. The bait box must be resistant to destruction by dogs and by children under six years of age, and must be constructed in a manner that prevents a child from reaching into the bait compartments and obtaining bait (Fig. 8). If bait can be shaken from the box when it is lifted or tipped, the box must be attached to stakes in the ground or otherwise immobilized. Clearly label all bait boxes with appropriate warnings, and store unused bait in a locked cabinet or other areas inaccessible to children and domestic animals.

(See "Poisoning" for additional information.)

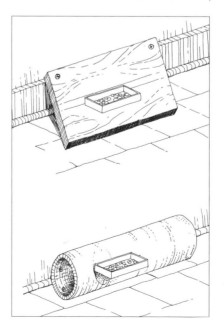

Clearly label all bait boxes "Caution—Poison Bait" as a safety precaution. Some poison-bait labels or situations may require the use of approved tamper-resistant bait boxes. If so, be sure to secure these boxes to buildings by screwing or otherwise securing them to walls or floors in a way that will not permit a person or animal to knock them over or shake the bait out.

Poison bait can be purchased packaged in a variety of formulations. Most baits contain one of several anticoagulants that prevent blood clotting, causing rats to bleed internally and die. (See "Pets, Wildlife, and Rodent Control" for important information.)

Check bait boxes periodically to make sure the rats are taking the bait and that the bait is fresh. Rats rarely feed on spoiled bait. Because rats fear new objects, the bait may not be eaten for a few days or a week.

Remove and properly dispose of all uneaten bait at the end of a control program. Also, collect and properly dispose of any dead rats found while baiting. Pick them up, using a sturdy plastic bag inverted on your hand, and seal them in the bag for disposal with household garbage, or bury them in a location where pets or scavengers will not easily dig them up.

Dried rat carcasses and skeletons indicate a past control effort using poison bait. Many fresh carcasses are an indication that poison bait is currently being used in the area.

Figure 10. Where children, pets, and non-target species cannot access poison bait, place it along the rat's route and in such a way that the rat will feel safe when consuming it. (Drawings by Jenifer Rees.)

Flooding or Smoking Rats Out

Norway rats may be drowned or flushed from their burrows with water from a garden hose. The entire tunnel system will need to be quickly and completely flooded to evict its tenets, which can be dispatched with a shovel or caught by a dog. Be careful when attempting to flood out a rat near a building—doing so could damage the foundation or flood the basement or crawl space. Concentrate the effort in late winter and early spring, before rats give birth. Fill all holes with dirt when done.

Smoke or gas cartridges are registered and sold for the control of burrowing rodents. When placed into the burrows and ignited, these cartridges produce toxic and suffocating smoke and gases. Norway rat burrows may extend beneath a residence and have several open entrances, however, permitting toxic gases to permeate the dwelling. For this reason, and because fire hazards are associated with their use, smoke and gas cartridges are not recommended for rat control around homes.

Shooting

Controlling small, isolated groups of Old World rats with a pellet gun is effective. For safety considerations, shooting is generally limited to rural situations and is considered too hazardous in more populated areas, even if legal.

Follow-up

It is difficult to permanently rat-proof a structure—when houses settle new opportunities for re-entry open up, and rats can chew a hole into or burrow under an unprotected structure at any time. Therefore, keep watch for any new signs of rats (droppings, food damage, gnawing damage, burrow entrances, etc.).

Public Health Concerns

Among the diseases that can be spread from Old World rats to humans are bubonic plague, salmonella (food poisoning), leptospirosis, and tularemia.

If a person is bitten or scratched by a wild rat, immediately clean the wound by thoroughly scrubbing it with soap and water. Flush the wound liberally, using tap water. Contact a physician and the local health department. This same precaution applies if a person has handled an obviously sick rat that may be harboring a contagious disease. (Children are particularly inclined to handle lethargic small mammals.)

If you can place a large bucket over the rat and secure the bucket with a heavy object, the animal can then be held for inspection by a health official.

Concerning the regulations for rat control, city, county, or state departments of health have regulations concerning rat infestations that reach levels of public health concern and generally have the authority to mandate control. Restaurants and fast-food outlets must meet certain conditions of sanitation and landlords must keep rodents under control so that the premises are considered habitable.

Legal Status

Because legal status and other information concerning rats change, contact your local, state, or provincial wildlife office, or visit their Web site for updates. (See Appendix E for contact information and where to access the state and provincial laws mentioned below.)

Oregon: Old World rats are not under the jurisdiction of the Oregon Department of Fish and Wildlife. All woodrats are classified as unprotected wildlife. The killing of Old World rats and woodrats is unrestricted. Woodrats may be captured from the wild and held as pets without permits, but they must be maintained in a humane manner (OAR 635-044-0132).

Washington: Old World rats are not considered to be wildlife and are therefore not regulated by the Washington Department of Fish and Wildlife. The killing of Old World rats is unrestricted.

Woodrats are considered wildlife and may be trapped or killed year-round without a permit to prevent or stop damage.

British Columbia: Old World rats are not included under the definition of wildlife and do not fall under the regulatory authority of the Ministry of Water, Land, and Air Protection.

Bushy-tailed woodrats are listed under Schedule "B" of the Wildlife Act. They may be captured or killed for the purposes of protection of property by the person that owns or occupies the property or his or her spouse, parent, guardian, or child.

CHAPTER 17

Opossums

P rior to European settlement of North America, the Virginia opossum (*Didelphis virginianus,* Fig. 1 and back cover) was found only in Central America and the southeastern United States. During the 1900s, its range expanded northward and westward.

Virginia opossums, also known as "possums," first arrived in the Pacific Northwest in the early 1900s as pets and novelties. Some of these animals, or their offspring, later escaped from captivity or were intentionally released.

With few natural predators, the absence of hunting, and an abundance of food and shelter, opossums have adapted well to living close to people in urban and suburban environments. Except for higher elevations, opossums now occupy most human-occupied habitats in western Oregon and Washington, and southwestern British Columbia.

Opossums are marsupials (mammals with a pouch in which they carry their young), a primitive group of mammals found most commonly in Australia. Kangaroos, koalas, and wombats are other well-known marsupials. Opossums are the only marsupial in North America. All other mammals are placentals, which means their young develop within a saclike membrane called the placenta inside the mother's uterus, rather than in an exterior pouch.

In Australia and elsewhere, many species of marsupial have been out-competed and even driven to extinction by more modern mammals. Yet, the opossum has adapted to the changing environment in the Western Hemisphere, and continues to thrive.

Opossums are inhibited animals, especially in daylight or under artificial light, but are by no means stupid. Results from some learning and discrimination tests rank opossums above dogs and more or less on a par with pigs in intelligence.

Facts about Opossums
Food and Feeding Habitats

- Opossums lived during the time of the dinosaurs and one reason for their continued survival is their ability eat nearly anything.
- Foods include fruits, nuts, grains, insects, slugs, snakes, frogs, birds, bird eggs, shellfish, mice, and carrion (dead animals).
- Around human habitation, opossums also eat garbage, pet food, birdseed, poultry, and handouts.
- A study of an urban opossum population in Portland, Oregon, found that small mammals (dead and alive) were the most important food in winter and spring, slugs in summer, and fruits in fall.
- Because opossums eat many road-killed animals, including other opossums, they often become road kill themselves.
- Because opossums accumulate little body fat for winter and don't store food, they must forage year-round.

Den Sites

- Opossums will den nearly anywhere that is dry, sheltered, and safe. Den sites include burrows dug by other mammals, rock crevices, hollow stumps, logs and trees, woodpiles, and spaces in or under buildings.

Figure 1. Opossums measure 2 to 3 feet (60–90 cm) in length, a third of which is a round, scaly, sparsely haired tail. The head is conical, tapering to a slender, elongated snout tipped by a pink-colored nose. The face is light gray to white, whereas the general color of the fur from neck to rump is grayish white. Because of its body shape, a small opossum is sometimes mistaken for a large rat. (From Christensen and Larrison, Mammals of the Pacific Northwest: A Pictorial Introduction.*)*

- Their fur doesn't provide much insulation, so opossums fill their dens with dried leaves, grass, and other available soft material to form well-insulated nests. Nest materials are carried in their coiled tail.
- To avoid predators, opossums move to a different den every few days. (A male opossum followed by radio tracking used 19 different dens in five months.)
- A female with young or an opossum "holed up" during a cold spell will use the same den for a greater length of time.

Reproduction

- Opossums are successful as a species due in part to the size and frequency of litters.
- In western Oregon, the breeding season begins in late December or early January, and there are three peaks in breeding activity: January to February, mid-April to early May, and October to mid-November.
- Being marsupials, opossums give birth to undeveloped young. Only 12 days following breeding, five to ten bumblebee-sized pups crawl into their mother's pouch, where each firmly attaches to a teat.
- Opossum pups find nourishment, warmth, and safety in the pouch. When closed, it is so well sealed that if the female swims, the pups remain dry.
- At 60 to 70 days old, the house-mouse-size young begin to leave the pouch for brief periods, returning to suckle.
- At 80 to 90 days old, the young begin to ride on their mother's back with their feet and tail firmly attached to her fur (Fig. 1). (Contrary to myth, a female opossum never carries her young on her tail.)
- At 3½ months of age, the young begin to leave the den to feed on their own, and soon disperse to establish their own territories.

Mortality and Longevity

- Opossums have high mortality rates at all ages. They are killed by dogs, coyotes, foxes, raccoons, bobcats, eagles, hawks, and owls, with young opossums being the most vulnerable.
- Car kills in the fall and in winter conditions account for many opossum deaths.
- Opossums rarely live a full two years in the wild.

Viewing Opossums

Opossums are nocturnal, spending the day in dens or other protected spots. However, they can be seen at any time of day, especially in winter when food is scarce. At night, opossums forage in areas near their current dens, but can travel up to 2 miles (3.2 km) in search of food.

Opossums are solitary animals, and except during breeding season or a female with her young, they are rarely seen together. Opossums do not hibernate.

Although they can climb and are good swimmers, opossums prefer to amble about on the ground. With a top speed of about 4 miles (6.4 km) per hour, when "running," opossums appear to be walking quickly, with the tail rotating in circles for balance. When idle, opossums constantly groom themselves, much as house cats do.

A nighttime walk along a path bordering a stream or wetland, or down an alley lined with trashcans,

Early Encounters

Captain John Smith (of Pocahontas fame) first adopted the name "opossum" in Western culture in 1608. It derives from the Algonquin Indian name of the animal, *apasum*, meaning "white animal." The spelling has undergone a few changes over the years (opassom, ouassom, opussum, apossumes, ospason, opuson, oppassum, apossum) with the current spelling being settled upon in 1787.

John Smith's description of the opossum is certainly not the first, but it is the best-known and most quoted, early description: "An Opassom hath an head like a Swine, and a taile like a Rat, and is of the bignes of a Cat. Under her belly she hath a bagge, wherein shee lodgeth, carrieth, and sucketh her young."

Table 1. Introduced mammal species that are successfully maintaining populations in Oregon, Washington, and British Columbia

(Adapted from Johnson and O'Neil, Wildlife-Habitat Relationships in Oregon and Washington, *Oregon State University Press, 2001.)*

Species	Reason	Status	Origin
Virginia opossum *Didelphis virginiana*	Aesthetics, pet escapees, fur trapping	W WA, W OR and NE OR, SW B.C.	E United States
Red fox *Vulpes vulpes*	Fox hunting, fur farming, escapees	Widespread in W OR & WA, less so in E OR and WA, south-central B.C.	Holarctic
House cat *Felis catus*	Pet escapees, pest control	Widespread in OR, WA, B.C.	Eurasia, Africa
Domestic dog *Canis familiaris*	Pet escapees	Occasional occurrences	Eurasia
Burro *Equus asinus*	Escapees or released when no longer needed	Small population in SE OR	Africa
Horse *Equus caballus*	Escapees or released when no longer needed	Moderate population in SE OR, central B.C.	Asia
Feral pig *Sus scrofa*	Hunting, escapees	Very small, localized populations	Eurasia
Axis deer *Axis axis*	Aesthetics or escapees	Small, localized population	India
Fallow deer *Dama dama*	Aesthetics or escapees	Small, localized populations	Europe
Mountain goat *Oreamos americanus*	Aesthetics, hunting	Moderately abundant in Olympic Mtns. native to N Cascades and Rocky Mtns.	Alaska to WA, Cascade and Rocky Mtns.
Eastern gray squirrel *Sciurus carolinensis*	Aesthetics	Localized, urban/suburban areas of OR, WA, and southern B.C.	E United States
Fox squirrel *Sciuris niger*	Aesthetics	Localized, urban/suburban areas of E WA, W and NE OR, and southern B.C.	E United States
House mouse *Mus musculus*	Stowaway, then range expansion	Widespread, urban/suburban areas in OR, WA, and B.C.	Europe
Norway rat *Rattus norvegicus*	Stowaway, then range expansion	Localized, urban/suburban areas OR, WA, and B.C.	Asia
Black rat *Rattus rattus*	Stowaway, then range expansion	Localized, urban/suburban areas in OR, WA, and B.C.	Asia
Nutria *Myocastor coypus*	Fur farming, escapees, vegetation control	Localized, mostly in W OR & WA	South America
European rabbit *Oryctolagus cuniculus*	Aesthetics, hunting	Abundant island populations in WA and B.C., other localized populations	Europe
Eastern cottontail *Sylvilagus floridanus*	Hunting	Widespread, locally abundant	E United States

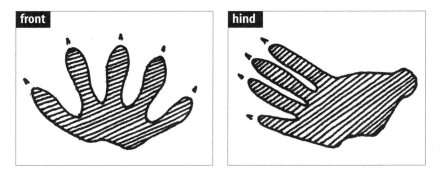

Figure 2. The opossum's front tracks are about 2 inches (5 cm) in diameter and hind tracks are slightly larger. The opossum's long tail often leaves drag marks in snow or mud. (Drawings by Kim A. Cabrera.)

will occasionally turn up an opossum searching for food. Strong but not agile climbers, opossums can be observed climbing trees to escape, search for food, rest, or to look for dens. Their tails are able to wrap around and grasp tree limbs and can support the animal's full weight for short periods. (Contrary to myth, opossums do not hang upside down by their tails when sleeping.)

Trails

Opossums readily use trails made by other wildlife or humans near creeks, ravines, and wetlands. Like raccoons and foxes, opossums use culverts as a safe way to cross under highways and roadways.

In developed areas, trails occur along buildings and fences. Wear marks and hairs may be found around the edges of entry points where opossums are entering a building or crawling under a fence. Opossum hair is long and silver to gray in color.

Tracks

Tracks can be found in mud, snow, or fine soil; also on deck railings, downspouts, and other surfaces that opossums use to gain access to structures (Fig. 2). The opossum's opposable hind thumbs create a unique print, pointing as much as 90 degrees away from the direction of travel.

Droppings

Opossum droppings are not easily found, but can be seen along trails they use and near favorite feeding spots. Opossum droppings vary in appearance according to the animal's diet and may resemble the droppings of house cats and small domestic dogs, coyotes, and foxes. Firm droppings are pointed on the ends and 1 to 3 inches (2.5–7.5 cm) long.

Calls

Opossums are among the most silent animals that live in the Pacific Northwest, but when frightened or threatened they growl and hiss.

Preventing Conflicts

In urban areas, opossums are beneficial as rodent and carrion eaters. They also clean up uneaten food that might otherwise attract mice and rats. However, in rural areas the impact of non-native opossums preying upon native invertebrates, small mammals, amphibians, reptiles, ground-nesting birds, nestlings, and eggs is of concern to wildlife biologists.

Although they may overturn garbage containers, eat garden fruit and vegetables, and enter chicken houses or other structures, opossums rarely cause serious problems. As long as they are kept out of human homes, not cornered, and their interaction with pets is limited, opossums are not dangerous.

If an opossum finds its way into your house, stay calm, close surrounding interior doors, leave the room, and let the animal find its own way out through the pet door or an open door or window. If necessary, gently use a broom to corral the opossum outside.

Do not corner an opossum, thereby forcing it to defend itself. If the opossum appears sick or injured, call a wildlife rehabilitator or your local wildlife office (see "Wildlife Rehabilitators and Wildlife Rehabilitation" in Chapter 20).

The most effective way to prevent conflicts with opossums is to modify your home and property so it will not attract these animals. For ways to do this, see "Preventing Conflicts" in Chapter 21.

For information on trapping opossums, see Appendix A.

Public Health Concerns

Although opossums might carry several diseases of significance to humans, their role in the transmission of any of these diseases is uncertain. Anyone handling a dead or live opossum should wear rubber gloves, and wash his or her hands well when finished.

There is convincing evidence that the parasite that causes **Equine Protozoal Myeloencephalitis** (EPM), a disease in

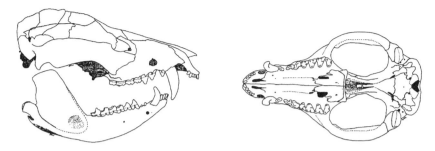

Figure 3. Their 50 teeth (more than any other mammal in North America) give opossums a menacing look when threatened. (From Verts and Carraway, Land Mammals of Oregon.*)*

"Playing Possum"

The opossum is a slow runner and when threatened will usually growl, hiss, and bare its teeth or try to escape by climbing the nearest tree. However, when caught out in the daylight with little chance of escape, or when attacked, the opossum will "play possum." This is a surprisingly effective defense commonly seen in insects. In such cases the opossum will fall on its side, curl its body, open its drooling mouth, and excrete droppings—all to give the appearance of being dead.

While the opossum is in this state, which lasts several minutes or several hours, no amount of prodding will produce a response. Though it appears to be in a catatonic state, its metabolic processes are as high as when the animal is fully alert.

When the opossum believes the danger has passed, it will begin to wiggle its ears in an effort to pick up sounds. If it thinks the danger has passed, it will pick up its head and look around. If danger persists, the opossum will play dead again.

Although generally gentle and placid, opossums have 50 teeth and will use them to protect themselves, or their young (Fig. 3). So avoid close encounters.

horses, is carried by the opossum. EPM is an infection of the central nervous system; the neurologic signs that are most apparent in horses include dizziness, weakness, and spasticity.

While there are no guaranteed methods of preventing exposure to this parasitic organism, horse owners can minimize risks by making facilities less attractive to opossums. Remove or seal up food that opossums might find attractive, such as cat food left out, grain sitting in buckets, feed in uncovered bins, and garbage in open cans. If feed has been left exposed, check it for droppings before serving it to your horses. Droppings need not be fresh to be dangerous; the parasite can live outside a host and remain potentially infectious for as long as one year.

The **rabies** virus does not exist in Pacific Northwest opossum populations, and for some unknown reason opossums rarely get rabies elsewhere.

If a person is **bitten or scratched,** immediately clean the wound by thoroughly scrubbing it with soap and water. Flush the wound liberally. A physician should examine all wounds caused by an opossum.

Legal Status

Because legal status, trapping restrictions, and other information about opossums change, contact your local, state, or provincial wildlife office, or visit their Web site for updates. (See Appendix E for contact information and where to access the state and provincial laws mentioned below.)

Oregon: Because opossums are non-native, possession of opossums has been prohibited in Oregon. People may not possess them as pets, rehabilitation of injured or orphaned opossums is not allowed, and the relocation of trapped opossums by wildlife-control professionals is prohibited. The killing of opossums is unrestricted, although a hunting or trapping license is required.

Washington: The opossum is unclassified and may be trapped or killed year-round; no permit is necessary. Although trapping opossums is not illegal, there may be restrictions on what devices may be used in some jurisdictions.

It is unlawful to release an opossum anywhere within the state, other than on the property where it was legally trapped, without a permit to do so (RCW 77.15.250 and WAC 232-12-271).

British Columbia: The opossum is identified as "Wildlife" and is protected under Section 34 of the Wildlife Act. However, it is also listed under Schedule "C" and may be captured or killed anywhere or at any time in the province. No permit is required but a person must have land owner/occupier permission on private property.

CHAPTER 18

Pocket Gophers

In some areas, the name gopher is applied to a variety of mammal species including ground squirrels and moles. True pocket gophers are burrowing rodents that get their name from their fur-lined cheek pouches, or pockets. These pockets are used, like a squirrel's, for carrying food. However, the pockets on a gopher open on the outside and turn inside out for emptying and cleaning.

Pocket gophers are well-equipped for a digging, tunneling lifestyle, with large-clawed front paws, small eyes and ears, and sensitive whiskers that assist with movement in the dark (Fig. 1). Their pliable fur and sparsely haired tails—which also serve as a sensory mechanism—help gophers run backward almost as fast as they can run forward. Their large front teeth are used to loosen soil and rocks while digging, as well as to cut roots (Fig 2).

The pocket gopher's short fur is a rich brown or yellowish brown, but also may be grayish or closely resemble the local soil color.

There are five species of pocket gophers in the Pacific Northwest (Table 1). They occur from almost sea level to above timberline, and inhabit grassy areas including lawns, meadows, highway rights-of-way, rangelands, orchards, and clearcuts.

Pocket gophers can be a problem for homeowners, but they actually benefit the soil and vegetation in many areas. Unfortunately, the positive effects are not as visible as the mounds pocket gophers create in lawns and pastures.

Facts about Pocket Gophers
Food and Feeding Habits

- Unlike moles, which mostly eat insects and other invertebrates, pocket gophers only eat vegetation.
- Gophers eat roots, bulbs, and other fleshy portions of plants they encounter while digging underground.
- Gophers also eat the leaves and stems of plants around their tunnel entrances and can pull entire plants into their tunnels.
- In areas with a snowpack, gophers will gnaw on bark several feet up a tree or shrub.
- Because gophers obtain sufficient moisture from their food, they don't need a source of open water.

Figure 1. Pocket gophers are stout-bodied rodents with small ears and eyes and large-clawed front paws. Their large front teeth are used to loosen soil and rocks while digging, as well as to cut and eat roots. (From Christensen and Larrison, Mammals of the Pacific Northwest: A Pictorial Introduction.*)*

The Benefits of Pocket Gophers

A typical pocket gopher can move approximately a ton of soil to the surface each year. This enormous achievement reflects the gopher's important ecological function.

Their tunnels are constantly being extended, then gradually filled up with soil as they are abandoned. The old nests, toilets, and partially filled pantries are buried well below the surface where the buried vegetation and droppings become deep fertilization. The soil thus becomes mellow and porous after being penetrated with burrows. Soil that has been compacted by trampling, grazing, and machinery is particularly benefited by the tunneling process.

In mountainous areas, snowmelt and rainfall are temporarily held in gopher burrows instead of running over the surface, where they are likely to cause soil erosion.

Surface mounds created by gophers also bury vegetation deeper and deeper, increasing soil quality over time. In addition, fresh soil in the mounds provides a fresh seedbed for new plants, which may help to increase the variety of plants on a site.

Many mammals, large birds, and snakes eat gophers and depend on their activities to create suitable living conditions. Salamanders, toads, and other creatures seeking cool, moist conditions take refuge in unoccupied gopher burrows. Lizards use abandoned gopher burrows for quick escape cover.

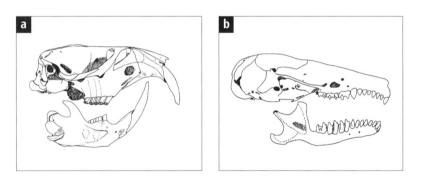

Figure 2. Lateral views of the Camas pocket gopher (a) and the Townsend mole (b) show the differences in their teeth. A pocket gopher's teeth are adapted for gnawing on plant material, and a mole's long jaws and 44 teeth are adapted for consuming small insects. (From Verts and Carraway, Land Mammals of Oregon.*)*

Reproduction and Social Structure

- Depending on the species, pocket gophers breed from early spring to early summer, resulting in one litter of three to ten young per year. Botta's and Townsend pocket gophers may produce an additional litter.
- The nest chamber is located in the pocket gopher's burrow system, is about 10 inches (25 cm) in diameter, and is lined with dried grass.
- The young develop quickly, remain in the nest for five to six weeks, and then wander off above ground to form their own territories.
- Pocket gophers are solitary except during the breeding season or when females have young with them.
- In south-central Oregon, densities of northern pocket gophers have been found to range from 2 to 20 gophers per acre, depending on food availability, species, and ages of the gophers.

Mortality and Longevity

- Coyotes, domestic dogs, foxes, house cats, and bobcats capture gophers at their burrow entrances; badgers, long-tailed weasels, skunks, rattlesnakes, and gopher snakes corner gophers in their burrows. Owls and hawks capture gophers above ground.
- A deep snowpack can result in high gopher mortality. If the snow melts rapidly it saturates the ground and floods the burrows.
- Pocket gophers live one to two years and the majority of the population consists of young adults.

Table 1. Pacific Northwest Pocket Gophers

The **Northern pocket gopher** *(Thomomys talpoides)* is the smallest and most widespread Pacific Northwest gopher, occupying much of eastern Oregon and Washington, and the southern interior of British Columbia. Adults measure 8 inches (20 cm) in length, including their 2-inch (5 cm) tail.

The **Mazama (Western) pocket gopher** *(Thomomys mazama)* occurs in western Oregon and is the only pocket gopher found in western Washington—in the Olympic Peninsula and the southern Puget Sound area. Adults measure 8 inches (20 cm) in length, including their 2½-inch (6.5 cm) tail.

The **Camas pocket gopher** *(Thomomys bulbivorus)* is found only in the Willamette River valley of Oregon and is the largest species in the Pacific Northwest. Adults measure 12 inches (30 cm) in length, including their 3½-inch (9 cm) tail.

The **Botta's pocket gopher** *(Thomomys bottae)* reaches its northerly limit in west-central Oregon. Adults measure 9 inches (23 cm) in length, including their 3-inch (7.5 cm) tail.

The **Townsend pocket gopher** *(Thomomys townsendii)* is found in scattered areas throughout southeastern Oregon. Adults measure 11 inches (28 cm) in length, including their 2½-inch (6.4 cm) tail.

Viewing Pocket Gophers

Although pocket gophers are active year-round and at all hours of the day, their underground lifestyle makes them difficult to observe.

If you are patient, you may be able to watch a pocket gopher feed above ground, or see their food being taken underground. (Of all the gopher species, the western pocket gopher spends the most amount of time above ground, generally at night and on overcast days.)

When sitting in a grassy area, keep your eyes and ears alert for the sight and sound of a wiggling clump of grass, wild flowers, or similar vegetation. You might see the entire plant slowly disappear below ground, and a few minutes later, the same gopher may venture a body length's distance from its tunnel opening to alertly feed or gather food.

When gathering food above ground, a pocket gopher will cut vegetation quickly, cram as much as possible into its external pouches (or pockets), and then disappear below ground. It may reappear in a few minutes, gather more food, and disappear to consume the food underground or store it away for later.

Pocket gophers live in extensive burrow systems, which they use for locating food, rearing young, storing food and droppings, and escaping predators. Burrow systems are a closely regulated microenvironment, and gophers will plug any openings in the system within 24 hours. Evidence of a gopher's burrow system includes mounds, soil plugs, and winter soil casts.

Mounds

As a pocket gopher tunnels, it loosens soil with its front legs. When digging becomes difficult, it bites off chunks of earth or roots with its incisors. The gopher then somersaults to turn around and push the loose earth and other debris to the surface bulldozer style, with its front feet and head. The excavated material is pushed out of the exit tunnel to the front, right and left, creating a fan-shaped or heart-shaped mound (Fig. 3). (See Table 2 for a comparison of pocket gopher and mole mounds.)

The capacity of pocket gophers to excavate tunnels is

Table 2. Differences between Pocket Gophers and Moles
Before setting out to control what you assume to be gopher damage,
be sure to properly identify the animal causing the damage.

Pocket Gophers	Moles
Small eyes are clearly visible.	Minute eyes are often not visible.
Muzzle is rounded.	Muzzle is long and tapering.
Orange, chisel-like pairs of upper and lower incisors are apparent.	The many small teeth are not apparent.
Mounds are crescent- or heart-shaped when viewed from above.	Mounds are round when viewed from above.
Soil plug is in the middle of the V shape or off to the side of the mound and may leave a visible depression.	Soil plug is in the middle of mound and may not be distinct.
No tunnels are visible from above ground.	Tunnels are often just beneath the surface, leaving a raised ridge.

Tunnels

Pocket gopher tunnels are 1¾ to 4½ inches (4.5–11.5 cm) in diameter, depending on the size of the gopher. Tunnels are 4 to 12 inches (10–30 cm) below ground, whereas the nest and food storage chamber may be as deep as 6 feet (1.75 m). Tunnels tend to be deeper in drier soils. Short, sloping, lateral tunnels connect the main tunnel system to the surface and are created for pushing dirt to the surface and access to foraging on the surface (Fig. 3).

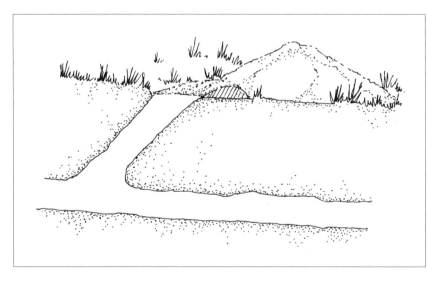

Figure 3. The side view of a portion of a pocket gopher's burrow system showing the mound, the short, sloping lateral tunnel, and the deeper main tunnel. (Drawing by Jenifer Rees.)

Soil Plugs

Tunnel exits made by a pocket gopher are marked by a 1- to 3-inch (2.5–7.5 cm) circle of disturbed soil, or a circular depression, called a "soil plug." Soil plugs occur where a gopher emerged to forage or deposit

phenomenal and heavy clay soil doesn't deter them. Gophers create several mounds per day, especially during the seasons when the soil is moist and easy to dig. In irriga-ted areas such as lawns, gardens, and pastures, digging conditions may be optimal year-round, and mounds can appear at any time.

soil, and then plugged up the hole on reentry. Plugs are found at mounds or along the course of the burrow system. Vegetation may be clipped around the soil plugs where a gopher was foraging.

Winter Soil Casts

Soil casts are created because pocket gophers commonly backfill their previously excavated tunnels with excess soil when they dig new tunnels. Casts are the result of this excess soil being backfilled into snow tunnels. When the snow melts, these then become apparent. Castings are nearly always fragmented and in short sections. The castings represent only a fraction of the snow tunnels produced, and only those backfilled with soil.

Preventing Conflicts

The ecological services of pocket gophers, which are substantial, are often not appreciated, particularly when the animals make their presence known by eating garden crops or damaging orchard or ornamental trees.

For homeowners and small-scale gardeners, gophers may be only an occasional nuisance in lawns and garden beds, and not a long-term problem or threat. Where these animals are not so numerous as to be causing heavy damage, they should be considered neutral.

If gophers are gnawing on water lines or wires, or are burrowing into dams and dikes, see Chapter 15 for management recommendations.

The following are suggestions for reducing conflicts. In cases where these methods are not practical, contact your local County

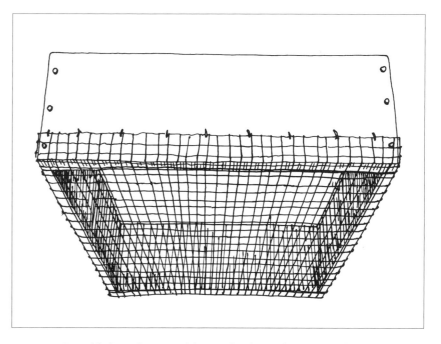

Figure 4. Raised beds can be protected from gopher damage by screening the bottom side with ½-inch (12 mm) mesh hardware cloth. (Drawing by Jenifer Rees.)

Extension Agent, Department of Agriculture, or Ministry of Agriculture (B.C.) for further information (see Appendix E for contact information).

Frightening devices and repellents: Although many devices are commercially available for use to frighten pocket gophers (vibrating stakes, ultrasonic devices, wind-powered pinwheels, etc.), gophers do not frighten easily. This is probably because of their repeated exposure to noise and vibrations from sprinklers, people and livestock moving about, and lawnmowers and other power equipment. ***Note:*** Be skeptical of commercial products and claims, and make sure the manufacturer offers a money-back guarantee if the product proves ineffective.

No repellents currently available will reliably protect lawns or other plantings from pocket gophers. Mothballs, garlic, spearmint leaves, predator urine placed

in tunnels—and a perimeter of mole plant or castor bean planted around gardens—have all provided mixed results. Such control strategies may be experimented with where gophers are an occasional problem, but not a long-term threat.

Barriers: Constructing a barrier to keep pocket gophers from tunneling into an area can be labor-intensive and costly; however, this approach is recommended for small areas and areas containing valuable plants. Flower beds and nursery beds can be protected by complete underground screening of the sides and bottom. Raised beds with rock or wooden side supports will only require bottom protection (Fig. 4).

Wire baskets can be used to protect the roots of individual trees and shrubs. These can be purchased from nurseries or farm supply centers, or be homemade. Use a double layer of light-gauge

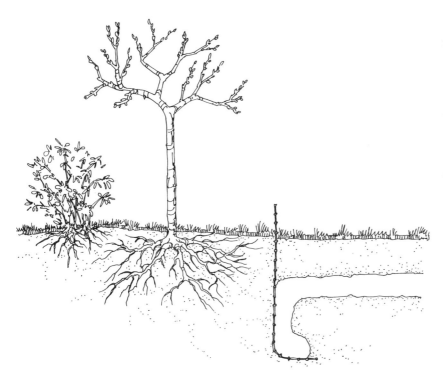

*Figure 5. A pocket gopher (and mole) fence should be installed at least 24 inches (60 cm) below ground (or down to the hardpan or bedrock) and 6 inches (15 cm) above ground. Use a 36-inch (90 cm) wide roll of ½-inch (12 mm) mesh hardware cloth. Before placing the hardware cloth perpendicularly in the trench, the bottom 6 inches (15 cm) should be bent outward at a 90-degree angle. Alternatively, fill a 10-inch (25 cm) wide by 24-inch (60 cm) deep trench with gravel that is 1 inch (2.5 cm) or more in diameter. **Note:** Such barriers will likely only work temporarily. (Drawing by Jenifer Rees.)*

wire, such as 1-inch (24 mm) mesh chicken wire for trees and shrubs that will need protection only while young. Leave enough room to allow for a few years of root development before the wire rots away.

Groups of bulbs (gophers are reported not to eat daffodil bulbs) and other plants needing long-term protection can be placed in baskets made from ½-inch (12 mm) mesh hardware cloth, available from hardware stores and building supply centers.

Large areas, such as vegetable gardens, can be protected using an underground gopher fence (Fig. 5) or a stone-filled trench. However, such a below-ground barrier will only slow the movements of

gophers for a time; sooner or later the barrier will be breached since gophers are capable of digging much deeper than 24 inches (60 cm).

To add to the life of underground barriers, spray on two coats of rustproof paint before installation. Above-ground parts can also be painted to blend in. Always check for utility lines before digging in an area.

Several types of barriers are effective at protecting above-ground parts of small plants, such as newly planted conifers (see "Preventing Conflicts" in Chapters 7 and 25).

Gophers may be deterred from chewing underground sprinkler lines or utility cables by sur-

rounding them with 6 to 8 inches (15–20 cm) of coarse gravel 1 inch (2.5 cm) or more in diameter.

Natural control: Because no population will increase indefinitely, one alternative to a pocket gopher conflict is to do nothing, letting the population limit itself. However, by the time gopher populations level off naturally, much damage may already be done to commercial crops.

A long-term way to help prevent conflicts is a combination of natural and active control. Predators—including snakes, dogs, coyotes, long-tailed weasels, and skunks—kill gophers. In addition, attracting barn owls and other raptors, which prey on young adult gophers when they disperse, may help control a gopher population, particularly in rural areas (see "Maintaining Hawk Habitat" in Chapter 33 for information). Encouraging these species, or not discouraging them, may help control the gopher population.

Predators alone won't keep a gopher population below the levels that cause problems in gardens and landscaped areas. Before removing every gopher, they will move on to hunt at more profitable locations. However, when combined with the other control techniques described here, natural control can contribute to overall control.

Lethal Control

Many species of pocket gophers in the Pacific Northwest are in decline and are, or soon will be, of conservation concern (see "Legal Status" and "Mazama Pocket Gopher Conservation"). The presence of a species of con-

1. Locate an area of recent gopher activity by looking for fresh mounds.

2. Locate the gopher's main tunnel using a probe constructed from narrow-diameter pipe, rebar, wooden dowel, or sharpened stick. The main tunnel can be found by probing 8 to 12 inches (20–30 cm) out from the soil-plug side of the mound. When the probe penetrates the 4- to 12-inch (10–30 cm) deep tunnel, there will be a noticeable drop of about 2 inches (5 cm).

3. Use a shovel or garden trowel to open the tunnel wide enough to set two traps facing opposite directions to intercept a gopher coming from either end of the tunnel. Scoop out a small depression for each trap so the gopher doesn't have to step up onto it to get caught. Make sure the opening around the trap is big enough so that the trap has room to close. Anchor traps with a stake and wire so they can be easily retrieved from the burrow and a predator cannot carry off the catch and trap.

4. After setting the traps, prevent light from entering the tunnel by covering the opening with a board, sod, cardboard, bucket, or large pot. Fine soil can be tamped around the edges to ensure a light-tight seal. **Note:** Some trappers feel this step isn't necessary and that gophers will enter the trap even with the tunnel open.

5. Mark the trap spot so it can be located easily and tamp down existing mounds to distinguish new activity.

6. Check the traps after 48 hours—capture in most cases occurs within 24 hours. Relocate traps if a catch is not made.

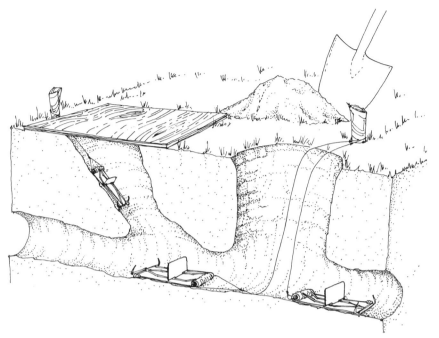

Figure 6. For most effectiveness, locate the main tunnel and place two traps, facing in opposite directions. Some trappers don't feel it's necessary to cover the tunnel with a board or soil. (Drawing by Jenifer Rees.)

cern in an area where you plan to take action—chemical, nonchemical, mechanical, or otherwise—could preclude use of this action. For information before moving forward with any type of control, contact your local fish and wildlife office, or Water, Land, and Air Protection office in B.C. (see Appendix E for contact information).

Because pocket gophers are territorial, removing one animal from an area may temporarily solve the problem. However, other gophers will eventually enter the area if attractive habitat is available.

Gophers can be lethally controlled at any time, but it is best to concentrate the effort from fall to early spring, before gophers give birth.

Before hiring someone to control gophers, read Appendix C, "Hiring a Wildlife Damage Control Company."

Trapping

Where legal, trapping can be effective method to control pocket gophers. Several types and brands of gopher traps are available from hardware stores and farm supply stores. Before choosing a trap to purchase, consider your ability to set the trap. Good hand strength is needed to set some gopher traps.

The trap most commonly used is a two-pronged pincher trap, such as the Macabee® trap, which is triggered when the gopher pushes against a flat vertical pan (Fig. 6). (Traditional Victor scissor-type® and Macabee® traps are not recommended for use on Camas pocket gophers because of that animal's large size.) Pacific Northwest trappers have had success using the Cinch® and the DK-2 Gopher Getter® trap on this larger species.

Trapping procedures are as follows:

7. When a gopher is caught, level the mounds and plug all but one small hole in that burrow system. Since pocket gophers usually maintain a closed burrow system, if a new animal moves into a vacant system or if more than one gopher is occupying the burrow, it will plug the small opening. A trap should then again be set in this system.

If a catch wasn't made, one or more of the following may apply:

• The trap was placed in the wrong location, or the trigger didn't work properly.
• The tunnel was overly disturbed.
• The gopher detected the trap. (When this happens, the gopher will burrow under it or fill the trap with loose dirt.)
• The gopher moved to another area in its territory.
• All the gophers in the area have been captured.

Poisoning

Poisoning pocket gophers is the least preferred method of control due to the possibility of poisoning non-target species, domestic animals, or humans.

If toxic gopher bait is used, follow all label directions. Place the bait well below the soil surface and in the main tunnels, taking care not to spill any on the ground surface. A funnel is useful for preventing spillage. After placing the bait in the main tunnel, cover the hole with sod, rock, board, or some other material to exclude light and prevent dirt from falling on the bait. Several bait placements within a tunnel system will increase success. Level existing mounds to distinguish new activity.

Gas cartridges (smoke bombs) are unreliable. Their effectiveness is probably compromised by the extensive nature of gopher tunnel systems and because gas diffuses in soil. Gophers will seal off their burrows in seconds when they detect smoke or gas.

Chewing gum placed in burrows is often suggested as a control technique, however it is ineffective. In a study at U.C. Davis, six gophers were offered Juicy Fruit® gum over a period of 12 days. Two gophers consumed nearly a stick of gum per day, and appeared healthy throughout the study and the 10-day post-treatment observation period.

Flooding

Pocket gophers can easily withstand normal garden or home landscape irrigation, but flooding can sometimes be used to force them from their burrows where they can be dispatched with a shovel or caught by a dog. The entire tunnel system will need to be quickly and completely flooded to evict its tenets. Five-gallon buckets of water poured in the hole will flood the area more quickly than a running hose.

Flooding has the greatest chance of succeeding if gophers are invading the property for the first time. Where they are already well established, their systems are too extensive.

For humane reasons, concentrate the effort in late winter and early spring, before gophers give birth. Be careful when attempting to flood out a gopher near a building; doing so could damage the foundation or flood the basement or crawl space.

Shooting

Since pocket gophers spend most of their time underground, shooting is impractical unless you have the time and patience to wait for one to forage at or near the surface. For safety considerations, shooting is generally limited to rural situations and is considered too hazardous in more populated areas, even if legal.

Follow-up

Once pocket gophers have been eliminated, monitor the area on a regular basis for signs of their return. Level all existing mounds so fresh mounds can be seen easily.

Be prepared to take immediate control action if new gophers do arrive. It is easier, safer, cheaper, and less time-consuming to control one or two gophers than to wait until the population increases.

Gophers are here to stay, and extermination is impractical, especially if your property borders wildlands or another area that historically has served as gopher habitat. While you may be able to remove the existing population of gophers or drive them elsewhere temporarily, other gophers will quickly move into vacated areas if conditions for them are present and nearby gopher populations are high.

Public Health Concerns

Gophers are not considered to be a significant source of any infectious disease transmittable to humans or domestic animals.

Legal Status

Because legal status, trapping restrictions, and other information about gophers change, contact your local, state, or provincial wildlife office, or visit their Web site for updates. (See Appendix E for contact information and where to access the state and provincial laws mentioned below.)

Oregon: Pocket gophers are classified as unprotected wildlife and the killing of pocket gophers is unrestricted.

Washington: The subspecies brush prairie pocket gopher *(Thomomys talpoides douglasi)* of Clark County is a state candidate species, and the Mazama (Western) pocket gopher *(Thomomys mazama)* of Thurston, Pierce, and Mason Counties is a state and federal candidate species. Because only remnant populations of these subspecies and species exist, people should not use lethal control in these areas.

Elsewhere, pocket gophers are unclassified and may be trapped or killed year-round without a permit. Although trapping gophers is not illegal, there may be restrictions on what devices may be used.

It is unlawful to release a pocket gopher anywhere within the state, other than on the property where it was legally trapped, without a permit to do so (RCW 77.15.250 and WAC 232-12-271).

Mazama Pocket Gopher Conservation: In the south Puget Sound area of Washington, many populations of Mazama pocket gopher have disappeared since the 1940s, and the species was recently listed as a candidate for protection under the federal Endangered Species Act.

Mazama pocket gophers continue to decline in numbers in part because of their small, local breeding populations. For many years the species has persisted by continually recolonizing areas after local extinctions have occurred; however, loss of habitat to development and other environmental changes have probably stopped much of this recolonization.

In addition, Mazama pocket gophers are trapped and poisoned because they sometimes damage tree seedlings and, like moles, their diggings can be considered an ugly nuisance to landowners who desire a perfect lawn. While large populations of some pocket gopher and mole species can recover, the small and isolated populations of the Mazama pocket gopher can be completely lost.

If Mazama pocket gophers are to persist in the south Puget Sound area, they will require protection and lands where management is compatible with their needs. In addition, because Mazama gophers occupy grassy areas near homes and private property, a heightened level of tolerance will be required from those people who share their territories.

British Columbia: Pocket gophers are not considered furbearing animals under the Wildlife Act, and there are no legal trapping or hunting seasons for these species. The Northern pocket gopher is, however, listed under Schedule "B," and so may be captured or killed for the purposes of protection of property by the owner or occupant of the property, or by his/her spouse, parent, guardian, or child.

The Northern pocket gopher, subspecies *T. talpoides segregatus,* is designated as a Red-listed subspecies, but can still be killed to protect personal property. See Chapter 1 for listing information on Canadian wildlife.

CHAPTER 19

Porcupines

Porcupines (*Erethizon dorsatum,* Fig. 1) are heavy-bodied, short-legged, slow-moving rodents with a waddling gait. They range in length from 24 to 36 inches (60–90 cm), including their 6- to 8-inch (15–20 cm) long tail. They vary in color from black to light yellowish brown.

Porcupines are best known for the barbed-tipped quills that cover most of their bodies. Contrary to common belief, the quills of the porcupine are not thrown, but do easily detach after contact with flesh or fur. A porcupine is unlikely to cause any harm unless it is provoked.

In addition to serving as a defense mechanism, the quills, along with fur, guard the animal from rain, snow, and the sun's rays. (See "The Porcupine's Protective Quills" for additional information.)

Porcupines occur in a variety of habitats throughout the Pacific Northwest, but are most common in forested areas. They do not occur on Vancouver and Queen Charlotte Islands in British Columbia, or on many islands in western Washington.

Figure 1. Porcupines are the second largest members of the rodent family in North America. Only the beaver is larger. Their fur, quills, and a thick layer of body fat keep them warm during the winter. (From Christensen and Larrison, Mammals of the Pacific Northwest: A Pictorial Introduction.*)*

Facts about Porcupines
Food and Feeding Habits

- Porcupines are plant-eaters (herbivores). During the spring and summer, porcupines eat buds, tender twigs, leaves, flowers, berries, nuts, and other vegetation, including garden crops.
- During winter, porcupines eat mainly evergreen needles and the inner bark (cambium layer) of trees. Woody shrubs, including blackberry, raspberry, and

elderberry, are also included in their winter diet.
- Supported by their buoyant hollow quills, porcupines will swim out to water lilies and other aquatic plant foods.
- A porcupine's plant diet is low in sodium (salt) and they seek out antlers, bones, salt-enriched soils, and other sources of sodium.

Den Sites
- Porcupines den in rock or earth caves, hollow trees, logs and logging debris, as well as in other animal burrows and under buildings.
- Den use by porcupines is not mandatory, and some animals remain outside all year. This is especially true in areas where evergreens are abundant and no other suitable crevices are available.
- Porcupines use the same dens and resting sites year after year.
- Porcupines are usually solitary, but may den together in cold winter areas.

Reproduction
- Porcupines breed in late fall, and after a seven-month gestation, one or occasionally two offspring are born in May or June.
- Mating is conducted by the conventional method of most mammals, while the female's quills are relaxed.
- The young are born with their bodies covered with long hairs and softened quills. Within hours the quills dry and serve as protection.
- The young stay close to their mothers throughout the summer, learning about den sites and food trees, but toward the end of summer they start to spend more time apart.

- By October, when females mate again, the young wander off to face the winter alone.

Mortality and Longevity
- Common causes of mortality include disease, winter stress, injury, and predation.
- Predators of porcupines include cougars and fishers, less often coyotes, wolves, bobcats, lynx, and great horned owls.
- Humans undoubtedly kill more porcupines with poisons, guns, traps, clubs, and automobiles than predators do.
- Porcupines rarely live more than three years in the wild.

Viewing Porcupines

Porcupines are solitary animals, and except during breeding season or for a female with her young, they are rarely seen together.

Porcupines do not hibernate; however they often remain in their dens for several days during bad weather. They generally forage during the night, but on a warm winter day they may be seen wandering on the ground or climbing in a tree.

Porcupines use their long curved claws and bristled tail for climbing. Stiff, backward-pointing bristles on the tail underside are pressed against tree bark, anchoring the animal against the bark and preventing downward sliding. By moving the tail up and down prior to each claw-hold, and feeling the path below, the tail is used as a quick guide stick for crawling down a tree.

Wherever you spot a porcupine it will be moving slowly, and, having poor eyesight, it probably won't see you. If you stay downwind and quiet, you can usually

observe and photograph the animal at your leisure.

Before engaging in a fight, a porcupine will warn its attacker by displaying its quills and emitting threatening vocalizations. It will use its quills only as a last resort for defense.

Trails and Tracks
Porcupines use regular trails through vegetation and snow, often leaving a distinct path from their dens to their feeding areas. The tail quills leave a pattern on the ground between the prints that resembles the marks of a broom on dirt.

The porcupine's slightly pigeon-toed tracks made in tall grass will probably be too faint to read, but the furrow made by the wide, low-slung body may be more than a foot wide and remain visible for several hours after the porcupine's passing.

Droppings
Porcupine pellets are usually oblong, yellowish brown, and deposited in large numbers. Each pellet is ¾ to 1 inch (2–2.5 cm) long, about ⅜ inch (1 mm) in diameter, and sometimes slightly curved. In summer, when more succulent vegetation is available, the pellets may be joined together by thin strands.

Look for droppings at the base of a tree in which the porcupine feeds in or in a den where the porcupine rests. The droppings have a rather sweet, resinous odor, and a large accumulation can be smelled several yards (meters) away.

Because pellets are deposited on deer trails, which porcupines frequently use, they're often mistaken for deer droppings.

The Porcupine's Protective Quills

The best-known characteristic of the porcupine is its needle-sharp quills, numbering about 30,000 per individual. The quills are modified hairs 1 to 5 inches (2.5–13 cm) long, generously mixed with the fur on the porcupine's back and tail. The longest quills are located on the porcupine's rump; the shortest are found on the face. As porcupines climb up and down trees, their bellies and underarms continually rub against the trunk, resulting in an absence of quills in these areas. The quills are tipped with tiny barbs that overlap like roof shingles; these are what prevent the quills from being easily pulled out once they are embedded in the skin of the attacker.

When threatened, the porcupine draws up the skin on its back to lift the quills and point them in all directions. The porcupine tries to keep its back toward the attacker, waving its barbed tail back and forth in a threatening manner (Fig. 2).

An encounter with a porcupine can be painful. Some unfortunate carnivores have also starved to death because a mouthful of quills prevented them from eating. Once embedded, the hollow, airtight quills expand as a result of being warmed by the victim's body heat, causing the barbs to dig in more firmly. The immediate result is intense pain, followed by throbbing and swelling. In an effort to remove the quills, the predator may cause the barbed quills to work deeper into tissue. If stuck with someone else's quills—during mating or a conflict between males—porcupines are adept at removing them with their teeth and paws.

Quills have other uses besides defense: Porcupines spend a lot of time in trees, and their quills help cushion their occasional falls. The air-filled quills also insulate them from winter cold, and act like a raft, making the animal buoyant when it swims to feed on aquatic plants or to escape an enemy. When the porcupine is relaxed, the quills lie flat and point backward, protecting the animal as it passes through underbrush.

Predators have different means of killing porcupines. The fisher *(Martes pennanti)* circles around the porcupine and makes repeated swift attacks at its face and exposed head with its sharp teeth and foreclaws. After a half-hour or more of repeated blows, the porcupine tires and dies. A fisher may also drive a porcupine out on a small limb to get at its soft, unprotected underparts. The porcupine is then eaten, leaving an empty, quill-covered skin. Lynx, wolves, and coyotes also use the circling method described earlier.

When an automobile strikes a porcupine, the quills may be embedded in a tire, work their way in, and cause the tire to go flat. If you run over a porcupine, it is wise to stop immediately and withdraw any quills while they can still be easily pulled out.

Porcupine quills are used by Native Americans to make quill boxes, jewelry, and other artwork. Natives used to kill porcupines for their quills, but today they corner the animal and then tap its back with a Styrofoam paddle to collect needed quills.

See "Public Health Concerns" for information on how to deal with confrontations between porcupines and domestic dogs.

Den Sites

A porcupine den is easy to recognize because the porcupine leaves its droppings in the den; over time these accumulate into a thick layer in the bottom. The freshness of droppings is a clue as to how recently the den has been used.

Feeding Areas

Often a porcupine will chew bark from the upper portions of a tree trunk that is next to a limb where the animal sits while eating. The bark will be neatly chewed away, leaving large patches of exposed wood. On large, heavily barked trees, pieces of the outer bark,

Figure 2. When threatened, the porcupine will keep its back toward the attacker, waving its barbed tail back and forth in a threatening manner. (From Larrison, Mammals of the Northwest: Washington, Oregon, Idaho, and British Columbia.*)*

½ to 1 inch (1.5–2.5 cm) wide, will be found at the base. Prominent horizontal tooth marks, ⅛ inch (3 mm) wide, are evidence of porcupine use.

Porcupines also chew off branches up to ½ inch (1.3 cm) in diameter. The chewed twig ends are distinctive; they are cut at a diagonal and show tiny parallel rows of bites. Uneaten twigs and branches provide important winter food for deer, elk, rabbits, hares, and other mammals.

Note that tree squirrels also bite off twigs, dropping them purposely to the ground so they can remove the seeds to store in their winter larders. Porcupines tend to use larger trees than squirrels do, and if the feeding activity occurs only at night, chances are it's a porcupine . . . or a mountain beaver (see Chapter 14).

Rest Trees

Porcupines spend a lot of time resting in trees called "rest trees." These trees may be deciduous or conifer species, but a common characteristic is a dense crown that provides cover and protection from predators. Use of a rest tree is made evident by the presence of broken quills, recent droppings, and the pungent odor of urine at the base of the trunk.

Calls

Porcupines make moans, whines, squeaks, and snorts, depending upon the circumstance. Grunts, groans, and high-pitched cries may be heard for ¼ mile (0.4 km), especially in the fall mating season.

Preventing Conflicts

Porcupines are not aggressive animals but they can be very persistent when they find something they want to eat. Their loud gnawing noises can wake you from a sound sleep, and when scared away they often return shortly.

To obtain needed minerals in their diets, porcupines chew tool handles, boat paddles, shoes, steering wheels, and other objects impregnated with human perspiration. Tires, hydraulic lines, and electrical wiring coated with road salt may also be chewed. Plywood has a high glue content, and is often targeted by porcupines. Any object impregnated with urine also attracts porcupines because of its salt content.

When feeding on tree bark, porcupines can kill or damage trees, but damage is seldom significant and control is unnecessary. Porcupines can eat the terminal buds of newly planted conifers, but large-scale damage is rare.

To prevent conflicts or remedy existing problems:

Protect trees: To prevent porcupines from climbing trees, install a metal collar or a loosely fitting metal or heavy plastic cylinder around the tree above the expected maximum level of snowpack. (For an example, see "Preventing Conflicts" in Chapter 21.) Porcupines rarely cross from tree to tree above the ground.

Protect seedlings: Several types of small barriers are available to protect small plants and the growing tips of trees. (For detailed information, see "Preventing Conflicts" in Chapter 7.)

Commercially available or homemade chemical repellents designed for deer and rabbits may also repel porcupines. These are sprayed or painted on the plants subject to damage, and then renewed occasionally to remain effective. (For detailed information on repellents, see "Repellents" in Chapter 7.)

Protect crops: Fencing orchards, gardens, and other vulnerable areas can reduce porcupine damage. Porcupines climb wire and wood fences, but will be discouraged by two strands of overhanging wire installed at the top of the fence, at a 65-degree angle outward. Several electric-fence designs will also keep porcupines out of an enclosed area (see "Preventing Conflicts" in Chapter 21 for examples).

Two electrified wires, one 4 inches (10 cm) and the other 8 inches (20 cm) above ground, will also prevent porcupines from entering an area. (For detailed information on electric fencing, see "Electric Fences" in Chapter 7.)

Protect structures and other objects subject to gnawing: The smell and taste of common wood preservatives may repel porcupines when applied to exterior plywood. Avoid using wood preservatives that are metal-salt solutions, as these will attract porcupines.

Commercially available liquid taste repellent, such as Ropell®, may be effective at preventing damage to buildings, car hoses, and other equipment. Spray a liberal amount on all surfaces you have found them gnawing. (For detailed information on repellents, see "Repellents" and "Scare Tactics" in Chapter 7).

A salt lick may satisfy the porcupines' need for salt, and keep them from seeking it elsewhere.

Lethal Control

Nonlethal techniques to control problem porcupines are always the preferred approach. But in some cases of extensive damage, or where nonlethal techniques are ineffective on individual problem animals, lethal control techniques must be considered.

Removal of a porcupine by any means is probably a short-term solution since other porcupines are likely to move in if attractive habitat is available and other porcupines occur in the area.

In rural areas, you may be able to shoot a porcupine if it can be done safely and if local regulations allow it. There are no legally registered toxicants to use in controlling porcupines.

See Appendix A for information on trapping porcupines.

Public Health Concerns

Porcupines are not a significant source of any infectious disease that can be transmitted to humans or domestic animals.

If your pet has lots of quills, quills inside the mouth or throat, or one that has broken off under the skin, take it to the veterinarian as soon as possible to ensure that quills are removed correctly and with as little discomfort to the pet as possible.

If a veterinarian isn't available, cut off the ends to release pressure in the hollow quills. To soften quills and ease their removal, thoroughly wet them with a mix of 12 ounces of vinegar and two tablespoons of baking soda. Restrain the pet, then, with pliers, grasp each quill as close to the skin as possible and work it out steadily but slowly. Because of the barbs, yanking a quill can cause further damage and pain. Examine your pet's entire body for hidden quills so you get them all. Allow wounds to bleed as a method of self-cleansing, provided the bleeding is not excessive. Wash the wounds with warm, soapy water, rinse with warm water, and dab with antibiotic ointment. Watch the wounds for three or four days and take the pet to the veterinarian if you see any sign of infection.

Humans who have embedded quills should consult a physician to determine a course of treatment.

Legal Status

Because legal status, trapping restrictions, and other information about porcupines changes, contact your local, state, or provincial wildlife office, or visit their Web site for updates. (See Appendix E for contact information and where to access the state and provincial laws mentioned below.)

Oregon: Porcupines are classified as unprotected wildlife, so killing porcupines is unrestricted.

Washington: The porcupine is unclassified and may be trapped or killed year-round without a permit. Although trapping porcupines is not illegal, there may be restrictions on what devices may be used.

It is unlawful to release a porcupine anywhere within the state, other than on the property where it was legally trapped, without a permit to do so (RCW 77.15.250 and WAC 232-12-271).

British Columbia: Porcupines are not considered game animals under the Wildlife Act, and there are no legal trapping or hunting seasons for this species. The porcupine is, however, listed under Schedule "B" of the Designation and Exemption Regulation of the Wildlife Act, and so may be captured or killed for the purposes of protection of property by the owner or occupant of the property, or by his or her spouse, parent, guardian, or child.

Rabbits and Hares

Three species of rabbits are native to the Pacific Northwest and two others have been introduced to the area. The Pacific Northwest is also home to three species of hares, two of which are called "jackrabbits."

Rabbits differ from hares in that female rabbits give birth to blind, hairless young that require considerable attention for their first two weeks of life. Hares are born fully furred with their eyes open, and can hop about within hours of their birth. In addition, hares are generally larger than rabbits, with longer ears that allow them to detect enemies from greater distances, and longer legs to outrun them in open country.

Figure 1. The Nuttall's cottontail rabbit (also known as the mountain and Oregon cottontail rabbit and shown here) is found in eastern Oregon, eastern Washington, and southern British Columbia. The Eastern cottontail, which looks similar, was introduced to the Pacific Northwest as a game animal in the 1930s. (From Christensen and Larrison, Mammals of the Pacific Northwest: A Pictorial Introduction.*)*

Facts about Rabbits and Hares

Food and Feeding Habits

- From spring to fall, rabbits and hares eat grass, clover, wildflowers, weeds, and farm and garden crops. In winter, their diet shifts to buds, twigs, bark, conifer needles, and practically any green plant.
- In dry interior areas, Nuttall's cottontails climb sloping tree trunks to access green, dew-laden vegetation.
- Rabbits and hares re-ingest their droppings to further digest the material, a process called coprophagy. After returning to their shelters, they pass soft pellets containing undigested vegetation, and then eat these at a more leisurely pace, later passing hard pellets.

Nest Sites and Shelter

- Females of most rabbit and hare species create a shallow, bowl-like nest called a "form," and line it with leaves, grass, and fur plucked from their bellies.
- Female domestic rabbits and Pygmy rabbits excavate burrows for their shelter and den site.
- Where soil conditions and food supplies permit, domestic rabbits live in groups in large, complex burrow systems called "warrens."
- Nests and dens are located in or near brushy fencerows or field edges, brush piles, gullies containing shrubs, and landscaped areas with suitable cover.

Reproduction

- The breeding season for rabbits and hares begins in mid-February and can continue

Table 1. Pacific Northwest Rabbits and Hares

Native Rabbits

The **Mountain** or **Nuttall's cottontail** (*Sylvilagus nuttallii,* Fig. 1) averages 14 inches (35 cm) in length, is gray or brownish-gray, and has a relatively short, inconspicuous white tail. Also called the **Oregon cottontail,** it is found in sagebrush areas, weed and tall-grass patches, and orchards throughout eastern Oregon, eastern Washington, and in the Similkameen and Okanagan valleys of southern British Columbia.

The **brush rabbit** *(Sylvilagus bachmani)* is dark brown with a brown tail, and averages 13 inches (33 cm) in length. It occurs in and around brushy areas and blackberry thickets throughout western Oregon.

The **Pygmy rabbit** *(Sylvilagus idahoensis)* is the smallest rabbit in North America, measuring only 11 inches (28 cm) in length. It is slate-gray with a buff-colored tail and is found in the dense sagebrush and rabbitbrush areas of eastern Oregon and south-central Washington (where fewer than 50 of these rabbits exist).

Introduced Rabbits

The **Eastern cottontail** (*Sylvilagus floridanus,* back cover) was introduced to several areas in western Oregon, southwestern British Columbia, and eastern and western Washington as a game animal beginning in the 1930s. It averages 17 inches (43 cm) in length and is light brown in color; the white underside of its 2-inch (5 cm) tail is readily visible when the rabbit runs. It is commonly seen along roads, brushy fencerows, and blackberry thickets in and around areas where it has been introduced.

The **domestic rabbit** *(Oryctolagus cuniculus)* is the other introduced species. Also known as the European rabbit or Belgian hare, this single species is the ancestor of all domestic rabbits (about 80 varieties!). The domestic rabbit is considerably larger than other Pacific Northwest rabbits, measuring 20 to 30 inches (50–75 cm) in length. It has black, white, brown, or multicolored fur, and is most frequently seen in the San Juan Islands of northwest Washington where it was first introduced in 1900, although it is spreading into other areas where it has been released. Feral populations exist on Sidney, James, and Triangle Islands in British Columbia; small localized populations also exist on Vancouver Island and the lower mainland.

Native Hares and "Jackrabbits"

The **snowshoe hare** *(Lepus americanus)* is found in forested areas throughout most of the Pacific Northwest except the coastal islands of British Columbia. It has large feet, relatively small ears, and ranges from 15 to 17 inches (38–45 cm) in length. Also known as the **varying hare,** it is reddish-brown in summer and white in winter (except for its nose and ear tips, which are black year-round). At low elevations the snowshoe hare remains brown throughout the year.

The **white-tailed hare** *(Lepus townsendii,* Fig. 2) occurs in grasslands and lower mountains in eastern Oregon, eastern Washington, and perhaps the Okanagan Valley of southern British Columbia. Adults average 23 inches (58 cm) in length, have large ears, and are grayish brown in summer and white or nearly all white in winter.

The **black-tailed hare** *(Lepus californicus* Fig. 2) is found in sagebrush and grasslands of eastern and southwestern Oregon, and eastern Washington. Adults average 21 inches (53 cm) in length, are grayish to brownish, and have large ears tipped with black. They do not have a white winter coat.

Both the white-tailed hare and the black-tailed hare are referred to as **jackrabbits.**

through late summer. Famous for their reproductive abilities, rabbits have a 30-day gestation (pregnancy) period, and have several litters containing four to eight young each year.

- For about two weeks, the mother rabbit stays away from the nest so as not to show predators the location of the young, returning only at dusk and dawn to nurse and lick her young clean.
- At two weeks of age the young rabbits begin to eat vegetation; at four to five weeks old they are feeding alongside their mother.
- Hares give birth in 35 to 47 days, depending on the species. They have two litters per year (occasionally three), with litter sizes varying from one to seven.
- Hares are born fully furred, with the ability to eat and move about with their mother within hours.
- The number of rabbits and hares in a given area will increase and decrease in a cycle connected with predator populations and food availability.

Mortality and Longevity

- Adult rabbits and hares are preyed on by large hawks and owls, domestic dogs, coyotes, foxes, and bobcats. Vehicles also kill many rabbits and hares.
- The young are eaten by mink, skunks, long-tailed weasels, gopher snakes, and domestic cats; young

in nests are vulnerable to weed-wackers, lawn mowers, hay mowers, and other agricultural equipment.
- Rabbits and hares may live two or more years in the wild, but when predators are numerous and weather conditions are extreme, they seldom live more than a year.

Viewing Rabbits and Hares

Rabbits and hares are most active at dawn and dusk. Because their eyes shine yellow or red in a flash-light beam, they are easily seen at night. Rabbits and hares do not hibernate.

During the day, rabbits and hares rest under overhanging limbs, brambles, brush piles, old farm implements, and junk. In extreme cold or heat, they may seek greater protection by bur-rowing into snowdrifts or using another mammal's unoccupied burrow.

Rabbits' and hares' survival depends on their sitting still for long periods, a trait they've likely developed to avoid being seen by predators who watch for move-ment. Indeed, the most common way to find a rabbit's or hare's resting site is by accidentally scar-ing the animal from it.

Rabbits and snow-shoe hares establish safe "bolt areas," then move

Releasing Unwanted Pet Rabbits

Every few years the population of domesticated rabbits explodes in many city parks and "wild" areas around urban centers due to released or abandoned pet rabbits breeding.

Pet bunnies are purchased at Easter and other times and later released when people lose interest in them, move, feel the animal is too messy, or when it is no longer cute and cuddly. Unfortunately, domestic rabbits are not prepared to "live off the land."

Because the release often happens in or close to residential areas, these rabbits quickly begin feeding in gardens and flower-beds, causing conflicts with property owners. Disoriented rabbits crossing roads in search of their own territories cause traffic distractions and accidents. Biologists are concerned that domestic rabbits can introduce diseases into wild rabbit populations or mate with wild rabbits, undermining the vigor of wild populations.

Domestic animals should never be abandoned to fend for themselves. If you have a pet rabbit and no longer want it, take it to an animal adoption center or find a home for it by advertising or putting up signs in local pet shops and animal clinics.

Figure 2. Both the black-tailed hare and the white-tailed hare are referred to as jackrabbits. (From Christensen and Larrison, Mammals of the Pacific Northwest: A Pictorial Introduction.*)*

out in a radius from those safety points to forage, always keeping bolt areas nearby. Animals on the move are highly stressed because they are insecure until they know the safe places they can bolt to.

Traveling in 12-foot (3.7 m) leaps, the white-tailed hare can maintain a speed of 35 mph (55 km/h), with spurts up to 45 mph (75 km/h).

Trails

Rabbits and hares commonly feed and travel along human paths, rural roads, and roadsides.

Heavily used rabbit and hare trails look like 4-to 5-inch (10–12 cm) wide paths in grass, dirt, decaying vegetation, and snow. In tall grass, trails are more tunnel-like. Tall blackberry thickets are a favorite shelter, and nicely pruned little tunnels can be seen leading into these thickets.

Tracks

Rabbit and hare tracks show four toes on each print, with the hind tracks being about two and a half times as long as the front. The claws occasionally register, and on hard ground this may be the only sign of a passing rabbit or hare.

Droppings

Rabbit and hare droppings are easy to identify in any season and are found in groups of five to ten scattered on the ground in feeding areas. Droppings are generally spherical, sometimes slightly oblong or irregular, but never acorn- or capsule-shaped like those of deer and elk. Rabbit droppings are about ⅜ inch (1 cm) in diameter; those of hares are slightly larger. They are composed of light brown, sawdust-like material.

Figure 3. The front and hind feet of the snowshoe hare are bigger that those of cottontail rabbits. The hare's rear toes can spread to more than 3½ inches (9 cm) wide; their abundant hair also helps the animal run on top of snow, while its predators sink in. (From Christensen and Larrison, Mammals of the Pacific Northwest: A Pictorial Introduction.*)*

Feeding Areas

Look for the clean-cut, angled clipping-off of flower heads, buds, and young stems up to ¼ inch (6 mm) in diameter. Evidence of gnawing can be found on the stems of woody plants, blackberry canes and other brambles, and on fallen twigs and branches (Fig. 4).

A rabbit's feeding area can also be located by looking for grass, clover, plantain, and other weeds kept cropped to within an inch (2.5 cm) or two of the ground.

In areas covered with snow, young plants may be clipped off at snow height, and the smooth, thin bark of trees and shrubs may be completely girdled.

Twig clipping by rabbits and hares is sometimes confused with deer browsing. Deer damage can be identified easily: It occurs above 2 feet (60 cm) and, because deer have no upper front teeth and must twist and pull when browsing, they leave a ragged break. Rabbits clip twigs off cleanly, as if

with a knife. (See "Feeding Areas" in Chapter 7 for figures.)

Dust Baths

Rabbits and hares roll in dust baths to help rid themselves of external parasites. These are small areas of sand or dry soil, about a foot (30 cm) in diameter, that have been cleared of vegetation, possibly through the animals' scratching or through repeated use.

Calls

The large feet of the snowshoe hare make a good sound thump, and often when a family group is together, one will be lookout and thump a warning when something suspicious approaches. A rabbit or hare captured by a predator emits a loud, high-pitched scream.

Preventing Conflicts

A rabbit's or hare's appetite can get it into trouble with gardeners, landscapers, orchardists, and foresters throughout the year. The following are suggestions for preventing conflicts. For best results, use control methods at the first sign of damage. Once rabbits and hares are used to feeding in an area, all control tactics become less effective.

Fences: Fences provide the most long-term and effective way to protect plantings from rabbit and hare damage. For rabbits, a 2-foot (60 cm) tall fence, constructed with 1-inch (2.5 cm) mesh chicken wire and supported by sturdy stakes or posts every 4 to 6 feet (1.25–1.75 m) will exclude the animals from an enclosed area. Larger hares (jackrabbits) ordinarily will not jump a 2-foot (60 cm) fence unless chased by dogs or otherwise frightened. Increasing

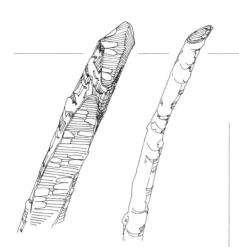

Figure 4. Signs of rabbits and hares include the clean-cut, angled clipping-off of flower heads, buds, and young stems, and gnawing on the stems of woody plants, blackberry canes and other brambles, and on fallen twigs and branches. (Drawing by Jenifer Rees.)

the above-ground height to 3 feet (90 cm) will prevent them from jumping a fence. Where deep snow is common, fences will need to be higher, or adjusted to exclude animals during winter.

Rabbits and hares are more likely to go under a fence than over it. To prevent this, place the bottom of the fence 6 inches (15 cm) underground, stake the bottom of the fence flush to the ground, or line the bottom of the fence with rock, bricks, fence posts, or similar items. Anther option is to create a 1-foot (30 cm) wide wire apron on top of the ground on the animal side of the fence. Be sure to secure the apron firmly with stakes (Fig. 6).

Electric netting (a type of electric fence) is also suitable for rabbit and hare control. It is designed for ease of installation and frequent repositioning. Electric netting is intended for temporary use at any one site, making it ideal for seasonal gardens. Daily inspections will be necessary during the first three

Figure 5. Brush piles for rabbits and other wildlife can be made in many different ways. In areas with an abundance of trees, "half-cuts" can create instant cover. (From: Link, Landscaping for Wildlife in the Pacific Northwest. *Drawing by Jenifer Rees.)*

Brush Piles for Rabbits, Hares, and other Wildlife

If your yard lacks mature shrubs, thickets and tangles, or other natural shelter, you can attract rabbits, songbirds, and other wildlife almost immediately by building a shelter of brush. Brush piles are fun to build, and the necessary materials are often found on site.

Wildlife use brush piles in different ways: The inside network of strong twigs protects rabbits and other inhabitants from sun, rain, and predators. During strong winds, birds that would ordinarily use an evergreen tree for evening shelter may instead use a brush pile located out of the wind. The outside, where sticks protrude from the pile, provides places for birds to perch, sing, preen, and catch insects. If the base of the pile contains large limbs or logs, salamanders, snakes, and lizards may hibernate under them. Ants, worms, beetles and other insects will live and feed in the rich soil beneath a pile.

For optimum wildlife watching, locate your pile where you can easily see it through a window or from an outdoor seating area. Other good locations include:

• An unused corner of your backyard or any wild portion of the landscape, particularly near a hedgerow, thicket, or group of mature trees or shrubs.
• Near a pond, irrigation canal, or birdbath.
• In view of a constructed blind used for photography or wildlife viewing.
• In an area that was recently cleared of blackberries or other brambles.
• In the shade (in areas with little rain and hot summers); in the sun (in cooler areas).

Places not to locate a brush pile for wildlife include: Where it could create a fire hazard, near a heavily traveled road, and where there is the strong possibility of attracting unwanted wildlife, such as rattlesnakes and rats.

The larger the brush pile the larger the number of wildlife species that will use it. However, a loose heap of limbs and branches 3 feet (90 cm) high and 5 feet (1.5 m) wide is adequate for most rabbits, hares, and songbirds. To create some permanent access points into the pile, use a few 6-inch (15 cm) diameter (or smaller for smaller creatures) plastic, concrete, or ceramic pipes, about 18 inches (46 cm) long, in the bottom layer.

A large, well-constructed brush pile will last many years if maintained. As time passes, top branches will collapse and brush will settle. When this happens, add branches to the top or sides, making sure small animals can still get to the center of the pile.

weeks after installation. If maintenance is poor, rabbits can do considerable damage to electric netting fences by chewing through the wires. This could result in sections becoming inoperative, which may be difficult to detect. *Note:* Before purchasing the product, ask the representative about the effects of electric netting on frogs, toads, and baby rabbits. (See "Preventing Conflicts" in Chapter 7 for information on electric fences.)

The lower 2 to 3 feet (60–90 cm) of an existing fence or gate can be covered with 1-inch (2.5 cm) wire mesh to exclude rabbits and hares. Attach the protective wire to the fence at enough points to prevent sagging, and follow the above recommendations to prevent rabbits and hares from pushing through from underneath. Use tight-fitting gates and keep them closed as much as possible.

Inspect the fence regularly to make sure animals have not dug or pushed their way under it, or worked their way over it. Once a rabbit gets into a fenced area, it may not be able to get out without being directed to a gate or other opening.

Other barriers: In some cases, protecting individual plants may be more practical than excluding rabbits and hares from an entire area. Newly planted vegetables can be protected using commercially sold cloches or 1-gallon plastic milk containers that have the bottom cut out. Placed over the seedlings, they provide protection from animals as well as late frosts.

One-inch (2.5 cm) mesh chicken wire can be cut and formed into cylinders and placed around plants needing protection

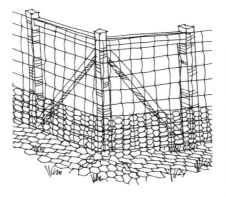

Figure 6. A fence provides the most long-term and effective way to protect plantings from rabbit and hare damage. (Drawing by Jenifer Rees.)

(Fig. 7). Bury the bottom of the cylinders 3 inches (7.5 cm) below the soil line and brace them away from the plants so animals cannot press against the cylinder and nibble through the mesh. Inspect these barriers regularly to keep the area inside the barriers clean of leaves, weeds, and other debris, which can hide damage caused by mice and voles.

Individual plants can also be protected with a variety of commercially available protectors, including nylon mesh and plastic tubes; aluminum foil has been double-wrapped around small trees with effective results. (See "Preventing Conflicts" in Chapters 3 and 25 for figures and additional examples.)

Plant rabbit-resistant plants: Protecting vulnerable plants from rabbit and hare damage within a fence, and landscaping with rabbit-resistant plants elsewhere makes an effective combination. A walk or drive through the neighborhood can give you an idea of what plants are less palatable to these animals. *Note:* When preferred foods become scarce, there are few species that rabbits will not eat.

The following list of rabbit-resistant (or close to it) plants (Table 2) is a general guide.

Rabbits sometimes will browse the plants listed and sometimes will avoid plants not listed.

Scare devices and repellents: If fencing and other barriers are impractical or undesirable, or if damage is so slight that fencing is not cost-effective, scare devices can provide **temporary control.** Visual scare tactics include mylar tape and mylar party balloons located above areas of potential damage, and pinwheels and other devices located at ground level. However, urban and suburban rabbits are often unafraid of such devices and other rabbits quickly get used to them. Ultrasonic units, which rely on sound waves to repel rabbits and hares, have not been proven effective.

A dog can help keep rabbits away, especially if it is awake at night. For information on keeping a dog outside, see "Scare Tactics" in Chapter 7 and Appendix D.

Plants may also be protected with commercially available or homemade taste repellents that render the treated plant inedible. Research has shown that repellents with putrescent whole-egg solids can reduce browsing by rabbits.

Apply repellents before damage occurs and reapply frequently, especially after a rain, heavy dew, sprinkler irrigation, or when new growth occurs. In all cases, follow the label directions for the repellent you are using. Many repellents cannot be used on plants or plant parts to be eaten by humans. (For recipes and detailed information on repellents, see "Preventing Conflicts" in Chapter 7.)

Other nonlethal control methods: Encouraging natural predators of rabbits and hares—or at least not interfering with them—may aid in

Table 2. Rabbit-resistant (or close to it) Plants for Pacific Northwest Landscapes
Plants are listed alphabetically by their botanical name.

Garden Annuals
Snapdragon, *Antirrhinum* spp.
Borage, *Borago officinalis*
Calendula, *Calendula officinalis*
Lobelia, *Lobelia erinus*
Lupine, *Lupinus* spp.
Stock, *Matthiola* spp.
Flowering tobacco, *Nicotiana* spp.
Marigold, *Tagetes* spp.

Garden Perennials
Bear's breeches, *Acanthus* spp.
Yarrow, *Achillea* spp.
Monkshood, *Aconitum* spp.
Lady's mantle, *Alchemilla mollis*
Columbine, *Aquilegia* spp.
Wild ginger, *Asarum caudatum*
Butterfly weed, *Asclepias tuberosa*
Astilbe, *Astilbe* spp.
Bergenia, *Bergenia* spp.
Bellflower, *Campanula* spp.
Daisy, *Chrysanthemum* spp.
Cyclamen, *Clyclamen* spp.
Coreopsis, *Coreopsis* spp.
Bleeding heart, *Dicentra* spp.
Bishop's hat, *Epimedium* spp.
Fleabane, *Erigeron karvinskianus*
Euphorbia, *Euphorbia* spp.
Blanket flower, *Gaillardia* spp.
Hardy geranium, *Geranium* spp.
Christmas/lenten rose, *Helleborus* spp.
Daylily, *Hemerocallis* spp.
Coral bells, *Heuchera* spp.
Douglas iris, *Iris douglasiana*
Poker plant, *Kniphofia* spp.

Gayfeather, *Liatris* spp.
Flax, *Linum perenne*
Crown-pink, *Lychnis coronaria*
Catmint, *Nepeta* spp.
Peony, *Paeonia* spp.
Beard-tongue, *Penstemon* spp.
Cape-fuchsia, *Phygelius capensis*
Solomon's seal, *Polygonatum* spp.
Lungwort, *Pulmonaria* spp.
Black-eyed Susan, *Rudbeckia hirta*
Pincushion flower, *Scabiosa caucasica*
Sedum, *Sedum* spp.
Blue-eyed grass, *Sisyrinchium* spp.
False Solomon's seal, *Smilacina racemosa*
Fringecups, *Tellima grandiflora*
Foamflower, *Tiarella trifoliata*
Starflower, *Trientalis latifolia*
Trillium, *Trillium* spp.
Johnny-jump-up, *Viola tricolor*
Zauschneria, *Zauschneria californica*

Herbs
Chives, *Allium schoenoprasum*
Hyssop, *Hyssopus officinalis*
Lavender, *Lavandula* spp.
Lemon balm, *Melissa officinalis*
Mints, *Mentha* spp.
Oregano, *Origanum vulgare*
Rosemary, *Rosmarinus officinalis*
Sage, *Salvia officinalis*
Thyme, *Thymus* spp.

Ground Covers/Sub-shrubs
Kinnikinnik, *Arctostaphylos uva-ursi*
Ajuga, *Ajuga* spp.

Point Reyes creeper, *Ceanothus gloriosus*
Cotoneaster, *Cotoneaster* spp.
Heather, *Erica* spp.
Wild strawberry, *Fragaria* spp.
Sweet woodruff, *Galium odoratum*
Wintergreen, *Gaultheria* spp.
Twinflower, *Linnea borealis*
Honeysuckle, *Lonicera* spp.
Creeping Oregon-grape, *Mahonia repens*
False lily-of-the-valley, *Maianthemum dilatatum*
Oxalis (wood sorrel), *Oxalis oregana*
Emerald carpet, *Rubus calycinoides*
Lamb's ears, *Stachys byzantina*
Piggyback plant, *Tolmiea menziesii*

Sub-shrubs
Cascade Oregon-grape, *Mahonia nervosa*
Salal, *Gaultheria shallon*
Santolina, *Santolina* spp.
Germander, *Teucrium chamaedrys*

Bulbs and Corms
Nodding onion, *Allium cernuum*
Crocosmia, *Crocosmia* spp.
Fairywand flower, *Dierama* spp.
Gladiola, *Gladiolus* spp.
Iris, *Iris* spp.
Daffodils, *Narcissus* spp.
African corn-lily, *Ixia maculata*

reducing plant damage. It is common to provide perches for owls and hawks in some commercial areas to control rabbits, hares, and small rodents (see "Maintaining Hawk Habitat" in Chapter 33).

Another form of control is to remove brush piles, weed patches, rock piles, and other debris where rabbits and hares live and hide. Before doing this, consider the potential impact on other desirable wildlife species. (Removing cover will probably have little effect on jackrabbits because they can use cover that is often great distances from feeding sites.)

Live Trapping Rabbits and Hares

Trapping and moving wild rabbits and hares several miles away has appeal as a method of resolving conflicts because it is perceived as giving the "problem animal" a second chance in a new home.

Unfortunately, the reality of the situation is quite different.

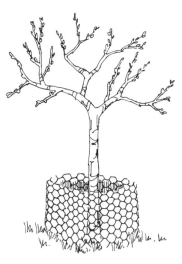

Figure 7. A cylinder of chicken wire, hardware cloth, or drainage pipe can protect young trees from rabbit damage. Tree prunings can be left as a decoy food source during the winter, as rabbits much prefer twigs and buds to tree bark. (Drawing by Jenifer Rees.)

Wildlife Rehabilitators and Wildlife Rehabilitation

Wildlife rehabilitation involves caring for injured, ill, displaced, and orphaned wild animals—from bats to wolves to eagles to woodpeckers—with the goal of releasing physically fit and psychologically sound animals back into their natural habitat. Each animal is examined, diagnosed, and treated through a program of veterinary care, hospital care, feeding, medicating, physical therapy, exercising, and prerelease conditioning.

For rehabilitation to be deemed successful, released animals must be able to truly function as wild animals. This includes being able to recognize and obtain the appropriate foods, select mates of their own species and reproduce, and show the appropriate fear of potential dangers (people, cars, dogs, etc.). To accomplish this, releases are planned for appropriate weather, season, habitat, and location.

Some animals brought into wildlife rehabilitation centers, of course, are not releasable. Some of these animals can provide valuable research information and some are suitable as educational aids; others may need to be euthanized.

Some people advocate for "letting nature take its course," indicating that injured, ill, and orphaned wild animals should be allowed to meet their natural fate. However, records indicate that the majority of distressed animals handled by rehabilitators are suffering not because of "natural" occurrences, but because of human intervention. Some of these are accidental, some are intentional, and many are preventable—such as those by vehicles, mowers, pets, high-voltage wires, firearms, traps, poisons, and oil spills.

Permitted, trained rehabilitators are a link in the network of people and agencies working with wildlife. Some are involved in research, captive propagation, and reintroduction projects. Many are involved in public education, exposing both children and adults to biological facts, ecological concepts, and a responsible attitude toward all living things.

Because most rehabilitators are swamped with injured and orphaned animals during spring and summer months, they sometimes cannot take in animals for care. Many wild animals found by the public do not need to be rescued. The public can help by always consulting a rehabilitator, delaying intervention if possible, and working toward a solution that does not necessitate the handling of the animal.

Wildlife rehabilitation is a profession that is licensed by state and federal agencies. Most rehabilitators, however, are volunteers and pay any expenses out of their own pockets. Typically, their capability (both financial and timewise) is limited and the demand is great, given all the calls from the public for assistance. Your local wildlife office keeps a list of rehabilitators and can tell you which ones serve your area, or you can look under "Animals" or "Wildlife" in your phone directory.

The animals typically become disoriented, which results in them getting hit by a car or eaten by a predator. If they remain in the new area, they may cause similar problems there, or transmit diseases to other animals in the area. If a place "in the wild" is perfect for rabbits and hares, they are probably already there. It isn't fair to the animals already living there to release another competitor into their home range to the detriment of both of them.

In many cases, moving rabbits and hares will not solve the original problem because others will replace them and cause similar conflicts. Hence, it is often more effective to use the above recommendations for making the site less attractive to rabbits and hares, than to constantly trap them.

Trapping also may not be legal in some cities; check with local authorities. Transporting animals without the proper permit is also unlawful in most cases (see "Legal Status").

Biologists might want to release native rabbits or hares if there was a need for population recovery. An event leading to this scenario might be a major disease outbreak. The justification would be if there was sufficient habitat to support a viable population in an unoccupied area. Such a project would have to be under the supervision of a federal, state, or provincial wildlife biologist.

For additional information on trapping rabbits and hares, see Appendix A.

Lethal Control

Lethal control may become necessary when all efforts to dissuade problem rabbits and hares fail.

However removing these animals rarely gives a lasting solution since survivors have larger litters, and others will resettle favorable habitats.

Shooting has traditionally been the primary form of controlling rabbits and hares and is effective in eliminating small, isolated groups of animals. For safety considerations, shooting is generally limited to rural situations and is considered too hazardous in more populated areas, even if legal. (See "Legal Status" for important information.)

Public Health Concerns

Rabbits, hares, voles, muskrats, nutrias, and beavers are some of the species that can be infected with the bacterial disease **tularemia** (see "Public Health Concerns" in Chapter 3). Tularemia can be acquired through ingesting undercooked rabbit meat or handling a dead or sick animal. Anyone handling a rabbit or hare should wear rubber gloves and wash his or her hands well when finished.

Legal Status

Because legal status, trapping restrictions, and other information about rabbits and hares change, contact your local, state, or provincial wildlife office, or visit their Web site for updates. (See Appendix E for contact information and where to access the state and provincial laws mentioned below.)

Oregon: Rabbits are classified as a "predatory animal" under the jurisdiction of the Oregon Department of Agriculture, which allows any rabbit to be controlled that is or may be "destructive to

agricultural crops, products and activities" (ORS 610-002). Any person owning, leasing, occupying, possessing or having charge of or dominion over any land or place, has the authority to control rabbits through any appropriate legal means.

The Pygmy rabbit and the white-tailed hare or jackrabbit are listed as State Sensitive species and cannot be indiscriminately killed, hunted, pursued, taken, caught, or held in possession either dead or alive. All other species of rabbits and hares are unprotected.

There are no seasons or bag limits, but a hunting license is required if pursuing rabbits or hares in a non-damage situation. Because the Eastern cottontail is not native to Oregon, it is not legal to hold this species as a pet, or to rehabilitate or release it back into the wild.

Washington: The Nuttall's cottontail rabbit, Eastern cottontail rabbit, and snowshoe hare are classified as game animals. A hunting license and open season are required to hunt them.

A property owner or the owner's immediate family, employee, or tenant may kill or trap the above species on that property if they are damaging crops or domestic animals (RCW 77.36.030). In such cases, no permit is necessary for the use of live (cage) traps. However, a special trapping permit is required from the Department of Fish and Wildlife for the use of all other types of traps.

The domestic rabbit is unclassified and may be trapped or killed year-round and no permit is necessary; however, the above restrictions on trapping devices still apply (RCW 77.15.192).

The Pygmy rabbit is a Federal Endangered Species and cannot be hunted or trapped. The white-tailed jackrabbit and black-tailed jackrabbit are Species of Concern and a special permit is required to kill them in damage situations.

It is unlawful to release wildlife anywhere within the state, other than on the property where it was legally trapped, without a permit to do so (RCW 77.15.250 and WAC 232-12-271).

British Columbia: Nuttall's cottontail and the white-tailed hare are both considered species at risk. Nuttall's cottontail is Blue-listed and the white-tailed hare is Red-listed (see "Legal Status" in Chapter 1 for listing requirements).

Nuttall's cottontails and the white-tailed hares that cause property damage may not be captured or killed without first obtaining a permit from the regional Fish and Wildlife Office.

Snowshoe hares are designated as small game. They may be legally hunted with a license during open season as defined by regulation. Snowshoe hares are also listed under Schedule "B," which means they may be captured or killed to protect a property owner or occupant, or his/her spouse, parent, guardian, or child (Wildlife Act, Designation and Exemption Regulation 2 and 3). No permit is required in such cases.

The Eastern cottontail and the domestic rabbit are listed under Schedule "C" and may be captured or killed anywhere at any time. No permit is required, but a person must have landowner/occupier permission on private property.

CHAPTER 21

Raccoons

The raccoon *(Procyon lotor)* is a native mammal, measuring about 3 feet (90 cm) long, including its 12-inch (30 cm), bushy, ringed tail. Because their hind legs are longer than the front legs, raccoons have a hunched appearance when they walk or run. Each of their front feet has five dexterous toes, allowing raccoons to grasp and manipulate food and other items (Fig. 1).

Raccoons prefer forest areas near a stream or water source, but have adapted to various environments throughout the Pacific Northwest. Raccoon populations can get quite large in urban areas, owing to hunting and trapping restrictions, few predators, and human-supplied food.

Adult raccoons weigh 15 to 40 pounds (6.75–18 kg), their weight being a result of genetics, age, available food, and habitat location. Males have weighed in at over 60 pounds (27 kg). A raccoon in the wild will probably weigh less than the urbanized raccoon that has learned to live on handouts, pet food, and garbage-can leftovers.

As long as raccoons are kept out of human homes, not cornered, and not treated as pets, they are not dangerous.

Facts about Raccoons
Food and Feeding Habitats

- Raccoons will eat almost anything, but are particularly fond of creatures found in water—clams, crayfish, frogs, fish, and snails.
- Raccoons also eat insects, slugs, dead animals, birds and bird eggs, as well as fruits, vegetables, nuts, and seeds. Around humans, raccoons often eat garbage and pet food.
- Although not great hunters, raccoons can catch young gophers, squirrels, mice, and rats.

Figure 1. Sharp, nonretractable claws and long digits make raccoons good climbers. Like squirrels, raccoons can rotate their hind feet 180 degrees and descend trees headfirst. (Cats' claws don't rotate and they have to back down trees.) A young raccoon is shown here on a branch of an oak tree. (From Christensen and Larrison, Mammals of the Pacific Northwest: A Pictorial Introduction.*)*

- Except during the breeding season and for females with young, raccoons are solitary. Individuals will eat together if a large amount of food is available in an area.

Den Sites and Resting Sites

- Dens are used for shelter and raising young. They include abandoned burrows dug by other mammals, areas in or under large rock piles and brush piles, hollow logs, and holes in trees.
- Den sites also include wood duck nest-boxes, attics, crawl spaces, chimneys, and abandoned vehicles.
- In urban areas, raccoons normally use den sites as daytime rest sites. In wooded areas, they often rest in trees.
- Raccoons generally move to a different den or daytime rest site every few days and do not follow a predictable pattern. Only a female with young or an animal "holed up" during a cold spell will use the same den for any length of time.
- Several raccoons may den together during winter storms.

Reproduction and Home Range

- Raccoons pair up only during the breeding season, and mating occurs as early as January to as late as June. The peak mating period is March to April.
- After a 65-day gestation period, two to three kits are born.
- The kits remain in the den until they are about seven weeks old, at which time they can walk, run, climb, and begin to occupy alternate dens.
- At eight to ten weeks of age, the young regularly accompany their mother outside the den and forage for themselves. By 12 weeks, the kits roam on their own for several nights before returning to their mother.
- The kits remain with their mother in her home range through winter, and in early spring seek out their own territories.
- The size of a raccoon's home range as well as its nightly hunting area varies greatly depending on the habitat and food supply. Home range diameters of 1 mile (1.6 km) are known to occur in urban areas.

Mortality and Longevity

- Raccoons die from encounters with vehicles, hunters, and trappers, and from disease, starvation, and predation.
- Young raccoons are the main victims of starvation, since they have very little fat reserves to draw from during food shortages in late winter and early spring.
- Raccoon predators include cougars, bobcats, coyotes, and domestic dogs. Large owls and eagles will prey on young raccoons.
- The average life span of a raccoon in the wild is 2 to 3 years; captive raccoons have lived 13 years.

Viewing Raccoons

Raccoons can be seen throughout the year, except during extremely cold periods. Usually observed at night, they are occasionally seen during the day eating or napping in a tree or searching elsewhere for food. Coastal raccoons take advantage of low tides and are seen foraging on shellfish and other food by day.

Figure 2. Because raccoons manipulate and moisten food items in water, there is a misconception that raccoons "wash" their food before eating it. However, when water is not available, raccoons use many of the same motions in handling food. (Oregon Department of Fish and Wildlife.)

Tips for Attracting Raccoons

Raccoons can be attracted to your property by providing natural food sources and living spaces. Suggestions include:

• Keep as much wooded property in a natural condition as you can. Den trees, and potential den trees, should be given special protection.

• Where natural dens are scarce, den boxes can be installed on trees in wooded areas near water.

• Plant and protect native trees and shrubs that provide nuts, acorns, and fruits at different times of the year. These will be eaten while on the plants and after they have fallen.

• Protect creeks, streams, and marshes on and near your property from destruction and pollution.

• Build any size pond and stock it with fish.

• Keep domestic dogs indoors or fenced.

Trails

Raccoons use trails made by other wildlife or humans next to creeks, ravines, ponds, and other water sources. Raccoons often use culverts as a safe way to cross under roads. With a marsh on one side of the road and woods on the other, a culvert becomes their chief route back and forth. Look for raccoon tracks in sand, mud, or soft soil at either end of the culvert (Fig. 3).

In developed areas, raccoon travel along fences, next to buildings, and near food sources.

Tracks, Scratch Marks, and Similar Signs

Look for tracks in sand, mud, or soft soil, also on deck railings, fire escapes, and other surfaces that raccoons use to gain access to structures. Tracks may appear as smudge marks on the side of a house where a raccoon shimmies up and down a downspout or utility pipe.

Sharp, nonretractable claws and long digits make raccoons good climbers. Like squirrels, raccoons can rotate their hind feet 180 degrees and descend trees headfirst. (Cats' claws don't rotate and they have to back down trees.) Look for scratch marks on trees and other structures that raccoons climb.

Look for wear marks, body oil, and hairs on wood and other rough surfaces, particularly around the edges of den entrances. The den's entrance hole is usually at least 4 inches high and 6 inches wide (10 x 15 cm).

Droppings

Raccoon droppings are crumbly, flat-ended, and can contain a variety of food items. The length

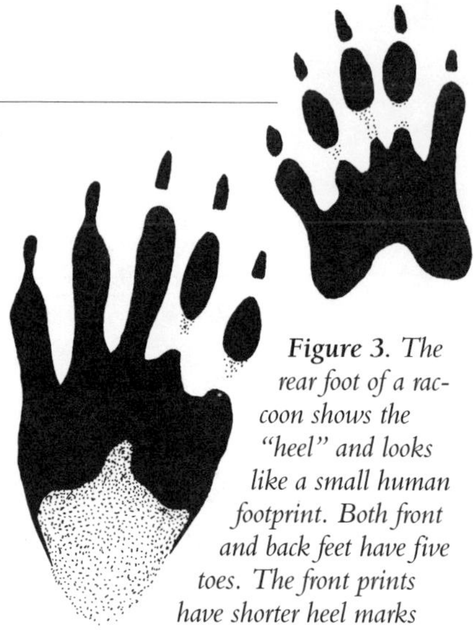

Figure 3. The rear foot of a raccoon shows the "heel" and looks like a small human footprint. Both front and back feet have five toes. The front prints have shorter heel marks and are 2 to 3 inches (5–7.5 cm) long; the hind tracks are 3 to 4 inches (7.5–10 cm) long. (From Pandell and Stall, Animal Tracks of the Pacific Northwest.)

is 3 to 5 inches (7.5–13 cm), but this is usually broken into segments. The diameter is about the size of the end of your little finger.

Raccoons leave droppings on logs, at the base of trees, and on roofs (raccoons defecate before climbing trees and entering structures). Raccoons create toilet areas—inside and outside structures—away from the nesting area. House cats have similar habits.

Note: Raccoon droppings may carry a parasite that can be fatal to humans. Do not handle or smell raccoon droppings and wash your hands if you touch any. (See "Public Health Concerns.")

Calls

Raccoons make several types of noises, including a purr, a chittering sound, and various growls, snarls, and snorts.

Preventing Conflicts

A raccoon's search for food may lead it to a vegetable garden, fish pond, garbage can, or chicken coop. Its search for a den site may lead it to an attic, chimney, or crawl space. The most effective way to prevent conflicts is to modify the habitat around your home so as not to attract raccoons. Recommendations on how to do this are given below.

Don't feed raccoons. Feeding raccoons may create undesirable situations for you, your children, neighbors, pets, and the raccoons themselves. Raccoons that are fed by people often lose their fear of humans and may become aggressive when not fed as expected. Artificial feeding also tends to concentrate raccoons in a small area; overcrowding can spread diseases and parasites. Finally, these hungry visitors might approach a neighbor who doesn't share your appreciation of the animals. The neighbor might choose to remove these raccoons, or have them removed.

Don't give raccoons access to garbage. Keep your garbage-can lid on tight by securing it with rope, chain, bungee cords, or weights. Better yet, buy garbage cans with clamps or other mechanisms that hold lids on. To prevent tipping, secure side handles to metal or wooden stakes driven into the ground. Or keep your cans in tight-fitting bins, a shed, or a garage. Put garbage cans out for pickup in the morning, after raccoons have returned to their resting areas.

Feed dogs and cats indoors and keep them in at night. If you must feed your pets outside, do so in late morning or at midday, and pick up food, water bowls, leftovers, and spilled food well before dark every day.

Keep pets indoors at night. If cornered, raccoons may attack dogs and cats. Bite wounds from raccoons can result in fractures and disease transmission.

Prevent raccoons from entering pet doors. Keep indoor pet food and any other food away from a pet door. Lock the pet door at night. If it is necessary to have it remain open, put an electronically activated opener on your pet's collar. *Note:* Floodlights or motion-detector lights placed above the pet door to scare raccoons are not long-term solutions.

Put food in secure compost containers and clean up barbecue areas. Don't put food of any kind in open compost piles; instead, use a securely covered compost structure or a commercially available raccoon-proof composter to prevent attracting raccoons and getting exposed to their droppings. A covered worm box is another alternative. If burying food scraps, cover them with at least 8 inches (20 cm) of soil and don't leave any garbage above ground in the area—including the stinky shovel.

Clean barbecue grills and grease traps thoroughly following each use.

Eliminate access to denning sites. Raccoons commonly use chimneys, attics, and spaces under houses, porches, and sheds as den sites. Close any potential entries with ¼-inch (12 mm) mesh hardware cloth, boards, or metal flashing. Make all connections flush and secure to keep mice, rats, and other mammals out. Make sure you don't trap an animal inside when you seal off a potential entry (see Appendix B, "Evicting Animals from Buildings"). For information on securing chimneys, see "Raccoons in Dumpsters and Down Chimneys."

Prevent raccoons from accessing rooftops by trimming tree limbs away from structures and

Raccoons Too Close for Comfort

If a raccoon ever approaches too closely, make yourself appear larger: stand up if sitting, shout, and wave your arms. If necessary, throw stones or send the raccoon off with a dousing of water from a hose or bucket.

If a raccoon continues to act aggressively or strangely (circling, staggering as if drunk or disoriented, or shows unnatural tameness) it may be sick or injured. In such a case, call a wildlife rehabilitator (see Chapter 20, "Wildlife Rehabilitators and Wildlife Rehabilitation"), your local wildlife office (see Appendix E for contact information), or the state patrol.

If aggressive raccoons are routinely seen in your area, prepare your children for a possible encounter. Explain the reasons why raccoons live there (habitat, food sources, species adaptability) and what they should do if one approaches them. By shouting a set phrase such as "Go away raccoon!" when they encounter one, instead of a general scream, children will inform nearby adults of the raccoon's presence. Demonstrate and rehearse encounter behavior with the children.

If a raccoon finds its way into your house, stay calm, close surrounding interior doors, leave the room, and let the animal find its way back out through the open door, window, or pet door. If necessary, gently use a broom to corral the raccoon outside. (Do not corner a raccoon, thereby forcing it to defend itself.)

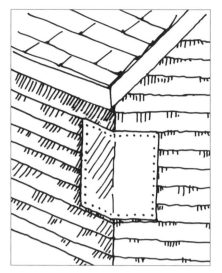

Figure 4. Raccoon access to rooftops can be eliminated by installing sheets of aluminum flashing, at least 3 feet (90 cm) square, around the corners of buildings.

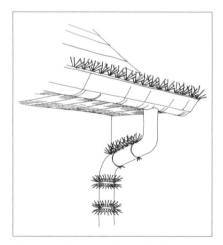

Figure 5. Commercially available metal or plastic spikes can help keep raccoons off of buildings. (Drawings by Jenifer Rees.)

by attaching sheets of metal flashing around corners of buildings (Fig. 4). Commercial products that prevent climbing are available from farm supply centers and bird-control supply companies on the Internet (see Fig. 5 and Appendix F for resources). Remove vegetation on buildings, such as English ivy, which provide raccoons a way to climb structures and hide their access point inside.

Enclose poultry (chickens, ducks, and turkeys) in a secure outdoor pen and house. Raccoons will eat poultry and their eggs if they can get to them. Signs of raccoon predation include the birds' heads bitten off and left some distance away, only the bird's crop being eaten, stuck birds pulled half-way through a fence, and nests in severe disarray. *Note:* Other killers of poultry include coyotes, foxes, skunks, raccoons, feral cats, dogs, bobcats, opossums, weasels, hawks, owls, other poultry, and disease.

If a dead bird is found with no apparent injuries, skinning it may determine what killed it. If the carcass is patterned by red spots where pointed teeth have bruised the flesh but not broken the skin, the bird was probably "played with" by one or more dogs until it died.

To prevent raccoons and other animals from accessing birds in their night roosts, equip poultry houses with well-fitted doors and secure locking mechanisms. A raccoon's dexterous paws make it possible for it to open various types of fasteners, latches, and containers.

To prevent raccoons and other animals from accessing poultry during the day, completely enclose outdoor pens with 1-inch (2.5 cm) chicken wire placed over a sturdy wooden framework. Overlap and securely wire all seams on top to prevent raccoons from forcing their way in by using their weight and claws. To prevent raccoons from reaching in at ground level, surround the bottom 18 inches (45 cm) of the pen with smaller-mesh wire.

See Figure 6 for examples of how to prevent raccoons from climbing enclosures. (See "Pre-venting Conflicts" in Chapter 23 for examples of how to prevent raccoons from digging into enclosures.)

Fence orchards and vegetable gardens. Raccoons can easily climb wood or wire fences, or bypass them by using overhanging limbs of trees or shrubs. See Figures 6 and 7 for examples of ways to prevent raccoons from climbing fences and accessing crops at ground level. Wire fences will need to have a mesh size that is no wider than 3 inches (7.5 cm) to keep young raccoons out. (See Chapter 7 for information on electric fencing.)

Protect fruit trees, bird feeders, and nest boxes. To prevent raccoons from climbing fruit trees, poles, and other vertical structures, install a metal or heavy plastic barrier (Fig. 8). Twenty-four-inch (60 cm) long aluminum or galvanized vent-pipe, available at most hardware stores, can serve as a pre-made barrier around a narrow support. (See "Predator Guards" under "Wildlife–Human Conflicts" in Appendix F and "Tree Squirrels and Bird Feeders" in Chapter 24 for more examples.) *Note:* Raccoons will attempt to use surrounding trees or structures as an avenue to access the area above the barrier.

Alternatively, a funnel-shaped piece of aluminum flashing can be fitted around the tree or other vertical structure. The outside edge of the flared metal should be a minimum of 18 inches (45 cm) away from the support. Cut the material with tin snips and file down any sharp edges.

Regularly pick up fallen bird-seed and fruit to prevent attracting raccoons.

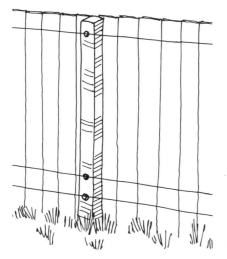

Figure 6. Install two electrified wires, 12 and 18 inches (30 and 45 cm) above ground and onto existing fence posts, poultry pen supports, and other structures, using the proper insulators. A single strand of wire may be sufficient, but two wires will provide added insurance against the animal making the climb. Run one or two electrified wires toward the top of the fence to prevent bobcats and other species from jumping the lower hot wires and making the climb. (Drawing by Jenifer Rees.)

Discourage raccoons from disturbing pond plants and other aquatic life.

Raccoons are attracted to ponds because they associate them with a food source. While a motion-activated light or sprinkler, or your shouting may scare off a raccoon, this is usually temporary. A raccoon, especially an urban raccoon, may run away the first night, walk away the second night, but, if there's no additional deterrent, by the third or fourth night the animal will be back with the light shining brightly or the sprinkler sprinkling strongly.

Always give fish a safe place to hide by constructing hiding places on the bottom of the pond. Use cinder blocks, ceramic drain tile, wire baskets, or upside-down plastic crates held in place with heavy rocks.

To prevent raccoons from disturbing aquatic plants in contain-ers, use containers that are too heavy or wide for raccoons to overturn. Securing chicken wire over the top of the containers will prevent raccoons from disturbing the soil inside.

Although it's awkward looking, small ponds can be completely covered with a barrier that can be left on permanently or removed daily. Since raccoons are nocturnal, be sure the pond is covered at night. Examples of barriers include one-inch (2.5 cm) mesh chicken wire laid over the surface and held in place with stakes—raccoons will walk on the barrier and try and go under it. (While black bird–netting is less conspicuous, raccoons and other animals can easily get entangled in it.) A wooden or PVC pipe frame covered with wire mesh can also be built to cover the pond. Maneuvering over pond plants with any of the above can be difficult.

An alternative frame can be constructed from heavy plastic lattice available from home improvement centers. Carefully cut the lattice so it fits in the pond; cut out pieces to accommodate any pond plants. Cover the lattice with bird netting (with the solid backing, animals are less likely to become entangled in the netting). The netting can be glued to the lattice using Shoe Goo® or other water-proof glue.

For larger ponds, stake 2-foot (60 cm) wide strips of chicken wire flat around the inside of the pond edge where raccoons are entering. (Cut the wire as needed to match the curvature of the pond.) Raccoons will have difficulty reaching over the wire, and will tend to not stand on it because of its instability. To camouflage and extend the life of the wire, spray it with dark-colored automobile undercoat paint or other rustproof paint.

Ponds with steep, 2-foot (60 cm) high side walls discourage raccoons from entering the water, but may be a safety hazard for small children and the elderly. These hazardous areas can be located away from paths and/or be heavily buffered with dense growths of tall marginal plants and shrubs.

Two electrified wires, 6 and 12 inches (15 and 30 cm) above ground and just back from the water's edge will deter raccoons (see Fig. 7 and "Preventing Conflicts" in Chapter 31 for examples). A single strand of wire may be sufficient, but two wires will provide added insurance against the animal making the climb. The wires can be hooked up to a switch for discretionary use; when you want to work near the wire, turn the system off. Where the barrier presents a safety problem, attach signs, short pieces of white

Figure 7. Install two electrified wires, 6 and 12 inches (15 and 30 cm) above ground around field crops and other areas needing protection. The fence can be hooked up to a switch for discretionary use; when you want to work near it, turn the system off. Where the fence presents a safety problem, install signs, short pieces of white cloth, or other material on the wire for visibility. (Drawing by Jenifer Rees.)

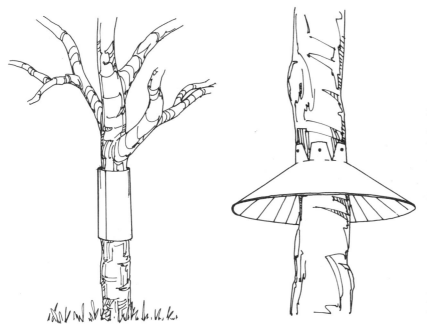

Figure 9. A raccoon guard can be secured around trees, pipes, posts, and other structures to keep raccoons from climbing. It can be made from a piece of aluminum flashing or sheet metal, held together with wire, nails, or screws, and painted to blend in. (Drawing by Jenifer Rees.)

cloth, or other material on the wire for visibility. (See Chapter 7 for information on electric fences.)

Prevent damage to lawns. Because worms and grubs inhabit areas just under well-watered sod, raccoons (and skunks) are attracted to these food sources. See Chapter 23 for ways to prevent conflicts.

Trapping Raccoons

Trapping and relocating a raccoon several miles away seems an appealing method of resolving a conflict because it is perceived as giving the "problem animal" a second chance in a new home.

Unfortunately, the reality of the situation is quite different. Raccoons typically try to return to their original territories, often getting hit by a car or killed by a predator in the process. If they remain in the new area, they may get into fights (oftentimes to the death) with resident raccoons for limited food, shelter, or nesting sites. Raccoons may also transmit diseases to rural populations that they have picked up from urban pets. Finally, if a place "in the wild" or an urban green space is perfect for raccoons, raccoons are probably already there. It isn't fair to the animals already living there to release another competitor into their home range.

Raccoons used to a particular food source, type of shelter, or human activity will seek out familiar situations and surroundings. People, organizations, or agencies that illegally move raccoons should be willing to assume liability for any damages or injuries caused by these animals. Precisely for these reasons, raccoons posing a threat to human and pet safety should not be relocated.

In many cases, moving raccoons will not solve the original problem because other raccoons will replace them and cause similar conflicts. Hence, it is more effective to make the site less attractive to raccoons than it is to routinely trap them.

Trapping also may not be legal in some cities; check with local authorities. Transporting animals without the proper permit is also unlawful in most cases (see "Legal Status"). See Appendix A for information on trapping raccoons.

Lethal Control

Lethal control is a last resort and can never be justified without first applying the above-described non-lethal control techniques. Lethal control is rarely a long-term solution since other raccoons are likely to move in if food, water, or shelter remains available.

If all efforts to dissuade a problem raccoon fail, the animal may have to be trapped (see Appendix A, "Trapping Wildlife").

While shooting can be effective in eliminating a single raccoon, it is generally limited to rural situations. Shooting is considered too hazardous in more populated areas, even if legal. (See "Legal Status" for important information.)

Public Health Concerns

A disease that contributes significantly to raccoon mortality is **canine distemper.** Canine distemper is also a common disease fatal to domestic dogs, foxes, coyotes, mink, otters, weasels, and skunks. It is caused by a virus and is spread most often when animals come in contact with the bodily secretions of animals infected with the disease. Gloves, cages, and other objects that have come in contact with infected animals can also contain the virus.

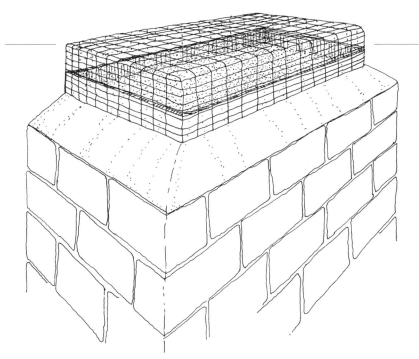

Figure 10. A commercially available chimney cap will prevent raccoons and other animals from entering the chimney. (Drawing by Jenifer Rees.)

Raccoons in Dumpsters and Down Chimneys

Raccoons are enticed by the food smells in dumpsters. When the lids are open they climb in and can't climb the slippery sides to get out. To help them escape, put a strong branch or board in the dumpster for the raccoons to climb out on.

If your disposal company leaves dumpster lids open, install a sign telling employees that it's vital to keep the lid closed so animals don't get trapped inside. Consider installing a totally enclosed trash-compacting dumpster. The trash is deposited in the front and regularly compacted.

In spring and summer, a female raccoon may be enticed into the dark, quiet, and secure environment of your chimney for a nesting place.

If you hear a large animal on the roof, or growls and whines coming from the chimney at night, there is probably a raccoon family inside. Using a powerful flashlight during the day, look for a raccoon down the chimney. (If spider webs are strung across the inside, you can be reasonably sure that no animal is using the chimney.)

The easiest solution to removing raccoons from a chimney is to wait for them to move on their own. After eight to ten weeks the female and young will leave and not return.

If raccoons need to be evicted, do not smoke them out and do not pour anything, including naphtha flakes or mothballs, down the chimney. Adult raccoons can easily climb out of a chimney, but the babies can't. The concentrated vapors can also damage the infant raccoons' mucous membranes and make an adult raccoon extremely agitated while attempting to flee from the vapors.

Instead, harass the adult female using the following methods until being there is no longer worth her effort. She will move her young to an alternate den, one by one, holding them by the back of the neck in her mouth. ***Note:*** Any time you try to evict any mother animal, there is a chance that she may leave some or all of the babies behind.

To encourage the female raccoon to leave:

1. Keep the chimney damper closed and put a loud radio tuned to a talk station in the fireplace.
2. With a short broomstick, pole, or board, bang on the underside of the damper as frequently as possible.
3. Wearing gloves, sprinkle dog, coyote, or male raccoon urine (available from farm supply centers, hunting stores, and the Internet) on a rag and wedge it in above the damper. If none of these natural repellents are available, place a bowl containing a cup of ammonia on a footstool just under the damper. If needed, open the damper 1/8-inch (3 mm). Most dampers are not airtight. Keep what deterrents you can in place 24 hours a day during a period of mild weather, and give the raccoons two to three nights to move out. On the night of departure there may be a lot of racket caused by the female raccoon's frequent climbing up and down the chimney as she retrieves her young.

In urban areas, harassment techniques may not work owing to raccoons' familiarity with humans. In such cases, call a wildlife damage control company and have them assess the situation (see "Hiring a Wildlife Damage Control Company" in Appendix C).

To make sure the eviction process was successful, shine a powerful flashlight down the chimney during the day and look for raccoons. Tap the chimney with a hard object and listen for any sounds of movement. If a young raccoon is left behind, it may be that the mother has abandoned it. In these rare cases it is best to hire a wildlife damage control company to remove the animal.

Once the raccoons are gone, promptly call a professional chimney sweep to remove any debris and to install a commercially designed and engineered chimney cap (homemade caps are often unsafe and may be a fire hazard). You can still have fires in your fireplace; however, the "cap" will keep raccoons and other wildlife out (Fig. 10).

The best prevention against canine distemper is to have your dogs vaccinated and kept away from raccoons.

Raccoons in the Pacific Northwest often have **roundworms** (like domestic dogs and cats do, but from a different worm). Raccoon roundworm does not usually cause a serious problem for raccoons. However, roundworm eggs shed in raccoon droppings can cause mild to serious illness in other animals and humans. Although rarely documented anywhere in the United States, raccoon roundworm can infect a person who accidentally ingests or inhales the parasite's eggs.

Prevention consists of never touching or inhaling raccoon droppings, using rubber gloves and a mask when cleaning areas (including traps) that have been occupied by raccoons, and keeping young children and pets away from areas where raccoons concentrate. (If washing raccoon droppings from a roof, watch where the liquid matter is going.) Routinely encourage or assist your children to wash their hands after playing outdoors. Unfortunately, raccoon roundworm eggs can remain alive in soil and other places for several months.

If a person is bitten or scratched by a raccoon, immediately scrub the wound with soap and water. Flush the wound liberally with tap water. In other parts of the United States raccoons can carry **rabies.** Contact your physician and the local health department immediately. If your pet is bitten, follow the same cleansing procedure and contact your veterinarian.

Legal Status

Because legal status, trapping restrictions, and other information about raccoons change, contact your local, state, or provincial wildlife office, or visit their Web site for updates. (See Appendix E for contact information and where to access the state and provincial laws mentioned below.)

Oregon: The raccoon is classified as a furbearer and hunting and trapping regulations apply. Property owners, persons lawfully occupying land, or their agents may trap raccoons at any time with a permit issued by Oregon Department of Fish and Wildlife (ODFW) if the raccoons are causing damage.

A raccoon may not be captured and held as a pet; however, a domestically raised raccoon may be imported from another state if accompanied by the proper documentation and a health certificate. A holding permit is required from ODFW to possess a raccoon.

Washington: The raccoon is classified as both a furbearer and a game animal (WAC 232-12-007). A hunting or trapping license is required to hunt or trap raccoons during an open season. A property owner or the owner's immediate family, employee, or tenant may kill or trap a raccoon on that property if it is damaging crops or domestic animals (RCW 77.36.030). In such cases, no permit is necessary for the use of live (cage) traps. However, a special trapping permit is required from the Department of Fish and Wildlife (WDFW) for the use of all other types of traps.

It is unlawful to release wildlife anywhere within the state, other than on the property where it was legally trapped, without a permit to do so (RCW 77.15.250 and WAC 232-12-271).

Except for bona fide public or private zoological parks, persons and entities are prohibited from importing raccoons into Washington State without a permit to do so (WAC 246-100-191).

British Columbia: Raccoons are classified as both a furbearer animal and a small game animal. They may be legally hunted or trapped with a license during open seasons as defined by regulation. Raccoons are also listed under Schedule "B," which means they may be captured or killed to protect property by the person who owns or occupies the property or his/her spouse, parent, guardian, or child (Wildlife Act, Designation and Exemption Regulation 2 and 3). No permit is required in such cases.

Importation of raccoons is not permitted except by educational institutions or scientific organizations if the Director of Wildlife is satisfied that the importation will not be detrimental to native wildlife or wildlife habitat (Wildlife Act, Permit Regulation 7).

CHAPTER 22

River Otters

River otters (*Lutra canadensis,* Fig. 1) have long, streamlined bodies, short legs, webbed toes, and long, tapered tails—all adaptations for their mostly aquatic lives. Their short thick fur is a rich brown above, and lighter, with a silvery sheen, below. Adult male river otters average 4 feet (1.25 m) in length, including the tail, and weigh 20 to 28 pounds (9–16.5 kg). Female adults are somewhat smaller than males.

Although seldom seen, river otters are relatively common throughout the Pacific Northwest in ponds, lakes, rivers, sloughs, estuaries, bays, and in open waters along the coast. In colder locations, otters frequent areas that remain ice-free in winter—rapids, the outflows of lakes, and waterfalls. River otters avoid polluted waterways, but will seek out a concentrated food source upstream in urban areas.

River otters are sometimes mistaken for their much larger seagoing cousin, the sea otter (*Enhydra lutris,* Fig. 2). However, male sea otters measure 6 feet (1.75 m) in length and weigh 80 pounds (36 kg). Sea otters are acclimated to salt water, and come to shore only for occasional rest periods and to give birth. In comparison, river otters can be found in fresh, brackish, or salt water, and can travel overland for considerable distances.

Figure 1. River otters are powerful swimmers with snakelike agility; their small eyes are adapted for seeing food items in murky or dark water. (From Larrison, Mammals of the Northwest: Washington, Oregon, Idaho, and British Columbia.*)*

Facts about River Otters

Food and Feeding Habits

- River otters are opportunists, eating a wide variety of food items, but mostly fish.
- River otters usually feed on 4- to 6-inch (10–15 cm) long, slowly moving fish species, such as carp, mud minnows, sticklebacks, and suckers. However, otters actively seek out spawning salmon and will travel far to take advantage of a salmon run.
- River otters can smell concentrations of fish in upstream ponds that drain into small, slow moving creeks, and will follow the smell to its origin, even in urban areas.
- River otters also eat freshwater mussels, crabs, crayfish, amphibians, large aquatic beetles, birds (primarily injured or molting ducks and geese), bird eggs, fish eggs, and small mammals (muskrats, mice, young beavers).
- In late winter, water levels usually drop below ice levels in frozen rivers and lakes, leaving a layer of air that allows river otters to travel and hunt under the ice.
- River otters digest and metabolize food so quickly that food passes through their intestines within an hour.

Den Sites

- River otters use dens for giving birth and for shelter from weather extremes. Birthing dens are lined with small sticks and shredded vegetation.
- Den sites include hollow logs, log jams, piles of driftwood or boulders, and abandoned lodges and bank dens made by nutria or beaver.

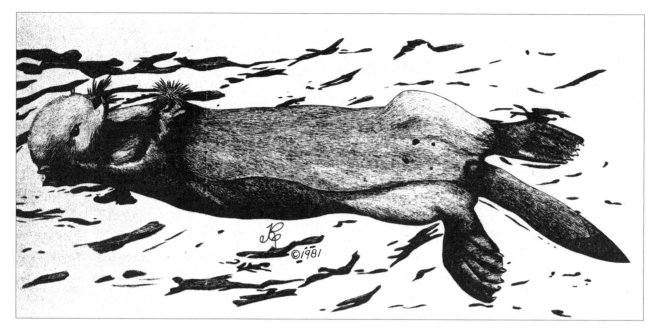

Figure 2. The river otter (shown here) is sometimes mistaken for its much larger seagoing cousin, the sea otter. (From Christensen and Larrison, Mammals of the Pacific Northwest: A Pictorial Introduction.*)*

- Dens are well hidden; those located at the water's edge will have an entry far enough below the surface to prevent it from being seen and/or frozen shut.
- River otters also den under boathouses, duck blinds, and other human structures up to ½ mile (0.8 km) away from water.

Reproduction and Family Structure

- River otters have what is called delayed implantation; the fertilized egg does not attach to the uterine wall for a period of time after breeding. Thus, gestation ranges from 285 to 375 days.
- Two to four pups are born March through May.
- Young otters begin playing at four weeks of age and learn to swim at about seven weeks of age.
- When eight to ten weeks of age, the pups begin exploring beyond their den and are introduced to solid food.

- In late fall, the pups leave to establish their own territories. During this time, wandering youngsters are seen far from water, traveling on land between lakes, ponds, and from one stream drainage to another.
- The basic social group for river otters is a female and her offspring. (Before and after breeding, male otters usually lead solitary lives.)

Mortality and Longevity

- Essentially safe from predators while in water, river otters are more vulnerable when they travel on land. Predators take mostly young river otters and include coyotes, bobcats, domestic dogs, cougars, and bears.
- Humans trap river otters to control fish predation in private ponds and commercial fish hatcheries and to prevent damage to private property. Where legal, river otters are trapped for their fur.

- The most significant impacts on river otter populations include reduced water quality from chemical pollution and soil erosion, and stream-bank habitat alteration by developments.

Viewing River Otters

River otters are active day and night; around humans they tend to be more nocturnal. Otters spend their time feeding and at what appears to be group play. They also dry their fur, groom themselves, and mark their territory by vigorously scratching, rubbing, and rolling on the ground.

River otters are active year-round, and, except for females with young in a den, are constantly on the move. They tend to follow a regular circuit that is covered in one to four weeks. Males can travel 150 miles (240 km) within a particular watershed and its tributaries in a year. A family may range 10 to 25 miles (16–40 km) in a season.

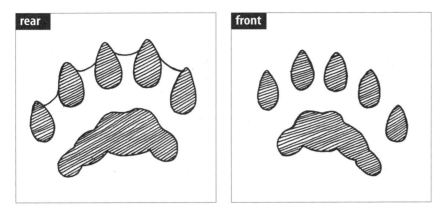

Figure 3. River otter tracks show five pointed toes around a small heel pad. Tracks are 3 to 3½ inches (7.5–9 cm) wide and 3 to 4 inches (7.5–10 cm) long. (Washington Department of Fish and Wildlife.)

To observe river otters, sit quietly on a high place (a bridge, overhanging bank or tree, or pier) above a known feeding area, trail, or slide. Find an angle from which you can avoid surface glare. A pair of polarized sunglasses and binoculars are useful. River otters are wary and their hearing and sense of smell are well developed. However, they are fairly near-sighted and they may not notice you if you stay still.

Never instigate a close encounter with river otters. They have been known to attack humans, and females with young are unpredictable.

Trails and Tracks

When traveling on land river otters walk, run, or bound. Bounding is their fastest gait. When bounding, the front and hind feet are brought toward each other causing the back to arch and the tail to be lifted off the ground. Otters make trails along the edges of lakes, streams, and other waterways. Trails often lead from one cove to another across a small peninsula, or alongside shallow rapids. Trails are 6 to 7 inches (15–18 cm) wide and may lead to slides or dens. Look for tracks in soft mud, damp sand, or fresh snow (Fig. 3).

Slides

Slides are a common sign of river otter presence and there are often several in a river otter's home range. Slides are about 1 foot (30 cm) wide and located at water's edge—frequently on islands in lakes, or in openings under bushes or brambles along creeks and streams. Slides are made in grass, dirt, sand, or snow. There is often a trail from the water's edge to the slide.

Droppings

River otters thoroughly chew their food, so their droppings contain only fine bits of fish scales, bones, and shells. The texture is oily, the smell fishy. Droppings are left on prominent spots at the water's edge, along trails, and near dens. Fresh river otter droppings are shapeless, slimy, and green; they darken with age.

Calls

River otter sounds include chirps (similar to a marmot), growls, whines, and, when alarmed, an explosive *hah!*

Preventing Conflicts

River otters are often blamed for preying on wild game fish, particularly trout. Nevertheless, studies indicate that the bulk of the river otter's diet consists of non-game-fish species.

However, river otters—particularly families containing young pups in spring—occasionally cause severe problems in fish hatcheries and private ponds. Otters also den under houses, decks, and other structures near water, and the smell of their droppings and discarded food remains can be unpleasant.

To prevent conflicts or remedy existing problems:

Eliminate access to feeding sites and other areas. Because river otters have heavy bodies and aren't jumpers, a 4-foot (1.2 m) high fence constructed with 3-inch (7.5 cm) mesh wire can keep them out of an enclosed area, such as where fish or aquaculture activities are concentrated. Because river otters are strong, fences should be sturdy and extend 6 inches (15 cm) below the surface to prevent otters from pushing under the fence. Alternatively, include a wire apron on the animal side of the fence to prevent otters from entering from underneath (see "Preventing Conflicts" in Chapters 6 and 14 for examples of such a barrier).

A double-wire electric fence, with wires set 6 and 10 inches (15 and 25 cm) above ground will also deter river otters. Such a fence can stand alone, or supplement an existing perimeter fence (see "Preventing Conflicts" in Chapter 21 for an example). A single wire can be used around docks and houseboats (see "Deer Fences" in Chapter 7 for information on electric fences).

Enhancing River Otter Habitat

People wanting to attract river otters to a property can do the following:

• Conserve wetlands and wooded areas along streams and rivers. Reducing soil erosion and preventing fertilizers and pesticides from washing into streams are important measures, even where river otters aren't likely to visit. Soil particles washed into a stream can settle when they reach slow-moving water, covering the rock, sand, or gravel that some fish need to lay their eggs and raise young. Fewer fish means less food for otters (Fig. 4).

• River otters like to use dens and wetlands created by beavers. Protecting the long-term health of the beaver population is an investment in the river otter's future.

• Retain large slash piles from forestry projects and preserve all large dead or dying trees (especially those containing cavities on the lower portion) within 500 feet (150 m) of waterways. These may be used as den sites.

• Stock your pond with fish (contact your local wildlife office for permits, recommended species, and places to purchase them). Give the fish places to hide or escape—a river otter can quickly deplete a small pond of its fish. See "Preventing Conflicts" for more information.

• Keep domestic dogs indoors or fenced. (See Appendix D for more information.)

Figure 4. Soil particles washed into a stream can settle when they reach slow-moving water, covering rock, sand, or gravel. Some fish that river otters eat need a clean substrate to lay their eggs and raise young. (Oregon Department of Fish and Wildlife.)

River otters are resourceful and will thoroughly investigate fence lines to find a way into a food source. They are known to use abandoned animal burrows as routes under fences. So, inspect fences regularly to make sure river otters have not dug or pushed their way under or worked their way over them.

Provide fish with hiding places. Give fish safe places to hide by constructing sturdy hiding places on the bottom of ponds using cinder blocks, ceramic drain tile, wire baskets made from leftover galvanized fencing, or upside-down plastic crates held in place with heavy rocks. In larger ponds, attach a group of cut conifer trees to a heavy anchor on the bottom of the pond.

Eliminate access to convenient denning sites. Close potential entries under porches, houses, sheds, and other structures with ¼-inch (12 mm) mesh welded-wire (hardware cloth), boards, or other sturdy material. Aluminum flashing, or aluminum or stainless-steel hardware cloth is recommended in saltwater areas since galvanized materials quickly corrode. (See Appendix B, "Evicting Animals from Buildings" for additional information.)

Eliminate noxious odors. Commercial odor-eliminators can be used to remove the smell of otter droppings and other debris under structures. Such products are available through hospital supply houses, drugstores, pet stores, and from the Internet using the keywords "Pest Control Supplies." If the smell is really bad, the beams and other areas under the structure may have to be cleaned with a bleach solution (1½ cups of household bleach in 1 gallon of water). Be careful of fumes.

Trapping and Lethal Control

Trapping or shooting river otters should be a last resort. Lethal control can never be justified without a serious effort to first prevent problems from recurring. Removing river otters by any means is a short-term solution since other otters are likely to move in if attractive habitat is still available.

(See Appendix A for information on trapping river otters.)

Public Health Concerns

Diseases and parasites associated with river otters are rarely a risk to humans. **Canine distemper,** a disease that affects domestic dogs, may be found in Northwest river otter populations (see "Public Heath Concerns" in Chapter 21). Have your dogs vaccinated for canine distemper to prevent them from contracting the disease.

Anyone handling a river otter should wear rubber gloves, and wash their hands well when finished.

Legal Status

Because legal status, trapping restrictions, and other information about river otters change, contact your local, state, or provincial wildlife office, or visit their Web site for updates. (See Appendix E for contact information and where to access the state and provincial laws mentioned below.)

Oregon: The river otter is classified as a furbearer and fur trapping regulations apply (OAR 635-067). Property owners, persons lawfully occupying land, or their agents may trap river otters at any time with a permit issued by Oregon Department of Fish and Wildlife if the animals are causing property damage.

Washington: The river otter is classified as a furbearer (WAC 232-12-007). A trapping license and open season are required to trap river otters.

A property owner or the owner's immediate family, employee, or tenant may kill or trap a river otter on that property if it is damaging crops or domestic animals (RCW 77.36.030). No permit is necessary; however, you must notify the Washington Department of Fish and Wildlife immediately after taking a river otter in these situations.

British Columbia: River otters are designated furbearing animals and may be legally trapped under license during open seasons as defined by regulation. There is no bag limit for river otter.

A person on their own property may trap a river otter outside of general open seasons if the river otter is a menace to a domestic animal or bird (Wildlife Act 26(2)). No permit is required in such cases.

CHAPTER 23

Skunks

Skunks are mild-tempered, mostly nocturnal, and will defend themselves only when cornered or attacked. Even when other animals or people are in close proximity, skunks will ignore the intruders unless they are disturbed.

Skunks are beneficial to farmers, gardeners, and landowners because they feed on large numbers of agricultural and garden pests. While young skunks are cute and kittenlike, they are wild animals and it is illegal to keep them as pets (see "Legal Status").

Two skunk species live in the Pacific Northwest: The **striped skunk** (*Mephitis mephitis,* Fig. 1) is the size of a domestic cat, ranging in length from 22 to 32 inches (55–80 cm), including its tail. Its fur is jet black except for two prominent white stripes running down its back. The striped skunk occurs throughout most lowland areas in Oregon and Washington. In British Columbia it occurs in the lower Fraser Valley, and the mainland east of the Coast Mountains. The striped skunk prefers open fields, pastures, and croplands near brushy fence-rows, rock outcroppings, and brushy draws. It is also seen—or its musky odor noticed—in some suburban and urban locations, particularly near sources of open water.

The **spotted skunk** (*Spilogale putorius,* Fig. 3), also known as the civet cat, ranges in length from 14 to 18 inches (35–45 cm), including its tail. Its fur is a black or grayish black, with white stripes on its shoulders and sides, and white spots on its forehead, cheeks, and rump.

The spotted skunk occurs throughout most of Oregon, west and southeast Washington, and southwest British Columbia. The spotted skunk and striped skunk use similar types of habitat, although the spotted skunk is more likely to be seen in and around forests and woodlands, and is not as tolerant of human activity as the striped skunk.

Facts about Skunks
Food and Feeding Habits

- Skunks will eat what they can find or catch. They have large feet, well-developed claws, and digging is their primary method used to obtain food.
- Animal foods include, mice, moles, voles, rats, birds and their eggs, and carcasses—also grasshoppers, wasps, bees, crickets, beetles, and beetle larvae.
- Skunks also eat fruits, nuts, garden crops, and scavenge on garbage, birdseed, and pet food.
- Skunks will roll caterpillars on the ground to remove the hairs before eating them. They will also roll beetles that emit a defensive scent, causing the beetle to deplete its scent before they eat it.

Den Sites

- Skunks use underground dens year-round for daytime resting,

Figure 1. While other animals try to camouflage themselves, the skunk advertises its identity with its bold coloration. Most predators need only one lesson to learn to avoid the skunk at first sight of black-and-white. As shown here with a badger, the striped skunk has two prominent white stripes running down its back. (From Christensen and Larrison, Mammals of the Pacific Northwest: A Pictorial Introduction.*)*

hiding, and birthing and rearing young.

- Dens are located under wood and rock piles, buildings, porches, and concrete slabs—also in rock crevices, culverts, drainpipes, and in standing or fallen hollow trees.
- Skunks may dig their own dens, but more often use the deserted burrows of other animals, such as ground squirrels and marmots.
- Dens are either permanent, or used alternately with other dens.
- Spotted skunks are excellent climbers and may use an attic or a hayloft as a den.
- Skunks do not hibernate; instead, they lower their body temperature and stay inside their dens during extreme cold, plugging the entrance with leaves and grass to insulate them from the cold.
- Female skunks sometimes share communal dens.

Reproduction

- Striped skunks breed from February through March. Spotted skunks breed from September through October and experience delayed implantation; the fertilized egg does not attach to the uterine wall for a period of time after breeding.
- In late April and May, females of both species give birth to four to five young in an underground nest lined with dried grass and other vegetation.
- At around 60 days of age, the mother leads her young out at dusk to forage and hunt. At three months old the skunks are almost full-grown and completely independent.
- Striped skunk families often remain together throughout the winter.

Mortality and Longevity

- Skunks have few predators—hungry coyotes, foxes, bobcats, and cougars, also large owls (which have little sense of smell). Domestic dogs will also kill skunks.
- Skunks also die as a result of road kills, trapping, shooting, and killing by farm chemicals and machinery.
- Striped skunks live three to four years in the wild; spotted skunks live half that long.

Signs of Skunks

Signs of skunks include their tracks, droppings, and evidence of their digging. A musky odor is another sign of their presence. A persistent smell and freshly excavated soil next to a hole under a building or woodpile indicates that a skunk may have taken up residence.

Skunks usually begin foraging after dark and are back in their dens before daylight. While striped skunks are sometimes seen during the day, spotted skunks seldom are—they may not even venture out on bright moonlit nights.

Skunks search for food along established routes and have a home range of less than 2 miles (3.2 km). Since they commonly patrol country roads looking for road-killed animals, vehicles often hit them.

When around skunks, avoid making loud noises, moving quickly, or taking other steps that could be interpreted by the skunk as a threat. If the skunk appears agitated, retreat quietly and slowly.

Skunks have poor eyesight and will often approach people who are standing still. If this happens, slowly move away from the approaching skunk.

Figure 2. Striped skunk tracks average 2 inches (5 cm) long by 1 inch (2.5 cm) wide. The tracks of spotted skunks are similar, but smaller. The long nails of the front foot are the skunk's identifying feature. (From Pandell and Stall, Animal Tracks of the Pacific Northwest.*)*

Tracks

Skunk tracks can be found in mud, dirt, or snow around den sites and feeding areas (Fig. 2). Skunk tracks look like domestic cat prints, except they show claw marks and five toes, not four. Unlike cats, skunks can't retract their claws, so each of their toe pads has a claw mark in front of it. Skunk tracks are also usually staggered, unlike domestic cat prints, which are often on top of each other.

Droppings

Look for droppings where skunks have been feeding or digging, or near a den. Droppings look like those of domestic cats and contain all types of food, from insect skeletons, to seeds or hair. Striped skunk droppings are ½ inch (1.3 cm) in diameter, 2 to 4 inches (5–10 cm) long, and usually have blunt ends. Spotted skunk droppings are similar looking, but half the size.

Getting Skunked

In other animals, musk is used for scent-marking and courtship. Only the skunks have turned musk into olfactory muscle. When an adult skunk or its young are threatened, they may emit a musky fluid from a nozzlelike duct that protrudes from the animal's anus. This fluid—nature's version of tear gas—can be discharged either in a fine mist or in a water-pistol-type stream. It has a stifling, pungent, often gagging odor that can persist for weeks and be detected over a mile away. On a still day, a skunk can discharge musk 12 feet (3 m) with good accuracy. On a windy day, spray may reach a person standing downwind 18 feet (5.5 m) away or more. Because even a few droplets of skunk spray smell so strongly, it doesn't take a direct hit to pick up the odor.

People's reaction to the odor varies greatly. Almost everyone finds it intolerable when in high concentration. Some people become violently ill. Low levels of the odor are still repugnant to most, while a few find them bearable or almost pleasant.

Because skunks have a limited supply of ammunition, they don't waste their defensive spray. A striped skunk can fire five to eight times before it has to reload, which takes about a week.

Fortunately skunks have various ways of warning when they are threatened, giving an intruder ample opportunity to back off. Dogs, however, tend to ignore this warning. That's why it's hard to find a human who has been sprayed, but easy to find a dog that has!

Contrary to popular myth, a striped skunk cannot spray over its back. When threatened it will stomp its front feet and, if the threat continues, it will make short charges with its tail raised in the direction of the threat. Next, the skunk will twist its hind end around so it is headed in the same direction as its snout. If the skunk continues to feel threatened, it will then spray.

Musk produced by spotted skunks is more pungent than that of striped skunks. However, they are less likely to spray, and will climb a fence post or a tree when threatened. When forced to, a spotted skunk will stand on its front feet with its back arched so that the spray is discharged forward (Fig 3).

The odor-bearing fluid, or musk, is amber in color, oily, and only slightly volatile. Therefore, it goes away "on its own" very slowly. However, it will go away eventually (perhaps in two to four months), even if nothing is done to get rid of the odor. This natural process is greatly slowed in areas with little ventilation and when the musk has penetrated porous materials.

If a person or pet is sprayed, the quicker you do something about it the more completely you can remove the odor. First, if eyes get irritated, flush them liberally with cold water. Next, because skunk spray is highly alkaline, counteract this by washing with mildly acidic substances such as carbolic soap, tomato juice, diluted vinegar, or the following home remedy:

- 1 quart of fresh, 3 percent hydrogen peroxide solution (old HP eventually turns into water)
- ¼ cup of baking soda (bicarbonate of soda)
- 1 teaspoon of a liquid soap that is known for its degreasing qualities

Always mix the solution in a large, open container. **A closed container can explode.** The mixture will bubble because of the chemical interaction between the baking soda and the hydrogen peroxide. Use the entire mixture while it is still bubbling. Wearing rubber gloves, apply the solution, work it into lather, and leave it on for 30 minutes.

Figure 3. The spotted skunk has white stripes on its shoulders and sides, and white spots on its forehead, cheeks, and rump. When forced to, it will adopt a defensive posture of standing on its front feet with its hind feet and tail in the air. (From Christensen and Larrison, Mammals of the Pacific Northwest: A Pictorial Introduction.*)*

Den Entrances

Look for a grass-free, smooth, 3 by 4 inch (7.5 by 10 cm) depression under a woodpile, shed, porch, or similar place. Generally, you will find only one entrance and a musky odor will be noticeable. Two-inch (5 cm) long black or white hairs may be found lodged in wood or other rough surfaces surrounding the entry.

Digs

Skunks dig in lawns and other grassy areas; usually several holes appear in the same few square yards (meters). When searching for insect grubs, skunks make small holes 1 to 3 inches (2.5–8 cm) in diameter and deep. (Such holes are similar to those made by gray squirrels.) Larger holes in rougher grass may be evidence of skunks digging for voles or other rodents. Skunks also tear apart logs and dig up nests of wasps and other insects in search of a meal.

Preventing Conflicts

Even though skunks possess a powerful spray defense, they will not spray unless surprised, cornered, harmed, or they need to protect their young. Young skunks are more likely to spray than more experienced skunks.

Occasional skunk sightings in a neighborhood need not be cause for alarm. Because skunks are nomadic, most concerns about them being under sheds, porches, and outbuildings are resolved in due time: skunks just move on.

The most effective way to prevent conflicts is to modify the habitat around your home so as not to attract skunks.

Do not feed skunks. Doing so may create undesirable situations for you, your children, your pets, and the skunks. Skunks that are artificially fed often lose their fear of humans. Artificial feeding also tends to concentrate skunks in a small area, and overcrowding can encourage diseases or parasites. Finally, these skunks might drop in on neighbors who do not want them around. These same neighbors might decide to have the skunks removed.

In addition, feed dogs or cats inside or clean up any spilled or uneaten food before dark, place indoor pet food or other food away from a pet door, and put food in secure compost containers. Also, regularly clean up bird feeding stations. (See "Preventing Conflicts" in Chapter 37 for detailed information on managing bird feeding stations.)

Prevent access to denning sites. Skunks frequently den under houses, porches, sheds, and similar places. Close off these areas with ¼-inch (6 mm) hardware cloth, boards, metal flashing, or other sturdy barriers. Make all connections flush and secure to keep mice, rats, and other small mammals out. Make sure you don't trap an animal inside when you seal off a potential entry (see Appendix B, "Evicting Animals from Buildings."). To prevent skunks from digging under a building or concrete slab, install a barrier (Fig. 4).

Remove access to shelter. Remove brush piles, lumber piles, and rock piles where skunks might live or hide. Before adopting this method, however, be aware that you are also eliminating habitat for other wildlife species, which you might want to attract.

Commercial preparations containing "neutroleum alpha," available from some pet stores, are also effective.

After washing with any remedy solution, follow with a long hot shower. Depending on the severity of the spray, you may have to repeat the process two or three times.

These solutions may be used to eliminate *most* of the skunk odor from people and pets. When washing a dog, wash the body first and then the head to keep the dog from shaking off the mixture. This will make the odor tolerable— only time will eliminate it.

Depending on the severity of the spray, clothing may be soaked in a weak solution of household chlorine bleach, ammonia, or products containing neutroleum alpha. If the clothing has been heavily sprayed, however, your best option may be to discard or burn it, because fabric will hold the skunk odor for a long time.

The above products may also be used to clean odor from inanimate objects. If the odor is inside or under your house, the area will need to be thoroughly aired out. Using fans will help.

Never use bleach or ammonia, at any dilution, on pets. Never use bleach or ammonia on materials you do not want to stain or discolor.

And remember . . . the best remedy is **Don't Get Sprayed!**

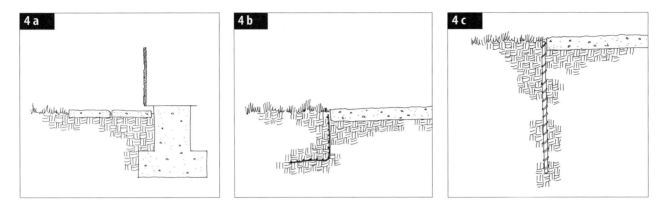

Figure 4. Various ways to install a barrier to prevent skunks (and other burrowing mammals) from digging under concrete slabs, decks, chicken coops, and similar places. To add to the life of any metal barrier, spray on two coats of rustproof paint before installation. Always check for utility lines before digging in an area. (Drawings by Jenifer Rees.)

a. *Lay large flat stones, concrete patio pavers, or ¼-inch (6 mm) hardware cloth (held in place with stakes) on the surface of the soil next to a concrete slab or wall. The barrier forces animals to begin digging farther out and they will most likely give up in the process.*

b. *Bend hardware cloth into an "L" shape and lay it in a trench so that the wire goes at least 1 foot (30 cm) below ground and 1 foot (30 cm) out from the concrete slab.*

c. *Excavate a 3 x 3 inch (7.5 cm x 7.5 cm) trench along the side of the slab or wall, and hammer 2-foot (60 cm) lengths of ⅜-inch (9 mm) rebar, spaced a few inches apart, into the ground. Cover the tops with concrete or dirt.*

Enclose ducks and chickens in a secure coop at night. A skunk may dig or otherwise find its way into a chicken coop and kill one or two small fowl, but if several chickens or ducks have been killed at one time, the predator is more likely a weasel, mink, fox, raccoon, or bobcat. If a skunk is eating the eggs of chickens or ducks, you will usually find eggs opened on one end with the edges crushed inward. A skunk cannot easily carry or hold chicken-sized eggs; therefore, the eggshells are rarely moved more than 3 feet (1 m) from the nest.

To prevent skunks from preying on fowl and their eggs, refer to the recommendations given under "Preventing Conflicts" in Chapter 21. To prevent skunks from digging under the coop or pen, create a barrier (Fig. 4).

Protect your pets. To keep pets from being sprayed, keep them inside at night.

Prevent damage to lawns. Because lawns—especially newly created ones—are often heavily watered, worms and grubs inhabit areas just under the sod, attracting skunks (and raccoons). Skunks tend to dig 1- to 3-inch (2.5–8 cm) deep holes only where a grub is located; raccoons tend to roll or shred the sod in their search. The use of pesticides to kill worms and grubs is not recommended because of their toxic effect on the environment, people, and animals.

To prevent digging, lay down 1-inch (25 mm) mesh chicken wire, securing the wire with stakes or heavy stones or heavy objects. Alternatively, sprinkle cayenne pepper or a granular repellent, such as Repel®—a commercial dog and cat repellent available at most pet stores or garden centers—over small areas.

Surrounding the area with a low chicken-wire fence used to prevent rabbit damage can protect large areas from striped skunks.

(See "Preventing Conflicts" in Chapter 20 for fence designs). Be sure to securely stake the bottom of the fence to the ground, but leave the top floppy to prevent climbing. A temporary, single strand of electric wire 5 inches (12 cm) above the ground will also deter skunks. (See Chapter 7 for information on electric fencing.)

Skunks in or Under Buildings

Occasionally a skunk will find a suitable den site in or under a building. Skunks normally occupy a den site for only two or three consecutive nights. However, during the mating and nesting season, females are attracted to warm, dry, dark, easily defended areas, and will remain longer if the setting remains favorable.

You may choose to let skunks occupy an area, such as under an outbuilding, if they don't pose a problem. Should you choose to remove the animals, a wildlife control company can be hired (see Appendix C, "Hiring a Wildlife Control Company"), or you can complete the process yourself (see Appendix B "Evicting Animals from Buildings.")

If a skunk finds its way into your house, garage, or other structure, stay calm, close all but one outside door, and let the animal find its own way out. If necessary, you can slowly encourage the skunk to move in a preferred direction while holding a large towel, or a large piece of plastic or cardboard in front of you. If the skunk appears agitated, retreat immediately. Don't use food as a lure—this will make the animal associate food with humans, and return for more. If the skunk appears sick or injured, call a nearby wildlife rehabilitator for assistance (see "Wildlife Rehabilitators and Wildlife Rehabilitation" in Chapter 20 for information).

Trapping Skunks

If all efforts to dissuade problem skunks fail, you may feel the need to trap the animals. Trapping skunks should be a last resort and can never be justified without first applying the above-described preventative measures. Trapping is also rarely a permanent solution since other skunks are likely to move into the area if attractive habitat is still available.

A wildlife damage control company can be hired to do the trapping, or you can do it yourself (see Appendix C, "Hiring a Wildlife Damage Control Company"). It is usually best to let someone with experience trapping skunks do the work. Because skunks

Removing Skunks from Window Wells and Similar Areas

Occasionally, striped skunks get stranded in window wells and similar areas. If this occurs, slowly lower in a rough board that is long enough to act as a ramp from the bottom to the top of the window well. Because striped skunks cannot climb a steep slope, the board should lean at no more than a 45-degree angle. You may need to attach wood cleats or a heavy piece of cloth so that the skunk can grip the board.

Crouch when approaching the area to stay out of sight of the skunk. If possible, have a second person on hand, with a vantage point high enough to see the skunk, and to warn you if the animal is becoming agitated.

Another method is to tie the board to the end of a long pole and lower it into the area. Once the board is placed, keep people and pets away from the area until nightfall, when the skunk should leave on its own. If the skunk doesn't leave, it probably can't get out because the board is positioned at too steep an angle.

Another approach is to place smelly cheese or cat food in the back of a small garbage can or a cat carrier. Slowly lower the can or carrier sideways into the window well with the open end facing the skunk. The skunk will smell the bait and go inside. When it does, slowly raise the skunk and carrier, elevator style, to ground level, keeping your hands on the outside so you don't risk being bitten. Leave the area and the skunk will amble out—probably after it is done eating.

Screen the top of the well to prevent the problem from reoccurring. Commercial well covers are available.

often live in groups, multiple traps are necessary to trap them out of an area. If you choose to do the trapping yourself, follow the steps listed in Appendix A, "Trapping Wildlife."

Public Health Concerns

The diseases or parasites associated with skunk populations in the Pacific Northwest are rarely a risk to humans.

Canine distemper, a disease that affects domestic dogs, is found in Pacific Northwest skunk populations (see "Public Heath Concerns" in Chapter 21). Have your dogs vaccinated for canine distemper to prevent them from contracting the disease.

Skunks may also be infected with the bacterial disease **tularemia** (see "Public Heath Concerns" in Chapter 3).

Never approach a skunk that appears to be ill, is overly friendly, or approaches you. If a person is **bitten or scratched** by a skunk, immediately scrub the wound thoroughly with soap and water. Flush the wound liberally with clean tap water. (In other parts of the United States skunks can carry **rabies**.) Contact your physician and the local health department immediately. If your pet is bitten, follow the same cleansing procedure and contact your veterinarian

to ensure that your pet has proper protection.

Anyone handling a skunk should wear rubber gloves, and wash their hands well when finished.

Legal Status

Because legal status, trapping restrictions, and other information about skunks change, contact your local, state, or provincial wildlife office, or visit their Web site for updates. (See Appendix E for contact information and where to access the state and provincial laws mentioned below.)

Oregon: Although not classified as furbearers, skunks are considered mammals with commercial fur value, so a trapping or hunting license is required to harvest them. Otherwise, skunks are not protected and can be killed without restrictions.

Capturing wild skunks and keeping them as pets is specifically prohibited (OAR 635-044-0015). Skunks may be imported as pets with the proper U.S. Department of Agriculture permits, but they are specifically prohibited from being sold, traded, or exchanged (ORS 498.029).

Washington: Skunks are not classified as game animals or furbearing animals (WAC 232-12-007). People can trap or shoot skunks on their own property

when the animals are causing damage to crops or domestic animals (RCW 77.36.030). No permit is required in such cases.

Except for bona fide public or private zoological parks, persons and entities are prohibited from importing skunks into Washington State without a permit from the U.S. Department of Agriculture and written permission from the Washington Department of Health. Persons and entities are also prohibited from acquiring, selling, bartering, exchanging, giving, purchasing, or trapping a skunk for a pet or for export (WAC 246-100-191).

British Columbia: Skunks are designated as both furbearing animals and small game. They may be legally hunted or trapped with a license during open seasons as defined by regulation. Both species may be captured or killed to protect property by the person that owns or occupies the property, or his/her spouse, parent, guardian, or child (Wildlife Act, Designation and Exemption Regulation 2 and 3). No permit is required in such cases.

Importation of skunks is not permitted except by educational institutions or scientific organizations if the Director of Wildlife is satisfied that the importation will not be detrimental to native wildlife or wildlife habitat (Wildlife Act, Permit Regulation 7).

CHAPTER 24

Tree Squirrels and Chipmunks

When the public is polled regarding suburban and urban wildlife, tree squirrels generally rank first as problem makers. Residents complain about them nesting in homes and exploiting bird feeders. Interestingly, squirrels almost always rank first among preferred urban/suburban wildlife species. Such is the paradox they present: We want them and we don't want them, depending on what they are doing at any given moment.

Although tree squirrels spend a considerable amount of time on the ground, unlike the related ground squirrels, they are more at home in trees. The Pacific Northwest is home to four species of native tree squirrels and two species of introduced tree squirrels (Table 1).

For information on chipmunks, see "Notes on Chipmunks."

Introduced Tree Squirrels

The **Eastern gray squirrel** (*Sciurus carolinensis,* Fig. 6) and **Eastern fox squirrel** (*Sciurus niger*) were introduced the Pacific Northwest in the early 1900s. Since then they have been repeatedly released in parks, campuses, estates, and residential areas. They are now the most common tree squirrels in urban areas, including the Lower Mainland and southern Vancouver Island in British Columbia.

Figure 1. The Western gray squirrel is the largest tree squirrel in the Pacific Northwest. It is found in low-elevation oak and conifer woods in western Oregon and parts of western and central Washington. (From Christensen and Larrison, Mammals of the Pacific Northwest: A Pictorial Introduction.*)*

Table 1. Native Pacific Northwest Tree Squirrels

The **Douglas squirrel, or chickaree** (*Tamiasciurus douglasii,* back cover) measures 10 to 14 inches (25–35 cm) in length, including its tail. Its upper parts are reddish- or brownish-gray, and its underparts are orange to yellowish. The Douglas squirrel is found in areas with large stands of fir, pine, cedar, and other conifers in the Cascade mountains and western parts of Oregon and Washington. In British Columbia it inhabits the southwest coast.

The **red squirrel** (*Tamiasciurus hudsonicus,* Fig. 4) is about the same size as the Douglas squirrel and lives in coniferous forests and semi-open woods in northeast Oregon and Washington, and also throughout British Columbia (except the southwest coast), including the islands. It is rusty-red on the upper part and white or grayish white on its underside.

The **Western gray squirrel** (*Sciurus griseus,* Fig. 1) is the largest tree squirrel in the Pacific Northwest, ranging from 18 to 24 inches (46–61 cm) in length. It has gray upper parts, a creamy undercoat, and its tail is long and bushy with white edges. This species is found in low-elevation oak and conifer woods in western Oregon and parts of western and central Washington.

The **Northern flying squirrel** (*Glaucomys sabrinus,* Fig. 2) is the smallest tree squirrel in the Pacific Northwest, measuring 10 to 12 inches (25–30 cm) in total length. It is rich brown or dark gray above and creamy below. Its eyes are dark and large, and its tail is wide and flat. These nocturnal gliders are surprisingly common, yet are seldom seen in their forest homes throughout the Pacific Northwest.

The upper parts of the Eastern gray squirrel are gray with a reddish wash in summer; its underparts are whitish. It's about 20 inches (50 cm) long, half of which is its prominent, bushy tail. The Eastern fox squirrel measures 22 inches (56 cm) in length, including a 9 to 10 inch (23–25 cm) tail. Its upper parts are usually dark grayish with a reddish cast, and the underparts are orange to deep buff.

The fur color of these two introduced squirrels can vary greatly. Some individuals, even whole populations, may be almost entirely black.

The increasing number of introduced Eastern gray squirrels is often said to be responsible for the decrease in Douglas squirrels in certain areas. However, given that these squirrels have different food and shelter preferences, it's likely that increasing housing and other development, and loss of coniferous forests is responsible for any decline in Douglas squirrel populations.

Facts about Tree Squirrels
Food and Feeding Habits

- Tree squirrels feed mostly on plant material, including seeds, nuts, acorns, tree buds, berries, leaves, and twigs. However, they are opportunists and also eat fungi, insects, and occasionally birds' eggs and nestlings.

- Squirrels store food and recover it as needed. Hollow trees, stumps, and abandoned animal burrows are used as storage sites; flowerpots, exhaust pipes, and abandoned cars are also used.

- Scientists credit flying squirrels with helping forest health by spreading species of fungi that help trees grow.

Nest Sites

- Tree squirrels construct nursery nests in hollow trees, abandoned woodpecker cavities, and similar hollows. Where these are unavailable, they will build spherical or cup-shaped nests in trees, attics, and nest boxes.

Figure 2. Flying squirrels can glide 150 feet (45 m) or more from a height of 60 feet (18 m). They don't actually fly, but glide downward through the air using skin flaps that stretch between their front and rear legs, forming two "wings." The tail is used as a rudder to help them keep on course. (From Christensen and Larrison, Mammals of the Pacific Northwest: A Pictorial Introduction.*)*

- At about three months of age, juvenile squirrels are on their own, sometimes remaining close to the nest until their parent's next breeding period.
- The second litter may stay with the mother in the nest through the winter until well after the winter courtship season.

Mortality and Longevity

- In trees, squirrels are relatively safe, except for an occasional owl or goshawk.
- On the ground, large hawks and owls, domestic cats and dogs, coyotes, and bobcats catch squirrels.
- Vehicles, disease, and starvation also kill squirrels.
- Most squirrels die during their first year; if they survive that, they live three to five years.

Viewing Tree Squirrels

Tree squirrels have many fascinating behaviors, and—except for nocturnal flying squirrels—they are commonly seen. Tree squirrels don't hibernate, but will remain in their nests in cold or stormy weather, venturing out to find food they stored nearby.

Squirrels are most active at dawn and dusk, but sharp eyes aided by a pair of binoculars can spot them moving among the treetops any hour of the day. On hot days, squirrels are less active and remain motionless on branches to enjoy whatever breeze is available.

Home ranges for tree squirrels are ½ acre to 10 acres (0.2–4 h). For the Eastern gray and Eastern fox squirrels living in city parks and suburban yards, home ranges average half an acre (0.2 h).

Flying squirrels can go at least three miles (5 kilometers) in four

- An alternate nest may be constructed in a tree for summer use. In areas with prolonged periods of cold weather, red squirrels may construct a winter nest underground, often in or near a food storage site.
- In urban areas, squirrels mostly nest in buildings and other structures.
- Nests contain leaves, twigs, shredded bark, mosses, insulation, and other soft material.

Reproduction

- Depending on the species, tree squirrels mate from early winter to late spring. One litter of two to four young is produced from March to June.

- All except flying and western gray squirrels may produce second litters in August or September.
- At about 30 days of age, the young are fully furred and make short trips out of the nest. At about 60 days of age, they begin eating solid foods and venture to the ground.

Notes on Chipmunks (*Tamias* spp.)

The smallest of the Pacific Northwest squirrels, chipmunks are active, vociferous, relatively small rodents that provide hours of enjoyment when watched in their surroundings. A series of alternating light and dark stripes cover their backs and upper sides from shoulders to rump. The sides of their heads are similarly striped.

People sometimes confuse chipmunks with ground squirrels; however, ground squirrels have longer bodies and live in open, grassy areas rather than wooded areas. Chipmunks also hold their tails up while running and ground squirrels do not.

Chipmunks live primarily on the ground, but spend time in shrubs and trees fleeing from danger, eating fruits or seeds, or nesting. Home ranges are from ½ to 3 acres (0.2–1.2 h).

Like tree squirrels, chipmunks are important in the dispersal of seeds because of their habit of storing them underground. Any buried seeds that are not consumed stand a better chance of germinating than those remaining on the surface. In this way, chipmunks assist in the spread of trees, shrubs, and other plants.

Figure 3. The yellow pine chipmunk, also known as the "Northwestern chipmunk," is a brightly colored species that inhabits the semi-open to open forests of central and eastern Oregon and Washington, and south and central British Columbia. (From Christensen and Larrison, Mammals of the Pacific Northwest: A Pictorial Introduction.*)*

Common Pacific Northwest Species

The Townsend's chipmunk (*Tamias townsendii*) is the largest species, at 11 inches (28 cm) in length (including its tail). This dark-colored chipmunk occupies coniferous forests in western Oregon, Washington, and British Columbia.

The slightly smaller **Siskiyou chipmunk** (*Tamias siskiyou*) inhabits the coniferous forests of southwestern Oregon and northern California.

The yellow pine chipmunk (*Tamias amoenus,* Fig. 3) measures 8 inches (20 cm) in length (including its tail). Also known as the "North-western chipmunk," this brightly colored species inhabits the semi-open to open forests of central and eastern Oregon and Washington, and south and central British Columbia.

The slightly smaller and mostly gray **least chipmunk** (*Tamias minimus*) tends to climb trees more than the other chipmunks, although it is rarely seen above 15 feet (4.5 m). The least chipmunk is restricted to the sagebrush areas of eastern Washington and Oregon, and the open forests of southeast British Columbia.

- **Nest sites and shelter:** Chipmunks nest, overwinter, and seek refuge in burrows located in hollow logs, rock crevices, under stumps, or in holes in trees. In areas blanketed with deep snow, chipmunks remain in their burrows until released by the spring sun. In areas with little or no snowfall, they may be active above ground throughout much of the winter, remaining in their nests only during the worst winter storms. Four to six young are born from early May through late June.

- **Food:** Chipmunks eat a wide variety of foods, including seeds, acorns, fruits, bulbs, fungi, insects, and some birds' eggs. Seeds and fruits form the bulk of their diet; mature seeds are eaten or stored for winter. Chipmunks have internal cheek pouches to carry food in, and they will regularly collect spilled food from under bird feeders if no domestic dogs or cats are around.

- **Signs:** Holes leading into burrows are about 2 inches (5 cm) in diameter, go directly into the earth, and show no evidence of excavation. Chipmunks tend to freeze or crouch when they are alerted by a sight or sound, and may not enter the burrow when they are being watched.

  Chipmunk droppings are very rarely seen except at spots where chipmunks regularly perch and feed. Individual droppings are 1/4 to 1/2 inch (6–12 mm) long and 1/8 inch (3 mm) wide.

  Their main call is a *chip,* a sound people often mistake for a bird's song. Another call is a lower-pitched chuck.

- **Predators:** Chipmunk predators include weasels, mink, spotted skunks, bobcats, domestic cats, and owls. They can be seriously wounded when they compete among themselves or with tree squirrels for food and space.

- **Conflicts:** Like squirrels, chipmunks occasionally dig up and eat flowering bulbs, such as crocus. Chipmunks also readily adapt to suburban gardens and often dig burrows under retaining walls and sidewalks, but their small burrows rarely cause structural damage.

  To prevent chipmunks from burrowing, use the recommendations provided under "Preventing Conflicts" in Chapter 23. Bulbs may be protected if planted beneath a ground cover of wire or plastic screen. This mesh should be 1/2 x 1/2 inch (12 x 12 mm) to allow plants to sprout, but prevent chipmunks from digging. Commercial repellents that prevent rodents from eating bulbs are available from nurseries and hardware stores.

  Chipmunks occasionally enter garages and storage buildings in search of food and shelter. Such problems can usually be solved by closing doors, sealing access points, and removing or securing stored bird seed or pet food.

  Live trapping chipmunks is not recommended because trapping a few individuals will not affect the neighborhood population. Trapping females during spring and summer months almost guarantees the death of their young, and releasing a chipmunk in new territory with no burrows will likely result in its death from predation, accident, or starvation.

  Because chipmunks are responsible for so little real damage, consider adopting a tolerant approach and sharing your land with them.

hours, soaring from tree to tree. Males are particularly prone to traveling, visiting different females in the spring. As proficient as they are in the air, flying squirrels are awkward on the ground. The large flaps of skin that make gliding possible obstruct walking.

Feeding Activity

In the fall, when Douglas squirrels and red squirrels are actively harvesting and storing food for winter, look for "cuttings" under oak, maple, walnut, hazelnut, and coniferous trees. Cuttings are made because seeds and nuts grow in clusters at the end of fragile, easily broken twigs, and squirrels have found that the easiest way to harvest them is to nip these twigs off the parent branch. The squirrels then climb to the ground, harvest the meal, or carry it off to a storage site.

Figure 4. Noisy sputterings and scoldings from the tree canopy call attention to the native Douglas squirrel, also known as the chickaree, or the similar size native red squirrel (shown here). (From Christensen and Larrison, Mammals of the Pacific Northwest: A Pictorial Introduction.)

Tips for Attracting Squirrels and Chipmunks

• Keep as much wooded property in a natural condition as you can.
• Include trees and shrubs that provide seeds, nuts, acorns, cones, and fruits at different times of the year.
• Leave dead or dying trees (snags) alone when possible. These provide nest sites and food-storage sites.
• Leave some tree or shrub prunings on the ground for squirrels to gnaw on during winter.
• Install a feeder and a nest box suitable for squirrels or chipmunks (see Appendix K for a plan of a nest box used by Douglas squirrels, red squirrels, Northern flying squirrels, and cavity nesting chipmunks). (Several flying squirrels will use a duck or owl box for hibernation, and an individual female will use a small box for raising her young.) Be careful when monitoring or cleaning these boxes, because any rodents could carry Hanta virus, which could infect you if dust from dried droppings and urine gets in your eyes, nose, or mouth ("see Public Health Concerns" in Chapter 8 for information).
• Keep domestic dogs and cats indoors or fenced.

A large pile of cone scales under a tree, called a "midden," generally indicates Douglas or red squirrels. In winter, holes in the snow may indicate where squirrels retrieved stored food.

Nest Sites

Winter is the time to spot the large, spherical nests built in deciduous trees. Nests are located 15 to 50 feet (4.5–15 m) high, and situated close to the trunk or a main branch.

Tracks and Scratch Marks

In urban areas, squirrels travel via rooftops and power lines, lawns, and concrete, leaving no visible trail. Tracks are also seldom visible on the soft forest floor. However in soft snow, the track pattern of a scampering squirrel can be seen as it leads from tree to tree. If you find tracks starting from an open area in the snow, a flying squirrel may have landed and scampered off.

The hind legs of squirrels are double jointed to help them run up and down trees and other objects. Their front claws are extremely sharp and help in gripping while climbing and traversing. Scratches may be found where squirrels access buildings via downspouts and painted surfaces. Look closely for ¼- to ½-inch (6–12 mm) long scratches in the paint that appear to have been made by a pin.

Droppings

Tree squirrel droppings are rarely obvious. Areas under a feeder or a nest are good places to check. Droppings are segmented, roughly cylindrical, and ¼ to ½ inch (6–12 mm) long, with a smooth surface. Coloration is typically black, but can be brown to red.

Calls

The red squirrel and the Douglas squirrel will announce an intruder's presence with much intensity. This territorial call sounds something like a rapid *tsik tsik tsik, chrrrrrrrr—siew siew siew siew.* The call of the Eastern gray squirrel—*que, que, que, que*—is usually accompanied by flicks of the tail. It makes other calls as well, including a loud, nasal cry.

The call of the relatively silent flying squirrel is a quiet, high-pitched, birdlike *tick tick.*

Preventing Conflicts

A tree squirrel's search for food may bring it to a bird feeder, back door, or a garden containing bulbs. Its search for a nest site may bring it into an attic or down a chimney. The most effective way to prevent conflicts is to modify the habitat around your home so as not to attract squirrels.

To prevent conflicts or remedy existing problems:

Don't feed squirrels. Tree squirrels that are hand-fed may lose their fear of humans and become aggressive when they don't get food as expected. These semi-tame squirrels also might approach a neighbor who doesn't share your appreciation of the animals, which would likely result in them dying.

Eliminate access into buildings. Repair or replace loose or rotting siding, boards, and shingles. When inspecting a building for potential access points, use a tall ladder to view areas in shadows. A pair of low-power (4x) binoculars can be a helpful inspection tool to use before making a dangerous climb. Inspecting the attic or crawl space during the day may reveal light

Figure 5. The Eastern gray squirrel is from the deciduous and mixed coniferous-deciduous forests of eastern North America, and was introduced into city parks, campuses, and estates in the Pacific Northwest in the early 1900s. (From Larrison, Mammals of the Northwest: Washington, Oregon, Idaho, and British Columbia.*)*

shining through otherwise unnoticed cracks and holes. Native squirrels chew holes 2 inches (5 cm) in diameter; Eastern gray and fox squirrels chew open baseball-size holes.

Cover the dryer vent with a commercial vent screen designed to exclude animals without lint clogging. Other vents can be covered with ¼-inch (6 mm) hardware cloth. Some roof-vent caps contain a flimsy, lightweight inner screen that a squirrel can easily penetrate. If the screen has been penetrated, it may be better to replace the whole vent cap with something stronger.

Because squirrels are excellent leapers, keep tree and shrub branches 10 feet (3 m) away from the sides and tops of buildings. To prevent squirrels from climbing a tree to access a building, install one of the barriers shown in Figure 6. Remove vines that provide squirrels a way to climb structures and hide their access points.

Prevent squirrels from accessing buildings via utility wires by installing 3-foot (90 cm) sections of 2- to 3-inch (5–7.5 cm) diameter plastic pipe barriers. Carefully split the pipe length-

wise with a saw, tin snips, or a sharp utility knife, spread the opening apart, and place it over the wire. The pipe will rotate on the wire and the squirrel will tumble off. **Do not attempt to install pipe over high-voltage wires.** Contact your local electricity/utility company for assistance.

To prevent squirrels from climbing the corner of a building, refer to the figure under "Preventing Conflicts" in Chapter 21.

Keep squirrels out of exhaust fans and chimneys (and help them out if they fall in). If a squirrel is trapped down an exhaust fan or a metal-lined chimney, you can let the squirrel exit through the house (see "Getting a Squirrel Out of the House") or drop a line down from above so the animal can climb out.

To help the squirrel exit from above, drop down a thick rope or cloth, such as strips of a sheet, so the squirrel can climb out. It is a good idea to tie knots in the rope or cloth 12 inches (30 cm) apart, to provide a secure climbing surface. You may have to tie a couple of lengths together to reach the

A Baby Squirrel May Need Help

Baby squirrels are often found on the ground after a storm has blown down a nest. If a baby squirrel is very small and has closed eyes, place the animal in a safe place below the nest tree, or along a trail used by local squirrels. Keep dogs, cats, and children away and stay completely out of sight. Within a couple of hours the mother will pick the baby squirrel up with her teeth and take it back to the nest. If she does not, the mother squirrel may have pushed the baby out for a reason, and it is best to let nature take its course.

As an alternative—or if the mother squirrel doesn't pick up the baby squirrel—call a wildlife rehabilitator for advice. Look under "Animal" or "Wildlife" in your phone book or search the Web for "wildlife rehabilitator." If a rehabilitator isn't available, follow the menu options provided on their phone message or on their Web site.

See "Wildlife Rehabilitators and Wildlife Rehabilitation" in Chapter 20 for more information.

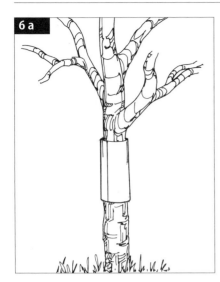

Getting a Squirrel Out of the House

To help a tree squirrel exit that accidentally gets caught inside a house or an enclosed fireplace:

• **Close all doors to any rooms in the house that the squirrel is not in.**

• **Close all of the curtains on the windows or the doors in the room where the squirrel is, or where the squirrel in the enclosed fireplace is. If necessary, use newspaper to cover incoming light sources.**

• **Leave the curtain open on the window or the door where you want the squirrel to exit.**

• **Open that window or door as wide as possible. Remove the screen, if necessary.**

• **Open the door to the fireplace or the cover from the exhaust fan so the squirrel can exit. The squirrel will escape toward the light.**

Note: If the squirrel is in the chimney and cannot climb out, then you must open the flue, so that the squirrel will drop into the fireplace.

Figure 6 (a) A squirrel guard can be secured around trees, pipes, posts, and other structures to keep squirrels from climbing. A barrier can be made from a piece of aluminum flashing or sheet metal, 24 inches (90 cm) wide and as long as the circumference of the support (allow plenty of material for the overlapping seam and tree growth). The barrier can be held together with wire, nails, or screws, and painted to blend in. The top of the barrier should be at least 5 feet (1.5 m) off the ground.

Figure 6 (b) Alternatively, a funnel-shaped piece of aluminum flashing can be fitted around the tree or other vertical structure. The outside edge of the flared metal should be a minimum of 18 inches (45 cm) away from the support. Cut the material with tin snips and file down any sharp edges. (Drawings by Jenifer Rees.)

bottom of the chimney. Tie something to provide weight to the bottom of the rope or cloth, such as a pair of pliers, or other small heavy object.

Lower the rope or cloth slowly, make sure this reaches the bottom, and then secure it at the top. Leave the area completely alone. The squirrel should climb out in 1 to 24 hours.

If the chimney is firebrick then the squirrel can climb out on its own. But if it falls through the flue into the fireplace it usually cannot get back up into the chimney. Open the fireplace door and place a board or branch leading

from the fireplace up to the flue. This way the squirrel can climb out on its own. Another option is to prepare the room as described under "Getting a Squirrel Out of the House."

Never leave the squirrel in the chimney or exhaust fan longer than 24 hours—it will die from dehydration. If needed, call a wildlife control company (see Appendix C).

After you are sure no animals are down the chimney, cap it with a commercially engineered chimney cap. Most hardware stores carry these, and chimney cover manufacturers are able to custom fit covers for unusual chimneys.

If a tree squirrel is nesting in a chimney, follow the recommendations given for raccoons under "Raccoons in Dumpsters and Down Chimneys" in Chapter 21.

Keep squirrels out of birdhouses.
Tree squirrels sometimes raid eggs or small nestlings in nest boxes for food. Introduced fox and Eastern gray squirrels are a worse problem than the native squirrels. Being larger they can reach farther down into a box to pull out the chicks, although all squirrels can simply chew their way into a box. More often, tree squirrels nest or over-winter in nest boxes intended for native birds. (For information on accommodating tree squirrels with their own nest boxes, see Appendix K and "Nest Structures and Den Boxes" in Appendix F.)

To prevent tree squirrels from climbing up a pole, tree, or other structure supporting a birdhouse, install a barrier (Fig. 6). To prevent squirrels from gnawing around the entry hole of a nest box intended for small birds, attach a pre-drilled metal plate (available from stores catering to the bird-feeding public). Alternatively, attach aluminum flashing or sheet metal to the front of the nest box, and drill an entry hole of the correct size through the flashing. Use a hole-saw bit that cuts through metal and wood, and file down all sharp edges.

Boxes with entry holes large enough to accommodate wood ducks or other large birds should have their entry holes blocked with a rag or other stopper until the desired bird species is seen or heard in the area. Alternatively, nest boxes can be set out with the top (or one of the sides used for cleaning out the box) left open until the desired bird is seen or heard. Immediately after the bird's nesting season, remove the boxes, or if that is not practical, plug the entrance holes or leave the lids off.

Alternatively, place nest boxes in the most open locations and positions available that will be unattractive to squirrels, but still attract the native birds appropriate to the site.

Keep squirrels out of fruit and nut trees. Install one of the barriers presented in Figure 6 and trim lower branches to 5 feet (1.5 m) above the ground. Barriers will not work if there are trees, fences, or buildings within 10 feet (3 m), because squirrels will leap from such objects to reach the food source.

Protect garden bulbs, plants, and seeds. Newly seeded areas and seedlings can be covered with a temporary wire cage or netting made from 1-inch (2.5 cm) mesh chicken wire. Where bulbs are being dug up, chicken wire can be laid down, securely staked at the edges, and lightly mulched to cover the wire for appearance sake. Commercial taste repellents that prevent squirrels from eating plant

material are available from nurseries and hardware stores.

Protect tree bark from being stripped or eaten. Tree squirrels strip the bark off redwood, redcedar, and certain other trees to line their nests. They seek the tender cambium layer of other trees for food. Protect individual trees by installing a barrier as presented in Figure 6, or by loosely wrapping vulnerable areas with 1-inch (2.5 cm) chicken wire. (For examples, see "Preventing Conflicts" in Chapter 3.)

Pepper spray or a commercial taste repellent such as Ropell® can be applied to the bark to prevent bark removal. Applications will need to be repeated in damp weather.

Tree Squirrels in the Attic

The most serious problem with tree squirrels occurs when an adult female squirrel enters an attic or similar space to nest. You may choose to let a squirrel stay if it isn't posing a problem. However, squirrels gather insulation for nests, create noise (especially on stormy days or nights when squirrels are less likely to be out for food), and may chew electrical wiring, causing electrical problems or fires.

Should you choose to remove the squirrel, a wildlife damage control company can be hired (see Appendix C, "Hiring a Wildlife Damage Control Company"). You can also do the work yourself by following the steps listed under "Evicting Animals from Buildings" in Appendix B.

Because attics can be difficult to access and maneuvering around

in them can be dangerous, it is recommended that a professional be hired when attics are involved.

If a squirrel has spent a prolonged amount of time in an area with exposed wiring, check your smoke detectors to make sure they are functioning in case of a fire. Also, inspect the area for wire damage or have an electrician inspect it.

Lethal Control

Trapping squirrels should be a last resort, and lethal control can never be justified without a serious effort to apply the above-described preventative control. Killing tree squirrels is also, at best, a short-term solution to any problem you may have. As long as you provide food or shelter and additional squirrels are in the area, other squirrels will move in to replace the ones that you have removed.

Shooting tree squirrels may be helpful if a small, localized population of introduced species is problematic. For safety considerations, shooting is generally limited to rural situations and is considered too hazardous in more populated areas, even if legal.

Don't trap a problem squirrel in a live trap, thinking that you can release it in another location. Doing so may be illegal (see "Legal Status"). In addition, once in the new location, the squirrel will likely die from hunger, stress, or territorial disputes with other squirrels. Relocated animals can also spread diseases to other squirrels. If they do survive, they are likely to create the same problems in their new home that they created in yours. Finally, in most cases if an area is optimum for squirrels—be it a city park or a wild area—chances are that enough

Tree Squirrels and Bird Feeders

Tree squirrels eat most types of birdseed and relish sunflower seeds. Once accustomed to a food source, they will be persistent at finding ways to reach it.

Many feeder designs on the market are advertised to be squirrel-proof. Some are but many are not. One popular design is a tube feeder enclosed in a cylindrical cage that allows only small birds to pass through freely. (The cage may also be purchased separately for use with an existing feeder.) However this approach also has drawbacks. Large holes may not exclude small native squirrels or immature non-native squirrels. If the feeder comes too close to the cage, adult squirrels will paw at the feeder ports,

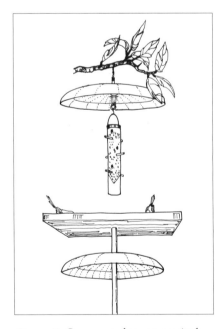

Figure 7. One way to keep tree squirrels off of bird feeders is to attach baffles above and/or below the feeder. The baffle on a pole should be attached at least 4 feet (1.22 m) above the ground and extend out at least 2 feet (60 cm). A baffle can also be hung above a feeder to prevent squirrels from climbing down to a feeder. (Drawing by Jenifer Rees.)

knocking or pulling food to the ground.

The best approach is to put a stop to squirrels accessing the feeder altogether. To prevent them from jumping to a feeder, place it at least 5 feet (1.5 m) off the ground and 10 feet (3 m) from the nearest large shrub, tree limb, deck, or other launching pad. To prevent squirrels from climbing wires, poles, or chains on which feeders are mounted, use one or more of the following techniques:

- Mount a commercially available dome-shaped "squirrel baffle" above and/or below the feeder to prevent squirrels from reaching it (Fig. 7). Adjust the baffle high enough above the feeder to prevent squirrels from reaching the feeder while holding on with their dexterous rear feet.
- Mount the feeder on a smooth pole made from metal or PVC pipe that is at least 5 inches (12.5 cm) in diameter (Fig. 8). The wide-diameter and smooth texture of the pipe will prevent squirrels from climbing.
- Suspend the feeder from a horizontal wire, placing a 3-foot (90 cm) length of 1-inch (2.5 cm) PVC pipe around the wire on either side of the feeder to function as a barrier. Similarly, soda bottles with a hole drilled in the bottom can serve as barriers.
- Coat the horizontal wire with a Teflon spray used to lubricate plastic wheels. The wire will need to be sprayed regularly, but the squirrels will eventually tire of spinning cartwheels around the wire. Always use heavy-gauge wire to hang the feeder. Squirrels have sharp teeth and strong jaws, and will chew through almost anything else, dropping the feeder to the ground.

Another way to discourage squirrels from eating birdseed is to offer birds foods that squirrels do not like. For example, squirrels will sometimes

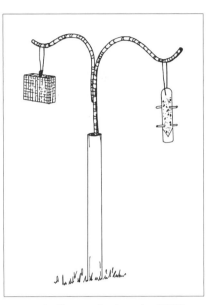

Figure 8. Create a barrier using PVC pipe that is at least 5 inches (12.5 cm) in diameter. The wide diameter and smooth texture of the pipe will prevent squirrels from climbing. (Drawing by Jenifer Rees.)

ignore safflower seed, millet, and plain suet (without nuts, peanut butter, or other additives). However, these foods may not be as popular with the bird species you want to attract, and squirrels will eat them when they are hungry enough.

Mixing dried pepper flakes into birdseed makes it less attractive to squirrels. (Give the seed a quick spray of Pam or other oil before mixing so the pepper flakes stick to the seed.) Birds are unaffected by the capsicum resin in peppers, but squirrels do feel the heat. Some people consider this practice inhumane, as squirrels may get capsicum in their eyes while grooming.

For additional information on managing bird feeders, see "Preventing Conflicts" in Chapter 37.

squirrels already live there. One or more of the squirrels is going to have to move or die.

For additional information, see "Trapping Wildlife" in Appendix A.

Public Health Concerns

Tree squirrels might carry diseases that could affect humans, but, as a practical matter, instances where squirrels have transmitted disease to humans are rare.

You may see a tree squirrel engaging in **unusual behavior,** such as repeatedly falling over or circling a small area. Such behavior can result from an injury, poisoning, or inflammation of the brain (encephalitis) caused by a parasite.

If a person is bitten or scratched by a tree squirrel, immediately scrub the wound with soap and water. Flush the wound liberally with tap water. Tree squirrels can carry **tularemia** (see "Public Health Concerns" in Chapter 3).

In other parts of the United States squirrels can carry **rabies.** Contact your physician and the local health department immediately. If your pet is bitten, follow the same cleansing procedure and contact your veterinarian. If you can place a large bucket over the squirrel and secure the bucket with a heavy object, the animal can then be held for inspection by a health official.

Legal Status

Because legal status, trapping restrictions, and other information about squirrels and chipmunks change, contact your local, state, or provincial wildlife office, or visit their Web site for updates. (See Appendix E for contact infor-mation and where to access the state and provincial laws mentioned below.)

Oregon: The Western gray squirrel is considered a game mammal and hunting regulations apply (OAR 635-067). The Douglas squirrel, red squirrel, and Northern flying squirrel are protected non-game species and may not be hunted. All chipmunks are protected nongame species and may not be hunted.

All native squirrels (except the Western gray squirrel) and chipmunks may be captured and held as pets, but only with a holding permit issued by Oregon Department of Fish and Wildlife (ODFW). Property owners, persons lawfully occupying land, or their agents may capture any native squirrel or chipmunk at any time with a permit issued by ODFW if the animals are causing damage.

The possession, importation, and sale of Eastern gray squirrels and fox squirrels is prohibited. In addition, the rehabilitation of injured or orphaned Eastern gray and fox squirrels is not allowed, and the relocation of trapped Eastern gray and fox squirrels by wildlife damage control companies is prohibited.

The killing of Eastern gray and fox squirrels is unrestricted, although a hunting license is required.

Washington: The Western gray squirrel is classified as a threatened species and cannot be hunted, trapped, or killed (WAC 232-12-007). The red squirrel, Douglas squirrel, Northern flying squirrel, least chipmunk, yellow pine chipmunk, Townsend's chipmunk, and red-tailed chipmunk are protected species and can be trapped or killed only in emergency situations when they are damaging crops or domestic animals (RCW 77.36.030). A special permit is required in such situations.

The Eastern gray squirrel and Eastern fox squirrel are unclassified and may be trapped or killed year-round and no permit is necessary. Although trapping these squirrels is not illegal, there may be restrictions on what devices may be used.

It is unlawful to release a squirrel or chipmunk anywhere within the state, other than on the property where it was legally trapped, without a permit to do so (RCW 77.15.250 and WAC 232-12-271).

British Columbia: Douglas squirrels, red squirrels, and Northern flying squirrels are designated as furbearing animals. Red squirrels may be legally trapped under license during open seasons as defined by regulation and there is no bag limit. There is no season for Douglas squirrels or Northern flying squirrels. Douglas squirrels, red squirrels, and Northern flying squirrels that are involved in property damage may not be captured or killed, outside of general open seasons, unless a permit is first received from the regional Fish and Wildlife Office.

Eastern gray squirrels and fox squirrels are listed under Schedule "C" and may be captured or killed anywhere or at any time in the province. No permit is required, but a person must have land owner/occupier permission on private property.

All species of chipmunks are designated as wildlife under the Wildlife Act and cannot be hunted, trapped, or killed unless under permit.

CHAPTER 25

Voles (Meadow Mice)

Voles, often called meadow mice or field mice, are rodents with stocky bodies, blunt noses, and small round ears that are hidden by fur (Fig. 1). They range in length from 5 to 10 inches (12.5–25 cm), including their tails.

More than 15 species of voles occur in the Pacific Northwest. Most live in grassland areas; however, some species are semi-aquatic and others are found living up in forest trees. Voles do not usually enter homes or other buildings.

West of the Cascade crest, you are most likely to find the **creeping vole** (*Microtus oregoni*), the **Townsend meadow vole** (*Microtus townsendii*, Fig. 1), or the **gray-tailed vole** (*Microtus canicaudus*) (occurs only in the Willamette valley, Oregon). East of the Cascade crest, the **montane vole** (*Microtus montanus*) is most commonly seen.

The above-mentioned species inhabit grassy or meadowlike habitats, areas or croplands adjacent to buildings, or gardens and landscaped sites with a protective cover of grass and other low vegetation.

Figure 1. The Townsend vole, like other voles, is distinguished from house mice and native mice of other genuses by its stocky body, more blunt snout, and shorter ears and tail. (Washington Department of Fish and Wildlife.)

Facts about Voles
Food and Feeding Habits

- Voles are herbivores (plant-eaters), consuming grasses, flowers, lichens, fungi, fruits, vegetables, bark, roots, and bulbs.
- Voles feed on the ground surface and dig out tubers, roots, and plant stems. Seeds, bulbs, and other foods may be concealed in runways for later use.
- Bones from dead animals seldom last very long in the wild because small rodents, including voles, gnaw on them for their nutrients.

Runway Systems and Nest Sites

- Grassland voles use underground runways or runways in vegetation that is thick enough to completely conceal their bodies.

Figure 2. Most small to mid-sized carnivorous mammals capable of catching voles eat them. Hawks, owls, and snakes also eat voles (montane vole shown here). (From Christensen and Larrison, Mammals of the Pacific Northwest: A Pictorial Introduction.*)*

- Nest sites are located under boards, rocks, logs, brush piles, hay bales, and in grassy tussocks.
- In areas with snow, nests are often constructed on the ground under a covering of snow.

Reproduction and Population Dynamics

- Voles living at low elevations along the coast breed from early spring to late fall.
- Voles can breed at a young age; female **gray-tailed voles** can breed when they are one month old.
- Litters sizes range from 1 to 13 young, with a gestation of about 21 days.

- Typical population density for grassland voles is 4 to 12 voles per ¼ acre (0.1 h) in old-field habitat. In peak years, population densities may reach 30 voles or more per ¼ acre (0.1 h) in favorable vole habitat.
- Vole population size tends to fluctuate widely over the course of years (4 to 8); however, cycles are generally not predictable. During this period of high density the number may be greater than 10 times that of a normal population.
- Vole populations tend to be lowest in early spring; the population increases rapidly through summer and fall.

Mortality and Longevity

- Most carnivorous animals capable of catching voles eat them: coyotes, foxes, bobcats, weasels, minks, skunks, raccoons, and domestic cats and dogs. Hawks, owls, and snakes also eat voles.
- Birds that aren't usually considered predators of mammals but that do eat voles include gulls, magpies, ravens, crows, and great blue herons.
- The age of a vole rarely exceeds 16 to 24 months and there is a high mortality rate due to natural causes among the young of less than two months of age.

Viewing Voles

Voles are active day or night, or both, depending on the species. Voles do not hibernate and are active throughout the winter.

Voles, depending on species, are not commonly seen because of their tall grass habitat. However, if you stand still within that habitat, you may see and hear voles dart from one site to another. Some species, such as the gray-tail vole, dig a more extensive burrow system and spend considerably more time below ground where they do most of their feeding.

Many times it is the vole's predator that is seen. Coyotes sniff around in fields and pounce on voles; hawks circle from above, and great blue herons wait to pierce voles with their sharp beaks.

One of the most common ways to see voles, or at least part of them, is in owl pellets. Great horned owls and barn owls may eat two or three voles in a day. Owls eat voles whole, and then regurgitate a pellet several inches long that contains the fur, skulls, and bones of several voles.

Runways and Tunnel Entrances

Vole runways are about 1½ inches (38 mm) wide. Vegetation in and next to well-traveled runways will be clipped close to the ground; long-used runways may be worn 2 inches (5 cm) into the ground. Vole droppings and small pieces of clipped vegetation can be found in the runways.

Some species, including creeping voles, create small sub-surface tunnels that appear as raised ridges, similar to those made by moles. Similar tunnels may also appear above ground in thick grass.

During the winter, montane voles make runways beneath the snow and feed on the snow-flattened grasses. After the snow melts, 1- to-2 inch (2.5–5 cm) "cords" are revealed. These cords are made of grass and debris stuffed into snow tunnels during the winter to make space elsewhere in the tunnel network.

Tunnel entrances above or underground are ¾ inch to 1½ inches (2–4 cm) in diameter. Entrances created underground may have mounds of dirt 3 to 4 inches (7.5–10 cm) wide next to them, but these are never as large as mole or gopher mounds. A good swimmer, the Townsend vole often constructs underwater entrances to its burrow system.

Droppings

Droppings are dark brown, oval-shaped, and shiny when fresh. They are slightly larger in size than those of house mice. Droppings are seen where a vole has been eating, and a section of runway may contain hundreds of pellets.

Feeding Sites

Evidence of gnawing voles can be seen above or below the soil line, up to 5 inches (13 cm) in either direction. Where snow cover is present, gnawing may extend higher. Close inspection of the exposed wood will reveal minute tooth-marks and small gashes at various angles, ⅛ inch (3 mm) in size. Feeding starts on one side of a seedling and progresses upward, so seedlings tend to be peeled only on one side at first. With later and more extensive girdling, all the bark on a plant may be removed up to its lower branches.

Note: Rabbits gnaw trunks higher up and leave larger, neater tooth-marks at consistent angles. When eating seedlings, rabbits leave the tips with 45-degree-angle cuts, deer leave ragged cuts, and mountain beaver leave 2-inch (5 cm) stubs. Pocket gophers girdle seedlings, but their much larger tooth marks are a clear indicator of their damage. Gophers also pull small seedlings underground or clip them at ground level.

Unlike moles or pocket gophers, voles don't create large dirt mounds at their tunnel entrances, so small tunnel entrances interconnected by trails are a sure sign of voles.

When the bark is soft in spring, deer and elk damage to conifer seedlings may resemble that of voles. They strip the bark of stem and lateral branches from bottom to top and often remove small needles in this process. Voles typically don't eat needles, and their damage doesn't extend as far up the main stem or include removal of lateral branches.

Voles defecate where they eat, so you may find piles of droppings at the base of damaged seedlings.

When the snow reaches 4 inches (10 cm) deep, voles will burrow up through it to reach grass seed-heads. Look for their 1½-inch (4 cm) diameter tunnels. In orchards, look under trees for hollow shells of apples and other fruits, eaten from the underside.

In newly planted orchards and landscapes where there is a large vole population, tree roots may be gnawed. During the growing season, a damaged tree will be leggy and thinly leafed. In winter when the tree is dormant, give it a wiggle. If it moves much, the entire root system is probably gone. If you pull the tree up, the undergound portions may look like they were run through a pencil sharpener.

Preventing Conflicts

Voles will tunnel through vegetable and flower gardens, feeding on roots, tubers, and bulbs—damage that is often incorrectly blamed on moles or gophers. Voles damage trees and shrubs when they gnaw on the bark or roots and eat seedlings and field crops. Called "bog rats" in Oregon, Townsend meadow voles burrow into bog dikes and eat the cranberries. Such damage is similar to that caused by muskrats (see "Preventing Conflicts" in Chapter 15 for management recommendations).

The following are ways to prevent conflicts and remedy problems. In cases where these methods are not practical, contact your local County Extension Agent, Department of Agriculture, or Ministry of Agriculture (B.C.) for further information (see Appendix E for contact information). Before hiring someone to control voles, read Appendix C, "Hiring a Wildlife Damage Control Company."

Habitat modification: Grassy cover encourages voles by providing food and protection from predators and environmental stresses. Removing this protection by regularly mowing, grazing, or tilling along ditch banks, rights-of-way, or field edges adjacent to vulnerable areas will cause the number of voles in that area to decline.

If feasible, mowed or otherwise managed strips can also serve as buffers around areas to be protected. The wider the cleared

Providing Vole Habitat

Grassy areas can be maintained specifically for voles. Fields and large lawns that were once routinely and completely mowed may, if left alone or mowed occasionally, eventually can have hawks, owls, and mammals hunting in them, and swallows, nighthawks, and bats swooping above them catching flying insects.

After checking city or county ordinances that could prohibit you from providing vole habitat, consider the following ideas to help you put a grassland management plan together on your property:

- Mow to a height of 6 inches (15 cm) to prevent harming voles (also newts, salamanders, and snakes). To achieve this height, mow with a powerful weed whacker rather than a lawn mower.
- Mow or weed whack only once a year to help control invading woody plants, such as scots broom and blackberry. Make this annual cut as late in the year as possible, preferably after the first frost, to reduce disturbance to voles and other wildlife. *Note:* Although infrequent mowing may save you time and maintenance costs, the extra height of the plants that results may require use of a special mower or heavy-duty weed whacker. If you attempt the job with an average lawn mower, first remove large unwanted woody plants by hand.
- Try cutting only a third of the grassy area at one time. Keep unwanted plants, such as thistles, under control with spot mowing or weed whacking. This minimizes disturbance to the voles and other wildlife using the uncut area, adds a variety of cut and uncut habitats, and helps prevent aggressive plants from gaining overall dominance.
- To enhance the shelter for voles, rake a portion of the cut grass into windrows (long piles) 2 to 3 feet (60–90 cm) wide and 12 inches (30 cm) high.

People seem to have an innate preference for a sense of order and purpose in a landscape. To prevent conflict among neighbors:

- Reduce expense, enjoy your efforts more, and engender less hostility from neighbors by taking small steps to transition your landscape into a more wild setting.
- You have good reasons to manage the grassland the way you do—let others know what they are. Educating your neighbors and discussing your project with local officials before you begin is essential. Signs can tell passersby that your yard is a special place and intended to be the way it is.
- Place a border between the grassy area and the sidewalk or a neighbor's property to make it clear that your yard is a product of intent and effort, not neglect (Fig. 3). The border can be a hedge, a series of low native plants, or a well-maintained path.
- Remember that, although you have a right to your "grassland," your neighbors have the right to their clipped lawns.

(Adapted from: Link, *Landscaping for Wildlife in the Pacific Northwest*.)

Figure 3. Keep a strip between the path and grassy areas mowed closely to indicate that plants in the adjoining grassland are left tall by design rather than from neglect. (Drawing by Jenifer Rees.)

strip, the less apt voles will be to cross it. A minimum width of 15 feet (4.5 m) and a maximum grass height of 6 inches (15 cm) are recommended, but even this can be ineffective when vole numbers are high.

Because voles prefer not to feed in the open, a 4-foot (1.2 m) diameter circle around the base of young trees or vines that is free of vegetation, or a buffer strip 4 feet (1.2 m) or more along a row of trees, may reduce problems.

Voles also use loose mulch for hiding and nesting cover, so reducing the depth of the mulch to 1 or 2 inches (2.5–5 cm) will discourage them. Rake up and compost or spread out tall cut grass to avoid a thatch buildup that the voles can hide under. Pick up any fallen fruit so the voles cannot feed on it.

Barriers: Individual young trees, shrubs, vines, and other plants can be protected with cylinders made from wire mesh (Fig. 4). Individual milk cartons or plastic soda bottles can also be cut at both ends to fit over small plants. To prevent voles from digging under the barrier, bury the bottom as deep as possible without damaging the plants' roots. Seedlings can be protected with rigid plastic tubes used to prevent damage from deer, rabbits, mountain beaver, or other rodents (see "Preventing Conflicts" in Chapter 7). Frequently inspect the barriers to make sure voles have not gnawed through or dug under them. Voles will use improperly placed barriers as protective cover for feeding or nesting.

Cheap but effective barriers for a large number of seedlings can be made using tinfoil. Cut 12-inch (30 cm) rolls of tin foil in half with a hacksaw. Wrap the base of the seedling at least twice with the foil and loosely crimp it. Don't use foil on seedlings less than 10 inches (25 cm) high—it can catch the wind and whip the seedlings, rubbing all the bark off at the base.

Small trees and shrubs are often surrounded by a weed mat or other material to reduce competition for nutrients, light, and water. Unfortunately, a weed mat can concentrate vole activity around the plant, resulting in it being girdled. For optimum protection, combine weed mats with other barriers.

Constructing a barrier to exclude voles from large areas can be labor-intensive and costly; however, this approach is recommended where high numbers of voles occur in areas containing valuable plants. Use a galvanized wire-mesh barrier that can stand alone, or be attached to the bottom of an existing fence (Fig. 5). A closely mowed or vegetation-free barrier on the outside of the fence will increase its effectiveness. See "Preventing Conflicts" in

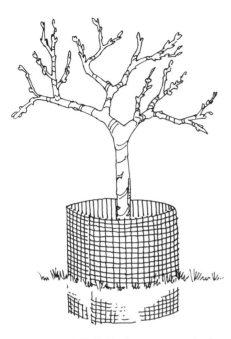

Figure 4. Individual young trees, shrubs, vines, and other plants can be protected with cylinders made from ¼-inch (6 mm) mesh hardware cloth. Bury the bottom of the barrier 4 to 6 inches (10–15 cm) below the soil surface to deter burrowing. If you extend the cylinder 18 inches (45 cm) above the ground you will also protect plants from rabbit gnawing and clipping. (Drawing by Jenifer Rees.)

Chapter 11 for additional information on ways to protect concentrated areas of valuable plants.

Natural control: Many species of wildlife eat voles, and having voles in your area may give you the opportunity to see one of them. Encouraging the vole's natural predators—or at least not interfering with them—also may aid in reducing vole damage. Although predators will not keep vole populations below the levels that present problems in gardens and landscaped areas, natural control can be combined with other methods to help prevent severe losses. Recently there has been interest in attracting barn owls and other raptors to an area for vole control using nest boxes and perching poles (see "Maintaining Hawk Habitat" in Chapter 33).

Repellents: The effectiveness of commercial and homemade repellents is highly variable and their use on voles is often not practical. They must be applied before damage occurs, and, because voles usually damage plants at or just beneath the soil surface, adequate coverage is difficult. Because rain, sprinklers, or even heavy dew may wash off repellents, they must be reapplied for continued protection. If commercial repellents are used, do not apply them to food crops unless such use is specified on the product label.

Electromagnetic or ultrasonic devices and flooding are not effective against voles.

Lethal Control

Removing voles from an area may temporarily solve the problem. However, lethal control is rarely a long-term solution because other voles will eventually enter the area if attractive habitat is available. It is important to understand that vole problems rarely can be resolved by a quick fix method, but that a continuing commitment to whatever solutions are adopted is required.

Lethal control of voles can be accomplished at any time, but it is best to concentrate the effort in late winter, before adults give birth.

Trapping

To trap voles, traditional mouse snap-traps can be used and no bait is needed when using traps with an expanded trigger (see "Lethal Control" in Chapter 8). Voles trigger these traps as they pass over them. Bait may also attract non-target animals, including domestic dogs.

Because voles seldom stray from their runways, set traps along these routes for best results. Look for their burrows and runways in grass or mulch. Place the traps at right angles to the runways with the trigger end in the runway. Examine traps daily and remove dead voles, or reset sprung traps as needed. Continue to trap in one location until no further voles are caught, then move the trap to a new location 15 to 20 feet (6 m) away. Destroy old runways or burrows to deter immigration of new voles to the site.

Because voles may carry infectious pathogens or parasites, do not handle them without gloves; you can use a plastic bag slipped over your hand and arm as a glove. Once the vole is removed from the trap, hold it with your bagged hand and turn the bag inside out while slipping it off your arm and hand. Be sure to keep small children and pets out of the area where you have set traps.

Poisoning

Check with the Cooperative Extension office or the Department of Agriculture before you attempt any chemical control method on voles (see Appendix E for contact information). Take necessary measures to ensure the safety of children, pets, and non-target animals; follow all product label instructions carefully. Some poison baits cannot be used for voles because of the potential risk they pose to predators such as cats and dogs; check the label carefully to ensure that the bait has voles or meadow mice listed. (See information on poisons under "Lethal Control" in Chapter 16 for more information.)

Burrow fumigants are not effective for the control of voles because the vole's burrow system

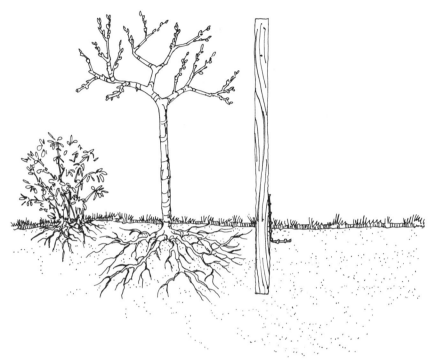

Figure 5. To protect large areas, bury ¼-inch (6 mm) mesh hardware cloth 8 inches (20 cm) deep in an L-shaped configuration. Extend the barrier at least 12 inches (30 cm) above ground to deter voles moving overland. To add to the life of the barrier, spray on two coats of rustproof paint before installation. Above-ground parts can also be painted to blend in. Always check for utility lines before digging in an area. (Drawing by Jenifer Rees.)

is shallow and has numerous open holes.

Public Health Concerns

Voles can be infected with bacterial diseases, including **tularemia,** which may be transmitted to humans. To avoid contracting a disease, wear rubber gloves if you need to handle a sick or dead vole, and wash your hands afterwards. A person who believes he or she may have contracted a disease should consult a physician as soon as possible, explaining to the doctor the possible sources of infection.

Legal Status

Because legal status, trapping restrictions, and other information about voles change, contact your local, state, or provincial wildlife office, or visit their Web site for updates. (See Appendix E for contact information and where to access the state and provincial laws mentioned below.)

Oregon: All but one of the species of vole in Oregon are classified as unprotected wildlife; the killing of these voles is unrestricted. The single protected species of vole is the white-footed vole, which is classified as a State Sensitive species (OAR 635-044-0130). White-footed voles may not be captured or killed. In the unlikely event white-footed voles were to cause property damage, property owners, persons lawfully occupying land, or their agents may capture and/or kill a white-footed vole under a permit issued by the Oregon Department of Fish and Wildlife.

Washington: The gray-tailed vole is under consideration for listing as a State Endangered, Threatened, or Sensitive species, in which case it cannot be trapped or killed (RCW 77.15.120 and 77.15.130). All other voles are unclassified and may be trapped or killed year-round without a permit.

British Columbia: All species of voles are listed under Schedule "B," which means they may be captured or killed for the purposes of protection of property by the person that owns or occupies the property or his/her spouse, parent, guardian, or child (Wildlife Act, Designation and Exemption Regulation 2 and 3).

Weasels and Mink

Weasels are powerful animals for their size and are able to successfully attack animals several times their weight. They have long, slender bodies, short legs, and a slightly bushy tail (Fig. 1). Their ears are short and rounded, and their eyes are small and black.

In areas where little or no snow accumulates during the winter, weasels essentially stay the same color throughout the year: The upper parts are a light yellowish brown and the undersides are creamy white to light yellow. In areas with significant amounts of snow, weasels are white in winter, except for a black tip on their tails that is thought to serve as a decoy, protecting weasels from lethal attacks by hawks and owls.

Weasels have a bounding gait. Because they bring their front and hind feet together, their back arches with each bound, giving their movement a resemblance to that of an inchworm.

Weasels are found in open woodlands, brushy fencerows, grasslands, and wetlands, as well as rock piles, woodpiles, and junk piles. An important requirement for weasels is a source of fresh drinking water, which they need daily.

There are three species of weasels in the Pacific Northwest (Table 1). Because the least weasel is rarely encountered, this chapter focuses on the more common species. For information on mink, see "Notes on Mink."

Facts about Weasels

Food and Feeding Habits

- To meet their high metabolic requirement, weasels eat almost half their body weight each day.
- Weasels hunt by moving rapidly and searching for prey in any burrow, tunnel, crevice, or hole where they can insert their head, long neck, and thin body.
- Ermines capture shrews, voles, mice, moles, and chipmunks; the larger long-tailed weasels capture rats, gophers, mountain beavers, and ground squirrels.
- Weasels also eat some insects, birds, and bird eggs.
- Movement triggers weasels to attack, so they will kill even if they are not hungry. Prey that is not eaten is often cached (stored) in a pile near the nest to be eaten later.

Figure 1. The long-tailed weasel occurs in riparian areas and brushy areas in forests, but it can also be found above timberline and far out into arid desert areas. (From Christensen and Larrison, Mammals of the Pacific Northwest: A Pictorial Introduction.*)*

Table 1. Pacific Northwest Weasels

The **long-tailed weasel** (*Mustela frenata,* Fig. 1) ranges in length from 12 to 22 inches (30–55 cm), including a 4½- to 7½-inch (12–19 cm) tail. It is present throughout Oregon and Washington and the southern and central mainland of British Columbia. It occurs in riparian areas and brushy areas in forests, but it can also be found above timberline and far out into arid desert areas.

The **ermine** or **short-tailed weasel** (*Mustela erminea,* Fig. 2) looks like the long-tailed weasel, but is noticeably smaller. It measures 7½ to 13½ inches (19–35 cm) in total length, including a 2- to 4-inch (5–10 cm) tail. The ermine occurs throughout the Pacific Northwest, except in dry interior areas and open pastures. It tends to frequent brushy areas and edges of forests rather than the forest interior.

The **least weasel** (*Mustela nivalis*) occurs in the northern and central parts of mainland British Columbia, and as far south as Ootsa Lake and Vanderhoof. Reaching a maximum length of 10 inches (25 cm), the least weasel well deserves its title of the smallest living carnivore.

Den Sites
- Den sites include hollow logs, trees or stumps, the burrows of their prey, and other natural cavities.
- Den sites also include areas in and under buildings, woodpiles, and rock piles.

Reproduction
- Weasels have a long gestation period of 205 to 335 days, with delayed implantation.
- Weasels breed in the summer, and four to eight young are born in April or May the following year.
- The young are born in a nest of interwoven vegetation, fur, and feathers located in the den.
- Young weasels grow and develop very fast; when six to eight weeks old, the young forage on their own; when four to five months old they become independent of their parents.
- Weasels lead solitary lives except during the breeding season and when females are rearing young.

Figure 2. The ermine, or short-tailed weasel, looks like the long-tailed weasel, only noticeably smaller, and its tail is one-quarter its body length. The females of both species are about 20 percent smaller than the males. (Oregon Department of Fish and Wildlife.)

Mortality and Longevity
- Causes of mortality include predation, disease, starvation, farm machinery, and vehicles (males are often hit when looking for mates in July and August).
- In areas where it is legal and profitable, humans trap weasels for their winter pelts.
- Predators of weasels include rattlesnakes, owls, hawks, coyotes, domestic dogs, foxes, mink, bobcats, and domestic cats.
- Where mink or fox populations are high, weasel populations are low, because mink and foxes both kill weasels and compete with them for food.
- Life expectancy of weasels is two to three years.

Viewing Weasels and Mink

Weasels and mink leave little sign and their presence often goes undetected. Weasels and mink are active year-round. Long-tailed weasels are most active in the morning and afternoon, ermine alternate periods of activity and rest throughout a 24-hour period, and mink are most active in the early morning and late afternoon. All make a circuit through their home range every 5 to 12 days.

Like other members of the mustelid family, weasels and mink have two anal glands that can discharge a disagreeable musk if the animals are frightened or disturbed. The pungent odor often hangs in the air long after the animal has vanished.

Notes on Mink

Anyone who spends a lot of time along streams, lakes, or visiting coastal estuaries and salt marshes will likely spot a mink (*Mustela vison,* Fig. 3).

Another member of the mustelid family, mink are long-bodied, muscular carnivores that have short legs and moderately bushy tails. An average male is 28 inches (71 cm) long, with the tail about one-third its body length. The luxurious fur of a mink is deep brown throughout the year, except for a whitish patch on its chin.

Although the mink's preferred prey are muskrats, it also eats rabbits, mice and other small rodents, fish, snakes, frogs, and young turtles. Fish (including spawning salmon) and marsh-dwelling birds provide additional food choices. Mink occasionally raid poultry houses.

On land, mink travel with a slow, arch-backed walk or a bounding lope, following the water's edge until something of interest draws them to swim and investigate. Mink are excellent swimmers and divers, aided by their partly webbed feet and dense, waterproof coats. Mink can also climb trees in pursuit of prey and to escape predators.

The shape of a mink's home range tends to be long and narrow, bordering the body of water along which the animal lives.

A mink's den is in a protected place near water. It will burrow under the roots of a live tree, stump, or a fallen tree, or use an abandoned muskrat or beaver den. All dens are temporary, as mink move frequently.

The female lines her den with plant fibers, feathers, and fur, and gives birth to 3 to 4 young in April or May. The family normally remains together until early fall, when each individual goes its own way.

Mink are extensively trapped for their fur, but are also preyed upon by great horned owls, bobcats, coyotes, wolves, and black bears. Mink depend heavily on aquatic areas and their main threat is the destruction of their habitat. The maximum life span for a mink is around three years.

Like weasels, mink occasionally exhibit surplus killing behavior—killing more than they can possibly eat—when presented with an abundance of food, such as in a coop full of chickens. Mink will place many dead chickens neatly in a pile.

To prevent conflicts with mink, use the management strategies described under "Preventing Conflicts" in this chapter.

Figure 3. Mink are long-bodied, muscular carnivores that have short legs and moderately bushy tails. (From Christensen and Larrison, Mammals of the Pacific Northwest: A Pictorial Introduction.*)*

Unlike skunks, however, they cannot spray musk a long distance.

Feeding Sites and Caches

Weasels tunnel through snow in search of small rodents living beneath it and escape predators in the same manner. Their tunnels are 1 to 3 inches (2.5–7.5 cm) in diameter, depending on the size of the weasel. Tracks on the snow's surface near the tunnel entrance will help confirm a weasel's presence.

A sign of a feeding weasel or mink is a partially eaten small mammal or fish. Because weasels and mink are small predators, they may eat only a small portion of their prey and store (cache) the rest. Caches are partially covered with grass, leaves, sticks, dirt, or snow. Weasels and mink will deposit droppings nearby as a territorial claim.

Any sign of feeding near the water, such as the remains of a fish, crayfish, or muskrat, is a clue to a mink's presence.

Trails and Tracks

The routes of weasels and mink seldom follow straight lines and have many loops and sharp turns. Because of their characteristic bounding gait, the track pattern shows two prints side by side that form a line of double tracks (Fig. 4). Tracks are most apparent in snow or mud.

Droppings

Weasel and mink droppings are not often seen, but when they are, they're usually along trails, on the tops of rocks or logs, next to partially eaten prey, or in a small pile a short distance from a den's entrance.

Weasel and mink droppings are long and thin, with tapered ends. They are usually made up of black, toothpaste-texture animal protein containing hair, bones, and feathers.

Mink droppings are 4 to 5 inches (10–13 cm) long and the diameter of a pencil. Long-tailed weasel droppings are 1 to 2 inches (2.5–5 cm) long and ¼ inch (6 mm) in diameter at their widest point. Ermine droppings are slightly smaller.

There may be more than one dropping at a location, each a different age, showing that a weasel or mink has passed by repeatedly. Piles of droppings are a form of territorial marking.

Den Sites

Den entrances are 1½ to 4 inches (4–10 cm) in diameter, depending on the size of the animal. The pungent aroma of musk, as well as scattered rodent fur and bones near entrances, can help identify their dens.

You may be able to coax a weasel or mink into poking its head out of a den by making quick, repetitive, squeaking noises like those made by a mouse or other rodent in distress.

Calls

Weasels and mink make a variety of sounds, including a repetitive low chatter. When excited or alarmed, weasels hiss or make a series of sharp barks.

Preventing Conflicts

Although weasels and mink occasionally kill poultry and game birds, the benefits they provide by killing rodents outweigh any harmful aspect. In a period of

Figure 4. Weasel and mink paws (shown here) are heavily furred. Long-tailed weasel tracks are about 1½ inches (3.8 cm) long and ¾ inch (2 cm) wide; ermine tracks are 1 inch (2.5 cm) long and ½ inch (1.3 cm) wide; mink tracks are 1¼ inches (3.2 cm) long and ¾ inch (1.9 cm) wide. (Washington Department of Fish and Wildlife.)

37 days, one female long-tailed weasel brought to her den 78 mice, 27 pocket gophers, 34 chipmunks, and 3 rats.

Commercial odor-eliminators can be used to remove the smell of weasel droppings in or under structures. Such products are available through hospital supply houses, drugstores, pet stores, and from the Internet using the keywords "Pest Control Supplies."

To prevent conflicts or remedy existing problems:

Enclose ducks, chickens, and rabbits at night in a secure coop. Weasels and mink quickly kill their prey by biting it through the skull or the back of the neck. Closely spaced pairs of canine teeth marks are the mark of a weasel or mink. Some dead birds may be missing heads, and mink will pluck the bird's feathers before eating it.

To prevent problems from occurring, follow the management

recommendations for small animals listed under "Preventing Conflicts" in Chapters 21 and 23.

Prevent access to denning sites.
These include spaces under porches, houses, sheds, and other structures. Weasels will use rat and other rodent holes to get under structures. See "Preventing Conflicts" in Chapters 16 and 21 for information on how to prevent and remedy conflicts.

Remove shelter. Remove brush piles, lumber piles, and rock piles where weasels and mink might live or hide. Before adopting this method, however, please be aware that you are also eliminating habitat for other wildlife species that you might want to attract. Always weigh all the consequences before manipulating natural habitats for wildlife.

Trap the animal. If all efforts to dissuade weasels or mink from using an area fail, the animal can be trapped. Such control can never be justified without a serious effort to prevent conflicts. Removal of weasels and mink by any means is a short-term solution since other animals are likely to move in if attractive habitat is still available. For information on trapping, see Appendix A, "Trapping Wildlife."

Public Health Concerns

Canine distemper, a disease that affects domestic dogs, is found in our weasel and mink populations. Have your dogs vaccinated for canine distemper to prevent them from contracting the disease. (For more information on canine dis-

temper, see "Public Heath Concerns" in Chapter 21.)

Legal Status

Because legal status, trapping restrictions, and other information about weasels, ermine, and mink change, contact your local, state, or provincial wildlife office, or visit their Web site for updates. (See Appendix E for contact information and where to access the state and provincial laws mentioned below.)

Oregon: Although not classified as furbearers, weasels are considered "mammals with commercial fur value," and a trapping or hunting license is required while someone is involved in these activities. Otherwise, weasels are not protected and can be killed without restrictions.

The mink is classified as a furbearer and fur trapping and hunting regulations apply. A license is required to trap or hunt mink during established open seasons. Property owners, persons lawfully occupying land, or their agents may trap mink and/or kill them under a permit issued by the Oregon Department of Fish and Wildlife if mink are causing damage.

Washington: Weasels and mink are classified as furbearers and a trapping license and open season are required to trap them (WAC 232-12-007).

Washington State does permit the owner, the owner's immediate family, employee, or a tenant to kill or trap weasels or mink on a property if a threat to crops or domestic animals exists. No per-

mit is necessary; however, you must notify your local wildlife office immediately after taking a weasel or mink in these situations.

British Columbia: All weasels and mink are considered furbearers. They may be legally trapped under license during open seasons as defined by regulation and there is no bag limit.

Mink and weasels involved in property damage may not be captured or killed, outside of general open seasons, without a permit first being received from the regional Water, Land, and Air Protection Office.

Mink and weasels may be trapped outside of general open seasons, by a person on their own property, if the animal is a menace to a domestic animal or bird (Wildlife Act 26(2)). No permit is required in such cases.

The *Mustela erminea anguinae* subspecies of short-tailed weasel found on Vancouver and Saltspring Islands is a Blue-listed subspecies, and there are no weasel trapping seasons on these islands. The *M. e. haidarum* subspecies, restricted to the Queen Charlotte Islands, is a Red-listed species and there are no open hunting or trapping seasons for *Mustela erminea haidarum*.

The long-tailed weasel subspecies *Mustela frenata altifrontalis* is a Red-listed species. A general weasel trapping season exists in the Fraser valley, and so *M. f. altifrontalis* may be legally trapped under license during open seasons as defined by regulation. See "Legal Status" in Chapter 1 for information on listing of Canadian animals.

Birds

ELVA H PAULSON

Canada Geese and Mallard Ducks

Canada geese *(Branta canadensis)* are among the most familiar birds in the Pacific Northwest. They are a source of recreation for bird-watchers and hunters and symbolize nature for many people. No one can miss the clear honking call of Canada geese when they fly over-head in their V-shaped formation.

Two groups of Canada geese populate the Pacific Northwest—migrating geese and nonmigrating (often called resident) geese. For a goose to migrate, it must be taught the flight path by its parents. Therefore, all following generations of nonmigratory Canada geese will also be nonmigratory, or resident geese, which will stay year-round in the vicinity where they were born.

Populations of resident Canada geese have dramatically increased over the past 25 years, particularly in urban areas where there are few predators, prohibitions on hunting, and a dependable year-round supply of food and water.

Canada geese are particularly attracted to mowed lawns around homes, golf courses, parks, and similar areas next to open water. Because geese and people often occupy these spaces at the same time of the year, conflicts arise. Many citizens enjoy the presence of geese, but others do not.

Seven subspecies of Canada geese breed or migrate through the Pacific Northwest. The taxonomy of Canada geese has been confused by the introduction of mixed subspecies, and will likely remain unclear for a long time.

The **Western Canada goose** *(Branta canadensis moffitti,* Fig. 1 and back cover) is the largest Pacific Northwest resident subspecies, referred to in the following chapter as Canada geese, or geese.

For information on mallards, see "Notes on Mallard Ducks."

Facts about Canada Geese

Food and Feeding Habits

- Canada geese graze while walking on land, and feed on submerged aquatic vegetation by reaching under the water with their long necks.
- Wild food plants include pondweed, bulrush, sedge, cattail, horsetail, clover, and grass; agricultural crops include alfalfa, corn, millet, rye, barley, oats, and wheat. Geese also eat some insects, snails, and tadpoles, probably incidentally.

Nests and Nest Sites

- Canada geese nest in areas that are surrounded by or close to water.
- Nest sites vary widely and include the shores of cattail and bulrush marshes, the bases of trees, the tops of muskrat lodges

Figure 1. The Western Canada goose has a black head and crown, a long black neck, and white cheek patches that connect under the chin. The adult gander (male) tends to be bigger than the goose (female) and averages 30 inches (75 cm) in length with a 60-inch (150 cm) wingspan. (Drawing by Elva Hamerstrom Paulson.)

Figure 2. The V-shaped flight formation allows each trailing bird to receive lift from the wingtip vortex of the bird in front of it, saving energy and greatly extending the range of a flock of birds over that of a bird flying alone. (Drawing by Jenifer Rees.)

and haystacks, and unoccupied nests of eagles, herons, and ospreys. Nests have produced successful broods of geese and ospreys in the same year.

- Other nest sites include planter boxes and nesting structures provided specifically for geese.
- The nest is a bowl-shaped depression approximately 1½ feet (45 cm) in diameter lined with grass, leaves, and goose down.
- A pair of geese may return to the same nest site in consecutive years.

Reproduction

- Canada geese usually begin nesting at three years of age.
- Adult pairs usually stay together for life unless one dies. Lone geese will find another mate, generally within the same breeding season.

- Between one and ten, but normally five to six eggs are laid in the nest in March, April, or May. Eggs are incubated by the goose (female) while the gander (male) stands guard nearby.

The female leaves the nest only briefly each day to feed.
- Eggs hatch after 25 to 30 days of incubation. The young, called goslings, can walk, swim, and feed within 24 hours.
- Both parents (especially the gander) vigorously defend the goslings until they are able to fly, which is at about ten weeks. The young geese remain with their family group for about one year.
- If the nest or eggs are destroyed, geese often re-nest in or near the first nest. Canada geese can raise one clutch per year.

Longevity and Mortality

- Predators of Canada geese and their eggs include humans, coyotes, raccoons, skunks, badgers, bobcats, and foxes, as well as gulls, eagles, crows, ravens, and magpies.
- Canada geese hatched in urban environments may have very low first-year mortalities due to the abundance of food and relative scarcity of natural predators.
- Canada geese can live more than 20 years in captivity; in the wild they have a much shorter life span.

Viewing Canada Geese

Geese are among the few water birds that will tolerate the environmental conditions found in urban areas. They are often the largest and most conspicuous bird species that people see.

Ducks and geese are often seen in a V-shaped formation when flying (Fig. 2). Such a formation allows each trailing bird to receive lift from the wingtip vortex of the bird in front of it, saving energy and greatly extending the range of a flock of birds over that of a bird flying alone.

French scientists taped heart monitors to the backs of pelicans (of a similar size to Canada geese) and observed them while in flight. They found that the birds' heart rates were lower and their wings flapped less frequently when they flew in formation than when they flew solo, again because the configuration creates an airstream that allows trailing birds to glide longer. The pelicans used 11 to 14 percent less energy in formation, savings that could be crucial for migrating birds.

Scientists have suggested that flying in V-formation may also be a way of maintaining visual contact and avoiding collisions.

Displays

Visit nearly any body of water in a nearby park (especially during the breeding period) and you will likely observe several obvious visual displays within a large active flock (Fig. 5).

Notes on Mallard Ducks

The mallard (*Anas platyrhynchos*, Fig. 3) is the most abundant duck in the Pacific Northwest. It is seen year-round on lakes, marshes, agricultural fields with standing water, and in bays, ditches, and city ponds. The recommendations for preventing conflicts associated with Canada geese also apply to mallards (including ducks and geese in swimming pools).

- **Identification:** In breeding plumage, the drake (male) is distinguished by its green head, white neckband, chestnut breast, and black, upcurled central tail feathers. The hen (female) is a much less colorful bird. Its back is mottled brown and its breast is heavily streaked with buff and darker brown. In early summer the drake goes through a molt (see "Molting") and assumes plumage like that of the hen.

    Because mallards are genetically capable of crossbreeding with other species of ducks, the resulting offspring can appear in a bewildering variety of patterns and colors. City park ducks often include groups with both mallards and hybrid ducks.

- **Food:** Mallards are dabbling ducks, which means they feed by tipping their tails up in the water and reaching their heads below the surface to find seeds, aquatic plants, and occasionally aquatic insects, fish eggs, and mollusks. Mallards also feed on land where they graze on grass and seeds.

- **Reproduction:** Mallard courtship starts in the fall, and by midwinter pairs have formed. Usually new pairs form every year, in part due to a high rate of predation on the nesting female. Mated pairs migrate northward together, heading for the female's place of origin. The drake stays with the hen until incubation

Figure 3. Perhaps the most familiar duck in the Pacific Northwest, the drake (male) mallard duck is a large and heavy bird. (Drawing by Elva Hamerstrom Paulson.)

is well under way, and then leaves to join a flock of other males.

Female mallards build their down-lined nests of leaves and grass at the edge of sloughs, lakes, or marshes, and sometimes on muskrat or beaver lodges, planter boxes, or specially designed nesting platforms. The nest may be placed 200 feet (60 m) from water, and even in a tree.

Females incubate eight to ten light-olive-green eggs, which hatch in about 25 days. To ensure that all the ducklings will hatch at approximately the same time, incubation does not start until the last egg has been laid. During the laying period and incubation the female plucks down from her belly to line the nest,

and to cover the eggs when she is away feeding. This supplies warmth and hides the eggs from marauding crows, magpies, and other predators, which are quick to find uncovered eggs. Mallard ducklings hatch from their eggshells fully covered with down, with their eyes open, and are able to walk and swim (Fig. 4).

The hen mallard is an excellent mother. Immediately after her ducklings hatch, she leads them to the nearest water, which can be a long and hazardous journey. If surprised by a human or a predator, she is likely to go flapping and squawking across the ground, as if injured. This faked injury may not fool a human, but hopefully lures predators away from the vulnerable young.

Figure 4. Mallard ducklings hatch from their eggshells fully covered with down, with their eyes open, and are able to walk and swim. (Drawing by Elva Hamerstrom Paulson.)

The ducklings first fly about two months after hatching, at which time the female finds a safe place to molt.

- **Territory:** A mallard's territory generally includes open water for feeding, cover for protection, and an area of open land called a "loafing area," used for preening and resting. Around the edge of a lake or pond the scarcity of loafing areas is often the main factor limiting the number of mallards that can establish territories.

     Three clues will help identify a mallard's territory:
     1. The male or pair will be seen in the same area day after day for about two weeks.
     2. When flushed into flight, the birds will return to the same spot after the danger is past.
     3. The birds will not allow other mallards on their territory, except on some occasions when first-year males are allowed in.

- **Voice:** The male utters soft, reedy notes, and the female a series of well-spaced, loud quacks.

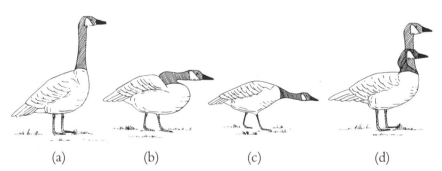

(a)   (b)   (c)   (d)

Figure 5. Some common displays of Canada geese. (From Stokes, A Guide to Bird Behavior, Volume 1.*)*

The **alert display (a)** is given when a goose is wary of some danger. The neck is vertical and straight and the head is horizontal.

The **bent-neck display (b)** is given in conflict situations with other geese. The neck is coiled back and the head is lowered and pointed toward the opponent. This display may be accompanied by a hiss.

The **head-forward display (c)** usually follows the bent-neck display and is an expression of increased threat. The goose extends its neck and holds the head low and points it toward the opponent. This display may be accompanied by a call.

The **head-pumping display (d)** is also given in conflict situations, and often precedes direct attack. The goose rapidly lowers and raises its head in a vertical pumping motion.

Nest Sites

Early in the breeding season, watch for a pair of geese quietly exploring an area. Later, listen for the honking call, which may be geese either greeting each other or engaging in a territorial squabble. Also, look for a lone male, feeding or resting, who is aggressive toward other geese or to you. Chances are its mate is on a nest nearby.

Because Canada geese are aggressive defenders of their nests and young, do not approach too closely; they may charge, and can inflict bruises with their beaks and wings.

Calls

The typical goose *ahonk, ahonk, ahonk* call is given during aggressive encounters, as a greeting, and when calling a mate. The call of

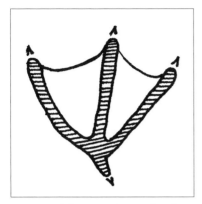

Figure 6. The Canada goose has four toes, but the hind toe is elevated and does not leave an imprint. (Drawing by Kim A. Cabrera.)

Molting

Like most waterfowl, adult Canada geese go through a complete molt every year. Molting is an opportunity for geese to replace their worn, frayed, or lost feathers with new ones. The molt takes 30 to 45 days and is completed by mid-July, a time when the adult geese are free from activities such as nesting, brood rearing, and migrating.

The young are still with the adults during the molt, and at this stage none of the family can fly—the young because they haven't grown their full flight feathers and the adults because they are replacing their flight feathers. Thus, the birds often move to areas that provide adjacent water for escape opportunities.

By late summer all of the family can fly, and they move to areas where there is abundant food, joining with other geese to form large flocks.

the male is thought to be lower than that of the female, and when a pair flies overhead, you may be able to distinguish the two sounds.

A hiss-call is given when geese are defending their territories, their nests, or their young, and is usually given only at close distances.

Tracks

Canada goose tracks are often seen on mudflats in conjunction with their sausage-shaped droppings. Their feet turn inward when they're walking. The foot's three main toes fan out in front and are connected by webs (Fig. 6). The claws are broad and blunt, and their imprint can usually be seen.

Droppings

Droppings are cylindrical and five to eight times longer than wide. Fresh droppings are greenish and coated with white nitrogenous deposits. Older droppings are darker.

Because geese have a rudimentary digestive system, they eat often and expel undigested remains in short order. Adult geese produce 1 to 3 pounds (0.45–1.4 kg) of droppings per day per bird.

Preventing Conflicts

Canada geese are extremely adaptable. They use food and other resources present in urban landscapes for nesting, raising young, molting, feeding, and resting. This has led to increasing conflicts between geese and people.

In parks and shorelines with short grass, large flocks of geese can denude areas of vegetation and litter them with their droppings and feathers. Public swimming

areas used by many geese have been closed to swimming (see "Public Health Concerns"). When nesting, geese can be aggressive toward humans, and may "attack" people who come near their nests or young.

In public areas with favorable habitat, it is rarely desirable, or possible, to eliminate geese entirely. Ideally, management programs should strive to reduce goose numbers and related problems to a level that a community can tolerate.

No single, quick-fix solution is likely to solve conflicts with geese. An integrated approach using several of the techniques described below in combination is required. Any approach to controlling geese ideally should be in place before the conflict starts—or quickly thereafter—as it is much more difficult to discourage geese after they have become attached to a site. After nesting has started, moving or scaring geese off a nest is illegal.

To prevent conflicts or remedy existing problems:

Stop feeding geese: When the diets of geese are no longer supplemented with handouts and they have to depend on the natural food supply, some or all the geese will move elsewhere.

In public areas, it is helpful to first install interpretive signs explaining the problems caused by feeding geese. Such signs might include the following in their text, preferably in the appropriate languages:

• Please don't feed the geese!
• Human food is not good for the geese because it lacks proper nutritional value.
• Feeding attracts more geese than the area can support naturally.

• Geese in high concentrations are more likely to get diseases and parasites.
• Geese droppings harbor parasites that can cause human health problems.
• Goose droppings increase algae growth that, in turn, results in fish kills.
• Goose droppings are unsanitary, unsightly, and contain parasites irritating to humans.
• Geese eat plants needed for ground cover and erosion control.
• Too many geese in one area may force the municipality to have them killed.
• Goose-management costs taxpayers money.

In order to prevent well-intentioned people from feeding geese, some localities may need to pass ordinances to regulate feeding and create authority to enforce such regulations.

Lawn management: Evolutionarily, Canada geese are tundra nesters that prefer to congregate on low vegetation adjacent to open water. Thus, areas of lawn next to water often attract geese. Large lawns provide food to graze on, room to take off and land, and an unobstructed sight line to scan for potential predators.

Although it can be expensive to transform a large lawn into something else—such as a play area or a landscape made up of plantings other than grass—it is the best long-term solution to human/goose conflicts. Such a transformation can occur over time and in phases; fencing or repellents may be necessary while the new landscape is getting established.

One important modification to a large area of lawn is to reduce its size to the point where geese no longer feel safe grazing on it. An open sight line (the distance from the geese to a place where a predator could hide) of less than 30 feet (9 m) will generally cause geese to move to a more comfortable grazing area.

Any size lawn can be made less attractive to geese by increasing its growth height to 6 inches (15 cm) and reducing the number of tender new shoots it produces. Stopping fertilizing and watering will reduce both the palatability of the lawn and the time it takes to maintain it. (The grass can be maintained at any height with a weed-whacker.) All of the lawn—or only a wide portion bordering a body of water—can be maintained this way.

Barriers

Barriers are most effective when geese numbers are low, when geese are molting (not flying), and when the barrier is in place before geese begin using the area.

Low barriers may not deter flying geese from entering an area. However, since geese typically do not land in an area that is less than 30 feet (9 m) wide, barriers, or lines of vegetation, can be used to break a site into smaller spaces. Low barriers can be combined with above-ground grids to prevent flying geese from accessing planted areas.

Plant Barriers

Geese have a fear of confinement you can take advantage of by the way you landscape. Shrubs, aquatic plants, and closely spaced groups of trees can be used to discourage geese if they block the birds' pathways to grazing areas and safety, and reduce the birds' sight lines to 30 feet (9 m).

For immediate results, plants should be at least 30 inches (75 cm) tall to prevent geese from seeing over them, and planted densely or in a staggered pattern to prevent geese from walking through gaps between the plants. Wide plantings (20 to 30 feet or 6–9 m) are more effective than narrow plantings. In wide plantings, winding footpaths prevent the geese from having a direct line of sight through the planted area, yet still provide shoreline access for humans (Fig. 7).

Where space is limited, one or two rows of shrub plantings can be combined with a fence, as described below. Ideally, the fence should be installed first and the shrubs planted as closely as possible to it so that as the shrubs grow, they envelope the fence.

Geese often gain access to grazing areas by simply walking out onshore from the adjacent body of water on which they have landed. Therefore, introducing a barrier of aquatic plants along the shoreline of a water body can create both a physical and a visual barrier to geese. Barriers of native aquatic vegetation that are at least 3 feet (90 cm) wide and include tall material, such as bulrush (*Scirpus* spp.), are most effective (Fig. 8).

If the limiting factor is the absence of an area on which to establish the new aquatic planting, constructing such an area can help. In man-made water bodies, cutting and filling can achieve a stable substrate on which to plant a barrier of aquatic plants. The water level of the pond, or other impoundment, can be temporally

Keep New Plantings in the Ground

Newly planted sites often suffer high plant mortality due to geese pulling small plants out of the ground. If still migrating, these geese would ordinarily arrive later and there would not be such pressure on the plants. To reduce this problem, or where barriers and other control tactics are not practical:

• Place large stones around the crowns of plants.

• Insert a metal staple (used to secure jute netting) over the crown of individual plants.

• Place long lengths of wood lath over the crowns of plants planted in a row. Secure the lath with metal staples or rocks.

• The above-mentioned devices will need to stay in place for two growing seasons—longer in areas where emergent plants are being established, or where there is a lot of pressure from resident geese.

• Another approach is to use large plant material (1-gallon containers instead of 4-inch pots or plugs). The larger root ball will have a better chance of getting established during the first few growing seasons.

• Drape bird netting over groups of new plantings; check netting daily for entangled birds.

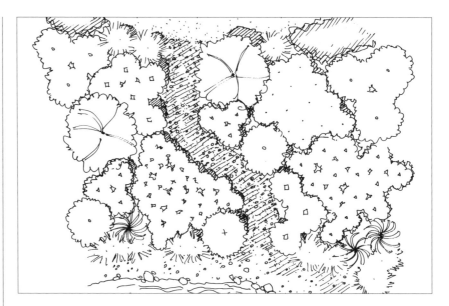

Figure 7. Plants should be planted densely or in a staggered pattern to prevent geese from viewing a passage through the planting. Wind paths through plantings to allow access for people, but not geese.

lowered to allow construction of the planting area. However, along natural water bodies, construction of a planting area can be more problematic—water levels may not easily manipulated, placing fill in deeper water is more likely to create unstable, slump-prone areas, and a permit may be required (contact you local wildlife office for permit information).

Fences

Fences can be made from woven wire, poultry netting, plastic netting, plastic snow fencing, monofilament line, or electrified wire. Fences should be **at least** 24 inches (60 cm) tall (3 feet may be better), firmly constructed, and installed to prevent the geese from walking around the ends.

Regardless of the material, lower openings should be no larger than 4 inches (10 cm) to prevent goslings from walking under or through the fence. Thus, a fence made from five monofilament lines (at least 20-pound test) should have lines set at 4, 8, 12, 18, and 24 inches (10, 20, 30, 45, and 60 cm) above ground.

Fences used in areas with tidal influence need to prevent geese entering the shore at all tide levels while not trapping fish. Turning field fencing upside down—moving the wider holes to the bottom—may accommodate fish passage.

Figure 8. In man-made water bodies, cutting and filling can provide a stable substrate on which to plant a barrier of aquatic plants. The water level of the pond, or other impoundment, can be temporarily lowered to allow construction of the planting area. (Drawings by Jenifer Rees.)

Figure 9. A low electric fence may be a temporary solution when geese have young or are molting. Flag the lines to warn people, and expect pets and wildlife to knock them awry. (Drawing by Jenifer Rees.)

Many electric fences are portable and can be set up in one or two hours and quickly taken down for storage when not in use (Fig. 9). The strands only need to be placed 4, 8, and 12 inches (10, 20, and 30 cm) above the ground. (See Chapter 7 for information on electric fencing.)

Flag all fences to warn people, and expect pets and wildlife to knock them awry.

Grids and Netted Rooms

A grid or network of multiple parallel lines of wire, stainless-steel cable, twine, rope, or monofilament (50 pound test) stretched 1 to 2 feet (30–60 cm) above a water body or other area will create a flight hazard and deter geese. There should be no more than 5 feet (1.5 m) of space between lines. If humans need to access the area under the grid, the grid can be installed high enough to accommodate them. To prevent geese from walking under the grid, install a perimeter fence as described earlier.

Attach separate lines to each vertical support (do not run the same length of wire through the entire grid) so that you will not have to rebuild the entire grid should one line break. Wherever two grid wires cross, tie the lines together to prevent rubbing and possible line breakage.

In places with large numbers of geese, and where funding is available, newly planted areas can be entirely enclosed in netting for the first few growing seasons. A netted room built high enough to allow access for maintenance can be constructed using wooden vertical supports sunk in the ground, horizontal steel cable supports, and heavy-duty bird netting. Such netting is commercially available from companies that specialize in bird control (see "Wildlife/Human Conflicts" in Appendix F). Previously used bird netting may be available from habitat restoration companies, as well as used gill netting from fisherman and fish hatcheries. The cost of new netting makes seeking out an alternative worthwhile.

Where long runs of steel cable are being installed to support netting, each line should get a separate length of cable, fitted at one end with an eyebolt, and at the other end with a turnbuckle. This will allow the cable tension to be adjusted or the cable to be removed if needed. The netting can be attached to the cable with nylon sting, wire, or hog rings. Hog rings and a special tool to attach the rings are recommended for large projects.

Note: All grids, netting, and fencing material should be regularly monitored for holes, trapped

wildlife, sagging, and overall effectiveness.

Harassment and Scare Tactics

Harassment and scare tactics are used to frighten Canada geese away from feeding, loafing, and resting areas where they are unwanted. Because geese learn that real physical danger isn't associated with harassment and scare devices, the birds will quickly learn to ignore them, no matter how effective these devices may be initially. Because of this, and to take advantage of geese being neophobic (fearful of novel objects), two important rules are: (1) never rely solely on one tactic, and (2) vary the use by altering the timing and location. Harassment and frightening devices are only as effective as the person deploying them.

Harassment and scare devices are available from the Internet, at over-the-counter bird-control businesses, and at some farm and garden centers (see "Wildlife/Human Conflicts" in Appendix F for vendors).

Harassment and scare tactics include:

Eyespot Balloons

Like most birds, geese rely more on vision than on their other senses to avoid danger, and so visual stimuli can be effective. Commercially available eyespot balloons are large, helium-filled balloons with a large, eye-like images. (Large colored spots on three sides of any helium balloon can suggest eyes.) Tether balloons on a 20- to 40-foot (6–12 m) monofilament line attached to a stake or heavy object. The bal-

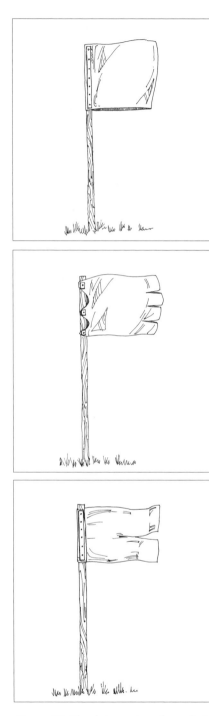

Figure 10. Flag designs using a large plastic garbage bag on a pole. Note the wooden battens installed to prevent the flags from ripping.(Drawings by Jenifer Rees.)

loons should be located where the wind will not tangle them in trees and utility lines, and should be repositioned at least once per day. Two balloons should be adequate for an average size yard.

Flags and Streamers

Flags and streamers work best in areas where there is a steady wind. A simple flag design uses plastic garbage bags mounted on tall poles (Fig. 10).

In addition, mylar tape can be made into 6-foot (1.8 m) streamers and attached to the top of 8 foot (2 m) long poles. Mylar tape is silver on one side, usually red on the other, and is very shiny and reflective.

A disadvantage of Mylar tape is that it is only effective in bright sunlight and wind. Poles with flags and streamers should be repositioned once per day.

Scarecrows

Scarecrows are only effective where geese view humans as dangerous predators, such as rural areas where they are hunted. Scarecrows can be made out of almost any material; however, the design should include movement, bright colors (red, blaze orange, or safety yellow), and large eyes. For maximum effect, the arms and legs should move in the wind, and the scarecrow should be moved once per day.

Geese occasionally will find a swimming pool an acceptable area. Large, blow-up toy snakes are reported to work as a type of scarecrow. Simply buy two or three of these, add weights (sinkers), and put them in the pool. Streamers made of mylar tape may also work if strung across the landing zone.

Noisemaking Devices

Devices that make a loud bang can scare geese, causing them to take flight. Promptness (beginning as soon after the geese arrive as possible) and persistence are the keys to success when using these devices.

Types of noisemakers include propane cannons, blanks, and whistle bombs. Propane cannons are stationary devices that explode propane gas at irregular intervals. Shell crackers and whistle bombs are shells that are fired from a shotgun or special pistol. When fired they either scream for a distance of 50 yards (45 m), or explode. Pyrotechnics should only be used by skilled individuals who understand the dangers that these tools can pose.

Loud auditory tactics generally require permits from area police departments and may be restricted in urban areas because of noise ordinances. When such devices are used, it is important that all organizations involved in the process be kept in communication. In addition, the surrounding neighborhood should be advised of what the process is trying to accomplish.

Where a small number of geese are a problem, construct a clapper by putting a hinge on two 24-inch (60 cm) 2 x 4s and smacking them together (see example in Chapter 6).

The more geese are exposed to these fear-provoking stimuli, the faster they will become accustomed to them and ignore them. For this reason, noisemakers should be used sparingly, and propane cannons should be set so that they fire only a couple of times per hour.

Lasers

Recent research conducted by the National Wildlife Research Center indicates that relatively low-power, long-wavelength lasers provide an effective means of dispersing geese, gulls, crows, and ravens under low light conditions, while presenting no threat to the animal or the environment. The lower power levels, directivity, accuracy over distance, and silence of laser devices make them safe and effective species-specific alternatives to noise-making devices.

Although researchers are not sure if birds see the same red spot as people, it is clear that in certain bird species the spot of laser light elicits an avoidance response. The birds view the light as a physical object or predator coming toward them and generally fly away to escape. *Note:* Lasers should never be aimed in the direction of people, roads, or aircraft.

At the time of writing, the cost of a laser device is still quite high. Check with dealers through the Internet and over the counter at bird-control businesses for current prices and instructions for use (see "Wildlife/Human Conflicts" in Appendix F for vendors).

Dogs

When directed by a handler, dogs are the method of choice for large open areas such as golf courses, airports, parks, agricultural fields, and corporate parks. In residential areas, parks with continuous public use, areas bisected by roadways, and large water bodies, dog use may not be appropriate.

Results are often immediate. After an aggressive initial use (several times a day for one or two weeks), geese get tired of being harassed and will use adjacent areas instead.

A dog can be tethered to a long lead (which may require relocating the dog and tether frequently to cover more area), be allowed to chase and retrieve a decoy thrown over a large flock of geese, or be periodically released to chase the birds (if this is not against leash laws).

While the wolflike gaze of border collies is frightening to geese, these dogs rarely harm them. These dogs can be purchased already trained, or be trained; however, it is also possible to hire a border collie "service."

Other breeds of dogs can also do the job. It is recommended that they be from proven working stock, preferably with prior experience with or exposure to live animals, particularly birds. (See Appendix D for information on the care of dogs.)

Chemical Repellents

Taste-aversion products and other chemical repellents are unobtrusive, may be applied directly to the problem area, and will not permanently harm the geese. Drawbacks to repellents include the high costs of covering large areas, the need for frequent application in rainy areas and during the growing season, odors associated with the few registered products, and their negative influence on the behavior of other wildlife.

If geese have used the area in the past, apply repellent before their return. Carefully read and follow all label and technical directions. (See "Wildlife/Human Conflicts" in Appendix F for vendors.)

Canada Geese "Relocation"

A control technique used early on to reduce the urban goose population was relocation, in which geese were corralled into pens, captured, and trucked to distant areas. The relocations have been only partly successful, at best. Some of the deported geese came back to the areas where they were captured as soon as they could fly again or by the following fall. When geese did stay in areas where they were released, their populations often grew, and so did conflicts with humans in the new area.

The roundups helped reduce goose numbers in specific areas during summer, the peak period of public use, but had little effect on the number of geese in these areas in the long term. Birds that returned were often joined by new birds, sometimes quickly bringing populations back to their former levels.

In many places the relocation methods were deemed to be politically unacceptable, and abandoned. One thing that was learned from these efforts is that if geese are removed from an area, other forms of control **must** be used after removal in order to keep the area from being reoccupied.

Lethal Control

If the above nonlethal control efforts are unsuccessful and the damage situation persists, lethal control may be an option. Lethal control techniques include legal hunting, shooting out of season by permit, egg destruction by permit, and euthanasia of adults by government officials.

Public Health Concerns

Canada geese are not considered to be a significant source of any infectious disease transmittable to humans or domestic animals, although their droppings are increasingly cited as a cause for concern in controlling water quality in municipal lakes and ponds.

Swimmers itch *(schistosome* or *cercarial dermatitis)* is caused by a parasite that can be spread by goose droppings, but does not mature or reproduce in humans. Recommendations to reduce the risk of swimmers itch are to: (1)

vigorously towel off immediately upon exiting the water (including under bathing suits), and (2) take a soapy shower immediately after exiting the water.

If you do get the itch, a topical rash cream should alleviate some of the itching, and the rash should clear up within a week. If you have concerns or questions, contact a physician.

Legal Status

Because legal status, hunting restrictions, and other information on Canada geese and mallards change, contact your local, state, or provincial wildlife office, or visit their Web site for updates. (See Appendix E for contact information and where to access the state and provincial laws mentioned below.)

Oregon and Washington: Canada geese and mallard ducks are protected under federal and state law and a hunting license and open season are required to hunt them. Where lethal control of Canada geese or mallards is neces-

sary outside of hunting seasons, it should be carried out only after the above nonlethal control techniques have proven unsuccessful and only under permits issued by the U.S. Fish and Wildlife Service. Currently, the only agency permitted for lethal removal is the U.S. Department of Agriculture's Wildlife Services.

British Columbia: Canada geese and mallard ducks are protected under the Federal Migratory Bird Convention Act and the Provincial Wildlife Act. A license and open season are required to hunt them. Any permit to control mallards outside of hunting seasons would need to be issued from the Canadian Wildlife Service (CWS), and would likely only be issued in very extreme cases. CWS issues egg-addling permits to organizations that show a need (usually parks, golf courses, etc.) to deal with "nuisance resident geese." Kill permits are sometimes issued for a limited amount of birds, and are usually restricted to golf courses or agricultural areas.

ELVA Paulson @2001

CHAPTER 28

Crows and Ravens

Crows and ravens belong in the Corvid family (which includes jays and magpies) and are considered to be among the most adaptable and intelligent birds. Its coal-black coloring, highly social behavior, and distinct call make the American crow *(Corvus brachyrhynchos)*, also known as the common crow, one of the most frequently seen and heard birds. Although most bird books recognize populations along the coast and around the Puget Sound to be a distinct species called the Northwestern crow *(Corvus caurinus),* some experts classify the smaller Northwestern crow as a subspecies of the American crow.

Crows will occupy almost any woodland, farmland, orchard, or residential neighborhood, as long as sufficient shelter and enough trees suitable for nesting are available. They seem to prefer lower elevations and moist places, including creeks, streams, and lakeshores. The Northwestern crow is found almost exclusively in very close proximity to Puget Sound and coastal marine waters.

In recent years, crow populations have expanded into urban and suburban areas. Their tameness becomes notable as they seek the plentiful food sources found on roadsides, parking lots, ferry landings, marinas, and other places where humans influence the landscape.

For information on ravens, see "Notes on Ravens."

Figure 1. The brain is especially well developed in corvids—crows, ravens, jays, and magpies—a family of birds considered intelligent because of their ability to adapt quickly to changing circumstances. (Drawing by Elva Hamerstrom Paulson.)

Facts about Crows
Food and Feeding Habits

- Crows are omnivorous and eat whatever is available—insects, spiders, snails, fish, snakes, eggs, nestling birds, cultivated fruits, nuts, and vegetables. They also scavenge dead animals and garbage.
- Crows are known to drop hard-shelled nuts onto a street, and then wait for passing automobiles to crack them. Similarly, along the coast they drop mussels and other shellfish on rocks

ELVA H PAULSON

Notes on Ravens

Ravens are found throughout the Pacific Northwest, except in major urban areas where competition from crows and a lack of nesting sites are probably too great. Ravens replace crows in mountainous areas, deserts, and rimrock areas; they are thus more common than crows east of the Cascade Range.

Ravens can be distinguished from crows by their larger size, much larger bill, and long, wedge-shaped tail (Figs. 2 and 3). In addition, a raven's flight pattern commonly includes soaring or gliding, while crows have a frequent, steady, "rowing" wing-beat with little or no gliding.

The raven can produce an amazing assortment of sounds. One study showed they have more than 30 distinct vocalizations. The trademark call of the raven is a deep, guttural *grock*. Other common calls include a high-pitched *aack* and a rolling, throaty rattle.

The breeding, nesting, and family structures of ravens and crows are similar. However, the courtship flights of ravens are most spectacular; male and female birds fly high and, with great exuberance, barrel-roll out of the sky together.

Ravens prey on a wide array of small animals, but are notorious scavengers. They are common visitors to garbage dumps, road kills, and the bodies of dead livestock. Ravens will hide or cache food supplies to eat later. Like crows, hawks, and owls, ravens regurgitate indigestible food in the form of pellets.

Long evoking strong emotion from humans, the raven has often played important roles in culture, mythology, writing, and stories. The spiritual importance of the raven to Alaska's Native people is still recognized. The Tlingit, Haida, Tsimshian, Bella Bella, and Kwakiutl tribes all viewed the raven as the creator of the world and bringer of daylight. The raven is also a significant social and religious component of Eskimo/Inuit culture.

Figure 2. Ravens are usually seen in pairs or small family groups, rather than the large flocks that crows often form. (Drawing by Elva Hamerstrom Paulson.)

to crack the shells and expose the flesh.

- Outside of the breeding season, crows travel as far as 40 miles (64 km) each day from evening roost sites to daytime feeding areas.
- Crows usually post "sentries," who alert the feeding birds of danger.

Nest Sites and Shelter

- Nests are built 15 to 60 feet (4.5–18 m) above ground in tall coniferous or deciduous trees. Nests are 1½ to 2 feet (45–60 cm) in diameter, and solidly built in the crotch of a limb or near the tree trunk.
- In areas that lack tall trees, nests may be placed lower in hedgerows or shrubbery. In urban areas, crows may nest on window ledges or the sides of buildings.
- Nests are constructed from branches and twigs, and are lined with bark, plant fibers, mosses, hair, twine, cloth, and other soft material.
- Hawks and owls inhabit old crow nests; raccoons and tree squirrels use them as summer napping platforms.

Reproduction and Family Structure

- Both sexes build the nest during a period of 8 to 14 days—beginning as early as mid-March and as late as mid-July—depending on latitude and elevation.
- The female incubates four to five eggs for 18 days, at times being fed by her mate or sometimes by offspring from the previous year.
- The chicks grow quickly and are out of the nest at around

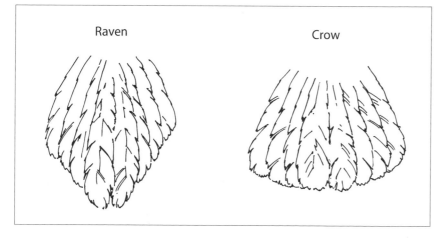

Raven   **Crow**

Figure 3. Ravens have wedge-shaped tails and crows have fan-shaped tails. This isn't very easy to see if the bird is sitting on the ground, but when it's flying overhead, you can often get a good look at the shape of the tail. (Drawing by Jenifer Rees.)

four weeks after hatching, although they continue being fed by the adults for about another 30 days.

• Frequently, one or more young crows remain with the parents through the next nesting season, or several nesting seasons, to help care for nestlings. This cooperative behavior during breeding includes bringing food to the nest and guarding the nestlings.

• In spring and summer, crows are usually seen in family groups of two to eight birds. During late summer, fall, and winter, crows gather from many miles to form communal night roosts.

Mortality and Longevity

• Adult crows have few predators—eagles, hawks, owls, and human hunters—with humans being their main predator.

• The causes of death of young crows still in the nest include starvation, adverse weather, and attacks by raccoons, great horned owls, and other animals.

• Mortality in the first year is about 50 percent, but adults live six to ten years.

Viewing Crows

Much of the time, crows are seen in small, noisy, family bands, spending the majority of their time in fairly restricted areas. For about a month during the nest-building, egg-laying, and incubation periods, breeding adult crows become uncharacteristically secretive and quiet. After the eggs have hatched, the parents become noisy defenders of their nest and later the young are heard wailing at their parents for food with an insistent, nasal *caw*.

In late summer through winter, crows are seen in large, raucous flocks that roam widely. In agricultural areas hundreds of crows may gather to forage in fields, while in cities, landfills and garbage dumpsters are crow favorites.

Interesting visual displays include male and female crows bobbing their heads up and down, and accentuating this by bowing. The wings and tail may also spread slightly and the body feathers may be fluffed. The bobbing display is usually performed in the presence of another crow in spring, and is possibly associated with courtship. Males may also engage in diving flight displays, chasing females.

Crows mob owls, hawks, and eagles throughout the year and are in turn mobbed by smaller birds. The loud, excited calls of crows are very characteristic and may lead you to sighting a local bird of prey.

Nest Sites

Even though crows are common, their nests are not easy to locate, except after deciduous trees lose their leaves. In addition to being secretive nesters, crows may partially construct a number of preliminary or decoy nests.

Crows return to the same nest territory year after year, often a few weeks before they start building. If a small group of crows remains in a particular area day after day, this may signal that nest building is about to begin.

Many larger twigs that form the base of the nest are broken directly off trees. If you see a crow hopping slowly about in some dead branches, continue to watch and you may see it break off a branch and carry it to the nest. This is the best time to try to find nests, as the birds are less secretive than during egg-laying and incubation.

Roost Sites

When the nesting period is over, the family group usually joins other groups of crows in communal night roosts. Roosts reach their highest numbers in late winter and may contain hundreds or even thousands of birds. Roost sites are generally located in groups of trees, often near water, and are used for many years if they aren't disturbed.

Communal roosting helps crows exchange information and find mates. Some birds, because of their age or familiarity with the surrounding landscape, are more efficient at finding food. Less-experienced members of a roost can follow other birds to known feeding sites. Communal roosting also helps crows remain safe and warm. Crows occupying the center of the roost are less exposed to predators and weather than those on the edges or those roosting alone.

Crows are believed to return to the same roost each night, and their behavior is often predictable. Each morning the roost breaks up into smaller flocks that disperse across the landscape to feed. In mid-afternoon, these smaller flocks start back toward the communal roost. They fly along the same flight lines each day and are joined by other flocks as they go. Often there are pre-roosting sites, where flight lines coincide and crows stop to feed before flying the final distance to the roost. Communication between groups of crows often takes place at these pre-roosting sites.

If you are near a flight line for as long as an hour, you will notice crows passing overhead, a few to several hundred at a time in the late afternoon.

Calls

The normal crow call is a loud *caw* or *awk*. The male also makes a dry, rattling call, very different from the normal call. If you are very fortunate you may hear the soft, almost melodious song of the crow.

Tracks and Trails

Crows spend a lot of time on the ground and tracks can be seen in snow, mud, or in wet sand at low tide (Fig. 4). Despite the fact that

Figure 4. The four toes in a crow's track are about the same length, and in good prints each toe leaves a claw mark at the end. Three thick toes point forward, and one long toe (equally thick) points back. The total length of a print is approximately 3 inches (7.5 cm). (From Pandell and Stall, Animal Tracks of the Pacific Northwest.*)*

"crow footed" is a term used to describe people who walk with their toes pointed inward, crows (and ravens) usually leave relatively parallel tracks.

Preventing Conflicts

Crows help control pest insects and "clean up" dead animals and garbage that has been scattered by other animals. Although crows prey on songbirds and their young, research suggests that they do not ordinarily have a significant impact on songbird populations. Robins, for example have evolved to have two to three clutches each year to make up for young lost to crows. However, because crows are intelligent, opportunistic, and protective of their young, and at times congregate in large numbers, they can create problems for people.

To prevent conflicts or remedy existing problems:

Keep crows out of the trash. Crows are often blamed for spilling garbage, trash, or grain that was actually spilled by other animals seeking food. To prevent other animals from making garbage available to crows, follow the suggestions under "Preventing Conflicts" in Chapter 21.

In addition, don't leave trash bags alongside a curb, in back of a pickup truck, or in an overfilled bin. Crows are early risers and will visit unattended garbage at first light or shortly thereafter. Therefore, overflow garbage bags should not be put out before sunrise on the morning of pickup. Ask your local restaurants and food chains to keep their garbage containers closed.

Keep crows away from crops. Protect fruit crops with flexible bird netting, which can be purchased in a variety of lengths and widths at garden and hardware stores or over the Internet from bird-control businesses (see Appendix F for vendors). Four-inch (10 cm) mesh will keep crows out, but not smaller birds such as robins and starlings. Tie the netting securely at the base of the shrub or onto the trunk of the tree to prevent crows from gaining access from below (see "Preventing Conflicts" in Chapter 38 for an example). Protect germinating corn plants and other crops with netting until plants are about 8 inches (20 cm) tall.

Visual scare devices, such as pie tins hung in trees, Mylar scare tape, Mylar balloons, scarecrows, or flags can be used to provide temporary protection (see "Preventing Conflicts" in Chapters 27

Figure 5. If you find an uninjured nestling that has fallen or been pushed out of its nest, replace it in the nest. It is not true that birds abandon their chicks if a person touches them. Birds have a poor sense of smell. (Washington Department of Fish and Wildlife.)

Baby Birds Out of the Nest

Sooner or later, no matter where you live, you'll come across a baby bird on the ground. You'll have to decide if you should rescue it or leave it to fend for itself. In most cases, it is best not to interfere. A baby bird that is featherless must be fed every 15 to 20 minutes from sunrise to sunset! This obviously requires a large time commitment on the part of the foster parent. The natural parents do a much better job at raising their young than we could ever do.

Finding fully feathered birds: If the bird is fully or partially feathered, chances are it doesn't need your help. As young birds develop they soon outgrow the limited space of a nest. The young birds, referred to as "fledglings" or "branchers" at this stage, typically leave the nest and move about on the ground and on low branches for a few days before they can fly. Their parents are nearby and continue to care for the birds, answering their demanding calls with regular deliveries of food. The scolding calls coming from the nearby tree are likely the adult birds, voicing their disapproval while they wait for you to leave.

Unless injured, the fledgling bird should be left where it is. Efforts should be made to keep cats, dogs, and curious children away from the bird so the mother can continue to feed it.

Unfortunately, this is when people often interfere and take a healthy bird out of the wild. Not only is this illegal in most cases involving native birds, but it also deprives the growing bird of essential care it needs from its parents.

Finding naked birds or birds with beginning feathers: If you find an uninjured nestling that has fallen or been pushed out of its nest (Fig. 5), replace it in the nest. (Note that this behavior is actually adaptive for some species. This way, only the strongest of the brood survive and go on to raise young themselves.) If the nest has fallen down (common after windstorms), replace the nest in a tree with the baby bird(s) in it. (It is **NOT TRUE** that birds abandon their chicks if a person touches them. Birds have a poor sense of smell.)

If you can't find the nest or accessing it is too dangerous, put the baby bird where its parents can find it but where it will be safe from cats. Use a small plastic berry basket, margarine tub, or similar container lined with shredded paper towels (no cotton products, which tend to tangle up in birds' feet). With a nail or wire, fasten the makeshift nest to a shady spot in a tree or tall shrub near where the bird was found. Next, place the nestling inside, tucking the feet underneath the body.

The parents will usually come back in a short time and will feed the babies in the container just as if it were the original nest. (Often, you will see the mother going back and forth between each "nest," feeding both sets of babies.)

Times when you should consider quickly getting the bird to a wildlife rehabilitator: (1) If the parents don't find the new nest within two hours, or if you are certain that the mother of a baby bird is dead; (2) if the bird is hurt or sick (unable to flutter wings, bleeding, wings drooping unevenly, weak or shivering), or the bird was attacked by a cat or dog, call a wildlife rehabilitator immediately. The longer the delay, the less chance the bird has of surviving. Your local wildlife office keeps a list of rehabilitators and can tell you which ones serve your area, or you can look under "Animals" or "Wildlife" in your phone directory. If a rehabilitator isn't available, follow the menu options over the phone or on the group's Web site for information on what to do. (See "Wildlife Rehabilitators and Wildlife Rehabilitation" in Chapter 20 for additional information.)

While waiting for a rehabilitator to arrive, pick the bird up with your gloved hands and place it in a well-ventilated, covered box or paper bag that is padded with paper towels.

Keep the baby bird warm and in a quiet, dark place until it can be picked up by a wildlife rehabilitator. If the bird is cold, put one end of the bird's container on a heating pad set on low. Or, fill a zip-top plastic bag, plastic soft-drink container with a screw lid, or a rubber glove with hot water. Wrap the warm container with cloth, and put it next to the animal. Make sure the container doesn't leak, or the animal will get wet and chilled.

Do not give the baby bird any liquids (they get all they need from their food and very often will inhale any liquid).

Wash your hands after contact with the bird. Wash anything the bird was in contact with—towel, jacket, blanket, pet carrier—to prevent the spread of diseases and/or parasites to you or your pets.

and 38 for detailed information).

One recent innovation is a motion sensor combined with a sprinkler that attaches to a spray hose. When a crow comes into its adjustable, motion-detecting range, a sharp burst of water is sprayed at the bird. This device appears to be effective by combining a physical sensation with a startling stimulus.

Keep crows out of nest boxes. Crows are capable of pulling nestlings out of nest boxes. They are most apt to snatch an older nestling that sticks its head out of the hole to accept food from its parents, but will also poke their heads into nest box entrance holes. This is a learned behavior that can result in individual predatory birds making the rounds of boxes and causing many losses of nestlings, and teaching other individuals to do the same.

To prevent this: Never put up a shallow box; there should never be less than 6 inches (15 cm) from the entry hole to the bottom of the box. Also, clean out used nests annually so the nesting birds do not fill the lower part of the box. Never put up a box designed with a perch or ledge under the hole; install wire mesh or similar predator guards on the entrance holes if necessary (see Appendix J for a sample of such a predator guard). Other predator guards are available from bird specialty stores and from the Internet.

Communal Night Roosts

The communal night roosts of crows create accumulations of droppings with the potential to spread disease. When and where

this poses a health risk to the public (as deemed so by a Public Health representative) or cannot be tolerated, steps need to be taken to remedy the problem.

Options include making the area temporarily off limits, routinely cleaning up the soiled area underneath the roost (see "Public Health Concerns"), or dispersing the flock by making the roost site undesirable to crows.

In Oregon and Washington, large-scale intervention strategies should be undertaken with the guidance of the Department of Agriculture. In British Columbia, contact the Ministry of Agriculture, Food, and Fish and the Ministry of Water, Land, and Air Protection (see Appendix E for contact information).

Methods to disperse crows from a night roost include:

Harassment Techniques

Harassment techniques include visual and audio stimuli and an assortment of other approaches to make crows uncomfortable enough to move elsewhere. If possible, act quickly when large numbers of roosting crows are detected. The birds will be more willing to abandon a roost site they have not been using long. *Note:* Most harassment techniques are effective only for a short time and the public may not like them because they cause crows to move elsewhere—such as a neighborhood park or someone's backyard containing large trees.

Visual scare devices include Mylar tape, eye-spot balloons, scarecrows, and laser devices. Visual harassment devices can provide effective short-term control, especially when they are used in combination with auditory

devices. (See "Preventing Conflicts" in Chapter 27 for detailed information.)

Audio scare devices include hazing with pyrotechnics such as cracker shells, blanks, propane cannons, and recorded crow distress and warning calls. (See "Preventing Conflicts" in Chapter 27 for detailed information on all the above except distress and warning calls.)

The main drawback with recorded calls is that crows ultimately learn they are not real and get used (habituate) to them. Because distress calls are given when a crow is being held by a predator, and alarm calls are given when there is a predator in the neighborhood, crows probably expect to see a predator whenever they hear one of these calls. If they do not, they may realize that something is not right and habituate more rapidly to the distress and alarm calls. For this reason, it is wise to pair the broadcast of these calls with a predator model, such as a "scarecrow."

When using any auditory scare device, change the area from which it is emitted, daily if possible. When using pyrotechnics, try to elevate them above the roost site.

Crows scare most easily when they are flying. They are most difficult to scare when perched in the protection of their roost. Therefore, audio devices should begin to be used when the first birds come in to roost, usually an hour and a half before dark. The same group of crows may circle around and come toward the roost many times, so scaring efforts need to continue until it gets dark.

Scaring should stop with darkness or the crows will become

accustomed to the sounds. If using recorded alarm calls, play them only 10 to 15 seconds per minute when the birds are coming in. When most of the birds are perched, play the call continuously until dark. If possible, early morning scaring should be used in conjunction with evening scaring, and should begin as soon as the first bird movement is detected in the roost, often just before daylight.

Success may not be achieved for several nights and will entail continuous efforts every evening and every morning. Because the crows may attempt to establish temporary roosts in other unsuitable locations, scaring efforts may be needed elsewhere until the birds move to an acceptable area. If crows are disturbed in their new roost site they will move back to the old one. Be prepared to resume efforts if they return.

Modify the Night Roost

Modifying the structure of the crows' night roost can discourage the birds from using it. This includes thinning up to 50 percent of the branches of roost trees, or removing trees from dense groves to reduce the availability of perch sites and to open the trees to the weather. A tree service company can remove tree limbs (see "Preventing Conflicts" in Chapter 38 for an example).

Other Techniques

Other techniques to disperse crows include using 4-inch (10 cm) mesh bird netting to create a barrier between the

Dive-Bombing Crows and Other Bird "Attacks"

Most aggressive behavior from birds is motivated by defense of their territory or young, or their search for handouts. Hummingbirds have been noted to buzz people wearing red, perhaps thinking that they were a group of nectar-rich flowers. Pigeons and swallows may appear to be attacking humans when actually they are returning to their nests in the eaves of buildings.

In the spring and summer crows and other birds establish territories, build nests, and rear young. During this period, adult birds may engage in belligerent behavior, such as attacking creatures many times their size. In this case, the birds are simply trying to protect their homes, their mates, or their young.

When possible, stay away from nesting areas with aggressive birds until the young are flying (three to four weeks after eggs hatch) and the parents are no longer so protective. (Do not attempt to "rescue" chicks found outside nests when adult birds are calling loudly nearby—see "Baby Birds Out of the Nest" for information.) If you must walk past a nest, wave your arms slowly overhead to keep the birds at a distance. Other protective actions include wearing a hat or helmet, or carrying an umbrella.

Other aggressive birds include ducks and geese found around city and suburban lakes, and gulls found near fishing piers, who become accustomed to being fed by humans and lose their natural fears. When a human appears at the spot where they usually are fed, the birds expect food and may approach without caution. Some of these birds become quite aggressive in their begging methods and may actually chase, hiss, and peck at the hapless human. This "conditioned response" technique was actually used to train the gulls and crows to chase the actors in the Hitchcock thriller *The Birds.* If this occurs, pick up small children and leave the area. If necessary, act aggressively toward the birds by waving your arms and shouting.

roost and the crows; spraying crows with water from a high-pressure hose (some cities have used a fire hose); installing a 360-degree sprinkler up in the roost tree; and lighting up the interior of the roost with bright fluorescent lights.

Lethal Control

Shooting is not an effective way to manage crow populations overall. The number of birds that can be killed by shooting is small relative to the size of the flock. However, shooting may be helpful where only a few birds are present, and in supplementing or reinforcing other dispersal techniques. First check the local ordinances regarding discharging firearms. For additional information regarding shooting crows, see "Legal Status."

Public Health Concerns

Although health risks from birds are often exaggerated, large populations of roosting crows may present risks of disease to people nearby. The most serious health risks are from disease organisms growing in accumulations of droppings, feathers, and debris under a roost. This is most likely to occur if roosts have been active for years.

Precautions need to be taken when working around large concentrations of crow droppings. See "Public Health Concerns" in Chapter 29 for information.

At the time of writing, West Nile virus, a virus carried by mosquitoes, has killed thousands of crows in the Northeast, South-east, and Midwest portions of the United States. Call your local Department of Health for updated information. Always wear gloves when handling dead or live birds.

Legal Status

Because legal status, hunting restrictions, and other information on crows and ravens changes, contact your local, state, or provincial wildlife office, or visit their Web site for updates. (See Appendix E for contact information and where to access the state and provincial laws mentioned below.)

Oregon: The crow is protected from being killed or captured at any time. The crow is not classified as a game bird, but there is an open season when it can be killed for sporting purposes and hunting regulations apply (OAR 635-054). When crows are found committing or about to commit depredations upon ornamental or shade trees, agricultural crops, livestock, or wildlife, or when concentrated in such numbers and manner as to constitute a health hazard or nuisance, they may be controlled without a federal or state permit.

The raven is protected under federal and Oregon law and cannot be hunted or trapped. A federal permit can be obtained from the U.S. Fish and Wildlife Service to use lethal means to control ravens when extreme damage is occurring on private property. Such a permit is only granted after all other non-lethal control techniques have proven to be unsuccessful.

Washington: The crow is classified as a predatory bird (WAC 232-12-004). A hunting license and an open season are required to shoot them. Under federal guidelines, individuals may kill crows without a hunting license or permit when they are found committing, or about to commit, depredations on agricultural crops, or when concentrated in such numbers and manner as to constitute a health hazard or other nuisance (16 U.S.C. Sections 703–712). The Code of Federal Regulations (CFR) is located at www.access.gpo.gov/nara/cfr/

The raven is classified as protected wildlife and cannot be hunted or trapped (WAC 232-12-011). Under federal guidelines, a permit can be obtained from the U.S. Fish and Wildlife Service to use lethal means to control ravens when extreme damage is occurring on private property. Such a permit is only granted after all other nonlethal control techniques have proven to be unsuccessful.

British Columbia: Crows and ravens come under the sole jurisdiction of the province. Both are identified as Wildlife and are protected under Section 34 of the Wildlife Act. However, crows are listed under Schedule "C" of the Wildlife Act and are exempted from protection (Wildlife Act, Designation and Exemption Regulation 2 and 3). Therefore, they, their eggs, nests, and young may be killed by the public without a permit (although municipal regulations governing firearms, etc., restrict methods in urban areas).

Ravens have full protection but in some regions hunting seasons and bag limits are used as a control method where there are conflicts with agriculture.

CHAPTER 29

Domestic Pigeons (Rock Doves)

The domestic pigeon *(Columba livia)* (also called the rock dove or city pigeon) was originally found in Europe, Northern Africa, and India. Early settlers introduced it into the eastern United States as a domestic bird in the 1600s. Since then, it has expanded throughout the United States to Alaska, across southern Canada, and south into South America.

Pigeons originally lived in high places—cliffs, ledges, and caves near the sea—that provided them with safety. Over time they have adapted to roosting and nesting on windowsills, roofs, eaves, steeples, and other man-made structures.

Pigeons typically have a gray body with iridescent feathers around their neck, a broad black band on their tail, and salmon-colored feet (Fig. 1). Breeders have created color variations, so the body color may also be white, tan, black, or a combination of several colors. Pigeons have a strutting walk and their call is a long, drawn-out *coo* that can be heard quite easily. When they take off, their wing-tips touch, making a characteristic clicking sound.

Two native birds, the **band-tailed pigeon** *(Columba fasciata)* and the **mourning dove** *(Zenaida macroura)* are sometimes confused with domestic pigeons. Band-tailed pigeons are similar in size but have a purplish head and breast, a dark-tipped yellow bill, yellow feet, and a small white crescent on top of the neck (Fig. 6). Mourning doves are smaller than domestic pigeons, have a long, pointed tail, large dark eyes, a dark bill, and a mournful *who-ooh, who-who-who* call.

Facts about Pigeons
Food and Feeding Habits

- Domestic pigeons mainly eat seeds and grains.
- Pigeons also eat insects, fruit, and vegetation, and scavenge food people provide for them—intentionally or unintentionally.
- While young birds of other species are fed a high-protein diet of insects, young pigeons are fed "pigeon milk"—a milky-white fatty substance regurgitated from both parents' crops.
- Pigeons feed on open ground such as that found in parks and squares, on rooftops, at food-loading docks and garbage dumps, and wherever people eat outdoors. They seem to prefer open feeding areas that permit a speedy getaway if a threat is detected.
- Unlike most birds that must tip their heads back to swallow water, pigeons can drink by sucking water directly from a puddle or other water source.

Figure 1. Pigeons were first domesticated around 4500 B.C. from stock inhabiting the sea cliffs of the Mediterranean. Since then, nearly 150 varieties have been developed, some for meat, some for fashion, and some for racing. The now extinct passenger pigeon (Columba migratoria), *originally from the eastern United States, is a different species. (Drawing by Elva Hamerstrom Paulson.)*

ELVA PAULSON

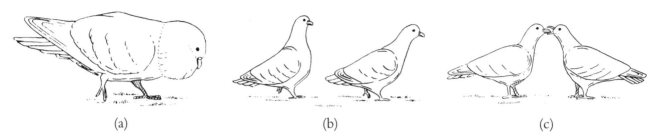

Figure 2. Displays associated with pigeon courtship. (From Stokes, A Guide to Bird Behavior.*)*

Nesting and Roosting Sites

- Nesting and roosting sites are protected from the elements and are situated on houses, barns, stadiums, and grain elevators, as well as bridges, wharfs, and cliffs.
- Nests in continual use become solid with droppings, feathers, and debris.

Territory

- Domestic pigeons don't migrate, but if removed from a nesting area, they have a good homing ability and can return from long distances. It is thought that this ability evolved to help them find their own nests on cliffs covered with large colonies of similar looking nests.
- Pigeons are gregarious and eat, roost, and nest in each other's company whenever possible.
- Usually only the immediate area around the nest site is defended against intruders.
- When pigeons are not involved in courtship behavior, caring for young, or eating, their day is spent cooing, preening, and sunbathing at their loafing and roosting sites. Sunbathing is common on cool mornings.

Reproduction

- Domestic pigeons mate for life unless separated by death or accident.
- Females usually lay two cream-colored eggs in a nest loosely constructed from twigs, feathers, and debris.
- Both male and female incubate the eggs, which hatch after 18 days.
- The young are independent at four to five weeks of age.
- Pigeons can raise four to five broods annually. Under optimal conditions, new eggs are laid even before the previous clutch has left the nest.

Mortality and Longevity

- Domestic cats are the main urban predators of pigeons, but opossums, raccoons, foxes, weasels, and rats all eat pigeons when they can access nests or catch adults.
- Urban-dwelling pigeons can also be an important food for peregrine falcons and Cooper's hawks. Crows sometimes eat juvenile pigeons.
- In captivity, pigeons commonly live up to 15 years, sometimes longer. In the wild (including urban areas) pigeons seldom live more than three years.

Viewing Pigeons

Many people find pleasure in viewing pigeons. Because they are one of the few animals that tolerate the environmental con-
ditions of an inner city, pigeons (and house sparrows) may be the only wildlife observed by many people living there. Look for flocks of pigeons in city parks and other places where the birds are accustomed to humans and gather to feed.

Within minutes of watching a large active flock of pigeons, you are bound to see several characteristic displays associated with courtship:

The **bow (a)** is performed primarily by males after landing near a flock or standing in front of a prospective mate. With neck feathers ruffled, the male lowers his head and turns in full or half circles.

Driving (b) occurs between members of a pair when other birds are present. The two birds will run in tandem with the male bird seeming to drive the female along.

Billing (c) occurs between members of a pair, often just prior to mating. A female puts her bill into her mate's open mouth, and the two move their heads rhythmically up and down together.

Preventing Conflicts

Pigeons are a major component of many urban and suburban wildlife communities. Most people don't object to them unless they are present in large numbers. In such cases, their droppings may ruin vegetation, produce an objectionable odor, and damage property such as park benches, statues, cars, and buildings. Large accumulations of droppings have been implicated in causing several fungal diseases (see "Public Health Concerns"). Droppings combined with nest materials and feathers may block downspouts and vents on buildings. Finally, pigeons carry a variety of parasites such as mites and lice. When they nest near windows, these small pests can find their way into homes and bedding.

The most effective way to prevent conflicts with pigeons is to modify your home and property so as not to attract them. Limiting available food and water may help, but pigeons find food and water in many places, even far from where they roost and nest. Where people feed pigeons in their backyards, parks, or lunch areas, education can help reduce the pigeons' food source; but this effort is often futile as there are always people who find the birds irresistible.

Occasionally a pigeon will get caught in a building. If this happens, turn off all inside lights and open all windows and other exits. The bird should leave on its own. If necessary, a broom or long pole with a T-shirt at the end can be used to direct the bird out an exit, or tire it to a point where it can be caught in a towel or similar item. If these methods fail or are impractical, a wildlife damage control company can be called to assist in the removal process. Call your local wildlife office for contact information or look under "Animal Control," "Pest Control," or

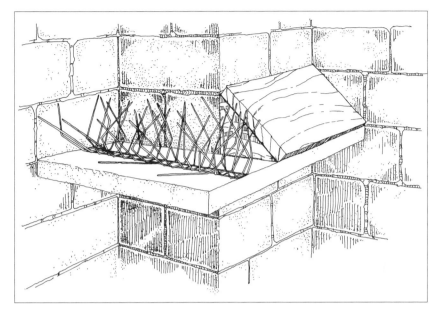

Figure 3. *The best way to prevent pigeons from roosting or nesting on ledges is to install barriers. (Drawing by Jenifer Rees.)*

"Wildlife Control" in your local phone book.

Other situations where it is wise to hire such a professional to remedy pigeon problems include: removing a large quantity of droppings from an old or well-used roost or nesting site, and installing netting or other barriers high on buildings.

Following the suggestions below can reduce problems caused by pigeons roosting or nesting in and around buildings.

Install barriers: The best way to keep pigeons from occupying ledges, window air-conditioning units, and similar sites is to install barriers. There are various ways to achieve this. *Note:* Established pigeons will fight any type of barrier put in place, especially if it is a popular nesting site. In such cases, the removal of pigeons prior to installing barriers is most effective.

• Install sheet metal, wood, Styrofoam blocks, or other materials at a 60-degree angle (Fig. 3).

• Place an outstretched slinky toy or a rolled-up piece of plastic mesh netting or chicken wire over the area.

• Place metal or plastic spikes, such as Catclaw®, Bird-B-Gone®, and Nixalite® (porcupine wire) where problems are severe or pigeons are persistent (Fig. 3). Metal coils (e.g., Bird Barrier®) function similarly. Electrified systems (Avi-Away®, Flock-Shock®, Flyaway®, VRS®) are designed to shock birds without killing them and thus exclude them from specific locations. These are commercial products available from farm supply centers and bird-control supply companies on the Internet (see Appendix F for resources).

• Tightly string single-strand steel wire (16–18 gauge) or monofilament line (80-pound test) between L-brackets installed at each end of the area used by pigeons. For increased tension, attach the wire to the L-brackets with turnbuckles. Install the

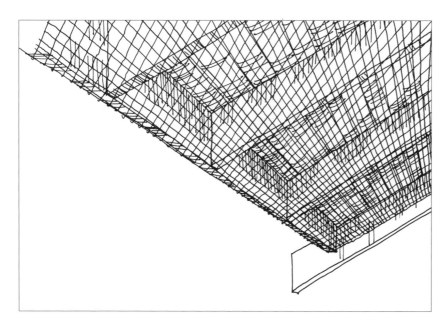

Figure 4. The undersides of rafters can be covered with bird netting to prevent pigeons from gaining access to roosting spots. (Drawing by Jenifer Rees.)

wire so it will come to the belly of the bird—about 2 inches (5 cm) high.

- Install bird netting to block off indoor roosting and nesting areas. If you can't reach a ledge from inside a building, netting can be attached to the roof, draped across the front of the structure, and then tightly secured to the base and sides of the building. Such netting is available from nurseries and hardware stores; professional quality material and associated hardware is available from bird-control companies and over the Internet (see Appendix F for resources). Two-inch (5 cm) mesh netting works well for pigeons, and it isn't as likely to trap small songbirds as the light, small mesh material. Install the netting so window washers can remove it or work under it.
- Cover the underside of rafters with bird netting to prevent pigeons from gaining access to roosting spots (Fig. 4).

Previously used bird netting may be available, as well as used gill netting from fishermen or fish hatcheries. The cost of new netting makes seeking out an alternative worthwhile.

- Holes in buildings can be boarded up or covered with quarter-inch (6 mm) galvanized wire mesh.
- Commercially available sticky products are not recommended. They attract dirt and may melt during hot weather. In addition to people's failure to use and monitor sticky products properly, these products can cause pigeons and smaller birds to suffer unnecessarily when they get stuck in them.

The cost of installing barriers may render them impractical on large buildings with extensive roosting sites. However, barriers are valid options for smaller areas. Always use care when working high above the ground and ensure that the barriers can't fall and injure a passerby.

Check the covered areas as needed for accumulated debris or nest material. Regularly remove falling leaves and other matter that can cover the barrier and reduce its effectiveness.

If pigeons are likely to drop nest material and other debris on top of the newly installed barrier, add an additional barrier on the landing site above the installation.

Harassment Techniques

Noisemaking devices and scare tactics have little permanent effect on pigeons, particularly at well-established roosting and nesting sites. However, harassment methods can be effective when installed before pigeons become accustomed to using an area. They may also be effective on small groups of pigeons. Various harassment techniques include:

- Continually remove pigeon nests to discourage the birds from nesting. Pigeons will leave an area after several unsuccessful attempts at nest building. This approach is most effective after barriers have been installed. When using a high-pressure spray, make sure the contaminated water doesn't spray where people are present. See "Public Health Concerns" for information on safely cleaning up bird droppings. When spraying is not possible, use a hook fastened to a long pole to remove the nests.
- Contact your local falconer's association to have a falconer come out to train their falcons weekly (search the Internet for "Falconers Association"). Trained falcons are especially effective at dispersing large flocks of pigeons and catching individual birds in large build-

ings. Most falconers will be reluctant to use their birds of prey near highways and other high-traffic areas.

- Ultrasonic devices have been tested by university, government, and private independent researchers, and were found to have no effect on pigeons.

(See "Harassment and Scare Tactics" under "Preventing Conflicts" in Chapter 27 for additional information.)

Lethal Control

If all efforts to dissuade problem pigeons fail and they continue to be a human safety concern, they may have to be trapped. Trapping is rarely a permanent solution since other pigeons are likely to move in if attractive roosting and nesting sites are still available.

Small-scale traps are available from the Purple Martin Conservation Association and other enterprises over the Internet (see Appendix F for resources). Check the trap at least twice a day for non-targeted birds.

Do not trap pigeons and release them elsewhere, because they will easily return or cause problems somewhere else. If you cannot humanely kill them yourself, find a falconer or wildlife rehabilitation center that will accept live pigeons to feed to hawks. (See Appendix A for information on euthanization.)

Shooting has been effective in eliminating small isolated groups of pigeons. For safety considerations, shooting is generally limited to rural situations and is considered too hazardous in more populated areas, even if legal. Where shooting is legal and safe, .22 CB caps work well; so does

any semi high-powered pellet rifle with a pellet velocity of 800 fps or more.

Public Health Concerns

The most common health concerns associated with starlings, crows, and pigeons involve disease that could result from inhalation exposure to large accumulations of droppings. **Histoplasmosis** cases are not reported in the Pacific Northwest, but do occur in other parts of the United States. **Psittacosis,** caused by the organism *Chlamydia psittaci,* can be related to exposure to pigeons or their droppings.

When working in or cleaning up areas where large amounts of bird (or bat) droppings occur, follow these precautions to minimize risk from disease organisms in the droppings:

Pigeons and Bird Feeders

Because domestic pigeons are relatively large birds, you can discourage them by using tube-type feeders that have small (or no) perches, and small feeding ports. Most other feeder birds can manage fine without perches (see an example under "Sick Birds at the Feeder" in Chapter 37).

Some commercially available caged feeders designed to frustrate squirrels will let smaller birds in but also keep pigeons out. Wire mesh placed over a platform feeder prevents larger birds such as jays, starlings, and pigeons from accessing seeds (Fig. 5).

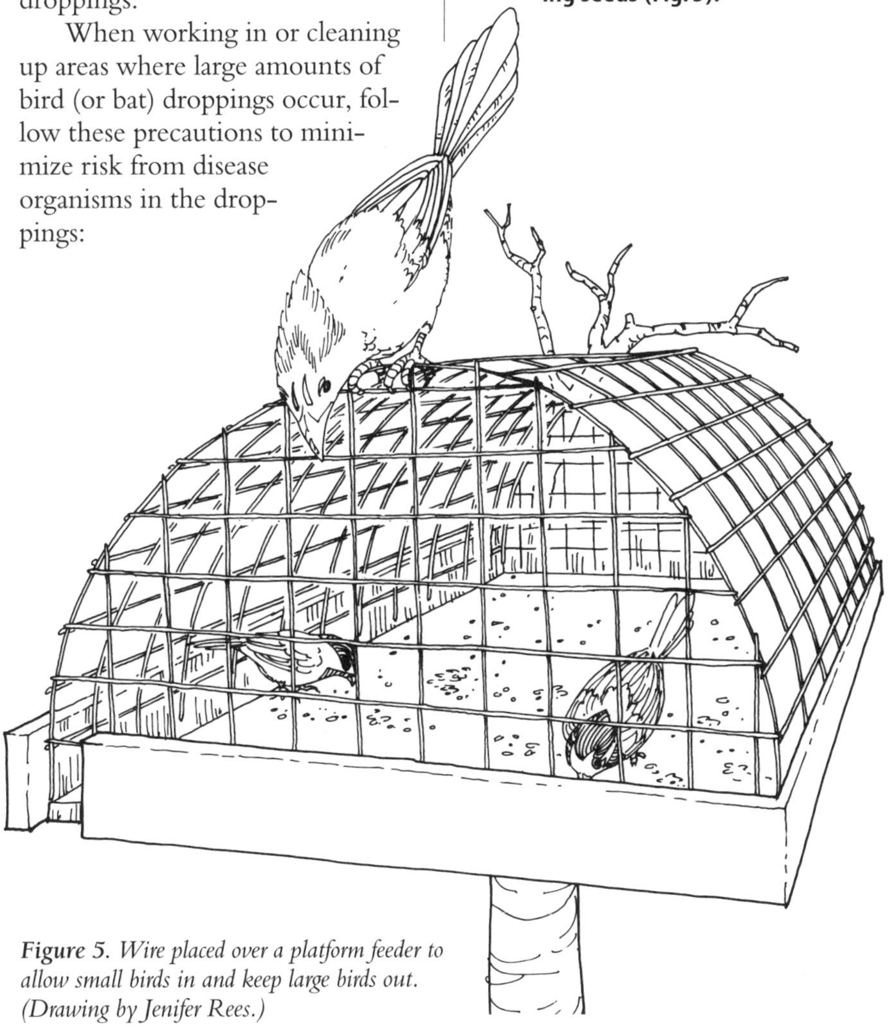

Figure 5. Wire placed over a platform feeder to allow small birds in and keep large birds out. (Drawing by Jenifer Rees.)

- Wear a National Institute for Occupational Safety and Health (NIOSH) approved full-face respirator with a high-efficiency particle air (HEPA) filter for screening particles down to 0.3 microns
in size. Simple dust and particle masks will not provide adequate protection. Make sure the respirator is properly fitted to your face before work starts. Contact the manufacturer for specific information. Respirators are available for purchase in larger hardware, paint, and home supply outlets.
- Wear disposable protective gloves, a hat, coveralls, and boots. When finished, and while still wearing the respirator, remove this protective clothing and place it in a plastic bag. If you wear nondisposable coveralls, be sure to put them in a plastic bag after you are finished wearing them, and keep the bag tied until you are ready to wash them. Wash the coveralls separately from other

clothing before you wear them again.
- Wet down the droppings to keep spores from becoming airborne, and keep the droppings damp for as long as you are working with them.
- Put droppings into sealed plastic garbage bags.
- Dispose of trash bags (disposal should be permissible through standard trash pickup).
- Wash or shower after you have removed your protective clothing.

Legal Status

Oregon: The domestic pigeon is an introduced species and is not protected in Oregon. The nest, eggs, young, and/ or adults may be removed or destroyed at any time.

Washington: During the writing of this publication, the legal status of domestic

pigeons is being revised. Contact your local Department of Fish and Wildlife office for updated information. (See Appendix E for contact information.)

British Columbia: The domestic pigeon is identified as "Wildlife" and is protected under Section 34 of the Wildlife Act. However, domestic pigeons are also listed under Schedule "C," which exempts them from protection (Wildlife Act, Designation and Exemption Regulation 2 and 3). Therefore, they, their eggs, nests, and young may be killed by the public without a permit.

Figure 6. The band-tailed pigeon is similar in size to the domestic pigeon, but it has a purplish head and breast, a dark-tipped yellow bill, yellow feet, and a small white crescent on top of its neck. This native pigeon is protected under federal and state law. (Drawing by Elva Hamerstrom Paulson.)

CHAPTER 30

Eagles and Ospreys

T he bald eagle of the waterways and the soaring golden eagle of the open country are two of the Pacific Northwest's most magnificent birds of prey. Long valued for their aesthetic beauty, eagles are recognized for their biological importance as scavengers and predators in the natural environment.

The bald eagle *(Haliaeetus leucocephalus)* is so named for its snowy white head and tail (Fig. 1). The distinctive white adult plumage is not attained until the birds are four to five years of age. Lacking this conspicuous characteristic, immature birds can be confused with golden eagles.

The bald eagle usually lives and hunts near a body of water. It was adopted as the U.S. national symbol in 1782.

The golden eagle *(Aquila chrysaetos)* is named for the golden buff–colored feathers on the crown and nape of the neck (Fig. 2). The adult body color is usually dark brown, and the dark-tipped tail is either darkly barred or spotted. Adult plumage is acquired over a three- to four-year period and involves a gradual reduction in the amount of white coloration. Golden eagles are often confused with juvenile bald eagles.

Golden eagles are found both in open country (grassland, agricultural fields, desert scrub) and in open coniferous forests and woodlands in east and southwest Oregon, eastern Washington, and throughout British Columbia. Though uncommon, golden eagles are sometimes seen in the eastern Olympic Mountains, the San Juan Islands, and in the lowland forests of western Washington.

For information on the osprey, see "Notes on Ospreys."

Fact about Eagles
Food and Feeding Behavior

- Eagles use their keen eyesight, which is thought to be four times more acute than that of humans, to hunt and scavenge for food. A golden eagle can spot the movements of a rabbit more than a mile away.
- Fish and waterbirds (especially ducks and seabirds) are the main diet of bald eagles. A pair of bald eagles may cooperate in hunting small ducks and other waterbirds, taking turns diving

Figure 1. The bald eagle is the Pacific Northwest's largest resident bird of prey, with a wingspan of up to 7½ feet (2.3 m) and a weight of 8 to 14 pounds (3.6–6.3 kg). Females are larger than males. (Drawing by Elva Hamerstrom Paulson.)

ELVA HAMERSTROM *Paulson*

after the birds until the prey is exhausted.
- In winter, spawned salmon along rivers become the eagle's most important food.
- Bald eagles raid gull and seabird roosts and nesting colonies to prey on adults, nestlings, or eggs. They are also known to prey on great blue heron nestlings and roosting adult crows and their nestlings.
- Bald eagles also prey on small mammals, including rabbits, muskrats, and raccoons, as well as shellfish, carrion (animal carcasses), and the afterbirth of cattle, sheep, and seals.
- Golden eagles prey on rabbits and hares, ground squirrels, and, less frequently, on rats and mice.
- Golden eagles are known to also eat owls, hawks, magpies, grouse, pheasants, snakes, and carrion.
- While golden eagles are capable of killing young livestock (i.e., lambs, calves, etc.), such killings are rarely observed.

Range
- The foraging and nesting habitat of a pair of bald eagles ranges from about 2½ to 8½ square miles (4–13.6 km).
- Bald eagles that nest on relatively straight, featureless shorelines typically choose an area that includes about 1 mile (1.6 km) of shoreline on each side of the nest.
- Some bald eagles that breed in Washington migrate up to southeast Alaska and the northern interior of British Columbia for several months in the fall.
- During the nesting season a golden eagle usually forages within 4½ miles (7.2 km) of the nest.

Nests and Nest Sites
- Bald eagle pairs will use the same nest year after year, but may also maintain alternate nests within their breeding terri-

tory; golden eagles often use alternate nests in successive years.
- Most bald eagles nest near marine shorelines, but nests are also found along lakes, reservoirs, and rivers.
- Golden eagles build their nests on sheltered ledges of secluded cliffs near their hunting grounds, and they also nest occasionally in dead or live trees, and on artificial nesting structures.
- Eagle nests are large and constructed of sticks, with nest cups lined with soft materials such as grasses, shredded bark, downy feathers, and fur.

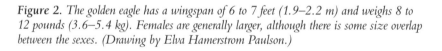

Figure 2. The golden eagle has a wingspan of 6 to 7 feet (1.9–2.2 m) and weighs 8 to 12 pounds (3.6–5.4 kg). Females are generally larger, although there is some size overlap between the sexes. (Drawing by Elva Hamerstrom Paulson.)

- New nests are typically 4 feet (1.25 m) (golden eagle) to 6 feet (1.85 m) (bald eagle) across and from 1 to 2 feet (30–60 cm) thick.
- Bald eagle nests that have been added onto for many years are as large as 10 feet (3 m) across and 10 feet (3 m) thick, because the occupants add new materials to the basic structure of branches and twigs each season.

Reproduction
- The breeding season for bald eagles begins as early as January; for golden eagles it begins around mid-March. Times can

vary according to geographic area and elevation.

- An average of two eggs are incubated for 35 to 45 days by both male and female eagles, with the female doing the majority of the incubation.
- Eggs are turned about every hour and are sometimes covered with soft nesting material on the brief occasions when they are left unattended.
- Nestlings first fly at 10 to 12 weeks (bald eagles) or 11 to 13 weeks (golden eagles) of age.
- Chicks then spend a month or so on their territory, still dependent on their parents, but become progressively independent as they develop hunting skills.
- Eagles are totally independent once they disperse from the nest site, at about 16 weeks of age. They do not travel in family groups and the young don't return to reproduce in their original nest.
- Where competition for food and nest sites exists, eagles typically begin breeding at around six years of age. Breeding may occur earlier where food and nest sites are abundant.

Mortality and Longevity

- Despite state and federal protection, a large percentage of adult eagle fatalities have human-related causes, including shooting, vehicle collisions, pesticides, lead poisoning, electrocution and/or collision with power lines, and a black market trade in eagle feathers and parts.
- Natural causes of eagle mortality include nestlings being killed by their nest mates, fights with other eagles, starvation during their first year, accidents, and predation.

- Animals that prey on bald eagle eggs and hatchlings include black bears, raccoons, gulls, red-tailed hawks, ravens, crows, and magpies.
- Once a wild eagle reaches maturity, its life expectancy is 15 to 20 years. The oldest wild bald eagle on record was 28 years. Captive bald eagles have lived to an age of at least 47 years.

Viewing Eagles

There was a time, not so long ago, when catching a glimpse of a majestic eagle would have been a rarity. While the solitary nature of golden eagles makes seeing one a rare event, today a combination of factors makes it possible to see large numbers of bald eagles.

There are indications that many bald eagles in urban areas have become fairly tolerant of human activity near nests. However, bald eagles in rural areas and golden eagles everywhere are sensitive to disturbance during nesting, particularly during the period of egg-laying through the early-nestling phase.

Adult eagles brood mostly when the eaglets are less than a month old. Brooding keeps the young warm, dry, and protected from predators. Humans approaching at this time may cause the adult eagles to abandon the nest or flush from it temporarily, leaving the developing young vulnerable to predators or cold. In addition, humans approaching the nest after four to five weeks can also be disruptive by keeping adults from delivering prey to the nest. Never approach a nest so closely that you cause an eagle to fly off the nest.

To prevent problems from occurring, nesting eagles should

be left alone or watched with binoculars or a spotting scope. Many designated areas have been set up for viewing eagles.

The following types of eagle behavior can be seen without putting the birds in jeopardy:

Courtship Behavior

Adult eagles go through a series of courtship behaviors to establish or reinforce a relationship known as a "pair bond," which often lasts until one eagle dies. The courtship of eagles can involve various chase displays in late winter and early spring.

You may see aerial chases between multiple pairs of bald eagles, and chases where one bird dives at another, with the attacked bird rolling over at the last second and extending its talons at the diver. During this "cartwheel display," the two eagles may touch briefly or lock talons and fall through the air, whirling. Such chases may last several minutes.

Golden eagles have a "roller-coaster flight display" in which the male bird ascends high in the sky, closes its wings, and plummets downward for several seconds. It then opens its wings, rises, and repeats the display. This display can be seen from a mile or more.

Territorial Behavior

Eagles defend their territories from other adult eagles that attempt to intrude during the breeding and nesting season. The adult pair tries to maintain exclusive occupancy of the territory by perching atop dominant trees and other structures, giving threat vocalizations, circling, and giving territorial chase.

Eagles occasionally fight, using their talons to grasp the

Figure 3. Ospreys are large birds with wingspans of 5½ feet (1.65 m). In flight, they are recognized by their long, narrow wings as they soar overhead. From a distance, an osprey at rest may resemble a bald eagle. The underside of an osprey's body and wings is white. A dark cheek patch contrasts with the white head. Female ospreys are slightly larger than males. (Oregon Department of Fish and Wildlife.)

Notes on Ospreys

The osprey (*Pandion haliaetus,* Fig. 3), also known as the fish hawk or sea hawk, is found along or near lakes, rivers, sloughs, and protected coastlines throughout the Pacific Northwest. Ospreys can be seen during the summer months soaring or hovering over water searching for fish, their main food source. When prey is sighted, an osprey dives steeply and plunges head and feet first into the water. It quickly resurfaces and, if it has made a catch, flies off.

Unlike other raptors, the osprey has four equal toes. The outer one is reversible, enabling the bird to seize its prey with two toes pointing forward and two pointing backward. (Most raptors have three toes pointing forward and one backward.) A long, sharp, curved claw on each toe, and short, rigid spikes on the sole of each foot, known as spicules, give the bird a firm grip on its slippery prey.

Ospreys occasionally also capture small mammals, birds, amphibians, and reptiles. Anglers sometimes complain of ospreys competing with them for fish; however, an osprey needs only one 10-inch (25 cm) fish per day, so it is doubtful they have much impact on fish populations. On occasion, a bald eagle will try to rob an osprey by harassing it in flight until it drops its prey.

Ospreys are migratory and generally spend their winters in Mexico and Central and South America, returning to the Pacific Northwest in mid to late April. A breeding pair probably mates for life and returns to the same nest area each year. Both adults help to build the 3 to 6 foot (90 cm–1.85 m) diameter stick nest, which typically is situated near water, or atop dead or live trees or rock pinnacles. Ospreys easily take to human-made structures and will nest on pilings, cross-members of power poles, cell-phone towers, on top of bridges, and on artificial nest platforms.

In mid-May the female lays two to four eggs, which are incubated by both parents for five weeks. Females closely guard their nestlings from the weather and predators while the male provides the family with food. At seven to ten weeks of age, the nestlings are nearly fully-grown and ready to leave the nest. Most ospreys have departed the Pacific Northwest by October. Ospreys usually remain south for their first several years, returning to their breeding grounds (north) during their third to fifth year.

Osprey calls are short, shrill whistles: *tewp, tewp, teelee, teelee.*

Like many other birds of prey that are high on the food chain, ospreys throughout the world had breeding problems during the 1950s and 1960s. These problems arose mainly from the then widespread use of organochlorine pesticides, especially DDT. DDE, a substance produced when DDT decomposes, caused thinning of the shells of the eggs, which consequently tended to break under the weight of the incubating adult. Since the use of these pesticides has been limited nearly everywhere, ospreys have begun to recover.

The osprey's adaptability to living near humans and using artificial structures for nesting, reinforced by access to artificial reservoirs and ponds stocked with fish, also have contributed to its recovery.

opponent while in flight. An eagle will roll over and extend its talons to meet an aggressor. Fights sometimes have fatal outcomes.

During egg-laying, incubation, and nestling periods, an eagle will chase, attack, and sometimes kill any crow, gull, hawk, or other bird that comes too near the nest.

Perching and Hunting Behavior

Both bald and golden eagles may be seen perched on large limbs in trees that provide a commanding view of their foraging areas. Perches are distributed throughout their nest territories and may be used daily. Golden eagles also use cliff ledges as perches.

Some bald eagles tend more toward scavenging than catching healthy prey. It is possible to watch an eagle in a perch tree, waiting for hours until the current brings in a floating meal.

However, bald eagles can be fierce predators. For example, an eagle will circle above and dive on a duck, causing it to go under water repeatedly until it is out of breath and easily taken.

Look for golden eagles perched near a hunting area in a location that has regular updrafts or thermals, which allow them to soar high above the hunting area to scan for prey. Attacks are generally made upwind so the bird has a high degree of maneuverability and aerodynamic control, and prey are forced to escape into the wind.

Golden eagles sometimes capture prey in the air, but usually fly to the ground and attack either during low- or high-speed glides, or low flapping flights.

Soaring

Although golden eagles are more often seen soaring high in open country, under suitable conditions bald eagles will also soar for long periods. During winter, soaring is usually seen in the afternoon after bald eagles have fed. Once an eagle has started this behavior, others often join in until a large flock is spiraling upward together, sometimes climbing to great heights in a loose formation called a "kettle." These kettles may consist of 25 to 50 bald eagles.

Winter Feeding

Bald eagles, including many Washington birds, generally leave their breeding grounds in early fall to seek late summer and fall salmon runs in Alaska and British Columbia. Hundreds of adult eagles from as far away as the northern Canadian provinces, Alaska, and Montana winter in relatively mild northwest Washington, where they rely particularly on spawning chum salmon as an annual food source.

Communal Night Roosting

During the winter, large numbers of roosting bald eagles can sometimes be spotted in stands of large coniferous trees. This is especially true during bad weather, when they may remain in the roost throughout the day. Big trees provide protection from cold, heavy rains, and high wind speeds and help birds conserve energy.

Many bald eagles roost singly and change roost sites frequently. Eagles may also roost in pairs or gather in congregations of as many as 500 individuals at locations that are used year after year. They leave the roosts in the morning and may start back to them as early as 2 p.m.

Caring for an Injured Eagle

Under federal, state, and provincial law, it is illegal for anyone to injure, harass, kill, or possess a bird of prey. Licensed rehabilitators are the only people legally permitted to transport and keep wildlife, including eagles.

If you find an injured eagle, contact a wildlife rehabilitation facility immediately. Your local wildlife office keeps a list of rehabilitators and can tell you which ones serve your area, or you can look under "Animals" or "Wildlife" in your phone directory. If a rehabilitator isn't available, follow the menu options over the phone or on their Web site for information on what to do. (For more information, see "Wildlife Rehabilitators and Wildlife Rehabilitation" in Chapter 20.)

Calls

Adult bald eagles give a series of harsh, squeaky chatters, and also a rapid series of chirping sounds, often trailing off at the end. Adult golden eagles give a high-pitched *kee-kee-kee* in a slow, measured series, and also a high scream or squeal. The nesting call of both species is given when the young see the adults flying toward the nest with food, or when the young are being fed or are fighting over the food on the nest. The call is a quiet *yeep* or a rising squeal to a shrill scream.

Preventing Conflicts

While golden eagles are capable of killing large domestic animals (i.e., lambs or calves), few such killings have been observed. Golden eagles commonly feed on animals that have already died from some other cause or on the animal's afterbirth. Since these birds are large and visible, however, many people assume incorrectly that they have killed what they are eating.

Some researchers suggest that golden eagles are beneficial to livestock interests because a large percentage of their diet is made up of rabbits, which compete with livestock for forage. Eight to 12 jackrabbits consume enough forage to support one sheep. The number of rabbits and rodents killed by golden eagles translates into a sizeable quantity of forage.

Although such incidents are rare, both bald and golden eagles have been known to attack domestic cats, dogs, and birds (mostly chickens). To prevent a problem, keep cats and small dogs indoors or under supervision in areas where eagles are commonly seen. See "Preventing Conflicts" in Chapter 33 for information on how to manage chickens and other fowl.

Public Health Concerns

Eagles and ospreys are not a significant source of any infectious disease that can be transmitted to humans or domestic animals.

Legal Status

Because legal status and other information about eagles and ospreys changes, contact your local, state, or provincial wildlife office, or visit their Web site for updates. (See Appendix E for contact information and where to access the state and provincial laws mentioned below.)

Washington and Oregon: Bald eagles and golden eagles in Oregon and Washington are protected by the Bald and Golden Eagle Protection Act and the Migratory Bird Treaty Act. The Bald Eagle Protection Act of 1940 (amended in 1962 to include golden eagles) protects eagles and their eggs and nests from "take," a term that "includes pursue, shoot, shoot at, poison, wound, kill, capture, trap, collect, molest, or disturb..." (16 USC 668–668d).

Ospreys are federally protected by the Federal Migratory Bird Treaty Act.

British Columbia: In British Columbia, eagles and ospreys come under provincial jurisdiction, as they are not covered by the Federal Migratory Birds Convention Act. They are protected under Section 34 of the Wildlife Act, which states that unless provided by regulation (i.e., a permit), it is an offence to possess, take, injure, molest or destroy a bird, its eggs, or its nest when active (contains young or eggs). However, for a few species, including the three discussed in this chapter, the nest is protected year-round whether active or not.

Bald Eagles in Washington State: Past, Present, Future

By 1980, shooting, trapping, poisoning, cutting of forests, commercial exploiting of salmon runs, and the use of DDT reduced the state's population of bald eagles to only 105 known breeding pairs. When white settlers first arrived in Washington, the early summer population of bald eagles may have been around 6,500.

The population has recovered dramatically with the ban on DDT use after 1972, and increased protection for eagles and bald eagle habitat. In the past 20 years, the population of nesting bald eagles has grown about 10 percent per year.

In 2001 there were 678 territories occupied and there are now some indications that the population has reached its maximum for habitat carrying capacity in some parts of western Washington. The population may still be increasing in northeastern Washington and along some western Washington rivers. If there is no decline in nesting habitat, productivity, or survival, the population may stabilize around 4,400 within Washington State.

After food sources, the single most critical habitat factor associated with bald eagle nest locations and breeding success is the presence of large trees, particularly conifers. Bald eagles need large trees capable of supporting their weight and their massive nests. Because the life expectancy of nests is 5 to 20 years, bald eagles need additional trees of similar size located nearby to serve as replacement nest trees if a nesting territory is to persist at the site.

Two-thirds of the known bald eagle nests are in trees on private property, the majority of which is shoreline property and thus highly desired for development and view management. Many trees left during construction of homes or commercial buildings will likely be removed when they become large enough to pose a threat to life or property should they fall. Other large trees, some of which are currently more than 300 years old, will succumb to disease. Therefore, in the future there may be fewer opportunities for bald eagles to find suitable spots to build their nests or perch while they watch for food.

The challenge for the future is to find a way to maintain appropriate stands of conifers in shoreline areas. These must include large, old trees as well as replacement nest trees that will provide nesting spots and screening from human activities continually, decade after decade.

CHAPTER 31

Great Blue Herons and Belted Kingfishers

The great blue heron (*Ardea herodias,* Fig. 1) is a large, grayish-blue wading bird with a long bill, neck, and legs. In flight, the great blue heron can be recognized by its long neck folded back on the shoulders, its long trailing legs, and its slow, deep wing beats. Adults can be recognized by the presence of a black plume. Males and females are identical in appearance.

Great blue herons are found year-round throughout the Pacific Northwest. They are at home in both salt and fresh water and are seen on lakes, ponds, rivers, marshes, and mudflats, as well as near irrigation ditches, and in farm fields and meadows.

For information on belted kingfishers, see "Notes on the Belted Kingfisher."

Figure 1. Great blue herons reach 4 feet (1.2 m) in height and have wingspans of nearly 6 feet (1.8 m). They have long necks and bills adapted for grasping prey. (Oregon Department of Fish and Wildlife.)

Facts about Great Blue Herons

Food and Feeding Habits

- Great blue herons are stand-and-wait-predators that remain motionless for long periods of time, waiting for prey to venture within striking range. (See "Feeding Habits of the Great Blue Heron.")

- Herons feed on a variety of prey, including fish, frogs, young birds and bird eggs, snakes, and insects, as well as mice, moles, gophers, and other small mammals.

- Herons feed during the day and at night in lighted areas, generally within 3 miles (4.8 k) of their colony. They tend to be solitary feeders, but where the food supply is abundant many can be found feeding together.

- Where the water freezes, heron populations concentrate along major rivers where food is available, or they hunt rodents on land.

Nest Sites and Nests

- Great blue herons nest in colonies, also called rookeries. Rookeries may contain a few or hundreds of nests.

- Rookeries are usually in isolated spots away from potential disturbance and near suitable feeding areas. Herons that have been frequently exposed to human disturbance may nest in large public parks and greenbelts.

- Herons nest in deciduous or evergreen trees and snags, usually near the top on vertical branches. Nests are usually constructed in the tallest trees available, on islands, or in trees with water around the base, presumably to reduce the risk of predation by mammals (Fig. 2).

- Where trees are absent, nests may be located on large shrubs, cliffs, and artificial structures.
- Nests are constructed from branches and twigs gathered from the ground, trees, and old nests. Nests are 25 to 40 inches (63–101 cm) in diameter and 12 inches (30 cm) or more thick.
- Rookeries may be used for decades; however, herons will relocate their colonies in response to increased predation on eggs and young, declines in food availability, human disturbance, and when nest trees fall.

Reproduction

- Adult herons can arrive on the nests as early as February. Nest building and repair usually begins in March or April.
- Three to five pale, greenish-blue eggs are incubated for 25 to 29 days by both sexes.
- Young birds first fly at around 60 days of age and leave the nest at 65 to 90 days, at which time they are similar in size to adults.
- Great blue herons have one brood per year; however, they may renest if their first clutch fails.

Mortality and Longevity

- Adult great blue herons don't have many predators. Bobcats and coyotes occasionally kill adults feeding at ground level.
- Mortality of the young is high: crows, ravens, gulls, eagles, and raccoons prey upon both the eggs and young. Heavy rains and cold weather at the time

of hatching also take a heavy toll.

- Six to eight years is the normal life span.

Viewing Great Blue Herons

People often observe great blue herons as the birds fly slowly and steadily—wings arching gracefully down with each beat, neck bent back, and feet trailing behind. They are also seen feeding—standing motionless and staring into shallow water, or walking with measured steps searching for prey.

Herons are great birds for the beginning birder to observe. In urban areas, they have acclimated to people so you can get a close view of their hunting behavior. A pair of binoculars or a spotting scope will allow for exceptionally close views of their black plumes and yellow eyes.

The nesting behavior of great blue herons is not often witnessed because, with some urban exceptions, they nest in colonies in fairly isolated areas. Nesting herons should be left alone, and area regulations and closures to protect colonies should be followed. Several studies have shown that human disturbance during the breeding season can cause adult herons to abandon the entire colony.

Great blue herons often congregate at mudflats and eelgrass beds during low tides from June to December, where they feed on small fish. Here you may have the opportunity to view many herons at once.

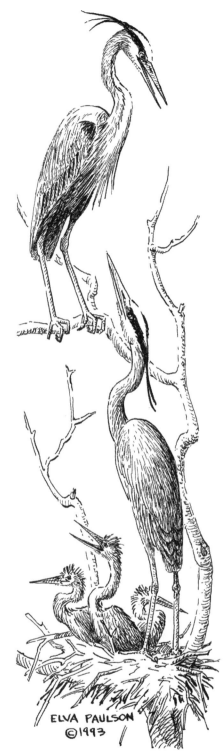

ELVA PAULSON
©1993

Figure 2. Great blue heron nests are usually constructed in the tallest trees available, on islands, or in trees with water around the base, presumably to reduce the risk of predation by mammals. (Drawing by Elva Hamerstrom Paulson.)

Feeding Habits of the Great Blue Heron

The great blue heron has two principal fishing techniques. The first consists of standing motionless, with its neck extended at an angle of about 45 degrees to the surface of the water. Only its head and eyes move to locate the prey. If no prey comes within range after a few minutes, the heron gradually moves a short distance away and takes up a similar position. When a potential meal comes close enough, the heron slowly folds its neck back and moves one leg in the direction of the prey. Suddenly, its entire body and neck unbend, its head plunges into the water, it catches the prey in its bill, and then swallows it outside the water, using a deft movement of the head to drop the prey headfirst into its gullet.

A similar technique is used to hunt small rodents in pastures, meadows, and similar habitats. Herons stab their prey using their bill like a long barbeque fork, or clamp on their prey, using their bill like barbecue tongs.

When its catch is too aggressive to be gulped down immediately or has dangerous spines, the heron drops it and grabs hold of it repeatedly and violently with its beak until it is dazed or the spines snap.

Other feeding techniques are observed more rarely. For example, herons in flight sometimes dive underwater to catch fish; others hover over the water and submerge their heads to catch fish; and some swim in deep water and feed on fish found near the surface.

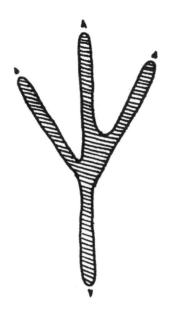

Figure 3. From front to back, a great blue heron's track is about 6 inches (15 cm) long and a claw imprint shows clearly at the end of each toe. The hind toe is well-developed for standing and walking. (Drawing by Kim A. Cabrera.)

Tracks

Great blue heron tracks are easily found in the mud or sand next to a feeding site (Fig. 3). Their tracks show four toes and the webbing may or may not appear, depending on the hardness of the surface.

Droppings and Pellets

Great blue heron droppings are semiliquid and mostly white. The ground beneath nests can become coated with droppings. Undigested material is coughed up as pellets, 2 to 3 inches (5–8 cm) in length, containing signs of fish, small rodents, and other prey. For information on pellets, see "What's in an Owl Pellet?" in Chapter 36.

Calls

The normal call of a great blue heron is a deep, hoarse *fraaaahnk* or *braak*. In aggressive situations or when frightened, the call is a short, harsh *frank frank frank taaaaaw*. Herons call in flight and on the ground, during the day and at night.

An adult arriving at the rookery usually gives a dull, guttural cry. The young can cry loudly and constantly when hungry and about to be fed.

Preventing Conflicts

Many people are willing to lose some inexpensive fish to great blue herons or kingfishers in exchange for the opportunity see one of these magnificent birds. However, some homeowners have expensive fish in their pond, or are fond of the cheap ones they've watched grow. Other people object to the mess herons sometimes leave behind.

Problems with herons and kingfishers can occur at any time of the year, although problems are greatest in spring and early summer, when the birds feed their young. Each day, an adult heron needs about 13 ounces (0.37 kg) of food, which is equivalent to three six-inch koi or 10 two-inch goldfish. They can take twice this amount when feeding their young. An adult kingfisher eats approximately 3.5 ounces (0.10 kg) of fish a day.

Herons and kingfishers generally visit ponds when everything is quiet, usually early in the morning or in the evening. Once they have found an easy food source, such as colorful fish in a shallow pond, they will return on several consecutive days until most of the fish have been taken. Feathers in the water, whitewash (droppings) or tracks near the edge of the water, and fish with wounds or old scars are typical signs of heron and kingfisher predation.

Notes on the Belted Kingfisher

The belted kingfisher (*Ceryle alcyon,* Fig. 4) is a Northern flicker-sized bird that is blue-gray above and white below, with a bushy crest, a large, daggerlike bill, and a short tail. It is usually heard before it is seen—its loud rattling *klek-klek-klek-klek-klek-klek* call travels far and is easily recognized, especially near water.

It is exciting to watch a kingfisher as it feeds. It sits on a perch or hovers in the air and then dives head first into the water to catch small fish. The birds teach their young to fish by dropping dead prey into the water for retrieval.

A kingfisher will often patrol a regular route along a stream or lakeshore, stopping at favorite exposed perches along the way. When flying from one perch to another, often a good distance apart, it utters its loud rattling call.

As its name implies, 3- to 4-inch (8–10 cm) long fish make up about 90 percent of the kingfisher's diet. In addition to fish, it eats crabs, crayfish, salamanders, lizards, mice, and insects.

The kingfisher frequents a variety of aquatic habitats throughout the Pacific Northwest, including rivers, marshes, lakes, coastal shorelines, tide pools, and beaver ponds; also backyard ponds and ponds in city parks. In areas where water remains open through the winter, kingfishers can be found year-round; otherwise they are migratory, leaving interior and higher elevation breeding grounds by September.

The kingfisher's breeding season begins in late April or early May. Both parents incubate the eggs and feed the young, which leave the nest about five weeks after hatching.

The kingfisher uses its toes and strong beak to chisel and shovel its nest cavity high in a sandy bank, resulting in a 3- to 8-foot (90 cm–2.4 m) long tunnel with a slightly enlarged space at the end. The nest site is usually free of vegetation, 1 to 3 feet (30–90 cm) below the top of the bank, and within 100 feet (30 m) of a dead limb, on which the bird can perch and see the nest.

When searching for a nest site, scan bank sides of a stream, ocean shore, or road. Look for a tunnel 3 to 4 inches (8–10 cm) in diameter. An active tunnel may have two slight grooves at the bottom of the entrance that the kingfisher has worn down with its feet.

For information on how to solve conflicts with kingfishers, see "Preventing Conflicts."

Figure 4. The belted kingfisher is a flicker-sized bird that is blue-gray above and white below, with a bushy crest, a large, daggerlike bill, and a short tail. The female has a second, rust-colored band on its upper belly. (Drawing by Elva Hamerstrom Paulson.)

Tips for Attracting Herons and Kingfishers

If you put out a bird feeder, you have to expect the birds to come and eat. A pond full of small fish makes a wonderful heron (and kingfisher) feeder. Keep the pond stocked with cheap feeder goldfish, give the heron a place to stalk them, and enjoy the show. To give fish a place to hide, one or two areas can be kept as hunting spots for herons and the other areas can be heavily planted to block access into the pond and provide cover for the fish.

To help preserve heron and kingfisher habitat:

• Preserve shoreline trees and any tall groups of trees by the water.
• Protect eelgrass beds because they provide great habitat for herring—a major food source for herons, kingfishers, salmon, seals, and other marine mammals.
• Protect wetlands.
• Keep pets, especially dogs, under control and away from great blue herons and rookeries.
• While boating or visiting the beach, give herons and heron rookeries plenty of space.
• Minimize development near heron rookeries.

For detailed management recommendations for great blue herons and their rookeries, contact your local wildlife office. (See Appendix E for contact information.)

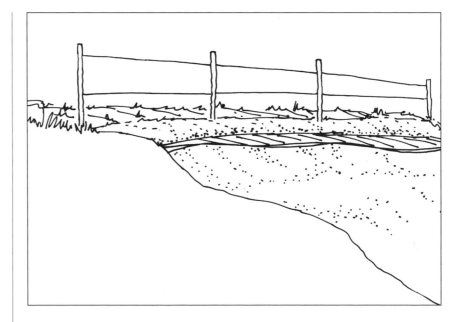

Figure 5. To create the least amount of visual distraction around a pond, install two strands of electrified wire to prevent herons from accessing fish. Make sure the barrier is far enough away from the pond edge to prevent a heron from sticking its neck through or over the fence to capture fish. However, keep the fence close enough to the edge to prevent the heron from landing between the fence and the pond. (Drawing by Jenifer Rees.)

There are several ways to reduce heron and kingfisher attacks on pond fish, all of which can be used with varying success. Any approach will be most successful when in place before the birds discover the easy food source.

Create a perimeter barrier. To prevent scaring fish, herons will not normally land directly in a pond. Instead, they will land and slowly walk toward the water. To prevent them from reaching the water, install a perimeter barrier using fencing or vegetation, or a combination of both.

To create the least visual distraction and most protection, install two strands of electrified wire, 8 and 18 inches (20 and 45 cm) off the ground, and back from the water's edge (Fig. 5). (Strong fishing line or a similar barrier may be tried at first, but herons are

likely to climb under or over it.)

The electrified wires can be hooked up to a switch for discretionary use; when you want to work near the wires, turn the system off. Be sure to keep the wires away from shoreline plants that can short the power out. (See Chapter 7 for information on electric fences.) For safety reasons, tie a sign, piece of white cloth, or other material on the wire for visibility where necessary.

Alternatively or in combination with a wire barrier, plant dense growths of tall marginal plants or shrubs around the pond to limit the herons' access to the water (Fig. 6).

Ensuring that the pond side is steep and the water is 8 to 12 (20–30 cm) inches below the edge of the pond will prevent a heron from reaching the fish (Fig. 7). (**Note:** Steep sides present safety

Figure 6. Dense growths of tall marginal plants or shrubs around the pond will limit the herons' access to the water. (Drawing by Jenifer Rees.)

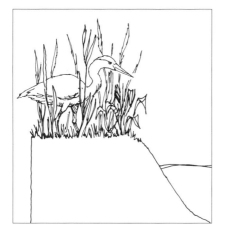

Figure 7. Ensuring the pond side is steep and the water is 8 to 12 (20–30 cm) inches below the edge of the pond will help keep herons from reaching fish. (Drawing by Jenifer Rees.)

problems for small children and the elderly, so keep steep sides away from walkways and other areas used by people.)

Create an overhead barrier. To prevent herons from landing in ponds and kingfishers from fishing from above, suspend netting (with a mesh size of 3 inches (7.5 cm) or less) above the entire pond surface. Such netting is commercially available from companies that spe-

cialize in bird control (see "Wildlife/Human Conflicts" in Appendix F). Previously used bird netting may be available from habitat restoration companies, as well as used gill netting from fishermen and fish hatcheries. The cost of new netting makes seeking out an alternative worthwhile.

Make sure the netting is at least 2 feet (60 cm) above the pond, taut, and cannot fall into the pond when a heron lands on it to spear the fish.

While such a barrier may not be the best looking solution, when combined with a fence or a perimeter barrier as described above, it is by far the most effective deterrent for most herons and kingfishers. (See "Discourage raccoons from disturbing pond plants and other aquatic life" in Chapter 21 for additional barriers.)

Herons can also be excluded by suspending parallel strands of steel wire (28-gauge) or monofilament line (50-pound test) over the pond. Wire or monofilament lines should be spaced no more than 12 inches (30 cm) apart. Where the attraction is high, such as in a fish-rearing pond, the spacing may need to be closer or a grid pattern set up. (See "Grids and Netted Rooms" under "Preventing Conflicts" in Chapter 27 for additional information.) Wire or line barriers should be installed at least 2 feet (60 cm) above an area needing protection. Prevent herons from entering under the barrier by installing a 2 foot (60 cm) high wire fence or a perimeter barrier as described above.

Where long runs of steel cable are being installed to support netting or to create a barrier, each line should get a separate length of cable, fitted at one end with an

eyebolt, and at the other end with a turnbuckle. This will allow the cable tension to be adjusted or the cable to be removed if needed. The netting can be attached to the cable with nylon string, wire, or hog rings. Hog rings and hog ring pliers are recommended for large projects.

Note: All grids, netting, and fencing material should be regularly monitored for holes, trapped wildlife, sagging, and overall effectiveness.

Provide hiding places for fish. Plant mat-forming aquatic plants, such as water lilies, for protection during the growing season. For year-round protection, construct hiding places on the bottom of the pond using cinder blocks, ceramic drain tile, wire baskets, or upside-down plastic crates held in place with heavy rocks. You can also deepen areas of the pond to at least 3 feet (90 cm)—too deep for the heron to reach.

Scare tactics. A variety of techniques can be used to frighten herons and, in some cases, kingfishers. Although law protects these birds, you may harass them without obtaining a permit as long as the birds are not nesting or touched by a person or an "agent" of a person (e.g., a trained dog).

There are a number of commercial scarers available that work in different ways. Some work on a "tripwire" basis, producing a loud noise and, in some cases, a visual deterrent, which scares the heron away. The Scarecrow® detects the presence of the heron using infrared detection, and scares it away by spraying a high-pressure jet of water.

Small rocks in a can or a "clapper" (see "Coyotes Too

Close for Comfort" in Chapter 6) can be effective noisemakers if you are lucky enough to be present when a heron or a kingfisher is around to scare away. If a heron is very reluctant to move or moves only a short distance away, it may be sick or injured. In such a case, call a wildlife rehabilitator (see Chapter 20, "Wildlife Rehabilitators and Wildlife Rehabilitation").

The effectiveness of scare devices is often short term as birds may quickly become accustomed to them. For detailed information, see "Harassment and Scare Tactics" in Chapter 27.

Plastic herons. Artificial plastic herons are very popular. Their success is based on the principle that herons are territorial and will not feed close to another heron. Unfortunately, this is not completely effective at any time, espe-cially in late winter and early spring when the herons are searching for a mate. In this case, it may actually attract herons to your pond.

Public Health Concerns

Great blue herons and kingfishers are not a significant source of any infectious disease that can be trans-mitted to humans or domestic animals. See "Public Health Con-cerns" in Chapter 29 for precau-tions to be taken when working around large concentrations of heron droppings.

Legal Status

Washington and Oregon: In Oregon and Washington, all heron and kingfisher species are protected under federal and state law. A fed-eral permit may be obtained from the U.S. Fish and Wildlife Serv-ice to use lethal means to control herons and kingfishers when extreme damage is occurring on private property. Such a permit is only granted after all other non-lethal control techniques have proven to be unsuccessful. (See Appendix E for contact infor-mation.)

British Columbia: In British Columbia, kingfishers come under provincial jurisdic-tion and are protected under the Wildlife Act, which states that unless provided by regulation (i.e., a permit), it is an offence to pos-sess, take, injure, molest or destroy a bird, its eggs, or its nest when occupied.

The great blue heron and its nests and eggs are protected year-round from persecution, hunting, and harassment.

CHAPTER 32

Gulls and Terns

Gulls are distributed from ocean beaches to the downtown cores of interior towns and cities, and from farmers' fields to fast-food outlets. The name seagull is a misnomer—not all gulls are associated with the sea.

Gulls are robust birds with webbed feet and slightly hooked bills. Like most seabirds, gulls have long, slender wings, naturally adapted for long periods of soaring on ocean air currents, alongside and behind large vessels, and over cliffs. There is no coloration distinction between males and females. Young gulls are usually all brown or gray and do not reach adult plumage for two to four years.

More than a dozen gull species spend all or part of the year in the Pacific Northwest (Table 1). The population of many species has risen dramatically in the last few decades. This increase is attributed to the expansion of cities and landfills, creation of dredge spoil islands and reservoirs behind dams, and the advent of large-scale agriculture.

Gulls have adapted well to humans and their environments, but because gulls and people often occupy the same areas at the same times of the year, conflicts have arisen.

For information on terns, a group of species closely related to gulls, see "Notes on Terns."

Figure 1. *Glaucous-winged gulls are one of the common gulls familiar to many in the Pacific Northwest. They are found along coastal saltwater bays and inland freshwater lakes, rivers, and marshes, also garbage dumps, strip malls, parking lots, airport runways, and lawns. (Drawing by Elva Hamerstrom Paulson.)*

Facts about Gulls

Food and Feeding Habits

- Gulls consume a wide range of foods, including the eggs and young of other seabirds, insects, earthworms, fish, shellfish, plant material, and carrion; human foods include waste discarded from fishing boats, garbage, and handouts.
- Gulls pick up food from the surface of water and from land. They seldom obtain food by diving or reaching underwater as terns do.
- Gulls drop mussels and other shellfish from heights of 20 feet (6 m) or more onto hard surfaces to break them open. Their heavy, slightly hooked bills help them extract the flesh and also tear up dead fish.
- Gulls can drink either fresh or salt water and they eliminate excess salt through two special glands on their heads.
- Gulls will fly 20 miles (32 km) or more to take advantage of a plentiful food source.

Nest Sites

- Gulls nest in small groups or large colonies. Nest sites include grassy bluffs, offshore rocks, and on islands and atop pilings in bays, lakes, and rivers.
- Glaucous-winged and Western gulls will nest on rooftops.
- Nests are a large bowl shape, built up a few inches in height. Nests are constructed by both sexes from grasses, seaweed, sticks, shells, fish bones, and feathers.
- Gulls defend a small area around the nest, usually only a few square meters in size.

Table 1. Some Common and Widespread Pacific Northwest Gulls

Glaucous-winged gull (*Larus glaucescens,* Fig. 1): Adults have white heads and bodies, yellow bills, brown eyes, pink legs and feet, and light gray backs and wing tips. (***Note:*** There is no black on the bird.) Adults are 23 to 26 inches (58–66 cm) long.

These gulls are mostly found along the coast and move southward to California in winter. Flocks follow fish runs (smelt and salmon) up rivers far inland.

Western gull *(Larus occidentalis):* Adults have snow-white heads and bodies, with dark slate-colored backs and wings, and black wing tips. (Their darker color and black wing tips are the best ways to distinguish them from glaucous-winged gulls.) They have yellow eyes, large yellow bills, and pinkish or flesh-colored feet. A breeding adult has a red dot near the tip of its lower mandible. Adults are 24 to 27 inches (61–69 cm) long.

They are mostly a coastal species, although in winter a few visit freshwater areas inland.

California gull *(Larus californicus,* Fig. 4): Adults have a dark gray mantle (the feathers of the back and folded wings), dark eyes and reddish eye rings, and yellow-green legs. The bill of a breeding bird has a red and black spot on the end of its lower mandible. Adults are 20 to 23 inches (51–58 cm) in length.

They nest in interior areas and migrate from late June into October toward the coast.

Ring-billed gull *(Larus delawarensis):* Adults are silvery gray on the backs and white on the heads, tails, and underparts. They have greenish-yellow feet and a narrow black ring around the yellow bill. Adults are 18 to 20 inches (46–51 cm) in length.

They nest in colonies in inland marshes, and on islands and rivers in inland areas, and also in coastal estuaries. They migrate with California gulls.

Mew gull *(Larus canus):* Adults are medium gray on the backs and white on the heads, tails, and underparts. They have yellow legs and feet and a thin, unmarked yellow bill. Adults are 16 inches (40 cm) in length.

They nest in British Columbia and Alaska and winter in large numbers in Washington.

Bonaparte's gull *(Larus philadelphia):* Adults have black heads in breeding plumage, and white heads during the winter, with a black spot behind the eye, gray backs, and white tails and underparts. They have pink legs and a black bill. Adults are 13½ inches (34 cm) in length.

They breed in British Columbia and Alaska and winter in Washington. They are our smallest regularly occurring gull and are notable for their light, buoyant flight, often picking food off the water's surface. They are frequently seen on ferry crossings on Puget Sound.

Reproduction
- Most Pacific Northwest gulls require three or four years to reach sexual maturity.
- Gulls begin to breed in late May or early June and produce one clutch per year.
- Both sexes incubate an average of three eggs for 21 to 29 days (depending on species).
- The young are tended by both parents, swim at an early age, and, depending on species, fly when 35 to 55 days old.

Mortality and Longevity
- Adult gulls are vulnerable to domestic dogs, foxes, seals, and sea lions. Gull chicks and eggs are eaten by ravens, crows, magpies, raccoons, domestic cats, minks, and rats. Bald eagles take chicks, juveniles, and adults.
- The most common predators of gulls' eggs and chicks are other gulls of a different species. Gulls will fly over the nesting grounds and swoop in to grab chicks or peck eggs. They are usually chased away by one of the parents before they can eat any of the eggs, but often the eggs are

Figure 2. In Washington, Caspian terns have shifted their preferred habitat from natural sites inland to coastal, human-altered sites (often islands made from dredged material). They have also shifted from nesting in small groups mixed with gulls to large colonies of only Caspian terns. (Drawing by Elva Hamerstrom Paulson.)

Notes on Terns

Gulls and terns are related and often mistaken for each other. With the exception of the black tern, their coloration is similar, but terns tend to be smaller (with the exception of the Caspian tern) and more slender and graceful than gulls.

Terns have long, pointed wings, sharply forked tails, and pointed red or black bills that are not hooked at the tip like those of gulls. Because of their tails, agility, and quickness in flight, terns have been called "sea swallows."

The three most abundant tern species in the Pacific Northwest include:

The **black tern** (*Chlidonias nigra*) is a mostly dark, robin-sized tern with a short, slightly forked tail. It is typically found in the interior areas of the Pacific Northwest.

The **Caspian tern** (*Sterna caspia*, Fig. 2) is a gull-sized tern with a black cap, grayish upper parts, white underparts, and a large red bill. It is mostly found in interior areas along large rivers and lakes, although large numbers are also seen in coastal areas.

The **Forster's tern** (*Sterna forsterii*) is a pigeon-sized, grayish-white tern with a black cap and a deeply forked tail. It breeds on freshwater marshes and lakes in interior areas.

One of the great wonders of the natural world is the migration of the **Arctic tern** (*Sterna paradisaea*), which flies off our coastal areas. Because it spends summers in the north and winters in the Antarctic, this bird experiences more daylight than any other creature. The 12,000-mile (19,200 km) flight between polar regions is completed twice a year. One Arctic tern, banded in the Arctic, was recaptured 11,000 miles (17,600 km) away less than three months later. The flight is even more remarkable when you consider that it was made far offshore by a bird that is not a particularly good swimmer.

Terns are quicker than gulls and adept hoverers. During flight, they carry their heads and bills pointed downward; gulls carry theirs pointed forward. Because their webbed feet are small, terns don't swim well and are seldom in the water longer than it takes to catch a billfull of fish.

Terns are expert fishers. They fly over water until they see a small fish, whereupon they plunge in after it. Terns also catch flying insects and other small invertebrates.

Terns are colonial nesters and are very territorial during the nesting season. To prevent close encounters with diving terns, avoid nesting colonies if at all possible.

Terns are less able to adapt to humans and are more dependant on special habitat than gulls. Many tern populations are declining in numbers as a result of habitat loss.

cracked and the developing chicks die.

- Other causes of gull deaths include oil spills and toxic substances in garbage dumps.
- Gulls live an average of eight years in the wild. The oldest recovered banded California gull was 27 years old.

Viewing Gulls

Gulls are among the few water birds that will tolerate environmental conditions found in urban areas. They are also one of the largest and most conspicuous bird species that people see in these areas. Visit almost any large body of salt or fresh water in the warm months and you will likely observe at least one of the species listed in Table 1.

In winter, gulls move about individually or in flocks. In spring, older birds that have bred before join with their mates and head to the breeding ground. Thus, in spring gulls will be seen in pairs. The male is slightly larger than the female and the two stay within a few feet of each other throughout the day. Look for the head-tossing courtship display given by both sexes (Fig. 3). Starting with its body in a horizontal plane and its head drawn in, the gull will repeatedly call and flip its bill upward. A similar display occurs when young gulls beg for food.

A good way to learn more about local gulls (and terns) is to join a local chapter of the National Audubon Society. Field trips, sharing sightings with others, and access to local bird experts are excellent ways to gain more knowledge about gulls and terns.

Figure 3. The head-tossing display is done by adult male and female gulls during courtship and by young birds begging for food. (From Stokes, A Guide to Bird Behavior.*)*

A field guide is helpful for identification of gull species. Illustrations should be in full color and show distinguishing traits. Clear, concise descriptions are necessary for identifying a bird in its various forms, including breeding and nonbreeding plumage, juvenile, and geographical variations. For those just starting to identify gulls, it is helpful to begin with adults and their more consistent plumages. Popular field guides are listed in Appendix F.

Tracks

Gull tracks show three of their four toes, with the webbing relatively straight between each toe. The feet turn slightly inward when a gull walks.

Droppings and Pellets

Gull droppings are semiliquid, primarily white, with indistinguishable contents. Their regurgitated pellets or castings contain bones, fish scales, garbage, and other foods. See "What's in an Owl Pellet?" in Chapter 36.

Feeding Areas

Cracked shells of shellfish that gulls drop onto rocks, roofs, and other surfaces from high in the air leave fragments in the area.

Preventing Conflicts

Gulls benefit people and the environment by "cleaning up" dead fish and other waste products, and some species are important in agricultural areas where they feed on crickets, grasshoppers, and other insects. However, because they readily habituate to humans and structures associated with people, they are often viewed as a problem.

Around outdoor food areas, gulls have been known to scare children into dropping or abandoning their food, or to steal it outright. People living along coastal areas object to gulls using sidewalks and parking lots to crack open shellfish, leaving shells and droppings scattered about.

Tolerance and common sense can usually solve these conflicts. However, at airports gulls are a potential danger to aircraft taking off and landing. At airports, trained falcons and hawks have been used to scare gulls, and cars are driven down the runway to disperse birds before planes take off. At other airports, gulls are shot after the proper permits are obtained.

The useful life of flat roofs where flocks of gulls nest and roost is thought to be half as long due to chemical erosion from gull defecation and water damage caused when feathers and nest material obstruct drainage. Droppings may pose slipping hazards on bridges, catwalks, handrails, docks, and other structures. Fish predation can also be a problem, as California and ring-billed gulls and Caspian terns feed upon young salmon.

If gulls are a problem, you can proceed with any of the following techniques yourself or hire a wildlife damage control company experienced in gull control (see Appendix C, "Hiring a Wildlife Damage Control Company"). In Oregon and Washington, large-scale coordinated intervention strategies should be under the guidance of the Department of Agriculture's Wildlife Services. In British Columbia, contact the Ministry of Agriculture, Food, and Fish, and the Ministry of Water, Land, and Air Protection (see Appendix E for contact information).

To prevent conflicts or remedy existing problems:

Stop feeding gulls. Clean up garbage, litter, human food wastes, and refrain from feeding gulls. (See "Preventing Conflicts" in Chapter 27 for information on interpretive signs for public areas.) Don't leave trash bags alongside a curb, in back of a pickup truck, or in an overfilled bin. Ask your local restaurants and food chains to keep their garbage containers closed.

Landscape modification. To discourage gulls from using park areas, lawns, and similar areas as resting or loafing areas, limit mowing. If the grass is maintained at a height of at least 12 inches (30 cm), gulls will be less likely to use the site because of their inability to visually detect potential predators. Grass height can be maintained with a weed-whacker.

Barriers. Prevent gulls from roosting on window ledges, pilings, light fixtures, and similar spots by installing the appropriate barriers. See Chapter 29 for detailed information on designs.

Exclude gulls from fish-rearing ponds, reservoirs, and rooftops

Figure 4. California gulls feed in agricultural lands, cities, and wetlands near their nesting areas. In agricultural areas, they feed primarily on small rodents. Insects, fish, eggs, and garbage are also part of the diet of this opportunistic feeder, which will fly 20 miles (32 km) to take advantage of a plentiful food source. (Drawing by Elva Hamerstrom Paulson.)

by suspending parallel steel wire (28-gauge) or monofilament line (50-pound test) over the area using 15-foot (4.6 m) spacing. To allow for maintenance, keep the lines 6 feet (1.5 m) above an area needing protection. Where the attraction is high, the spacing will need to be closer; or set up a 4 to 6 foot (1.2–1.8 m) grid pattern. Four-inch (10 cm) mesh bird netting and specially designed hardware, commercially available from bird-control companies, can be also be used to cover areas in extreme cases.

On rooftops or anywhere tall corner posts can be installed, use heavy cable on the outside frame and use smaller cable on the inside to create a grid pattern. A similar support structure can be used to hold bird netting in place. The netting can be attached to the cable with nylon string, wire, or hog rings. Hog rings and hog ring pliers are recommended for large projects.

Where long runs of cable are being installed to support netting

or to create a barrier, each line should get a separate length of cable, fitted at one end with an eyebolt and at the other end with a turnbuckle. This will allow the cable tension to be adjusted or the cable to be removed if needed.

Any barrier needs to be checked daily for sagging, broken lines, and entangled birds. (See "Grids and Netted Rooms" under "Preventing Conflicts" in Chapter 27 for more information.)

Harassment and scare devices. Harassment and scare devices are designed to frighten gulls away from problem sites. Although state and federal laws protect gulls, you may harass them without obtaining a federal or state permit as long as the gulls are not nesting or touched by a person or an agent of a person (e.g., a trained dog).

Frightening devices that have been used successfully against gulls include propane cannons, shell crackers, whistle bombs, and broadcasts of gull distress calls. Dead gull decoys placed in dead gull postures can be used, especially in conjunction with other frightening devices.

When directed by a handler, dogs are the method of choice for large open areas such as golf courses, airports, parks, agricultural fields, and corporate parks.

See "Preventing Conflicts" in Chapter 27 for detailed information these and other harassment techniques and scare devices.

Public Health Concerns

Although health risks from birds are often exaggerated, large populations of roosting gulls or terns may present risks of disease to people nearby. The most serious health risks are from disease organisms growing in accumulations of droppings, feathers, and debris under a roost. This is most likely to occur if roosts have been active for years.

Precautions need to be taken when working around large concentrations of gull or tern droppings. See "Public Health Concerns" in Chapter 29 for information.

Legal Status

Washington and Oregon: All species of gulls and terns in Oregon and Washington are protected under federal and state law and cannot be hunted or trapped. A federal permit must be obtained from the U.S. Fish and Wildlife Service to use lethal means to control gulls and terns when extreme damage is occurring on private property. Such a permit may be granted only after all other nonlethal control techniques have proven to be unsuccessful. (See Appendix E for contact information.)

British Columbia: In British Columbia, all gull and tern species are protected under the federal Migratory Bird Convention Act and the provincial Wildlife Act. No hunting or trapping of gulls and terns is allowed. Any permit to control gulls and terns would need to be issued from the Canadian Wildlife Service and would likely only be issued in very extreme cases. (See Appendix E for contact information.)

CHAPTER 33

Hawks

H awks belong to a group of birds called raptors, or birds of prey, which includes hawks, eagles, falcons, ospreys, and owls. Raptors are powerful birds with sharp, curved talons for capturing prey and strong hooked beaks used for grasping and tearing flesh. All raptors have an excellent sense of hearing, and their eyesight is the best in the entire animal world. Not only do they have a "zoom-like" focusing ability, their visual acuity (the ability to see clearly) is many times that of humans.

The Pacific Northwest provides a wide variety of habitats for raptors, including seven species of hawks that fall into two main groups: accipiters and buteos (Table 1).

Accipiters are the forest-dwelling hawks, characterized by short, rounded wings that enable them to accelerate rapidly, and a long, rudderlike tail for steering around trees. Their flight pattern consists of several rapid wing beats, then a short gliding flight, followed by more rapid wing beats. The Cooper's hawk (*Accipiter cooperii,* Fig. 1) and sharp-shinned hawk (*Accipiter striatus,* Fig. 2 and back cover) are the two most frequently observed accipiters in the Pacific Northwest.

The buteos are known as the broad-winged or soaring hawks. All buteos have long, wide wings and relatively short, fan-like tails. These features enable them to soar over open country during their daily travels and seasonal migrations. The red-tailed hawk (*Buteo jamaicensis,* Fig. 4) is the most common and widely distributed buteo in the Pacific Northwest.

Because of their broad distribution and association with humans, the hawks this chapter focuses on are the Cooper's hawk, sharp-shinned hawk, and red-tailed hawk. Where noted, information is also applicable to other hawks.

Figure 1. Adult Cooper's hawks are bluish-gray with brownish-orange horizontal bars on the breast. An adult female is 14 to 20 inches (35–50 cm) tall with a 28-inch (71 cm) wingspan. The male is approximately 20 percent smaller. The tail is longer and rounder (curved like the letter C) than that of the sharp-shinned hawk. Both Cooper's and sharp-shinned hawks have tails with three black bands. (Drawing by Elva Hamerstrom Paulson.)

Facts about Hawks
Habitat and Territory

• Cooper's and sharp-shinned hawks occur in areas with trees and large shrubs, including greenbelts, brushy edges of woodlands, shrub thickets and fencerows, as well as residential communities, cemeteries, and golf courses.

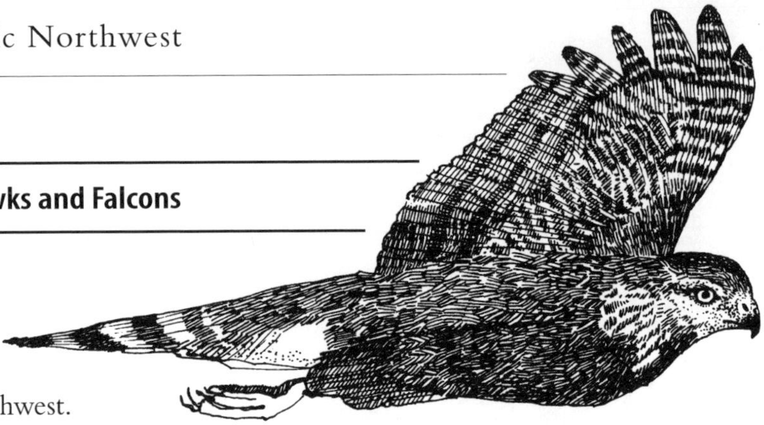

Table 1. Pacific Northwest Hawks and Falcons

Accipiters (woodland dwellers with long tails and short, rounded wings)

Northern goshawk (*Accipiter gentilis*): Occurs year-round throughout the Pacific Northwest. Occasionally seen at the edges of clearings and hunting below the treetops in coniferous forests.

Cooper's hawk (*Accipiter cooperii*, Figs. 1 and 3): Seen year-round throughout the Pacific Northwest in a wide variety of habitats, including forested urban areas.

Sharp-shinned hawk (*Accipiter striatus*, Figs. 2, 3, and back cover): Seen year-round throughout the Pacific Northwest in a wide variety of habitats, including forested urban areas.

Buteos (soaring hawks with broad tails and rounded wings)

Red-tailed hawk (*Buteo jamaicensis*, Figs. 4 and 5): Occurs year-round throughout the Pacific Northwest in a variety of habitats, preferring wooded areas near open fields. Commonly seen on the roadside perched on poles or midway up large trees.

Swainson's hawk (*Buteo swainsoni*): Seen in spring and summer east of the Cascade mountains and in British Columbia in prairie areas, open plains, cultivated fields, and clearcuts.

Rough-legged hawk (*Buteo lagopus*): Occurs in winter in open country, such as fields, prairies, and marshes throughout the Pacific Northwest. Tends to perch on smaller branches and shorter tree tops than the red-tailed hawk. Seen perched on wires more than on posts.

Ferruginous hawk (*Buteo regalis*): Seen in small numbers in spring and summer east of the Cascade mountains and even more rarely in British Columbia's dry, open country.

Harriers (low fliers with long-wings and tails)

Northern harrier (*Circus cyaneus*): Occurs year-round throughout the Pacific Northwest. Often seen gliding over marshes and grasslands where it hunts by taking birds and small mammals by surprise. A good field mark is its broad white rump. Formerly called "marsh hawk."

Falcons (fastest fliers, with long, pointed wings)

American kestrel (*Falco sparverius*): Occurs year-round throughout the Pacific Northwest in grasslands, farmlands, and urban areas. Commonly perches on telephone wires. Formerly called "sparrow hawk."

Merlin (*Falco columbarius*, Fig. 7): Occurs year-round throughout the Pacific Northwest in a wide variety of habitats, particularly areas with shorebirds. Also seen in urban areas since it preys on songbirds. Formerly called "pigeon hawk."

Prairie falcon (*Falco mexicanus*): Occurs year-round throughout the hot, dry, open terrain in the interior Pacific Northwest; occasionally seen in coastal areas during the winter.

Peregrine falcon (*Falco peregrinus*): Seen year-round throughout the Pacific Northwest, including cities, where it feeds on starlings and pigeons. Formerly rare but increasing in numbers.

Gyrfalcon (*Falco rusticolus*): Rare. A small number are seen each winter in the Pacific Northwest. Largest falcon in our region.

Figure 2. The sharp-shinned hawk is the smallest member of the accipiter group. A typical female is 13½ inches (34 cm) long with a wingspread of 25 inches (63 cm). The male is approximately 20 percent smaller. Adults are bluish-gray with brownish-orange horizontal streaks on the breast. Because of their similar size and appearance, it is quite difficult to distinguish a male Cooper's hawk from a female sharp-shinned hawk. (Oregon Department of Fish and Wildlife.)

Figure 3. Cooper's and sharp-shinned hawks perch quietly in trees or move stealthily from perch to perch. When prey is sighted, they make a dash, reach out with their long legs, and grab the prey animal with powerful, long, narrow toes equipped with sharp talons. (From Proctor, Manual of Ornithology.*)*

- Red-tailed hawks occur in open to semi-open coniferous, deciduous, and mixed woodlands; woodland edges; grassy roadsides; rangelands; and agricultural fields with scattered large trees.
- The size of a Cooper's, sharp-shinned, or red-tailed hawk's territory is variable and is determined by the abundance of food and the number of available perches and nest sites. Hawks will fly several miles away from the nest in search of food. (See "Migration Sites" for additional information.)

Food and Feeding Habits

- Cooper's and sharp-shinned hawks are often seen near pigeon and starling roosts and in the vicinity of bird feeders. In addition to pigeons and starlings, they eat robins, finches, juncos, sparrows, and similar size birds.
- Less frequently, Cooper's and sharp-shinned hawks eat rodents, frogs, lizards, and insects. Cooper's hawks will kill domestic pigeons and occasionally small poultry.
- Red-tailed hawks are opportunistic predators and will eat anything from large insects, lizards, snakes, frogs, and turtles to voles and rodent-like mammals up to the size of a rabbit—occasionally taking poultry. They will also eat dead animals.
- Red-tailed hawks search for prey from an elevated perch and are seen atop snags, utility poles, fence posts, and road signs, often gliding down to catch their meal.

- Some studies conclude hawks see reflected ultraviolet light. Therefore, they perch near areas well marked by urine trails of rodents whose urine reflects ultraviolet light.

Reproduction

- The breeding season for Cooper's, sharp-shinned, and red-tailed hawks begins as early as February, depending on latitude and altitude. Pairs raise one brood each year.
- The eggs are incubated by both sexes (mostly by the female) and hatch after 30 to 35 days (sharp-shinned), 36 days (Cooper's), or 32 to 35 days (red-tailed).
- The young grow rapidly and begin to fly at 23 days of age (sharp-shinned), 30 days (Cooper's), or 42 days (red-tailed).

Figure 4. The red-tailed hawk is a large, stocky bird. An adult is 18 to 25 inches (45–63 cm) long with a 40-inch (101 cm) wingspan. Red-tailed hawks occur in various color patterns; typical birds have dark heads, whitish breasts with a darker bellyband, a whitish lower belly, and rust-colored tails. Young birds are duller and more streaked, and lack the rust-colored tails of the adults. Some red-tailed hawks are nearly all dark brown. (Drawing by Elva Hamerstrom Paulson.)

Figure 5. *Humans approaching an occupied hawk's nest (red-tailed hawk shown here) may cause an adult hawk to abandon the nest or flush from it temporarily, leaving the developing young vulnerable to predators or cold. (Drawing by Elva Hamerstrom Paulson.)*

- Out-of-nest care by the parents (fledgling phase) is 21 to 28 days (sharp-shinned), 30 to 40 days (Cooper's), and 42 to 46 days (red-tailed).
- Juvenile red-tailed hawks may continue to associate with their parents for up to six months after they leave the nest.
- Hawks will return to the same breeding territory for several years, thus staying together as a pair. If one of the pair dies, another hawk will acquire the mate and territory.

Nests and Nesting Sites

- Hawks nest in coniferous forests or mixed stands of coniferous and deciduous trees, often at a forest or woodland edge.
- The nest is located 15 to 90 feet (4.5–27 m) above ground in the crotch of two or more large limbs, or where limbs meet the trunk of the tree. In treeless areas, red-tailed hawks nest in rocky cliffs, large shrubs, or utility pole crossbars and towers.
- Red-tailed hawks reuse an old nest or build a new one, sometimes using a squirrel or crow's nest, or another hawk's nest as a base. Cooper's and sharp-shinned hawks prefer to build new nests, which can result in several nests being located within 100 to 200 feet (30–60 m) of one another.
- The bulky nests have an outside diameter of 28 to 38 inches (70–96 cm) (red-tailed) or 20 to 25 inches (50–63 cm) (Cooper's and sharp-shinned); old red-tail nests that have been used for years may be three or more feet (90 cm) thick.

Mortality and Longevity

- Causes of mortality among young hawks include predation and unfavorable changes in weather. Nestlings are most vulnerable during incubation and brooding, and when they first learn to fly.
- Adult hawks, eagles, owls, and crows occasionally kill adult and young hawks; raccoons, black bears, and other climbing predators sometimes reach nests to eat eggs or young birds.
- Hawks also die from disease, illegal shooting, or flying into windows, vehicles, power lines, or poles.
- As many as 70 percent of hawks die during their first winter.
- In the wild, a red-tailed hawk can live seven years or longer; captive hawks live 20 years or more.

Viewing Hawks

Because Cooper's and sharp-shinned hawks are secretive and inhabit wooded areas, they are mostly seen during their spring and fall migration, often chasing songbirds near bird feeders. Red-tailed hawks are observed year-round because they soar in the open, perch in conspicuous places, and have loud, distinct calls.

Hawks are often sighted when crows, jays, magpies, or other birds discover them in their territory. The birds will defend their domain by diving and calling repeatedly at the perched or flying hawk, called "mobbing." Look and listen for this behavior and see if you can locate the "invader."

The sight of humans easily disturbs nesting hawks. Adult hawks brood (keep their young covered), mostly when the young are less

Maintaining Hawk Habitat

For people wishing to maintain hawk habitat on their property, things to include are:

• Multiple-acre patches of coniferous trees—good nesting areas for accipiters.
• Young stands of coniferous trees at various stages of growth—good hunting areas for Cooper's and sharp-shinned hawks.
• Quiet, protected areas away from human activity—good for all hawk species.
• Protected areas near water with big trees—good for all hawk species.
• Tall snags (dead or dying trees over 10 feet or 3 m tall)—good perch sites and nest sites for kestrels.
• Tall live trees—good nest and perch sites for several hawk species.
• Hedgerows and thickets bordering fields—good for accipiters and red-tailed hawks.
• Large unmowed or infrequently mowed grassy areas—good for red-tailed hawks, Northern harriers, kestrels, and other species that eat rodents and large insects such as grasshoppers. (See "Providing Vole Habitat" in Chapter 25 for management information.)
• Artificial nest boxes for kestrels (see Appendix F for a list of resources).
• Artificial perches and platforms—good for red-tailed hawks (Fig. 6).

Hawks (and owls) use tall dead trees and branches as places to rest, look for prey, and feed once prey is caught. The tree's height provides the birds with a wide visual range, easy takeoff, and greater attack speed when hunting. Where tall snags or dead branches don't exist or can't remain because of safety constraints, perch poles can be installed for owls and hawks. This may be as simple as adding extensions to

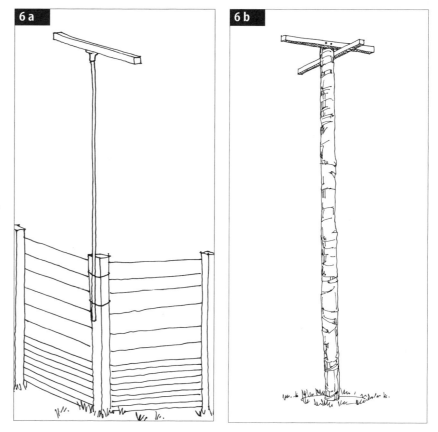

Figure 6. *Two perch poles built for hawks and owls. (Drawings by Jenifer Rees.)*

fence posts and sprinkler risers in grassy fields, orchards, and woodlots.

If attaching a metal perch pole to a fence post or other structure, bolt a 2 x 2 x 24 inch (5 x 5 x 60 cm) wooden crossbeam to a floor-type flange that is screwed onto the upper end of an 8 foot (2.5 m) section of ¾ inch (1.9 cm) galvanized steel water pipe (Fig. 6a). The galvanized pipe can be secured to the fencepost with bailing wire and be painted to reduce metal glare and to protect it.

If erecting a wood perch pole, it's best to use 4 x 6 inch (10 x 15 cm) or thicker timber, 16 to 20 feet (4.5–6 m) long, buried at least 3 feet (90 cm) in the ground (Fig. 6b). The timber should be well tamped or set in concrete to stand firm. If a crossbeam (owls use these more than hawks) is bolted at the top, it should be at least 2 x 2 inch

(5 x 5 cm) to ensure adequate hold for the birds. If necessary, diagonal struts can support it. Orient the crossbar so that birds can land on it against the prevailing wind. Where wind directions vary, use two crossbars.

In general, one perch pole every 200 feet (60 m) around a field or per acre (0.4 h) will provide the birds with enough perches for optimum hunting. Remember that a perch is only as useful as the surrounding habitat. Management of open habitat is essential if breeding and hunting grounds for hawks and owls are to be maintained or created.

Caring for an Injured Hawk

Under federal, state, and provincial law, it is illegal for anyone to injure, harass, kill, or possess a bird of prey. Licensed rehabilitators are the only people legally permitted to transport and keep wildlife, including hawks.

If you find an injured hawk, contact a wildlife rehabilitation facility immediately. Your local wildlife office keeps a list of rehabilitators and can tell you which ones serve your area, or you can look under "Animals" or "Wildlife" in your phone directory. If a rehabilitator isn't available, follow the menu options over the phone or on their Web site for information on what to do. (For more information, see "Wildlife Rehabilitators and Wildlife Rehabilitation" in Chapter 20.)

How to Report a Bird Band

Bird banding is one of the most useful tools in the modern study of wild birds. Wild birds are captured and marked with a uniquely numbered band or ring placed on the leg. The bander records where and when each bird is banded, how old it is, what sex it is, and other information. The bander sends these data to the North American Bird Banding Program, which is jointly administered by the U.S. Department of the Interior and the Canadian Wildlife Service.

Information from bands subsequently found and reported to the Bird Banding Program provides data on the distribution and movements of species, their relative numbers, annual production, life span, and causes of death. Such information increases the knowledge and understanding of birds and their habits and assists in their management and conservation.

If you see a live or dead hawk or other bird with a band on its leg, you can provide a vital service by reporting it. The preferred method of reporting bird bands is to call the toll-free number 1-800-327-BAND (2263). The operators will need to know the band number as well as how, when, and where the bird or band was found. Please do not use this number to call about other matters.

than a month old. Brooding keeps the young warm, dry, and protected from predators. Humans approaching at this time may cause an adult hawk to abandon the nest or flush from it temporarily, leaving the developing young vulnerable to predators or cold. To prevent problems from occurring, nesting hawks should be left alone from January through May, or watched from a distance with powerful binoculars or a spotting scope.

A good way to learn more about local hawks is to join a local chapter of the National Audubon Society or other local naturalist group. Field trips, sharing ideas and sightings with others, and access to local bird experts are excellent ways to broaden your knowledge of these animals.

A field guide is helpful to identify hawks. Illustrations should be in full color and show distinguishing traits. Clear, concise descriptions are necessary for identifying a bird in its various forms, including breeding and nonbreeding plumage, sex, age, and geographical variations. Popular field guides are listed in Appendix F.

Some zoos and wildlife rehabilitation centers have educational demonstration programs where hawks can be seen up close.

Feeding Sites

Evidence of a Cooper's or sharp-shinned hawk's recent feeding site will be an accumulation of feathers on a stump, log, or rock. This is where the hawk carries its prey to pluck off the feathers before bringing it to the nest or eating it. An indiscriminant scattering of feathers on the ground occurs where the hawk has plucked off its prey's feathers from a fence post of a branch above.

When a red-tailed hawk eats a rabbit-size mammal, there will be tufts of hair around the partially eaten animal.

Pellets and Droppings

Contrary to popular belief, owls are not the only birds that produce pellets. All birds of prey produce pellets to expel indigestible foods. When hawks consume feathers or tufts of fur along with their meal, they are not digested but are bundled into compact "pellets" and later regurgitated. Pellets are most commonly found under a hawk's favorite perch.

Unlike owls, which ingest entire animals, hawks tear much of the flesh from their victims and swallow very few bones. Thus, hawk pellets do not provide as much information about their diet. For more information, see "What's in an Owl Pellet?" in Chapter 36.

A hawk's normal droppings are semiliquid and primarily white, often causing a "whitewash" under a nest or perch site.

Calls

Calls are directed toward intruders, such as humans and other hawks. They are also given while the birds are engaged in aggressive encounters, courtship, and by young begging for food. Birds also use contact calls to locate their mate and keep track of their young.

Common calls include:

Cooper's hawk: a rapid series of loud *kaks* or a high-pitched *pee-a, pee-a, pee-a.*

Sharp-shinned hawk: a rapid series of short, harsh, high-pitched *kik-kik-kik-kik* sounds similar to the call of the Northern flicker; also a shrill squeal.

Red-tailed hawk: a high-pitched, descending, scream-like *keeeeer;* also a loud, piercing series of chirps. The *keeeeer* cry is so dramatic it is regularly used as a background sound in movies and TV shows. Steller's jays and starlings imitate the calls of red-tailed hawks.

Red-Tailed Hawk Displays

Look for red-tailed hawks soaring in late winter and early spring. If you watch them for 15 to 30 minutes you are likely to see several different displays associated with courtship and territorial behavior. Displays include the following:

The **talon-drop** is done by the male when soaring over the female. The hawk extends his legs down and may attempt to touch

Hawks at the Bird Feeder

A large concentration of songbirds around your feeder can attract a Cooper's hawk, sharp-shinned hawk, merlin (Fig. 7), or other raptor, especially during the winter. The occasional foray of a hawk or falcon into a backyard wildlife sanctuary should be welcomed rather than treated as a problem.

Predation is a natural part of a well-functioning ecosystem. Hawks weed out the unfit and thus help maintain the overall health of the songbird population. Healthy songbirds can protect themselves by taking cover quickly. To facilitate this, you can provide trees, shrubs, thickets, and brush piles (see "Brush Piles for Rabbits, Hares, and other Wildlife" in Chapter 20).

If you have a persistent hawk or other bird of prey hunting around your feeding station on a regular basis, remove the feeders for a few days and the hawk may move on.

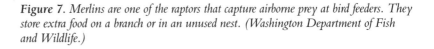

Figure 7. Merlins are one of the raptors that capture airborne prey at bird feeders. They store extra food on a branch or in an unused nest. (Washington Department of Fish and Wildlife.)

or hit the female with his talons. The display is often accompanied by a call.

The **dive-display** is a steep dive that starts from a soaring or undulating flight position and is usually directed toward the nest area. The display stops short of the ground or tree canopy.

Rare to see is a **death spiral,** which may occur during territorial encounters with intruding adults. Two hawks lock talons, spread their wings, and spin as they fall earthward, sometimes to the ground.

Migration Sites

Some Cooper's, sharp-shinned, and red-tailed hawks spend the winter months in warmer regions of the Pacific Northwest or farther south. Spring migration begins in February and lasts until mid-March; fall movement begins in late August and ends in late October, depending on the region.

It is at these times that hawks are likely to be seen around a bird feeder, flock of chickens, or other food source that attracts them while they are moving through your area. Several areas in the Pacific Northwest are popular sites for viewing large concentrations of migrating hawks. Contact your local chapter of the Audubon Society for information about viewing sites and the best time to visit.

Preventing Conflicts

Once juvenile hawks have left the care of their parents in summer and fall, they enter a period where they must quickly perfect their hunting skills or starve. During this time, chickens and other domestic birds can seem very attractive to these inexperienced hunters. Once hawks discover an easy supply of food, they can be very persistent.

Cooper's hawks will attack small farm and hobby birds; sharp-shinned hawks are less of a problem due to their small size. Red-tailed hawks are too slow to catch a bird in flight, but they occasionally prey on pigeons and free-ranging fowl and domestic rabbits.

Cooper's and sharp-shinned hawks occasionally eat songbirds that are attracted to bird feeders. Thus, they bring out mixed emotions in people, especially bird lovers. A hawk will usually kill only one bird per day. (See "Hawks at the Bird Feeder.")

Hawks pluck birds and mammals, leaving piles of feathers or fur at plucking spots or on the ground. The plucked feathers can be used to determine whether a hawk actually killed a bird or was simply feeding on one that had died of other causes. If the feathers have small amounts of tissue clinging to their bases, they were plucked from a cold bird that died of another cause. If the base of the feather is smooth and clean, the bird was plucked shortly after it was killed.

While trying to catch small birds, a hawk will occasionally get caught inside a building. If this happens, turn off all inside lights and open all windows and other exits. The hawk should leave on its own. If necessary, a long pole with a T-shirt or flag at the end can be used to direct the bird out an exit. If these methods fail, a licensed falconer, wildlife rehabilitation center, or a wildlife damage control company can be called to assist in the removal process. Call your local wildlife office for contact information.

Injured or cornered wildlife will react by defending themselves. Hawks do this by using their powerful talons and occasionally their beaks. Do not attempt to capture a hawk without first talking to a wildlife professional.

To prevent conflicts or remedy problems:

Enclose domestic animals. Free-roaming chickens, ducks, turkeys, and pigeons are all potential hawk meals because they are conspicuous and often concentrate in areas that lack adequate escape cover.

By far the best defense is to house domestic birds in a durable, fenced enclosure that will allow the birds to safely eat and loaf outside during the day. Such a structure can be constructed with a wooden framework that is **entirely** covered with 1-inch (2.5 cm) poultry wire or similar netting. This outdoor run can be permanent and attached to a coop or other building, or be a portable and moved periodically. (See Chapter 21 for information on how to exclude raccoons and other mammals from such enclosures.)

Some commercial chicken farms protect their fenced chickens from large hawks and eagles by installing wires in a grid fashion from the top of the fence and over the chickens. See "Barriers" under "Preventing Conflicts" in Chapter 32 for information. Such a barrier would not likely work at keeping the smaller Cooper's and sharp-shinned hawk out.

Where a permanent enclosure isn't practical or desirable, escape cover should be provided. Birds have natural defenses at the sight of a hawk and will quickly squeeze under a nearby building, old car, shrub, or other area. Escape cover can be made of planks, plywood, or chicken wire placed over logs, rocks, or bricks. It should be at least 7 x 7 feet (2 x 2 m) wide and long and the cover should be 8 inches (20 cm) off the ground.

Please understand that you cannot expect to fully protect free-range birds from hawks and an occasional loss of a bird is to be expected.

Change your routine. Hawks will quickly learn the routines on a property if they are successful at catching prey on site. If a problem occurs, people flying pigeons or allowing other birds to feed unattended should vary the routine. Use this technique anytime a hawk is spotted nearby; be particularly attentive during the spring and fall migration (see "Migration Sites").

Install scare devices. A variety of devices can frighten a problematic hawk. Increasing human activity in the area will keep most hawks at a distance. Yelling and clapping hands, firing a gun loaded with blanks (it is illegal to shoot any type of hawk), and banging cans together are all effective when a hawk is seen nearby. A "clapper" can be constructed by putting a hinge on the ends of two, 24-inch (60 cm) 2 x 4s and smacking them together. Any hesitation on the hawk's part will cut its odds of catching a targeted bird.

More extreme scare devices include projectile explosives fired from a special pistol (see Appendix F for resources). Some government entities consider these firearms and require permits before you can shoot them. Even if your local government does not consider them firearms, treat them as potentially dangerous.

The "hawk globe" is basically a round mirror designed to scare an attacking hawk. If it is placed in the flight path the hawk uses, an attacking hawk will see its reflection and retreat, giving domestic birds a second chance. Because hawks hunt on their own, they may avoid returning to a place where they perceive competition from another hawk.

Scare devices reduce losses rather than eliminate them. Those who use these devices must be willing to tolerate occasional losses. If predatory birds are hungry, they quickly get used to, and ignore, frightening devices.

Public Health

Hawks are not a significant source of any infectious disease that can be transmitted to humans or domestic animals.

Legal Status

Because legal status and other information about hawks change, contact your local, state, or provincial wildlife office, or visit their Web site for updates. (See Appendix E for contact information and where to access the state and provincial laws mentioned below.)

In **Oregon** and **Washington,** all species of hawks and falcons are protected under federal and state law and cannot be hunted or trapped, except by licensed falconers with appropriate licenses and permits. A federal permit may be obtained from the U.S. Fish and Wildlife Service to use lethal means to control raptors when extreme damage is occurring on private property. Such a permit is only granted after all other non-lethal control techniques have proven to be unsuccessful.

In **British Columbia,** hawks come under provincial jurisdiction and are protected under Section 34 of the Wildlife Act, which states that unless provided by regulation (i.e., a permit), it is an offence to possess, take, injure, molest or destroy a bird, its eggs, or its nest when occupied.

CHAPTER 34

House Sparrows and Native Sparrows

The commonly seen house sparrow (*Passer domesticus,* Fig. 1) is not a true sparrow, but is one of two Eurasian species in the family Passeridae that are now found in North America.

House sparrows were introduced to various places in the United States for a variety of reasons: sentiment, aesthetics, and with hopes they would control problem caterpillars. The first successful release was in a New York City cemetery in the spring of 1853. These 50 pairs of sparrows were from England, from which we get the bird's other common name, the English sparrow.

Through subsequent releases, westbound trips in railroad grain cars, and other ways, the first house sparrows were spotted in the Pacific Northwest in the late 1890s. The house sparrow is now considered to be the second most prolific introduced bird species, being outnumbered only by the European starling. Remarkably, its numbers have fallen since the 1920s. The decline has been attributed to competition for nest sites with European starlings, the lack of available grain as automobiles and trucks replaced horses, and a reduction in the number of farms.

The house sparrow prefers human-altered habitats in cities and around houses. It is adept at soliciting handouts at fast-food restaurants, sidewalk cafes, and parks. The house sparrow seldom enters areas very far from human activities.

For information on sparrows native to the Pacific Northwest, see "Notes on Native Sparrows."

Facts about House Sparrows

Food and Feeding Habits

- House sparrows prefer to eat seeds and grains located on the ground. They will also feed on fruits, flowers, buds, handouts, and whatever else is edible and available to them.
- Young in the nest are fed a high-protein diet of insects.
- House sparrows pick insects off the grills of parked cars and hunt for them at night under streetlights.

Nest Sites

- House sparrows nest wherever they can find suitable cavities or crevices in or near buildings. Sites include ledges, attics, light fixtures, utility poles, nest boxes, and old bird nests.
- In an area with no available cavities, house sparrows nest in trees or shrubs, often in small colonies.
- Nests are untidy, bulky structures of straw, plant stems, paper, string, cloth, and similar debris. Nests are lined with feathers, hairs, and other soft material.
- Nest building is done by both sexes and can begin as early as mid-February.

Figure 1. House sparrows are chunky birds, 5 to 6 inches (13–15 cm) long. Males in breeding plumage have streaked chestnut backs, chestnut heads, light gray cheeks and underparts, and gray crowns. Males have black throats and breasts which are more prominent during breeding season. The female and young birds are mostly brown overall, have a gray breast, buff eye stripe, and a streaked back. (Drawing by Elva Hamerstrom Paulson.)

Table 1. Introduced bird species that are successfully maintaining populations in Oregon, Washington, and British Columbia *(See Chapter 17 for introduced mammal species and credits.)*

Bird	Reason(s) for introduction	Origin	Status
Mute swan *Cygnus olor*	Aesthetics, escapes	Eurasia	Limited numbers in OR, WA, and BC
Gray partridge *Perdix perdix*	Hunting, brood stock sale	Eurasia	Scattered populations in NE OR, E WA, and S BC
Egyptian goose *Alopachen aegyptiacus*	Escapes from a wild animal park (Wildlife Safari) in Winston, OR	Africa	Approximately 250 geese in Umpqua Valley, OR
Chukar *Alectoris chukar*	Hunting, brood stock sale	Eurasia	Scattered populations in E OR, E WA, and S BC
Ring-necked pheasant *Phasianus colchicus*	Hunting, brood stock sale	Eurasia	Widespread, common, declining in some areas
White-tailed ptarmigan *Lagopus leucurus*	Aesthetics, hunting	SE Alaska, W Canada into WA	Small numbers in OR
Northern bobwhite *Colinus virginianus*	Hunting, brood stock sale	E United States	Localized, small numbers
Wild turkey *Meleagris gallopavo*	Hunting	E United States, S-central United States	Widespread, moderate numbers in E WA, E OR, and SE BC
Rock dove *Columba livia*	Aesthetics, racing, messengers, food	Eurasia	Widespread, common
European starling *Sturnus vulgaris*	Aesthetics, then range expansion	Eurasia	Widespread, common
American black duck *Anas rubripes*	Hunting	Eastern United States	Small populations in Puget Sound area and BC
Crested mynah *Acridotheres cristatellus*	Aesthetics, range expansion	SE Asia	Small numbers in Seattle and Bellingham, WA, and Vancouver BC
House sparrow *Passer domesticus*	Aesthetics, sentiment, and pest control, then range expansion	Eurasia	Widespread, common
Monk parakeet *Myiopsitta monachus*	Escapes	South America	Small numbers in the Portland, OR, area

Notes on Native Sparrows

The "little brown birds" that you see and hear around parks and gardens are often our native sparrows. Most species are about 6 inches (15 cm) or less in size. Sparrows as a group are seed-eaters, and all have short, stout bills adapted for seed-cracking. Several species come to bird feeders that offer seed.

During the breeding season, adult sparrows consume more animal matter in the form of spiders, mosquitoes, caterpillars, and other insects to provide protein necessary for their increased activity. Nestlings are fed almost entirely on such animal matter.

Because many sparrows nest in low brush and forage on and near the ground, and because the young develop rapidly and often leave the nest before they're able to fly, they are especially vulnerable to predation by house cats. This is an especially good reason to keep your cats indoors, besides the many health benefits enjoyed by indoor cats. (For more information, see Appendix D, "The Impact of Domestic Cats and Dogs on Wildlife.")

More than 20 species of sparrows occur in the Pacific Northwest. The following are some that are common and easily identified:

The **white-crowned sparrow** (*Zonotrichia leucophrys,* Fig. 2) is easily recognized, at least in adult form, by the black-and-white striping on its head. Its song is a slow, nasal series sounding like *oh-gee, kitty-did-scare-me* or *see see little-little me.* Or *see-me pretty-pretty-me.*

The **golden-crowned sparrow** (*Zonotrichia atricapilla*) looks similar to the white-crowned sparrow, but instead of white head stripes it has a dull-yellow cap heavily bordered with black. The descending, three-note whistled song is a *dee-er, dee-er, mee.*

The **song sparrow** (*Melospiza melodia*) is the quintessential "little brown bird," with its brown-streaked breast and brown head. As the specific name *melodia* implies, the song is quite musical. It begins with a *chit-chit-cha-wee,* followed by various trills.

The **savannah sparrow** (*Passerculus sandwichensis*) has a distinctive yellow eyebrow stripe that can be seen at close range. It is uniformly streaked both on the back and on the light-colored breast. Its song is a thin, insectlike *chip-chip-chipa-bzzz-bzzz.* Or *zeet-zeet-zeet zeetle-zee-ti-zeet.*

The **chipping sparrow** (*Spizella passerina*) has gray underparts, a cinnamon-colored cap, a white stripe over the eye, and a black line through the eye. Its call is an evenly pitched trill that sounds like a series of "chips," hence, its common name.

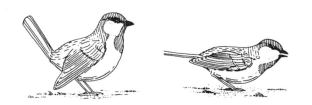

Figure 3. Two visual displays given by house sparrows. (From Stokes, A Guide to Bird Behavior.*)*

Figure 2. The white-crowned sparrow is a distinctive bird with bold black-and-white stripes on its head. (Drawing by Elva Hamerstrom Paulson.)

- Male house sparrows select and aggressively defend their nest sites. They take over nest boxes and other nest sites occupied by native birds, including purple martins, bluebirds, and violet-green swallows.

Reproduction

- The breeding season for house sparrows begins in mid February and ends in early June. However, because house sparrows often raise three broods a year, their breeding season can continue into late August.
- Both parents help incubate the three to six eggs for 10 to 14 days, and both feed the young.
- The young begin to fly at 15 days of age, are cared for out of the nest for about a week, and then flock with other juvenile birds.

Mortality and Longevity

- Sharp-shinned hawks, Cooper's hawks, merlins, pygmy owls, and barn owls take house sparrows in flight, particularly around winter bird-feeding stations.
- Tree squirrels, striped skunks, weasels, raccoons, jays, and crows take some eggs and nestlings. Domestic cats catch adult and juvenile birds feeding on the ground.
- Cold winters kill many adult house sparrows. The overheating of metal farm buildings leads to mortality of the young in the nest.
- Mortality is highest during the first year; few house sparrows survive more than three years.

Viewing House Sparrows

Except for juvenile birds that move out of an area to establish new territories, house sparrows do not travel far. They will stay in an area year-round as long as food, water, and nest sites are available. House sparrows survive winter by roosting in groups at night in heated or sheltered structures, including attics, soffits, grain elevators, barns, and outbuildings.

Because these birds are among the few animals that flourish in the inner city, house sparrows (and domestic pigeons) may be the only wildlife observed by people living there.

Look for flocks of house sparrows in parks, malls, and other places where the birds are accustomed to humans and gather to feed. Within minutes of watching, you are likely to see several displays associated with courtship or territory (Fig. 3).

The **hop-and-bow** display is done by male house sparrows. The male hops about in front of a female with his head held high, wings drooped slightly, tail feathers up and fanned, and its rump feathers fluffed. The display is accompanied by a chirp-call.

The **head-forward** display is performed by the male or female as a close-distance threat to another house sparrow or other bird. It is often seen at feeders. The bird places its body horizontally with head forward and wings slightly spread. Its bill may be open.

Since house sparrow nests are used almost year-round, you can often see some type of nest-related activity—repair work, new construction, and territoriality. Watch for both male and female house sparrows picking up objects off the ground and breaking twigs off of shrubs. Follow a bird carrying nesting material and it will lead to a nest site. While walking on a sidewalk you may see a male house sparrow perched on a sign, rail, or planter box, chirping repeatedly. (The house sparrow's call is a monotonous series of nearly identical *chirps.*) This is the typical behavior of a male attempting to attract a mate to a nest he has constructed nearby.

Preventing Conflicts

House sparrows are among the most common visitors to bird feeders and nest boxes. They provide enjoyment to those who enjoy watching them and clean up bread crumbs and other edible matter from the streets and sidewalks—perhaps curbing the numbers of rats and mice in urban areas. However, introduced wildlife species often become

problems, and the house sparrow is no exception.

Because house sparrows are aggressive, social birds that remain in the same area year-round, they often out-compete native bird species for food and nest sites. Because they form roosts and search out food sources and nest sites in and around structures, they can come into conflict with people.

The following are suggestions on how to prevent and remedy problems that might arise. If you lack the knowledge, ability, or interest to deal with the problem, see "Hiring a Wildlife Damage Control Company" in Appendix C.

House sparrows nesting in buildings and other structures. House sparrows are adept at finding nest sites in any nook or cranny. Their bulky, flammable nests are a potential fire hazard and can clog gutters and drainpipes. Their droppings contaminate stored food. House sparrows can also create noise problem for people, as their chatter begins at first light.

Prevent house sparrows from nesting in existing structures by sealing all existing holes wider than ¾ inch (2 cm) in diameter. Use wood, ¼-inch (6 mm) hardware cloth, aluminum flashing, or another suitable material. Lightweight material, such as plastic netting or window screening, will not keep determined house sparrows out. Replace loose siding and any broken windows.

Install commercially available vent guards to prevent house sparrows from entering exhaust vents and dryer vents. If necessary, cover the ends of elevated drainpipes with ¼-inch (6 mm) hard-

Figure 4. Before and after pruning of a small deciduous tree to reduce its attractiveness to roosting house sparrows in winter. (Drawings by Jenifer Rees.)

ware cloth during the nesting season. All screening should be checked periodically to make sure it isn't clogged.

Prevent house sparrows from roosting on walls covered with vegetation by removing the vegetation or draping bird netting over the area. In new construction, avoid creating small cavities or spaces with access from the exterior into which house sparrows can enter and nest.

House sparrows can be evicted from buildings and other sites any time of year. State, federal, or provincial laws do not protect this species. A stick with a 2½-inch (6 cm) angle bracket screwed to it can be used to remove nests. The nesting material should be collected and removed to prevent the birds from using it for a new nest. Take immediate steps to prevent the sparrows from rebuilding, which they will attempt repeatedly throughout the breeding season.

If the birds are caring for young, one approach is to wait until the young can fly out of the nest, then remove all nesting materials and cover all openings. (See Appendix A for information on euthanization.)

House sparrows roosting inside or on buildings. Large accumulations of house sparrow droppings can build up to such an extent that they smell or become a health hazard.

Covering the underside of rafters with bird netting prevents house sparrows from gaining access to roosting spots. For detailed information on its placement, see "Preventing Conflicts" in Chapters 29 and 39.

Heavy plastic or rubber strips hung in open doorways of buildings have been successful in excluding house sparrows while allowing people, machinery, or livestock to enter. Hang strips with about ½-inch (1.25 cm) gaps between them.

Occasionally a house sparrow will get caught in a building. If this happens, turn off all inside lights and open all windows and other exits. The bird should leave on its own. If necessary, a broom or long pole with a T-shirt at the end can be used to direct the bird out an exit, or tire it to a point where it can be caught in a towel or similar item. If these methods fail and in large public buildings, a wildlife damage control compa-

ny can be called to assist in the removal process. Their approach may involve catching the bird in a mist net or shooting it with a pellet gun.

House sparrows at feeders. House sparrows are attracted to seed feeders, where their aggressive habits can keep more desired birds from approaching. Don't place birdseed on uncovered platform feeders or on the ground and offer black sunflower seeds and suet to birds instead of millet-based seed mixes. For detailed information, see "Preventing Conflicts" in Chapters 37 and 38.

House sparrows in nest boxes. Male house sparrows are aggressive fighters and will attempt to evict any nesting bird from a cavity in order to take over the site. For detailed information, see "Protecting Native Cavity Nesters from Starlings and House Sparrows" in Chapter 38.

House sparrows roosting in trees and shrubs. House sparrows use daytime roosts to perch and preen, and from which they call noisily and continuously. These roosts are used for an hour or two, are generally small, and are near a feeding area.

Where droppings accumulate under a roost in an area where it may pose a health risk to people, steps need to be taken to disperse the flock. Options include installing visual or auditory scare devices, or modifying the structure of the roost (Fig. 4). See "Preventing Conflicts" in Chapter 38 for information on these and other options.

House sparrows eating fruits and vegetables. House sparrows can damage crops by pecking seeds, seedlings, buds, flowers, vegetables, and maturing fruits. Protect crops with bird netting, which can be purchased in a variety of lengths and widths at garden and hardware stores. Tie the netting securely at the base of the shrub or on the trunk of the tree to prevent house sparrows from gaining access from below (see "Preventing Conflicts" in Chapter 38 for an example). Individual small branches containing fruit can be protected with an onion sack or similar mesh covering. Row crops, such as strawberries, can be completely covered during the fruiting season.

Alarm or distress calls have not been found to be effective for house sparrows. Frightening devices designed for other species will move sparrows from an area for a short period. House sparrows, however, adapt quickly to frightening devices and will not be repelled by sounds for any great length of time unless the sounds are varied and their location is shifted regularly.

Visual frightening devices can be helpful in some areas where crops are susceptible to damage for only a short period. For information, see "Preventing Conflicts" in Chapters 27 and 38.

Other Control Techniques

Trapping

There are more types of traps available for house sparrows than for any other bird. Despite the house sparrow's willingness to associate with humans, its suspicious nature and high intelligence (for birds) make trapping a difficult control technique. Even when it is successful, other house sparrows will quickly occupy the area if food and shelter remain. However, trapping may be helpful where a small, isolated population is problematic.

House sparrow traps and trap designs are available over the Internet (see Appendix F for resources). Check the trap every two hours for non-targeted birds. The male house sparrow is very easy to identify, but the females are quite similar to some species of native sparrows. Check a field guide to birds if necessary.

Do not set an in-box trap until a house sparrow has laid claim to that box. Once he has claimed it, he will not allow any other species inside. The male sparrow will be more likely to enter the box if a small amount of nesting material is left in the bottom or tucked in the entrance hole. However, be careful that the nesting material does not interfere with the trap.

To remove a trapped house sparrow from a box, place a clear garbage bag over the entire box and remove the trap with the bag still over the box. Once the trap is removed, the sparrow will fly into the bag. This is a much easier method than trying to reach into the box and catch the sparrow by hand.

Do not trap house sparrows and release them elsewhere, because they will easily return or cause problems somewhere else. If you cannot humanely kill them yourself, find a falconer or wildlife rehabilitation center that will accept live sparrows to feed to birds of prey. (See Appendix A for information on euthanization.)

Shooting

Shooting is not an effective way to manage house sparrow populations overall. The number of birds that can be killed by shooting is small relative to the size of the flock. However, shooting may be helpful in supplementing or reinforcing other dispersal techniques, or if there is a smaller localized population that is problematic. First check local ordinances; using air guns or low-powered firearms may be restricted.

Public Health Concerns

House sparrows are not a significant source of any infectious disease that can be transmitted to humans or domestic animals. See "Public Health Concerns" in Chapter 29 for precautions to be taken when working around large concentrations of house sparrow droppings.

Legal Status

In **Oregon, Washington,** and **British Columbia,** house sparrow nests, eggs, young, and/or adults may be removed or destroyed at any time and no permit is required. All other native sparrows and sparrowlike songbirds are protected by law and cannot be hunted or trapped without a permit.

CHAPTER 35

Magpies and Jays

Magpies and jays are in the Corvid family, which includes ravens and crows. These social birds are bold and gregarious, and adapt well to living around humans. Many feel that corvids rank among the most intelligent of birds.

The black-billed magpie (*Pica pica,* Fig. 1) is easily distinguished from other birds by its striking black-and-white color pattern. It has an unusually long tail (at least half its body length) and short, rounded wings that show flashes of white in flight. The feathers of the tail and wings are iridescent, reflecting a bronzy-green to purple. The juvenile magpie is similar to the adult, but has less iridescence and a shorter tail.

Magpies primarily occur east of the Cascade mountains in Oregon and Washington and in the Southern Interior Ecoprovince in British Columbia. They are typically found close to water in open areas near scattered trees and thickets.

Since the mid 1900s, when they were regularly trapped or shot, magpies have increased in abundance. They are now common birds of rural, suburban, and even city parks and gardens.

For information on jays, see "Notes on Jays."

Figure 1. The black-billed magpie is a black-and-white, pigeon-sized bird with an unusually long tail. (Drawing by Elva Hamerstrom Paulson.)

Facts about Magpies

Food and Feeding Behavior

- Black-billed magpies are resourceful opportunists. They flip items over to look for food, follow predators with hopes of scavenging scraps, pick insects from the backs of cattle and sheep, and steal food from other birds.
- Magpie diets include mostly animal matter: insects, small mammals, small wild birds and their hatchlings and eggs. They also eat seeds, fruits, and nuts.
- Magpies often congregate on recently killed large animal carcasses and eat many road-killed small mammals and birds. They can use scent to find food—an unusual trait for birds, which generally have very little sense of smell.
- Food may be stored in trees, shrubs, and in shallow pits magpies dig in the ground.

Reproduction and Family Structure

- Adult magpie pairs stay together year-round and for life unless one dies, in which case the remaining magpie finds another mate.
- The breeding season for magpies is from late March to early July.
- The female incubates six or seven eggs for 16 to 18 days. The male feeds the female throughout incubation.
- Young fly three to four weeks after hatching, feed with adults for about two months, and then fly off to join other juvenile magpies.
- Magpies form loose flocks throughout the year; winter

Figure 2. In 1987 the Steller's jay was declared the provincial bird of British Columbia. The Steller's jay is often incorrectly called a blue jay, which is a common bird in the Eastern United States but only a recent arrival in the Pacific Northwest. (Drawing by Elva Hamerstrom Paulson.)

Notes on Jays

Jays are members of the Corvid family, which includes magpies, ravens, and crows. Five species of jays occur in the Pacific Northwest (see below).

Jays are generally noisy, aggressive birds, although inconspicuous when nesting or robbing nests. All jays are strong flyers and have rowing-like wing beats. They mob and dive-bomb predators, including much larger owls and hawks, throughout the year.

Jays eat a wide variety of foods. Foods of animal origin include insects, snails, small frogs and salamanders, small snakes, mice, carrion, and the eggs and young of game birds and songbirds; plant matter in the diet includes grains, acorns, fruits, and nuts. At feeders, jays are fond of peanuts, sunflower seeds, and suet. Food may be stored for later use.

Nests are located in shrubs and trees, often near water. Egg laying occurs from early March through June, with the peak occurring in April.

Jays remain in the same area year-round or migrate to lower altitudes in fall. Less common is a north-south migration.

Steller's and scrub jays have bad reputations among walnut and filbert growers, and because of this special conditions apply to their legal status (see "Legal Status").

Wire mesh placed over a platform feeder prevents larger birds, such as jays, from accessing seeds (see "Pigeons and Bird Feeders" in Chapter 29 for an example).

Techniques for solving other problems associated with jays are similar to those given for magpies.

The following species of jays occur in the Pacific Northwest:

The **Steller's jay** (*Cyanocitta stelleri,* Fig. 2 and back cover) is a striking bird with deep blue and black plumage and a tall, shaggy crest. It ranges from sea level to timberline throughout the wooded areas of the Pacific Northwest. It is an unafraid visitor to yards, gardens, and orchards, but a wary species in wilder settings. In 1987 the Steller's jay was declared the provincial bird of British Columbia.

Its call is a raspy *yaaak-yaaak,* but it imitates musical calls of other birds. Its *keeeer* call can have an uncanny resemblance to the scream of the red-tailed hawk.

The **Western scrub jay** (*Aphelocoma californica*) is a slender jay with a long a tail and large beak. It has a crestless head, an olive-gray back, and a white throat outlined in blue. Its head, tail, and wings are blue.

Typically a more southern species, the Western scrub jay has expanded its range and today is seen as far north as southwestern Washington in groves of oaks, shade trees, and the dense shrubbery of residential areas.

Its calls are harsh, raspy, and varied, often in series of ones or twos. In flight, its call is a long series of *check-check-check* notes.

The **blue jay** (*Cyanocitta cristata*) has bright, colorful, contrasting plumage. Its back is mostly blue, and the underside is mostly white or light gray. It has black facial markings and a crest that is less prominent than that of the Steller's jay.

The blue jay is a common bird in the Eastern United States, but before the mid-1970s was unknown to the Pacific Northwest except as a rare vagrant. Since then it has been turning up more frequently in cities, towns, and resorts. It is assumed that the abundance of winter food at feeders is aiding this advance.

Its call is a harsh *jaay* or a liquid *queedle.* It also imitates red-tailed hawks.

Steller's and scrub jays are often incorrectly called blue jays.

Other jays include the **gray jay** (*Perisoreus canadensis*), formerly called the "Canada jay" and popularly known as the "Camp Robber." It occupies coniferous forests, mostly above 2,000 feet (600 m).

The related **Clark's nutcracker** (*Nucifraga columbiana*) is found at higher elevations, and can also be seen near camps and picnic sites begging and stealing food scraps.

The **pinyon jay** (*Gymnorhinus cyanocephalus*) is restricted to the ponderosa pine forests on the eastern slopes of the Cascade Range and into the adjacent juniper/sage flats of Oregon.

congregations may include several hundred individuals.

Nest and Roost Sites

- Black-billed magpies nest once a year, but will re-nest if their first attempt fails.
- Magpies nest individually or in loose colonies, frequently toward the top of deciduous or evergreen trees or tall shrubs.
- Nests are built by both sexes over a 40- to 50-day period. Old nests are repaired and used, or a new nest is built on top, with older nests reaching 48 inches (120 cm) deep by 40 inches (100 cm) wide.
- Nests are loose accumulations of branches, twigs, mud, grass, rootlets, bark strips, vines, needles, and other materials, with branches and twigs constituting the base and framework. The nest cup is lined with fine rootlets, grass, and other soft material.
- Nests almost always contain a hood or dome of loosely assembled twigs and branches, and usually have one or more side entrances.
- Other bird species, including small hawks and owls, often use old magpie nests.

Mortality and Longevity

- During the first half of the 20th century, black-billed magpies were considered detrimental to game-bird populations and domestic stock, and were systematically trapped or shot. Many also died from eating poison set out for coyotes and other predators.
- In 1933, 1,033 magpies were shot in Washington's Okanagan valley by two teams of bounty hunters.

- Today, adult magpies have few predators, although large hawks occasionally take some. Nestlings die from starvation, adverse weather, and attacks by raccoons, owls, and other animals.
- Pesticides used on livestock are of some concern, since magpies perch on livestock and eat those pests that are being poisoned.
- The life span of a magpie in the wild is four to six years.

Viewing Magpies

The magpie is probably the most conspicuous bird seen throughout interior areas of the Pacific Northwest (Fig. 3). Where they are not harassed, magpies can be extremely bold; when harassed, they become elusive and secretive.

During the breeding season, magpies are often seen in areas that combine woods or thickets for nesting, water for drinking and bathing, and open areas for feeding.

During the nonbreeding season, particularly winter, magpies are most numerous in environments influenced by people, such as livestock feedlots, barnyards, roadsides, garbage dumps, railway loading yards, and grain elevators. Individual magpies may wander widely.

Magpie migration is mainly elevational, from low to high, although some north-south movements occur. Where winters are severe, all magpies leave the high country.

Magpies are not swift fliers and elude predators and danger by flitting in and out of areas with trees or diving into heavy cover.

If you locate a colony of magpies, find a spot where you can view an expanse of trees and meadow with binoculars. Sit down and

wait. It won't be long before the magpies will ignore you and go about their daily activities. If you notice that your presence is annoying the birds, you are too close. Back away, sit down quietly, and wait for them to return to the area.

Nest Sites

Magpies build huge, domed nests. In the winter when deciduous trees are bare, the large nests are easily seen. Because of their size, you may first think they are hawk nests, but look for the telltale dome, an adaptation that protects the young from predators.

Roost Sites

From midsummer through fall and winter, magpies gather together in the evening and form communal roosts. Roosts are located in dense thickets of shrubs and trees near major food sources. The birds use deciduous thickets until the arrival of colder temperatures, at which time they move into conifers. Sometimes old magpie nests are used as night roosts.

The portions of food magpies can't digest are bundled into compact "pellets" and later regurgitated. Pellets are usually found under or near the roost. (See "What's in an Owl Pellet?" in Chapter 36.)

Tracks and Trails

A magpie's tracks show four medium-wide toes, three facing forward and one back. The claws are long and their marks are detached from the footprint.

Like virtually all other corvids, magpies walk with a strut, and hop quickly when rushed. The walking stride is about 6 inches (15 cm) and a tail print may be visible.

Figure 3. Black-billed magpies are typically found close to water in open areas near scattered trees and thickets. (Drawing by Elva Hamerstrom Paulson.)

Calls

Typical calls include a rapid, nasal *mag? mag? mag?* or *yak yak yak.*

Preventing Conflicts

Magpies help control pest insects such as grasshoppers and tent caterpillars, and also "clean up" dead animals and garbage scattered by other animals. Other species of birds and mammals often use unoccupied magpie nests.

Although magpies prey on songbirds and their young, research suggests that they do not ordinarily have a significant impact on songbird populations. However, because magpies are intelligent, opportunistic, and at times congregate in large numbers in close proximity to humans, conflicts can occur.

Perhaps the magpie behavior that is most annoying to farmers and ranchers is the picking of wound sites on the backs of healthy horses and livestock under certain conditions. However, this is not a common occurrence, except possibly during severe winters when food for magpies is scarce.

The following are suggestions on how to prevent and remedy problems that might arise. In cases where these methods are not practical, contact your local wildlife office (see Appendix E for contact information).

Crop damage: Magpies can damage local crops of fruit, nuts, and grain. For information on how to prevent or reduce small-scale problems, see "Preventing Conflicts" in Chapter 38. For large-scale problems, see "Communal Night Roosts" in Chapter 28.

Lawn damage: Because new lawns are well irrigated during the growing season, worms and grubs collect under the sod, attracting magpies—and skunks and raccoons. If small sections are being damaged, lay chicken wire over the area and secure it with stakes, stones, or by some other means. Other techniques are described under "Preventing Conflicts" in Chapter 23.

Small farm-animal damage: Free-roaming chickens, ducks, and pigeons and their eggs and young are susceptible to magpie predation.

Housing domestic birds in a durable, **completely** covered coop will exclude not only magpies but also hawks and owls. The structure can be constructed by attaching 1-inch (2.5 cm) chicken wire or bird netting to a solid framework (see "Enclose domestic animals" in Chapter 33 for additional information).

Magpies at feeders: Magpies are attracted to suet feeders and their aggressive habits can deplete food supplies and keep smaller birds from approaching. Because magpies have trouble clinging upside down, a suet feeder that requires the birds to clasp the feeder in this position will discourage magpies. Your local bird specialty store can also give you information on feeder designs to deter these birds.

Public Health Concerns

Magpies and jays are not a significant source of any infectious disease that can be transmitted to humans or domestic animals. See "Public Health Concerns" in Chapter 29 for precautions to be taken when working around large concentrations of magpie droppings.

Legal Status

In **Oregon** and **Washington,** magpies and jays in are protected under the Federal Migratory Birds Treaty Act (16 U.S.C. 703–712). However, under the Code of Federal Regulations, "a Federal permit shall not be required to control . . . magpies, when found committing or about to commit depredations upon ornamental or shade trees, agricultural crops, livestock, or wildlife, or when concentrated in such numbers as to constitute a health hazard or other nuisance . . ."

In addition, people engaged in the commercial production of nut crops in certain counties in Washington and Oregon may, without a permit, trap or shoot scrub jays and Steller's jays when they are found committing or about to commit depredations to nut crops on the premises between August 1 and December 1.

In **British Columbia,** magpies and jays are protected under Section 34 of the provincial Wildlife Act; however, because they are exempted from protection under Schedule "C," they can be trapped or killed by the public on Crown and private land without a permit.

Owls

Most owls are nocturnal predators, with hooked bills and needle-sharp talons (claws). They have wide wings, lightweight bodies, and feathers specially designed to allow them to silently swoop down on prey. Depending on the species, adult owls hoot, screech, or whistle.

More than a dozen species of owls live in the Pacific Northwest (Table 1). The great horned owl (*Bubo virginianus,* Fig. 1) is the most widely distributed owl in both the Pacific Northwest and North America, occupying dense forests, open woodlands, clearcuts, deserts, and urban environments, including golf courses, cemeteries, and parks with adjacent woodlots.

This chapter focuses on the great horned owl because it is the species that is most often seen and heard. However, except where noted, information about this species applies to all other owls.

Figure 1. *The large-headed, neckless silhouette and large ear tufts or "horns" of the great horned owl are hard to miss. An adult female is shown here with three nestlings. (Drawing by Elva Hamerstrom Paulson.)*

Facts about Owls

Food and Feeding Habits

- Great horned owls primarily eat small mammals such as rabbits, skunks, and rodents. They also eat a variety of birds, including quail, ducks, and smaller owl species.
- To a lesser extent, great horned owls eat reptiles, amphibians, fish, and insects.
- Owls have keen hearing and keen vision in low light, both adaptations for hunting at night. Since their eyes don't move in their sockets, they rotate their heads on their flexible necks.
- Great horned owls use a sit-and-wait approach, watching from a perch and swooping down on passing prey to seize it with their talons.

Reproduction

- Great horned owls are early nesters. Pair formation can occur from mid-January to mid-March, depending on the north-south range and elevation.
- The male owl chooses a nest site and attempts to attract a female by frequent hooting.
- The female incubates one to four eggs for 30 to 36 days.

ELVA PAULSON ©1993

The male provides her food and guards the nest.

- The young remain in the nest for about six weeks, and then climb out onto nearby branches. They begin taking short flights at seven weeks, and can fly well at nine to ten weeks.
- Both parents feed and tend the young for several months, often as late as September.

Nests and Nesting Sites

- Great horned owl nests are located high up in trees—generally in the crotch of a branch next to the trunk. Where suitable nest trees don't exist, owls will use rock ledges, power-line towers, haylofts, and nest boxes.
- Great horned owls make little if any effort to construct nests or even to repair suitable existing ones. Rather, they customarily usurp the previous year's nest made by a red-tailed hawk, crow, magpie, great blue heron, or tree squirrel.
- The same nest is seldom used by owls more than once, because trampling by the young usually reduces any nest to a disintegrating mass of sticks.

Mortality and Longevity

- Adult great horned owls may be killed or seriously injured when attacking prey. It is, for instance, common to find these owls riddled with porcupine quills or reeking of skunk scent.
- About 50 percent of the young that leave the nest die within their first year from starvation, vehicle collisions, and various other causes.
- The oldest banded great horned owl recovered in the wild was 13 years old. Captive birds can live almost 30 years.

How Owls Hunt at Night

Owls locate faint sounds with remarkable accuracy. The best studied of these nocturnal predators is the barn owl (Fig. 2). Experiments conducted by neurobiologists in pitch-black, sound-proofed rooms have demonstrated that barn owls can locate and capture prey by sound alone.

An owl's eardrums are larger than those of any other bird, and allow it to have the best hearing of all birds. The barn owl's sensitive hearing is enhanced by its facial disc—a concave surface of stiff feathers that serves as a dish receiver, channeling sounds into the ears. Once a sound is detected, the owl orients toward it and accurately pinpoints its location.

The cue used to determine whether a sound comes from the right, left, or straight ahead is the very small difference in time that it takes for a sound to reach each ear. When the sound source is dead ahead, there is no difference in time.

The variety in intensity of sound received by each ear is also used to localize a sound vertically. Barn owls (and most other owls) have asymmetrical openings to their ears—the opening is higher in one ear than the other. Thus, a sound coming from above will seem slightly louder in the ear with the higher opening; if a sound is equally loud in both ears, the owl senses that the source should be at eye level.

An owl's hearing system is not the only reason that it can hunt successfully in the dark. Owls have extraordinary sight. Unlike other birds, their eyes are positioned in the front of the head, which permits binocular vision and depth perception. In addition, familiarity with the local environment, such as the heights of favorite perches, seem to be essential to the owl's ability to successfully catch prey.

(Adapted from Ehrlich, et al., *The Birder's Handbook*.)

Figure 2. The barn owl has a heart-shaped face and dark eyes. (Washington Department of Fish and Wildlife.)

Table 1. Common Owls of the Pacific Northwest

Some owls are uncommon or unlikely to be seen on your property due to their habitat preferences. These include the great gray owl (mountains), burrowing owl (shrub-steppe areas), and the threatened spotted owl (old-growth forests). The following owl species are seen or heard around wooded rural properties, agricultural areas, and large urban parks.

The **great horned ow**l (*Bubo virginianus,* Figs. 1 and 5) is easily identified by its large ear tufts or "horns." It's also called the cat owl because the tufts look like cat ears.

The great horned owl stands 20 inches (51 cm) tall and has a 48-inch (122 cm) wingspan. It is dark brown with black spots above; the underparts are pale brown with heavy, dark brown bars. Some subspecies are paler. All have large yellow eyes. Great horned owls can turn their heads 270 degrees either way when facing forward, but they can't turn their heads 360 degrees.

The **barred owl** (*Strix varia,* Fig. 3) is similar in size to the great horned owl, but has dark eyes, a thicker-appearing neck, and no ear-tufts. The barring pattern on the neck and breast is crosswise and lengthwise on the belly. The barred owl is primarily a bird of eastern and northern U.S. forests and is a recent arrival to the Pacific Northwest.

The barred owl is distinguished from the **spotted owl** (*Strix occidentalis,* Fig. 4) by its streaked belly (spotted owls have a combination of spots and bars). Barred and spotted owls are closely related and occasionally mate where their ranges overlap.

The **barn owl** (*Tyto alba,* Fig. 2), with its white, heart-shaped face, no ear-tufts, dark eyes, and golden-buff plumage with ghostly light underparts, is distinctive and unique. It has a 39-inch (100 cm) wingspan, and its long legs give it a height of up to 20 inches (50 cm). The barn owl has a preference for rats and other rodents and earned its name by readily nesting in barns, silos, and sheds.

The **Western screech owl** *(Otus kennicottii)* is a small owl, with a height of 10 inches (25 cm) and a 20-inch (51 cm) wingspan. Adults are dark brown or gray with small ear-tufts. A year-round resident throughout the Pacific Northwest, it nests in tree cavities, including old woodpecker holes, but will also use nest boxes. Screech owls favor areas near water, and eat rodents, small birds, and large insects, as well as the occasional fish, crayfish, or amphibian.

The **Northern saw-whet owl** *(Aegolius acadicus)* is 8 inches (20 cm) tall with an 18-inch (46 cm) wingspan. It has dark brown plumage, brown and white vertical breast streaks, and no ear-tufts. This seemingly tame, fearless little owl moves from higher elevations to valleys in winter. Like the screech owl, it nests in tree cavities and will use a nest box. It prefers dense coniferous and broad-leaved forests next to water. It eats small mammals, including shrews and mice, and will also catch small birds and bats.

The **Northern Pygmy-owl** *(Glaucidium gnoma)* is our smallest owl, at 7 inches (18 cm) tall, including its long tail. Because of its size, long tail, proportionally small head, and daytime hunting behavior, the Pygmy-owl is often misidentified or overlooked as just another brown bird in the brush. Watch for it near your winter feeder, where this fearless hunter may attempt to take small birds or mammals. It has sharply streaked undersides, but the most telling marks are the black patches on the back of its head that mimic eyes to deceive predators.

Figure 3. The barred owl is a medium-large owl with brown eyes and no ear-tufts. (Washington Department of Fish and Wildlife.)

Viewing Owls

Visual encounters with owls are relatively rare, because they spend most of the day perched high in trees, inside tree cavities, or in nest boxes. Due to its size, the Northern pygmy-owl almost always goes undetected. The screech owl camouflages itself by stretching tall, holding its wings close to its back, and appearing to be a dead stub on a tree branch.

You are more likely to hear an owl than to see it. If you remain quiet, you can sometimes spot a calling owl with your flashlight. Make every effort not to disturb an owl during its late winter to spring nesting season, a critical time in its yearly cycle.

Owls can also be viewed when crows, jays, magpies, or other birds discover them in their territory. The birds will defend their domain by diving and calling repeatedly at the perched or flying owl, an activity called "mobbing." Look and listen for this behavior and see if you can locate the "invader."

A good way to learn more about owls that live around you is to go on an owl walk with members of your local Audubon Society. Field trips, sharing ideas and sightings with others, and having access to local bird experts are some excellent ways to gain more knowledge.

A field guide to birds is helpful to identify owl species. Popular field guides are listed in Appendix F.

Roost Sites

Owls roost in places that offer maximum concealment during daylight hours, choosing trees with dense foliage. Conifers are favored when present; in deciduous forests, owls will use trees that hold clusters of dead leaves over winter.

During the day, scan tall trees for the silhouette of an owl. Also, look for the whitewash of droppings on branches and rocks, and owl pellets on the ground. At dusk or at night, look or listen for an owl roosting in the area.

Pellets and Droppings

Typically, owls ingest entire animals—including feathers, fur, teeth, and bones. The undigested material is bundled into compact pellets and later regurgitated. Pellets are usually found under or near the owl's favorite roost. (See "What's in an Owl Pellet?")

Pellets range from ½ inch to 4 inches (6 mm–10 cm) long, depending on the owl's size and its diet. Pellets, shiny and black when new, turn gray with age.

Owl droppings are semiliquid and primarily white; a whitewash can sometimes be seen under a nest or roost site.

What's in an Owl Pellet?

An owl pellet is a clod of fur or feathers and bone—the indigestible remains of the animals an owl has eaten. Because it swallows small prey whole and is able to digest only the fleshy parts, the owl regurgitates the remaining solid material as a compact pellet or casting. Where owls feed on insects, each regurgitated pellet contains the indigestible parts of the exoskeletons of numerous individual insects.

Although birds of many species regurgitate pellets, pellets from large owl species are especially suited for study because they are big enough to be examined without a microscope, and they contain the entire skeletons of small animals the owl has eaten. (Pellets of other raptors, such as eagles and hawks, are less useful since these birds tear much of the flesh from their victims, and do not swallow bones.) Because owl pellets accumulate in predictable locations, they are readily available for collection and examination.

Pellets last a long time in dry climates and in the protection of barns or other buildings. If they are soaked in warm water, carefully dissected, and examined under magnification, the identity of prey they contain can often be determined from the bones, teeth, and other remains.

The remains hidden inside a pellet usually represent the entire skeleton of every animal the owl has eaten during a night of foraging. There are almost always remains of two or more animals in each pellet.

Enjoy, and remember to wash your hands when done.

Attracting Owls

Things you can do to encourage owls to live or visit your property include:

• Retain multi-acre patches of coniferous and/or deciduous trees.

• Protect quiet, secluded areas near rivers, creeks, and lakes and away from human activity.

• Retain large dead or dying trees—over 20 feet (6 m) tall—as potential perches.

• Protect or plant hedgerows and thickets to attract small mammals that owls eat.

• Leave large grasslands alone or mow them only infrequently to provide habitat for small mammals that owls eat (see "Providing Vole Habitat" in Chapter 25 for management information).

• Manage mice and rat problems without poison baits, which can potentially kill owls (see "Preventing Conflicts" in Chapters 8 and 16 for management strategies).

• Install owl nest boxes for barn owls, Western screech owls, Northern Pygmy-owls, and Northern saw-whet owls. (See Appendix F for a list of resources.)

• Install perch poles (see "Maintaining Hawk Habitat" in Chapter 33).

Figure 4. The spotted owl is a medium-sized owl with brown eyes and no ear-tufts. It is gray-brown in color, with light spotting on the back and breast. They are slightly smaller than the closely related and similar-appearing barred owl. (Washington Department of Fish and Wildlife.)

Nest Sites

Look for the stick nests, originally built by large birds or squirrels, that great horned owls use. In deciduous trees these nests are easy to see in the winter, as the trees remain bare into the nesting season.

Calls

Owl calls are given at different times of day and year, depending on the species, and are associated with territorial behavior, courtship, or begging by the young. The following are the common calls given by each owl species:

Great horned owl: a series of four or five deep, resonant hoots given in various rhythms by different individuals: *hoo-hoo-hoo; hoo-hoo* ("who's a-wake, me too"). Calls are heard most in the early evening or predawn hours. They are given in all seasons, but commonly in fall and winter by the male as he advertises and defends his territory. The call may be answered in an unhurried way by another owl.

Occasionally two or more owls can be heard hooting, seeming to respond to one another. This is probably territorial hooting between males, since females are silent except for the few weeks of courtship.

Juvenile great horned owls beg with a high, scratchy *reeeek* well into the summer. The call is similar, but usually shorter and less rasping, than the barn owl's call.

Barred owl: a clear-voiced series: *hoo-hoo-hoo, hoo, hoo-hoo-hoo-hoo-a-aw.* Given in words: "who-cooks-for-you, who-cooks-for-you-a-all," ending with a descending note.

Western screech owl: a slow but accelerating series of short mellow whistles, *pwep pwep pwep pwep pwepwepwepepepep,* that is slightly lower at the end. Also a two-part trill, with the second part longer. Other calls infrequently heard include a soft bark and a short chuckle.

Northern saw-whet owl: low, whistled *toots* (about two per second): *toit toit toit…* or *poo poo poo.* Also a wheezy, rising, catlike screech: *shweeee.*

Northern Pygmy-owl: a soft, hollow *toot* (one note every two seconds). Also a high rattle or rapid trill: *tsisisisisisisi.*

Barn owl: a long hissing or raspy scream, *csssshhH* which sounds similar to a canvas being ripped. The call is similar to, but usually longer and more raspy, than the call for food made by juvenile great horned owls.

Preventing Conflicts

Because of their wide-ranging diet that includes rabbits, squirrels, chipmunks, and songbirds, great horned owls elicit mixed emotions in people, even wildlife-lovers.

Free-roaming chickens, ducks, turkeys, pigeons, small domestic rabbits, and similar animals are susceptible to owl predation. Birds are particularly vulnerable because they are usually conspicuous and concentrate in areas that lack brush or trees to hide in. Although rare, there have been reports of great horned owls preying on unattended puppies and small cats.

Figure 5. Great horned owls call mostly in the early evening or predawn hours. They call during all seasons, but are commonly heard in fall and winter when the male advertises and defends his territory. (Drawing by Elva Hamerstrom Paulson.)

Protect domestic animals by housing them in secure coops or houses. Birds can be conditioned to move into coops or houses by feeding them treats indoors at dusk.

Increased human activity in the area will keep owls at a distance. Yelling and clapping your hands or banging cans together are effective when an owl is seen nearby.

Another scare tactic is firing a gun loaded with blanks. If you can condition an owl to stay away from a specific area, it may stay in its nearby territory and keep other owls out.

See "Preventing Conflicts" in Chapter 33 for detailed information on scare tactics and housing birds.

Caring for an injured owl and bird bands: Under federal, state, and provincial law, it is illegal for anyone to injure, harass, kill, or possess a bird of prey. Licensed rehabilitators are the only people legally permitted to transport and keep wildlife, including owls.

If you find an injured owl, contact a wildlife rehabilitation facility immediately. Your local wildlife office keeps a list of rehabilitators and can tell you which ones serve your area, or you can look under "Animals" or "Wildlife" in your phone directory.

If a rehabilitator isn't available, follow the menu options over the phone or on their Web site for information on what to do. (For more information, see "Wildlife Rehabilitators and Wildlife Rehabilitation" in Chapter 20.)

If you see a live or dead owl with a band on its leg, you can provide a vital service by reporting it. See "Caring for an Injured Hawk"

and "How to Report a Bird Band" in Chapter 33 for information.

Dive-bombing owls: See "Dive-Bombing Crows and Other Bird 'Attacks'" in Chapter 28.

Public Health Concerns

Owls are not a significant source of any infectious disease that can be transmitted to humans or domestic animals. A few human *Salmonella* infections have resulted from handling owl pellets in school settings. Wash your hands after handling owl pellets.

Legal Status

Because legal status and other information about owls changes, contact your local, state, or provincial wildlife office, or visit their Web site for updates. (See Appendix E for contact information and where to access the state and provincial laws mentioned below.)

Oregon and Washington: All species of owls are protected under federal and state law and cannot be hunted or trapped.

British Columbia: Owls come under provincial jurisdiction and are protected under Section 34 of the Wildlife Act, which states that unless provided by regulation (i.e., a permit), it is an offence to possess, take, injure, molest or destroy a bird, its eggs, or its nest when occupied.

CHAPTER 37

Robins and Garden Finches

T he American robin (*Turdus migratorius,* Fig. 1), or robin, is one of the most familiar and widely distributed songbirds in the Pacific Northwest. It is equally at home in city parks and gardens, rural farms, woodland edges, and subalpine meadows. This North American "robin" is actually a thrush, and the English robin (*Erithacus rubecula)* of children's stories is in a completely different family of birds.

In late summer and continuing on up until the breeding season begins in spring, robins form nomadic flocks that roost together at night and feed together by day.

Robins remain in the same area year-round, or migrate short distances in the spring and fall. Often the robins you see in winter come from their northern breeding grounds, which may be 300 miles (480 km) away.

For information on finches, see "Notes on Garden 'Finches.' "

Figure 1. Adult male robins are dark gray above and brick red below. Their heads and tails are black and their beaks are yellow. Females are similar, but have duller coloring. Young robins have a freckling of white dots on their reddish fronts. Partial albino robins are uncommon, but are seen each year. (Washington Department of Fish and Wildlife.)

Facts about Robins
Food and Feeding Behavior

- During the breeding season robins mostly eat animal material, including earthworms, beetles, grasshoppers, ants, caterpillars, spiders, and snails (Fig. 2).
- Robins hunt on lawns, pastures, fields, and meadows, standing still with their heads cocked to one side as though listening for their prey, but actually discovering it by sight.
- With the decrease of available insects in fall and winter, robins feed on ripe fruits and berries in trees and shrubs.

Nest Sites

- Robins nest in deciduous and evergreen trees, shrubs, and hedges, as well as under bridges and on windowsills and other ledges.
- Robins nest early in the year. Their first nests are often placed in evergreens for protection, since deciduous trees and shrubs may not yet have leafed out.
- Females select the nest site and do the majority of nest building over a two- to six-day period.
- Nests are often placed in the crotch of a branch, or saddled on a branch next to the trunk.
- The nest is a bulky structure of twigs, weed and grass stems, and sometimes string or cloth. It contains a smooth inner cup of mud, with a thin lining of fine grasses.
- Robins often nest in the same area, or a nearby area, year after year.

Figure 2. During the breeding season, American robins forage primarily on soft invertebrates such as earthworms and ground-dwelling insects. Both parents feed the young. (Washington Department of Fish and Wildlife.)

Reproduction

- Breeding activity begins in early spring in lowland areas, later at higher elevations.
- The female incubates three to four glossy, light blue eggs for 12 to 14 days.
- The young leave the nest after 14 to 16 days and continue to be cared for by the parents for up to four weeks.
- Robins have two and sometimes three clutches of eggs each year. Nests may be used for multiple clutches; first-clutch nests may be built on top of nests from the previous year.

Mortality

- Robins have a high mortality rate, with up to 80 percent of the young dying each year.
- Tree squirrels, chipmunks, raccoons, magpies, crows, ravens, and jays eat robin eggs and nestlings.
- In winter roosting areas, great horned and barred owls take a toll on adult owls. Sharp-

shinned hawks catch adults in flight.
- Because robins feed on the ground, young and adult birds are vulnerable to attacks by domestic cats.
- In the 1950s and early 1960s, robins suffered from exposure to the insecticide DDT because they ate earthworms that accumulated high levels of DDT in their bodies.

Viewing Robins

Robins running over lawns in search of worms, perching and singing from utility wires, and bathing in rain puddles are familiar sights to most people. Robins sometimes nest on window ledges, beams under porches, in gutters, and on nest platforms provided for them (see Fig. 8 in Chapter 39).

After breeding season is over, robins gather for the night in communal roosts. Roosts are located in trees, under bridges, and in large open barns, and may contain a few birds or several hundred. In fall and winter, watch for the daily movement of robins to and from a roost after sunset and before sunrise. Robins generally remain in flocks through the winter, and the breakup of these flocks in spring signals the start of their breeding season.

Territories

The size of a robin's territory is one-third of an acre to several acres. The breeding pair spends most of their time there, on the nest or searching for food. The male actively defends the territory through all clutches. If another male intrudes, he will fly at the intruder to try to scare him away.

If that fails, he will dive-bomb the intruder and try to hit him chest to chest. This behavior is also seen when a male robin mistakes his image in a window for an opponent; homeowners often watch in amazement as the male robin beats himself silly against the glass, under the impression that he is attacking another robin.

You know you are in a robin's territory when a bird of either sex sounds its alarm call at your approach. Robins are particularly protective of their nest sites when young are in their nests. Nest predators, such as crows, will be mobbed by several robins in an area where there are a number of robin nests.

Nest Sites

When you see a robin perched or flying in midair with a wad of mud or grass in its beak, it's a sign of nest building. Another sign of nest building is a line of mud across the female's breast—she works mud into place with her feet and bill, molding it with her body. When foraging on a lawn, if a robin doesn't eat a worm or other prey immediately, but flies off with food in its beak, you can be fairly sure that it has young in a nearby nest.

Displays

If you watch robins over a period of 15 to 30 minutes in the spring, you are likely to see several different displays associated with courtship and territorial behavior (Fig. 4).

The tail-lift display is presented in situations of possible danger. The male or female robin lowers its head, raises its tail to a 45-degree angle, and repeatedly flicks its tail sharply while giving the *tuk tuk tuk* call.

Notes on Garden "Finches"

Several species of songbirds (finches, grosbeaks, crossbills, redpolls, siskins) are included in the Fringillidae or finch family of birds. These are small to medium-sized birds with short, conical beaks, short, slightly forked tails, bright wing and/or body markings, and "nervous" behaviors.

All birds in this group are seed specialists that feed in noisy flocks which, after breeding, may contain hundreds of birds, seen in undulating flight. Finches are heard and seen in thickets, weedy grasslands, and forest edges. All of them frequent birdbaths and feeders and are susceptible to some common diseases. (See "Sick Birds at the Feeder.")

All finches build shallow, saucer-shaped nests from bark, twigs, and moss and line them with plant down and feathers. Nests are placed in conifers, deciduous trees, or large shrubs. House finches will nest in hanging baskets and on buildings where suitable ledges or cavities exist.

You may want to purchase a field guide to birds, not only to identify the rare visitor but also to positively identify the ones you see every day, including the finches. (See Appendix F for a list of field guides.)

The following finches are commonly seen throughout the Pacific Northwest:

American goldfinch *(Carduelis tristis):* The breeding male is bright yellow with a black forehead and black wings and tail. The female and winter male are duller and grayer. The American goldfinch was designated as the Washington State Bird in 1951.

House finch *(Carpodacus mexicanus):* Adult males are bright red on the crown, breast, and rump. Females have a plain, unstriped head and heavy streaking on a light underside. Immature males are less highly colored, often orangeish or yellowish on head and breast.

Pine siskin *(Carduelis pinus):* A dark, streaked finch with a notched tail and small patches of yellow in the wings and tail. It is noticeably smaller than the other common finches, and has a more needlelike beak.

Evening grosbeak *(Coccothraustes vespertinus,* Fig. 3): A starling-sized, stocky finch with a very large, pale greenish or yellowish conical bill. The male has a brown head shading to yellow on the lower back, rump, and underparts, and a bright yellow forehead. The female is similar but grayer.

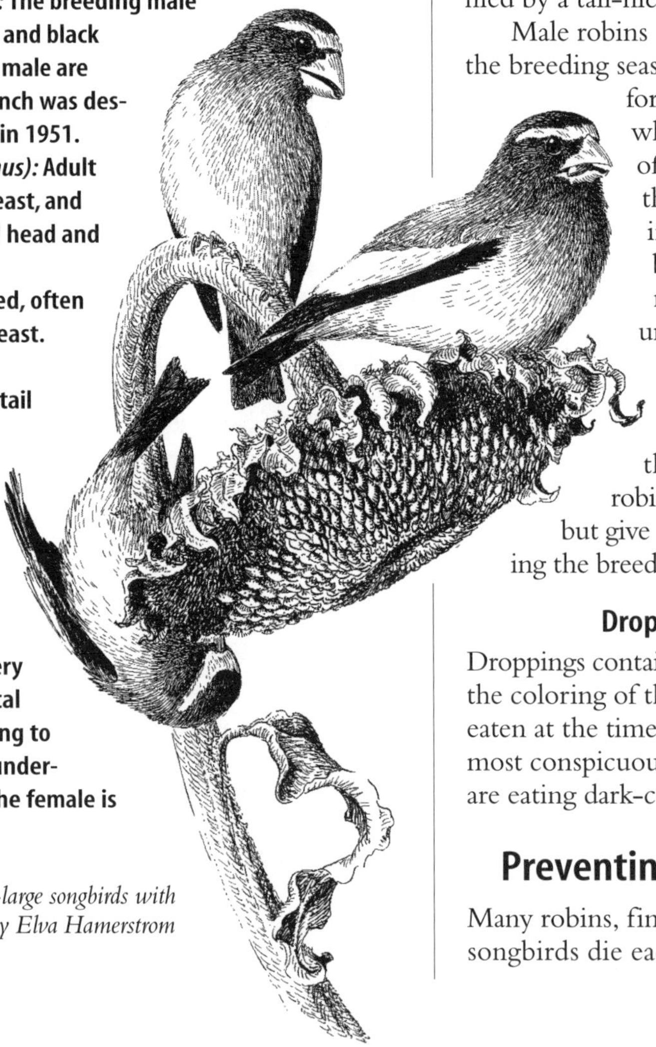

Figure 3. Evening grosbeaks are medium-large songbirds with very thick, seed-cracking bills. (Drawing by Elva Hamerstrom Paulson.)

The wing-droop display occurs just before or after an aggressive encounter. The wing-tips are lowered so they droop below the level of the tail, and the breast feathers may be puffed out.

Calls

The male's song is a series of rich caroling notes, rising and falling in pitch: *cheer-up, cheerily, cheer-up, cheerily.* It is sung early and late in the day during the breeding season. Some people confuse the song of the black-headed grosbeak with the song of the robin.

The *teeek teeek* or *tuk tuk tuk* call is given by either sex as an alarm call and in situations of possible danger. It is often accompanied by a tail-flick display.

Male robins stop singing after the breeding season and, except for a brief time when the shortness of daylight fools them into thinking it is time to breed again, do not sing again until the following spring. Alarm calls continue throughout the year. Female robins do not sing, but give alarm notes during the breeding season.

Droppings

Droppings contain seeds and have the coloring of the foods being eaten at the time. Droppings are most conspicuous when robins are eating dark-colored berries.

Preventing Conflicts

Many robins, finches, and other songbirds die each year due to

Figure 4. Two visual displays given by robins. (From Stokes, A Guide to Bird Behavior.*)*

collisions with windows and diseases they are exposed to at birdfeeders and birdbaths. Male birds also actively attack windows and other objects when they see their reflections and believe a rival bird has entered their territory.

Home gardens, commercial fruit-growing farms, vineyards, and orchards often attract migrating robins. For information on how to prevent or reduce losses from fruit-eating robins, see "Preventing Conflicts" in Chapter 38.

Birds and windows—Tips to ensure safe flight: Many robins and other birds are stunned, injured, or killed each year by flying into windows. This unfortunate event seems to occur because the birds have seen the reflection of landscape or sky in the glass, and have the illusion of space beyond the window. Problems with window collisions may increase after a robin has indulged in a binge of fermented berries, or when a hawk or other predator appears

Attracting Robins to Your Property

Ways to enhance your property for robins include:

- **Avoid using insecticides. Nearly 70 percent of the breeding birds in the Pacific Northwest (including robins) eat insects as a primary part of their diet during the nesting season.**
- **Protect and plant trees and shrubs that produce fruits and berries eaten by robins. Examples include salmonberry, madrone, mountain-ash, serviceberry, and hawthorn.**
- **Leave some "forest floor" in open soil, or mulched with leaf litter, to provide for ground foraging.**
- **Offer wild or cultivated fruits and berries on a platform feeder. Robins learn to take currants, raisins, small pieces of dates, and other dried or fresh fruits.**
- **Supplement the birds' supply of nest materials by offering straw and string and allowing muddy areas to remain for mud collecting (see "Attracting Swallows to Your Property" in Chapter 39).**
- **Install a nest platform where it is safe from house cats and can be observed from inside the house (see Appendix F for resources).**
- **Install a birdbath in an area where it can easily be observed and maintained (Fig. 5).**
- **Avoid pruning trees, shrubs, brambles, and other likely nesting spots in the spring and early summer when robins are nesting. If you must prune at this time, carefully examine the area for nests before you begin, and listen for an alarm call given by robins.**
- **Keep your cats indoors and discourage other cats from visiting your property (see Appendix D for information on cats).**

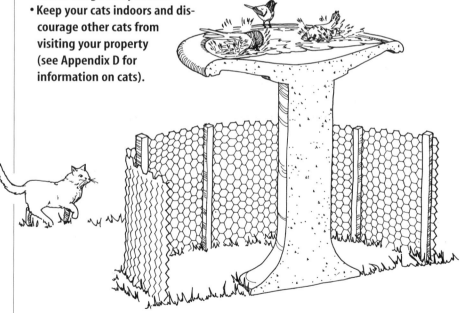

Figure 5. To bathe, birds need a gradual slope that allows them to wade in and find a comfortable depth. An 18-inch (45 cm) high wire fence can be added to keep cats from easily accessing the bathing birds. (Drawing by Jenifer Rees.)

suddenly and causes a bird, or flock of birds, to rush to escape.

Catalogs and stores selling bird-feeding supplies offer silhouettes of falcons or owls to be attached to windows to frighten birds or cover the reflection. But these silhouettes rarely accomplish either job. Birds quickly lose their fear of a silhouette, and because it covers only a small area, it has little effect on birds heading for other parts of the window.

For silhouettes to be effective, you must cover the outside surface of the window with them (the shapes really aren't important), or use other patterns placed no more than 6 inches (15 cm) apart.

It is important that whatever you place on the window be on the **outside surface;** anything on the inside of the glass will lose its effect because it won't interfere with the reflection. Other ways to prevent window collisions include:

- Create bird barriers by covering windows with black bird-netting, available from nurseries and hardware stores (Fig. 6). From inside the house, the netting will be barely noticeable, and will not impair bird-watching.
- Rub a bar of soap on the **exterior** surface of windows, using a design that leaves no area 6 inches (15 cm) or larger uncovered. Dusty windows also help to cut down on reflections.
- Turn windows into works of art by installing commercially available window film on the **exterior** of the windows to give the appearance of acid etching or sandblasting.
- Install **exterior** blinds or sun shields.

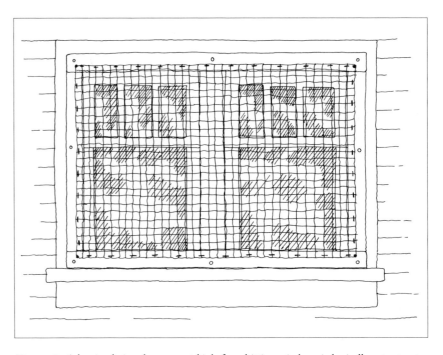

Figure 6. A barrier designed to prevent birds from hitting windows is basically a taut-net trampoline held out about 4 inches (10 cm) from the glass. Many variations for mounting the net are possible. The simplest is to use thumbtacks to attach black bird-netting from below the eaves to below the windows. Alternatives are to install 1 x 4 inch (2.5 x 10 cm) boards along the top and sides of the window frame. Stretch the netting over the boards, stapling as you go. You may also build a four-sided frame that you can put on over the window, much as you would with a storm window. (Drawing by Jenifer Rees.)

- Hang strips of Mylar tape, string, or other material no more than 6 inches (15 cm) apart on the **outside** of windows for the full width of the glass.

In addition:

- Make sure bird feeders and birdbaths are either less than 3 feet (90 cm), or more than 15 feet (4.5 m) away from windows. If birds are startled while using a feeder or birdbath that is close to a window, they can't build up enough momentum to injure themselves should they hit it. Alternatively, if birds are far enough away from the window when startled, they are less likely to make the mistake of flying into it.

- A bird that is able to see the landscape through a window that faces another window is likely to try to fly through the house, crashing into the glass. Closing curtains or blinds on one of the two windows can prevent this.

Note: These tips will not completely stop collisions, as the windows may still reflect the outside, giving the impression that it is possible to fly through them. But they do seem to help prevent at least some collisions.

Caring for birds that hit windows:
A bird that hits a window and falls to the ground may simply be stunned. On warm days, it is best to leave the bird alone; it will likely fly off after a few minutes.

However, if the weather is cool or if house cats are in your area, pick the bird up immediately. Stunned birds are subject to hypothermia and many cats recognize the sound of a bird striking a window and will quickly come investigate. Place the bird upright in the palm of your hand, cup your other hand over the bird, and hold it for about five minutes. When the bird starts moving, lift your hand and release it near a tree or large shrub so it will have a safe place to fully recuperate. Wash your hands immediately.

If the bird is large or doesn't revive within 20 minutes, place it in a brown paper bag or container with air holes and put it in a quiet place. Then, when you hear the bird moving, open the container outside near a tree or shrub and give it a chance to fly away.

If the bird doesn't fly off, contact a wildlife rehabilitation facility. Look under "Animal" or "Wildlife" in your phone book or search the web for "wildlife rehabilitator." If a rehabilitator isn't available, follow the menu options provided on their phone message or on their Web site. See "Wildlife Rehabilitators and Wildlife Rehabilitation" in Chapter 20 for additional information.

If you find a dead bird with a band on its leg, you can provide a vital service by reporting it. See "How to Report a Bird Band" in Chapter 33 for information.

Sick Birds at the Feeder

Birds that feed in large groups—notably birds in the finch family, such as pine siskins, goldfinches, and house finches—are particularly susceptible to a number of

Are Birds Attacking your Windows... or Vehicle?

Birds may fly into windows for a variety of reasons. Sometimes birds simply don't see the glass and attempt to fly through it. This can happen at any time of the year; however, window "attacking" birds are more common in spring because they become territorial during the breeding season.

Males in particular will drive away intruders with great ferocity. When they see their own reflection in a window, they may attack. Male grouse and wild turkeys have been seen attacking parked cars, presumably because they saw their reflection in the shiny surface of the car. Chickadees and other birds seasonally attack their reflections in windshields, hubcaps, and rearview mirrors. Male robins have attacked red objects, including socks, handkerchiefs, and other items hanging on a clothesline, and ornaments and discarded toys on the lawn. Apparently they mistake the red object for a trespasser.

Sometimes birds fly at windows from angles that would not permit them to see their own reflections. These confrontations seem more bumbling than belligerent, so perhaps they just involve birds that see reflected vegetation they want to investigate as a possible nest site.

Although the above behavior can be repeated for days or weeks, usually the bird does not injure itself seriously. What seems to be more bothersome is watching these disturbances! So what can you do to prevent them? Some people place small paper sacks over the mirrors of their vehicles when these are parked; using a protective cover for a vehicle also solves the problem. Where birds are striking windows, see the suggestions previously listed.

Fortunately, these remedies are generally only necessary during the spring breeding season. After this period of hectic romance, birds usually come to their senses.

diseases. Because individuals in these species tend to move around frequently, they increase the chance of infecting other birds. Also, some infected birds do not die from the disease, which increases the probability that they will transmit it to other individuals.

Five diseases have the potential of affecting bird species that frequent bird feeders:

Salmonella is the most common bird-feeder disease. The salmonella bacterium is the same one associated with a common form of food poisoning. Infected birds appear tame and might sit quietly for days in a sheltered spot. Often their feathers are fluffed out, and you might see a bird hold its head under its wings. Within a few hours after the symptoms become severe, birds fall over and die. Infected birds shed the salmonella bacteria in their droppings and via bird-to-bird contact.

Domestic cats and dogs that have picked up dead, salmonella-infected birds have spread the disease to humans, although this is rare.

The **avian pox virus** is most common in house finches. Infected birds may have wartlike growths on the face, legs, feet, or any unfeathered skin. The virus is spread by direct contact with infected birds, ingestion of food and water contaminated by sick birds, or contact with contaminated surfaces, such as bird feeders, birdbaths, and perching areas. Insects, especially mosquitoes, also carry the disease from one bird to another.

Avian pox growths on the face can become large enough to impair the bird's vision or eating ability and growths on the legs and feet can affect its ability to stand or perch.

Birds with **avian conjunctivitis** often have red, swollen, watery, or crusty eyes; in extreme cases the eyes are so swollen or crusted over that the birds are virtually blind. Although infected birds have swollen eyes, the disease is primarily a respiratory infection. Birds with this disease have trouble feeding, and you will see them remaining on the ground under the feeder, trying to find seeds.

The **aspergillus fungus** (mold) grows on damp feed and in the debris beneath feeders. Birds inhale the fungal spores and the fungus spreads through their lungs and air sacs, causing bronchitis and pneumonia.

Trichomoniasis is a disease caused by small parasites that can affect a wide variety of animals, including humans. The mourning dove and band-tailed pigeon seem to be particularly susceptible. The disease causes sores in their mouths and throats, and results in death from starvation or dehydration.

All the above diseases lead to death when birds become vulnerable to secondary infections, predation, starvation, dehydration, or exposure.

When it is necessary to handle sick or dead birds or their droppings, wear rubber gloves and finish up by thoroughly washing your hands. Prevent pets from getting access to sick or dead birds.

Preventing Disease Problems at Feeders and Birdbaths

Messy feeding areas are particularly conducive to the spread of infection as well as the growth of molds that can cause birds to become ill. A variety of strategies will help prevent disease-producing organisms from occurring or spreading.

Give birds space. Spread your feeders out to discourage crowding. Lots of birds using a single feeder looks wonderful, but crowding is a key factor in spreading disease. Crowding causes birds to jostle each other and creates stress, which may also make birds more vulnerable to disease.

Clean up waste. Rake the area underneath your feeder to remove droppings and old, moldy seed. Mount feeders above a hard surface, such as plywood or concrete, and sweep it regularly. Move feeders around periodically to keep droppings from collecting in one area.

Keep feeders in good repair, clean, and dry. To clean feeders, soak plastic feeders in a bucket (not the sink) and a solution of 10 parts water to 1 part bleach for at least 10 minutes. Scrub, thoroughly rinse, and dry well before rehanging. Another option is to run the feeder through a dishwasher cycle. Repeat every couple of weeks, more often if you notice sick birds at your feeder. Also, make sure your feeders keep the seeds dry.

Use fresh food. Do not serve moldy or damp seed to your birds. Disinfect any storage container that held spoiled food and also the scoop used to fill feeders from it.

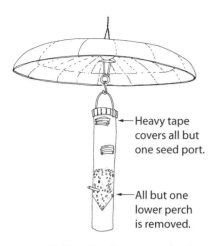

← Heavy tape covers all but one seed port.

← All but one lower perch is removed.

Figure 7. To reduce the amount of seed that falls to the ground, offer seed in a tube-style seed feeder. To reduce waste and prevent crowding at the feeder, cover all but one lower feeding port with duct tape or electrician's tape. (Drawing by Jenifer Rees.)

In addition:

- Prevent squirrels from accessing feeders and spreading seed on the ground. (See "Keeping Squirrels Away from Bird Feeders" in Chapter 24.)
- Cater to birds that create little or no mess around feeders—jays, chickadees, and nuthatches. Because birds in the finch family gather in large numbers and eat directly from the feeder (instead of flying away to eat elsewhere), they are prone to sharing diseases.
- Avoid mixed seed, especially mixes containing mostly milo or millet, which most birds, with the exception of juncos and a few others, don't prefer. Most

birds prefer black sunflower seeds.
- Offer shelled peanuts, hulled sunflower seed, or suet to eliminate hulls that would otherwise be dropped on the ground. (**Note:** Shelled nuts can quickly become rancid.)
- Use small feeders designed to allow only one or two birds to feed at a time.
- Offer seed in tube-style seed feeders and cover all the feeding ports but the lower one or two with duct tape or electrician's tape (see Fig. 7).

Diseases can also spread quickly and easily in an untended bird-bath. Change the water every few days. Change the water more often if many birds are using the bath. (Locating your birdbath near a faucet or hose will make refilling and cleaning easier.) Scrub birdbaths a few times each month with a plastic brush to remove algae and bacteria. (Rinse and thoroughly dry the brush following each use.) Never add chemicals to a birdbath for any reason.

Baby Birds Out of the Nest

Sooner or later, no matter where you live, you'll come across a baby bird on the ground. You'll have to decide whether you should rescue

it or leave it to fend for itself. In most cases, it is best not to interfere. The natural parents do a much better job at raising their young than we could ever do. A baby bird that is featherless must be fed every 15 to 20 minutes from about sunrise to 10 p.m.! This obviously requires a large time commitment on the part of the foster parent. See "Baby Birds Out of the Nest" in Chapter 28 for detailed information.

Public Health Concerns

Robins are not a significant source of any infectious disease that can be transmitted to humans or domestic animals. See "Public Health Concerns" in Chapter 29 for precautions to be taken when working around large concentrations of robin droppings.

Legal Status

In **Oregon, Washington**, and **British Columbia,** robins and all members of the finch family are federally protected. Any permit to lethally control these species would need to be issued from the U.S. Fish and Wildlife Service or the Canadian Wildlife Service, and would likely only be issued in very extreme cases.

CHAPTER 38

Starlings

It is hard to imagine now, but European starlings *(Sturnus vulgaris)* were purposefully introduced from Europe into this country. After two failed attempts, about 60 European starlings were released into New York's Central Park in 1890 by a small group of people with a passion to introduce all of the animals mentioned in the works of William Shakespeare. The offspring of the original 60 starlings have spread across the continental United States, northward to southern Canada and Alaska, and southward into Central America. There are now an estimated 150 million starlings in the United States.

In 1889 and 1892, the Portland Song Bird Club released 35 pairs of starlings in Portland, Oregon. These birds established themselves, but then disappeared in 1901 or 1902. The next sighting of a starling in the Pacific Northwest was not until the mid 1940s. Presumably these birds could be genetically linked to the 1890 Central Park introduction (Fig. 1).

It is difficult to reach a consensus on starlings. Some value the species for their creative adaptiveness and their odd beauty. Many hold a strong dislike of starlings because of their aggressive behavior at feeders and nesting sites, and their overwhelming flocks and roosting habits. There is only one thing on which agreement can be reached regarding starlings—they are ubiquitous.

Facts about Starlings
Food and Feeding Habits

- Starlings forage on lawns and other areas of short grass, such as pastures, golf courses, turf farms, and similar places.
- One of their very favorite foods is the large larva of the leatherjacket, or marsh cranefly *(Tipula paludosa)*, which eats the roots of grass plants. Leatherjackets (like starlings) are not native here, and were unintentionally introduced from Europe.
- Starlings have unique jaw muscles designed both to clamp shut and spring open, allowing them to use their bills to pry things open, including openings in the soil.
- Starlings also eat fruit, seeds and suet at bird feeders, and food scraps.

Figure 1. The European starling is a medium-sized, black songbird with short, triangular wings, speckled plumage, and a short tail. The adult in breeding plumage has a distinctive yellow bill and speckled black plumage with purple-green iridescence. The nonbreeding adult has a black beak and light spots. Juveniles are drab gray-brown overall. Males and females look alike. (Drawing by Elva Hamerstrom Paulson.)

ELVA HAMERSTROM Paulson

Nesting and Roosting Sites

- Starlings nest in suitable holes and crevices in buildings, utility poles, decaying trees, and holes in cliff faces, 6 to 60 feet (1.8–18 m) above ground.
- Males establish territories and choose nest sites, then attract females.
- Male starlings are very aggressive when claiming nest sites, taking over nest boxes and other cavities even while they are in use by such native birds as bluebirds, woodpeckers, and swallows.
- The nest is an untidy collection of grasses, bark strips, twigs, rope, and other debris. The nest cup is lined with feathers, mosses, or other soft material.
- In late summer and fall, starlings form large flocks and roost in large deciduous trees. In early winter, when trees lose their leaves, starlings roost in areas that provide protection from wind and cold, including coniferous trees, areas under bridges, and in grain terminals and barns.
- During the night, individual birds change their position in the roost to minimize energy loss, with older birds maintaining the "best" positions. (See "Roost Sites" for more information.)

Reproduction

- Starlings can be building nests, sitting on eggs, or caring for young anytime from mid-February to early July.
- Four to six slightly glossy, pale blue eggs hatch after an incubation period of 11 to 13 days.
- Both parents take turns with incubation during the day; at night only the female remains on the nest.
- The young begin to fly at 18 to 21 days of age, and out-of-nest care by parents lasts 2 to 4 days.
- A pair of adults can raise two broods per year. The female typically starts laying a second brood shortly after the first one fledges.
- Starling eggs (which are about the same size, shape, and color as robin eggs) often are found lying on the ground. It is believed that the females drop eggs if they are ready to lay, but the nest is not yet complete or has been taken over by another bird.

Mortality

- Adult starlings have few predators, although hawks and falcons occasionally catch them in flight.
- Loss of young starlings results from starvation, adverse weather, and predation by owls, raccoons, rats, domestic cats, and other predators.
- Humans, via control programs in agricultural areas, are probably responsible for most starling mortality.

Viewing Starlings

Starlings can be seen almost any time of the year in low elevations throughout the Pacific Northwest, particularly in areas associated with humans. They are among the few species of birds that tolerate high human density and poorly vegetated landscapes such as industrial sites. Starlings are normally absent only from heavily wooded areas, deserts, and areas above timberline. They appear to be partially migratory, but patterns vary regionally and individually.

Many birds move into valleys and urban areas during the winter.

Starlings are often observed walking or running along on lawns, stopping to probe for cranefly, moth, and beetle larvae with their powerful beaks. The short grass makes it easy for them to walk, locate food, and view potential predators.

The wings of starlings have a triangular shape when stretched out in flight. Their flight is direct and swift, not rising and falling, like the flight of many "black birds."

When starlings spot a perching hawk, falcon, or owl, they will "mob" it by flying around it and diving toward it, calling loudly. Dense flocks of starlings will also take flight and perform complex evasive movements in unison to avoid predators, such as falcons.

Huge, undulating flocks containing thousands of starlings can be observed during the winter months flying over towns, water, and fields.

Displays

Since starlings are widely distributed and abundant in populated areas, they make great subjects for bird-watchers interested in wildlife behavior (Fig. 2).

The **wing-wave** display (a) is performed when the male is perched; the bird spreads his wings and moves them in a rotating manner.

The **fluffing** display (b) is performed by males and females during aggressive encounters. The displaying bird faces another bird and puffs out all its feathers. The other bird may do the same.

Starlings have diverse calls and songs, such as whistles, high-pitched squeaks, and imitations of other birds' calls and songs,

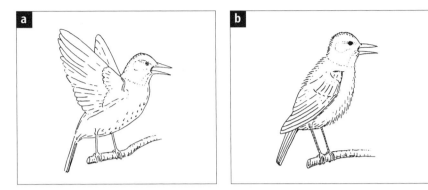

Figure 2. Two visual displays made by starlings. (From Stokes, A Guide to Bird Behavior.*)*

including those of bald eagles and other raptors. Just before pairing in spring and on warm fall days, the male commonly gives a squeal-call near the nest hole when a female flies by.

As long ago as the fifth century B.C., the Greeks and Romans kept starlings as caged birds and taught them to imitate human speech.

Roost Sites

Starlings roost on structures or in trees from late summer until the beginning of the breeding season. The number of birds using roosts can vary from a hundred to 150,000 or more. Roosts are largest in late summer, when composed of newly hatched young, their parents, and other birds that did not breed. The roosts become smaller, and may change location, in fall and winter when the adults migrate or return to breeding grounds.

Each sunrise, starlings leave their roost site and scatter across the land in small flocks to feed on nearby lawns, cultivated fields, golf courses, and similar places, as well as natural areas including wetlands, tidal flats, and debris-rich beaches. Starlings will fly 30 miles (48 km) to a productive feeding site.

Up to two hours before sunset, the starlings farthest from the roost site begin their return trip to the roost. The birds travel along established flight lines that are used day after day. Other small flocks join them and the flock size increases as it approaches the roost site. Some members will drop out and perch on pre-roosting sites such as trees, power lines, bridges, and towers, along the way. These pre-roosting areas are constantly changing in membership as birds leave and rejoin the main flocks.

Before sunset, all birds at pre-roosting sites will have left for the primary roost, where immense flocks will be swarming. The birds make spectacular dives into the primary roost, flutter about in search of a good perch, and settle down for the night.

Preventing Conflicts

The starling's long association with humans has strengthened its adaptive characteristics. Because these birds congregate in large numbers and aggressively search out food sources and nest sites in and around buildings, they can come into conflict with people.

The following are suggestions on how to prevent and remedy

conflicts that arise. In cases where these methods are not practical, contact the Department of Agriculture for more information. In British Columbia, contact the Ministry of Agriculture, Food, and Fish, and the Ministry of Water, Land, and Air Protection (see Appendix E for contact information).

Starlings nesting in buildings:
Starlings are adept at establishing nest sites in nooks or crannies in buildings. Nesting activity can damage buildings, create fire hazards, and clog gutters and drainpipes, causing water damage.

Prevent starlings from nesting in structures by sealing all potential points of entry. Although starlings have difficulty entering holes smaller than 1½ inches (4 cm) in diameter, house sparrows, bats and other small mammals can slip right in. Use wood, ¼-inch (6 mm) hardware cloth, aluminum flashing, or similar sturdy material. Light material, such as bird netting or rags, will not keep determined starlings out. Replace any loose shingles or siding, and repair broken windows.

Install commercially available vent guards to prevent starlings from entering exhaust vents and dryer vents. If necessary, cover the ends of elevated drainpipes with ¼-inch (6 mm) hardware cloth during the nesting season. All screening should be checked periodically to make sure it isn't clogged.

Prevent starlings from roosting on walls covered with vegetation by removing the vegetation or draping bird netting over the area. In new construction, avoid creating small cavities or spaces with access from the exterior into which starlings can enter and nest.

Starlings can be evicted from buildings and other sites any time of year. State, federal, or provincial laws do not protect this species. A stick with a 2½-inch (6 cm) angle bracket screwed to it can be used to remove nests. The nesting material should be collected and removed to prevent the birds from using it for a new nest. Take immediate steps to prevent starlings from rebuilding.

If the birds are caring for young, one approach is to wait until the young can fly out of the nest, then remove all nesting materials and cover all openings. (See Appendix A for information on euthanizing birds.)

For other ideas on how to prevent starlings from roosting in or on buildings, see "Preventing Conflicts" in Chapter 29.

Starlings at feeders: Starlings are attracted to both seed and suet feeders, and their aggressive habits can deplete food supplies and keep smaller birds from approaching. By choosing the right bird feed, style of bird feeder, or modifying an existing feeder, you can discourage starlings.

Because starlings have difficulty cracking the commercially available black sunflower seeds, these can be offered in feeders.

Because starlings have difficulty landing on a small perch, making the perches on a feeder smaller by sawing them, or removing the perches altogether, can keep starlings off (see "Preventing Conflicts" in Chapter 37 for an example). Most songbirds do not need a perch to access the seed.

Starlings may also be deterred by small feeders that swing and twirl whenever the heavy birds land on them (Fig. 3).

Figure 3. A hanging pine cone stuffed with peanut butter or suet will prevent starlings from accessing this simple feeder. Smaller birds will have no trouble landing and feeding. (From Link, Landscaping for Wildlife in the Pacific Northwest.*)*

Because starlings have trouble clinging upside down, a suet feeder that requires the birds to clasp the feeder from below will discourage starlings. Your local bird specialty store can give you information on suet feeder designs to deter these birds.

Wire mesh placed over a platform feeder will prevent starlings from accessing the seed (see "Preventing Conflicts" in Chapter 29 for an example). Don't place large amounts of birdseed on the ground or on an uncovered platform feeder.

Starlings roosting in trees: In fall and winter, the communal night roosts of thousands of starlings create accumulations of droppings below the roost. When a health official deems this a health risk to the public, steps need to be taken to disperse the flock. Options include installing visual and auditory scare devices, and thinning 30 to 50 percent of the branches of roost trees—or removing trees from dense groves—to reduce the availability of perch sites and to open the trees to the weather (Fig. 6). A tree service company can provide this service.

Experience has shown that the best results occur when the pruning of trees is combined with scare tactics. (See "Preventing Conflicts" in Chapters 27 and 28 for information on visual, auditory, and other scare devices.)

Starlings eating fruits and vegetables: A small flock of starlings can quickly ruin or remove the year's fruit or young vegetable crop.

Protect fruit crops with flexible bird netting, which can be purchased in a variety of lengths and widths at garden and hardware stores; professional quality materials and hardware are available from bird-control companies and over the Internet (see Appendix F for resources). Secure the base of the shrub or the tree to prevent starlings from gaining access from below (Fig. 7). Individual small branches containing fruit can be protected with an onion sack or similar mesh covering.

Row crops, such as strawberries, can be completely covered during the fruiting season. If the netting is to be used for several harvest seasons, it may be worth the extra effort to construct a frame to support the netting.

Scare devices, such as pie tins and commercially available Mylar

Protecting Native Cavity Nesters from Starlings and House Sparrows

Although starlings and house sparrows (Chapter 34) can be interesting to watch in highly built-up areas where few other bird species thrive, they are a serious problem in areas where native birds exist. These introduced species compete with native, cavity-nesting birds for nesting spots, which are becoming increasingly less plentiful as trees are cut down.

Male starlings and house sparrows are especially aggressive in their search for nest sites: They will peck holes in eggs laid by other birds, throw out their nesting material, and kill their young. Starlings will build nests on top of existing nests containing eggs, and can evict the larger wood duck from its nest boxes.

To prevent problems:

• Don't attract house sparrows and starlings. (Follow recommendations under "Preventing Conflicts" in this chapter and in Chapter 34.)
• Install nest boxes designed to exclude starlings and house sparrows (Figs. 4 and 5). Many native songbirds can use an entry hole smaller than the 1½ inches (3.8 cm) needed by starlings (Table 1), and the diamond-shape hole can be used where house sparrows are abundant (Fig. 5). Be alert to hole enlargement by flickers and rodents, and replace or add a new front with the proper hole size. To reduce the size of an existing entry hole, attach a piece of wood to the front of the existing box and drill the appropriate size hole. File down all rough edges. It is also possible to buy a pre-drilled metal plate that can be attached over the entry to a nest box.
• Don't install nest boxes that have perches. Perches are used by star-

lings and house sparrows, but are not necessary for native species.
• To prevent house sparrows from perching on the roof of a nest box and driving off the preferred species of birds, attach a piece of metal roof flashing (or a flattened can) to the nest-box roof, extending this vertically above the box (Fig. 4).
• When observing a house sparrow or starling building a nest in a nest box, repeatedly remove the nesting material, or plug the entry hole for a few days or longer to prevent them from entering. Carefully monitor the box throughout the breeding season for use by house sparrows or starlings.
• If house sparrows or starlings have laid eggs in a nest box, vigorously shake the eggs and return them to the nest. The adults will incubate them, but the eggs will not hatch. Because state, federal, or provincial laws don't protect these birds, it is legal to remove their nests and destroy the eggs or the birds themselves.
• Clean out nest boxes each year. When not cleaned out, birds will build new nests on top of old ones. This raises the new nest close enough to the entry hole so starlings or other predators can pull out the occupants.

• If you have just a couple of boxes, take them down each year (or block the entrance holes) after the breeding season and do not put them back up until the native species are seen or heard the following spring. *Note:* Since starlings and house sparrows do not migrate for the winter, they will be looking for nesting sites long before the migrants return in the spring.

Table 1. Entry hole dimensions needed by some small native cavity-nesting birds

Bird species	Diameter of entrance
Chickadees	1 to 1⅛ in. (25 – 28 mm)★
Tree swallows	1¼ in. (32 mm)
Violet–green swallows	1⅛ in. (28 mm)
House wren	1 in. (25 mm)★
Nuthatches	1¼ in. (32 mm)★
Western bluebird	1½ in. (38 mm)
Hairy woodpecker	1⅝ in. (42 mm)
Downy woodpecker	1¼ in. (32 mm)

★ *This species will also use the diamond-shaped entry hole shown in Figure 5.*

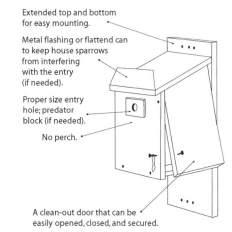

Extended top and bottom for easy mounting.

Metal flashing or flattend can to keep house sparrows from interfering with the entry (if needed).

Proper size entry hole; predator block (if needed).

No perch.

A clean-out door that can be easily opened, closed, and secured.

Figure 4. A nest box designed to provide a safe nesting site for native cavity-nesting songbirds. (From Link, Landscaping for Wildlife in the Pacific Northwest.*)*

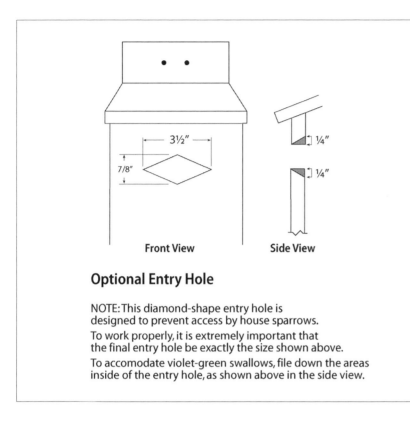

Optional Entry Hole

NOTE: This diamond-shape entry hole is designed to prevent access by house sparrows.

To work properly, it is extremely important that the final entry hole be exactly the size shown above.

To accomodate violet-green swallows, file down the areas inside of the entry hole, as shown above in the side view.

Figure 5. A diamond shape hole created to these exact measurements will exclude house sparrows. (From Link, Landscaping for Wildlife in the Pacific Northwest.*)*

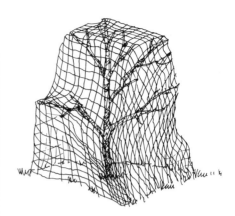

Figure 7. Protect fruit crops with flexible bird netting. Secure the netting at the base of the shrub or tree to prevent starlings from gaining access from below. (Drawing by Jenifer Rees.)

balloons or Mylar scare tape, are known to provide temporary protection. Suspend balloons at least 3 feet (90 cm) above trees or bushes, or from lines between posts. Use tethers at least 3 feet (90 cm) long.

Figure 6. Before and after pruning of a large coniferous tree to reduce its attractiveness to roosting birds. (Drawing by Jenifer Rees.)

Attach special red and silver bird-scare tape to stakes and stretch it 18 inches (45 cm) above the areas that need protection. Twist the tape several times before attaching it to stakes so that the visible interval of red/silver is 16 inches (40 cm). The tape should move freely, so that when a slight breeze blows it will flash in the sun. The space between tapes will have to be no more than 5 feet (1.5 m) to be effective.

Because most birds will fly into a strawberry patch, land on the ground between the plants and eat the ripe strawberries from there, scare devices placed above the patch are not effective. Instead, place the scare tape between the rows. The tape should sag slightly but should not be less than 3 inches (8 cm) or more than 5 inches (13 cm) from the ground.

Scare devices need to be moved weekly (daily if possible) so birds don't become accustomed to them; they are also most successful if put in place before the birds become a problem. Always harvest ripe fruit immediately.

Protect germinating corn plants and other crops with bird netting until plants are about 8 inches (20 cm) tall. Large plastic trash bags attached to 6 to 7 foot (1.8–2 m) wooden stakes, along with the above-mentioned scare tactics, can be used in areas with lots of air movement. Cracker shells and propane cannons may be needed in larger plantings (see Chapter 27 for information on these options). Ultrasonic devices are not effective at frightening starlings.

Other Control Techniques

Trapping

Research has shown that intensive trapping and euthanizing can temporarily reduce starling numbers and damage. This may be worthwhile in some situations, such as at a winter cattle-feeding operation or at airports. However, it has no effect on the number of starlings returning the next year unless it is done repeatedly and over 50 percent of the population is removed each time.

Small-scale traps are available from enterprises over the Internet (see Appendix F for resources). Check the trap every two hours for non-targeted birds.

Do not trap starlings and release them elsewhere, because they will easily return or cause problems somewhere else. If you cannot humanely kill them yourself, find a falconer or wildlife rehabilitation center that will accept live starlings to feed to birds of prey. (See Appendix A for information on euthanizing birds.)

Shooting

Shooting is not an effective way to manage starling populations overall. The number of birds that can be killed by shooting is small relative to the size of the flock. However, shooting may be helpful where only a few birds are present, and in supplementing or reinforcing other dispersal techniques. First check with the local ordinances regarding discharging firearms.

Public Health Concerns

Although health risks from birds are often exaggerated, large populations of roosting starlings may present risks of disease to people nearby. The most serious health risks are from disease organisms growing in accumulations of starling droppings, feathers, and debris under a roost. This is most likely to occur if roosts have been active for years.

Precautions need to be taken when working around large concentrations of starling droppings. See "Public Health Concerns" in Chapter 29 for information.

Legal Status

Starlings are exempted from protection in **Oregon, Washington, and British Columbia.** Their nests, eggs, young, and/or adults may be removed or destroyed at any time. No permit is required.

CHAPTER 39

Swallows and Swifts

Swallows are migratory songbirds that occur and breed in the Pacific Northwest from spring to fall. They are sparrow-sized birds with long, pointed wings and streamlined bodies developed for fast, acrobatic flight. They are seen swooping and flying over fields, orchards, lakes, and anywhere else that flying insects are abundant.

Seven members of the swallow family breed in the Pacific Northwest (Table 1). Of these seven species, barn and cliff swallows regularly build mud nests attached to buildings, a process that sometimes brings them into conflict with humans. Because of their close association with humans, this chapter focuses on these two species.

Swifts, which are also migratory and have slender bodies, long, pointed wings, and similar flight patterns, are sometimes mistaken for swallows. For information on swifts, see "Notes on Pacific Northwest Swifts."

Facts about Swallows
Food and Feeding Behavior

- Swallows are insectivores, catching a variety of insects in midair with their wide-gaped bills and expert flight. Barn swallows eat some berries, seeds, and dead insects from the ground, particularly during bad weather.

Figure 1. Barn swallow nests are made of mud pellets and some fibrous material and are often built under eaves, bridges, docks, or other man-made structures. (Drawing by Elva Hamerstrom Paulson.)

Table 1. Pacific Northwest Swallows

A field guide is helpful for identification of swallow species and learning about their distribution throughout the Pacific Northwest. Popular field guides are listed in Appendix F.

The **barn swallow** (*Hirundo rustica,* Fig. 1) is a distinctive bird with bold plumage and a long, slender, deeply forked tail. It has blue-black upper parts, a reddish throat and breast, and a rust or buff colored belly. Females are slightly duller and shorter-tailed than males.

Although they are still common in Washington, Breeding Bird Census data indicate that barn swallows have decreased significantly in the state since 1980.

The **cliff swallow** (*Petrochelidon pyrrhonota,* Fig. 2) looks somewhat like the barn swallow, but has relatively broad, round wings and a short, squared-off tail. The back, wings, and crown of the adult are a deep blue, and its belly is light colored.

The **violet-green swallow** (*Tachycineta thalassina,* Fig. 5) is well named because of the beautiful iridescent violet-green color of its back. It can be distinguished from the tree swallow by the white patches at the base of its tail. Compared to males, females are drab in color.

The **bank swallow** *(Riparia riparia)* is North America's smallest swallow. It is brown above and slightly darker on the wings than on the back and rump. It has a distinct dark breast-band and a white throat.

The **tree swallow** *(Tachycineta bicolor)* is an elegant bird, with white undersides, an iridescent blue-green back, and a moderately forked tail. It is most likely to be confused with a violet-green swallow, but the violet-green swallow has white patches extending up the sides of the rump that can be seen in flight.

The **Northern rough-winged swallow** *(Stelgidopteryx serripennis)* is plain brown above with a white belly and buffy throat and upper breast. The name "rough-winged" refers to the tiny serrations on the outermost wing feathers of this swallow, visible only when the bird is in the hand.

The **purple martin** *(Progne subis)* is the largest swallow in North America. The adult males are a solid, glossy, purplish-blue both above and below. The female is similarly colored, but less brightly. The **purple martin** is a rarity in our region, and few purple martin nesting colonies occur in the Pacific Northwest. Erecting the traditional apartment-type martin houses here has, to date, been a wasted effort. Other nest-box designs for purple martins, however, have been successful (see Appendix F for resourses).

- Swallows will fly several miles from their nest site to forage.
- Long periods of continuous rainfall make it difficult for adult swallows to find food, occasionally causing young birds to die.
- Swallows drink mid-flight; as they fly over water they dip their bill to the surface to drink.

Nests and Nest Sites
- Barn and cliff swallows construct nests formed from mud pellets that they collect in their beaks.
- Barn swallow nests are cup-shaped (Fig. 1); cliff swallow nests are gourd-shaped (Fig. 2). The interior of both these birds' nests contains an inner cup lined with grass, hair, and feathers.

- Historic nesting sites of both barn and cliff swallows include cliffs, walls of canyons, and vertical banks protected from rain.
- Today, barn swallows almost always build nests on eaves, bridges, docks, or other man-made structures that have a ledge that can support the nest, a vertical wall to which it can be attached, and a roof.

- Cliff swallow nests are built on vertical walls, natural or man-made, frequently with some sort of sheltering overhang. Freeways, bridges, barns, and other large buildings are regularly used.
- Barn swallows usually nest in single pairs; cliff swallows nest in colonies that may contain a dozen to over 500 nesting pairs.
- Barn and cliff swallow nests are prone to external parasite infestations. Colonies may not be reoccupied because of heavy infestations, and if parasite populations become too great, both species will prematurely desert their nests, abandoning their young.

Reproduction

- Time from start of nest building to departure of young is 44 to 58 days.
- Nest building is done by both sexes, and in mild years may begin as early as late March in Oregon and late April in British Columbia.
- Both parents take turns incubating three to five eggs, which hatch after 12 to 17 days.
- Brood parasitism is common among cliff swallows. Females will lay eggs in other females' nests and will also carry eggs in their beaks from their own nests to the nests of others.
- Both sexes care for the young, which begin to fly at 20 to 25 days of age.
- After learning to fly, the young remain in the nest, or near it, to be fed by parents and to roost at night. They leave the nest after a few days and will remain in the area for several weeks.
- Barn and cliff swallows can raise two clutches per year. Re-nesting will occur if nests or eggs are

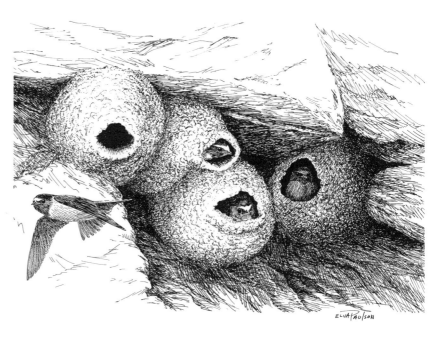

Figure 2. Cliff swallow nests are made of mud pellets and some fibrous material. The larger cliff swallow nest may contain 1,000 pellets or more, each representing one trip to and from the nest. (Drawing by Elva Hamerstrom Paulson.)

destroyed. For example, nests may fall because they were built too rapidly, or may crumble because of prolonged wet or humid weather.

Mortality and Longevity

- Young swallows may fall out of their nests or die from nest infestations of blowfly and other parasitic insects and mites.
- Other deaths of young occur from severe hot or cold temperatures, fallen nests, and predators, including crows, ravens, magpies, starlings, house sparrows, domestic cats, rats, and tree squirrels.
- Most barn and cliff swallows die during their first year, and rarely live longer than four years.

Viewing Swallows

The flowing flight of swallows can be enjoyed from dawn to dusk. Barn swallows are agile flyers that come to within inches of the ground to catch flying insects.

Cliff swallows glide, soar, and circle more than barn swallows do, and are often seen higher in the sky. When not in flight, swallows can be seen perched on utility wires, TV antennas, and on dead branches in large trees. Young swallows can be observed sticking their heads out of the nest, begging for food when a parent arrives.

Mud Sources

Barn and cliff swallows travel up to a half-mile (0.8 km) to gather mud from the edges of ponds, puddles, and ditches. Gathering mud and constructing nests are social activities for cliff swallows; even unmated swallows may build a nest that goes unused.

To find one of their mud sources, look for swallows landing on the ground—they rarely do this at times other than when nest building. The birds will remain on the ground for a minute or so and then fly off to a nest site.

Notes on Pacific Northwest Swifts

Swifts are one of the fastest flying birds. They seldom land, having legs that are small and weak. They have strong feet, however, and this allows them to cling to vertical surfaces, such as rock walls and chimneys (Fig. 3). They perform all bodily functions— including feeding, bathing, drinking, and mating—in the air.

Swifts resemble swallows but are grouped in the same taxonomic order as hummingbirds, rather than in the passerine order to which swallows belong.

Swifts measure 4 to 8 inches (10–20 cm) in length, with 12- to 15-inch (30–38 cm) wingspans. They have a streamlined, pale gray or brown body, and long, pointed wings that curve backward. Their scythe-like wings beat rapidly in shallow strokes followed by short glides.

Swifts are often overlooked because they spend much of their time chasing small insects high in the air. Like swallows, they are spring and summer residents in the Pacific Northwest, wintering in California, Mexico, and perhaps further south.

Three species of swifts breed in the Pacific Northwest: the black swift (*Cypseloides niger*), the white-throated swift (*Aeronautes saxatalis*), and the Vaux's swift (*Chaetura vauxi,* Fig. 3). The Vaux's swift is the most common. (The name is English, and the "x" is sounded.) Because the Vaux's swift occasionally roosts and nests in chimneys (like the chimney swift in the eastern United States), it is also the species most likely to attract the attention of humans.

Although Vaux's swifts use chimneys, they usually roost and nest in natural cavities with vertical entranceways, such as hollow trees. Unfortunately, Vaux's swift populations are declining as old-growth forest in the

Figure 3. The feet of the Vaux's swift have four hook-shaped toes with claws that can grasp onto a rough surface. These partly support the bird, while the stiffened tail feathers with their exposed spiny tips help hold it up. (Oregon Department of Fish and Wildlife.)

Northwest is destroyed and snags are removed from mature forests in an effort to prevent forest fires.

This swift's teacup-sized nest is made of small twigs or grass cemented to the inside of the chimney or tree with the bird's saliva. A few swift feathers may also become nest material. Clutches usually contain six eggs, and are incubated by both parents for 18 to 19 days. Both parents feed and tend the young, which first fly when about a month old. The young may continue to return to the nest site for a few days after learning to fly. Vaux's swifts raise only one brood each year.

Vaux's swifts in chimneys during late May to early August are almost always a breeding pair. If this should occur, delay annual cleaning until after young have left the nest. Although the noise of young birds begging for food may be heard, such noise is short-term and should be tolerated. The nest is

probably too small to be a fire hazard, but should always be removed after the birds have left in the fall.

Chimneys lined with metal should always be capped, as birds that enter these can easily become trapped and die. Chimney caps are available from hardware stores (see sample under "Preventing Conflicts" in Chapter 21).

As they prepare to migrate south in early fall, Vaux's swifts congregate— sometimes in the thousands—to use a single chimney as a night roost (Fig. 4). The nightly return of foraging birds is impressive, as they dart into the chimney at dusk with an uncanny synchronization that must be highly organized, even if it looks confused. In such cases, make sure the chimney flue is closed so that the birds cannot enter your house. Such use of a chimney is temporary. It should not cause any problems that cannot be solved by a routine cleaning after the birds move on.

The Marvel of Migration

Each autumn, almost half the bird species that breed in the Pacific Northwest migrate south to tropical Central and South America. This migration is one of the wonders of the natural world.

These birds, called "neotropical migrants," spend six or more months in southern locations before returning north in spring to mate and rear young. Most vireos and warblers winter in western Mexico and northern Central America, as do tanagers, black-headed grosbeaks, orioles, swifts, and violet-green swallows. Barn, bank, and cliff swallows and purple martins winter in Central and South America.

The famous swallows of Capistrano are cliff swallows, and, contrary to legend, they return to Capistrano in late February, considerably earlier than the fabled March 19.

The main reason for this seasonal migration is the lack of insects to eat during winter in the north. Avoiding cold temperatures is actually a less important reason for leaving.

Barn and cliff swallows begin their return to northern climes in late winter and early spring. Depending on weather conditions, they are first spotted in southern Oregon in late March or early April. They start appearing in British Columbia two to three weeks later.

Swallows migrate during the day, catching flying insects along the way. They will normally not move into areas unless flying insects are available for food, which occurs after a few days of relatively warm weather—60°F (16°C) or more.

Swallows are usually the first to begin the southern migration in mid-August to early September. They gather in large groups (sometimes as many as 2,000 birds) on telephone wires and other perches before departing.

Figure 4. A school in Portland, Oregon, is thought to have had up to 40,000 Vaux's swifts in one of its chimneys. It may be the largest swift roost in the western United States. (Drawing by Elva Hamerstrom Paulson.)

The collection site will be marked with numerous small holes made by the birds as they poke their beaks into the mud several times to get a good load. You may also see swallows flying with feathers or grass—materials used in the final stages of nest building.

Calls and Songs

The barn swallow's song is a series of twitters and gurgles. They emit a soft *wit wit* call when feeding with other swallows, and when approaching their nests. A louder

Attracting Swallows to Your Property

To best manage your property for swallows, protect nearby undisturbed wild areas, including wetland, lake, or grassy areas of any size used by swallows for drinking, mud collecting, and feeding. Also, retain as many large snags as possible—these dead or dying trees are used by swallows for perching and/or nesting.

Because nearly 70 percent of the breeding birds in the Pacific Northwest eat primarily insects, avoid using insecticides. Also, keep your cats indoors and discourage other cats from visiting your property (see Appendix D for information).

House sparrows sometimes take over empty swallow nests and have been known to drive off swallows from new nests (see Chapters 34 and 38 for information on managing house sparrows).

Supplementing the supply of nest materials and installing nest boxes are other ways to enhance your property for swallows. Swallow species that use nest boxes are violet-green swallows (Fig. 5), tree swallows, and purple martins; barn swallows will use nesting platforms (Fig. 8). (See "Nest Structures and Den Boxes" in Appendix F for resources.) Children enjoy providing nest materials and boxes, and they often observe animal behavior that adults don't notice.

Have nesting materials and boxes available for use by early April. Mud can be provided to barn swallows and cliff swallows (also robins and phoebes) by spraying a shaded area with water to create a mud puddle. Mud can also be offered in a tray—an inverted garbage can lid works great! Recess the lid into the ground, fill it with moist soil, and place it in an area where birds can see it. *Note:* Don't set this out in an area where house cats run freely. A mud-filled, elevated birdbath may serve as a safe mud-gathering site for swallows.

Make the soil the consistency of a mud-pie. Claylike soil with some humus is a good building material because plant fibers strengthen the nest. Swallows will also gather small downy feathers and 3-inch (7.5 cm) lengths of dry grass placed nearby.

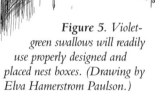

Figure 5. Violet-green swallows will readily use properly designed and placed nest boxes. (Drawing by Elva Hamerstrom Paulson.)

version of this call is given when there is possible danger near the nest, such as at your approach.

The cliff swallow's call is a low, soft, husky *verr* or *churr,* sounding like the squeaking of a door with rusty hinges. The song is a series of thin, strained, drawn-out rattling sounds that is shorter and simpler than the song of the barn swallow.

Preventing Conflicts

Many people enjoy swallows nesting on or around their homes. Colonies of cliff swallows on school grounds can provide excellent opportunities for study. The anticipation of the swallows' arrival in the spring is exciting, watching parents feeding their young is a wonderful sight, and swallows consume thousands of flying insects that are considered pests.

It has been speculated that one reason swallows choose to nest on door stoops, light fixtures, and porch fronts is because the close proximity to humans keeps crows and other predators away. The birds will even risk cat predation and human vandalism and nest close to the ground if the location is in a place frequented by humans.

The barn swallow's close association with humans in Europe goes back over 2,000 years. Thus, when you thwart a barn or cliff swallow's nesting effort, you may be denying the birds their only chance at successful reproduction.

To prevent conflicts or remedy problems:

Manage swallow droppings:
Conflicts with swallows occur when these birds nest close to humans, primarily because of the droppings and other debris they deposit.

When swallows first hatch, the parents eat their droppings, which keeps the nest clean and insect free. After a few days, the adults carry the droppings (which are encased in a fecal sac made from clean mucous membrane) away from the nest to prevent detection by predators. After about the twelfth day, the young back up to the edge of the nest and defecate out over the rim.

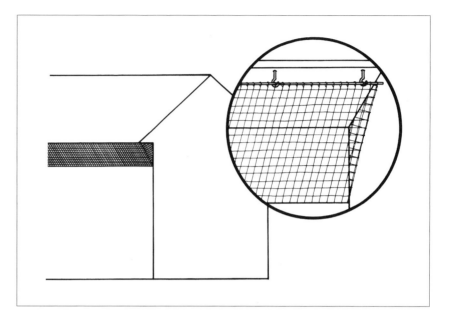

Figure 7. To deter swallow nesting on structures, attach bird netting or chicken wire from the outer edge of the eave down to the side of the building. Alternatively, create a small curtain of netting. (From Hygnstrom et al., Prevention and Control of Wildlife Damage.*)*

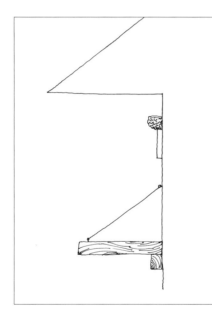

Figure 6. One way to deal with barn swallow droppings is to hang a board under the nest using eye screws and wire. Place newspaper or a piece of thin paneling on the board, and remove it when it needs cleaning. A longer board, or other structure, can be used under groups of cliff swallow nests. The board or other device should be a couple of feet below the nest and not so wide as to interfere with the birds' comings and goings. (Drawing by Jenifer Rees.)

Placing newspaper or some similar material where droppings accumulate can solve the problem. As necessary, the paper and droppings can be added to a compost pile, dug into the ground (droppings make wonderful fertilizer), or placed in the garbage. Similarly, a blanket or sheet can be used to cover a car or structure, and moved when needed.

Another solution is to install a board under the nest(s) to catch the droppings and debris (Fig. 6). Because of its close proximity to the nest, the board should be cleaned as needed to prevent infestations of insects and mites that may live in the accumulated debris. Before attaching the board, observe the swallows comings and goings to prevent installing something that could interfere with the birds accessing their nest.

Create a barrier: If for some reason swallows nesting on a building or other structure cannot be tolerated, a barrier can be installed. Again, because barn swallow populations have been on a decline for the past 20 years and cliff swallows have specific nesting requirements that are as yet unknown, preventing these species from nesting should be done only in extreme cases.

Barriers include any physical structure placed between the swallow and the structure. A permit is not required for this method if it is done before the birds arrive, during nest building when there are no eggs or young in the nest, or after the birds have left for the winter. If swallows have eggs or young in the nest, exclusion may not be used without a permit (see "Legal Status").

To prevent barn swallows from nesting on door jambs, window jambs, and other sites on the side of a building, cover the area with bird-netting or 1-inch (2.5–cm) mesh chicken wire. Drape the

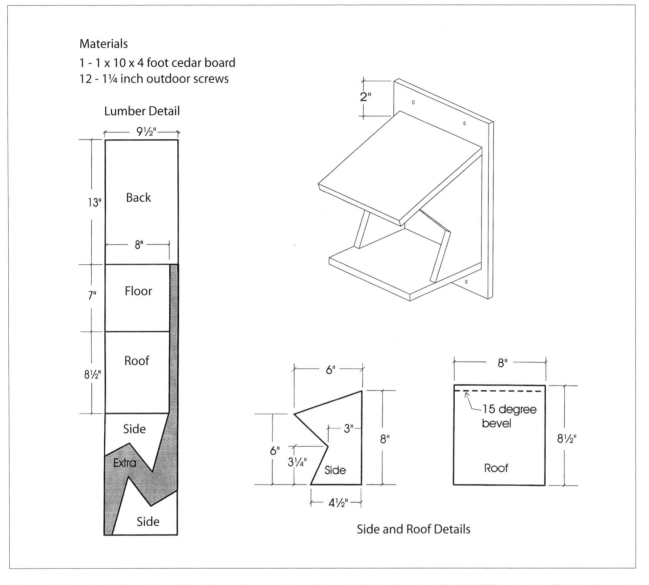

Materials
1 - 1 x 10 x 4 foot cedar board
12 - 1¼ inch outdoor screws

Lumber Detail

Back

Floor

Roof

Side

Extra

Side

Side

Roof

Side and Roof Details

Figure 8. A nesting platform designed for robins and barn swallows. (From Link, Landscaping for Wildlife in the Pacific Northwest.*)*

material from the outer edge of an eave down to the side of the building (Fig. 7). Remove wrinkles and folds that could trap or entangle swallows or other birds.

Bird netting and chicken wire are available from nurseries, hardware stores, and farm supply centers. Some pest-control companies sell a heavy-duty netting material with a larger mesh than common black netting used to protect fruit from birds. The netting is not as

likely to create problems for songbirds, which sometimes get caught in the smaller mesh netting. To find the product, search the Internet for "bird control supplies" or look in your phone book under "Pest Control."

Attach the barrier using staples, brass cup-hooks, adhesive-backed hook-and-loop Velcro, trash-bag ties, or other fasteners. To avoid unsightly rust stains, use only rust-resistant fasteners.

The barrier may also be first stapled to or wrapped once or twice around wood laths, which are then attached to the structure. This technique can also be modified to keep swallows from entering a breezeway, or similar sites, to nest.

Another technique is to hang a curtain of bird netting or chicken wire from the eave (Fig. 7). The curtain should be 3 to 4 inches (7.5–10 cm) from the wall and extend down from the eave 18 inches (45 cm) or more. A well-done application under the eaves is nearly invisible from 50 feet (15 m) because it is in a shaded area and gets obscured by the shadows.

A solution for small areas is to install aluminum foil, aluminum flashing, or heavy plastic over the spot where swallow nests are unwanted. The smooth surface will prevent swallows from adhering mud to the wall. Painting the area with a glossy latex paint may also be effective. It may be possible to offer barn swallows an optional nesting site by constructing a nest platform (Fig. 8).

Note: Hawk, owl, and snake models, noisemakers, revolving lights, red-and-silver flash tape, and hanging pie tins are unlikely to deter swallows.

Nest removal: At the first sign of nest building, remove the nest. *Note:* All swallows are protected under the law. You cannot disturb them once they lay their eggs in the nest (see "Legal Status").

Usually nests can be washed down with a water hose or knocked down with a pole. Because swallows are persistent at rebuilding nests, you will need to continually remove the nest mud for several days until the birds stop. Swallows are strongly attracted to old nests or to the remnants of deteriorated nests, so all traces of mud should be removed.

For information on what to do if young swallows fall from a nest, see "Baby Birds out of the Nest" in Chapter 28.

For information on what to do if you find a swallow wearing a bird band, see "How to Report a Bird Band" in Chapter 33.

Public Health Concerns

Swallows and swifts are not a significant source of any infectious disease that can be transmitted to humans or domestic animals.

See "Public Health Concerns" in Chapter 29 for precautions to be taken when working around large concentrations of swallow or swift droppings.

Legal Status

In **Oregon, Washington,** and **British Columbia,** swallows and swifts are federally protected. Any permit to lethally control these species would need to be issued from the U.S. Fish and Wildlife Service or the Canadian Wildlife Service, and would only be issued in very extreme cases. Some examples are concerns for aircraft safety from a nesting colony at an airport or potential food contamination from a colony over a loading area at a food-processing center.

In most cases a permit for lethal control will not be issued for swallows nesting on a residence or other buildings and causing aesthetic damage.

A permit is not required to remove swallow nests under construction that do not contain an adult, any new eggs or young, or nests abandoned after the breeding season. If an adult swallow is occupying a half-built nest, or a fully built nest without eggs, then the law protects it.

CHAPTER 40

Woodpeckers, Flickers, and Sapsuckers

Woodpeckers, including flickers and sapsuckers, are all members of the family Picidae. These unique and interesting birds play an important role in Pacific Northwest forests and wooded backyard ecosystems. They eat all life-stages of wood-boring insects that are inaccessible to most other forest birds. Northern flickers, or flickers, eat quantities of carpenter ants.

Holes that woodpeckers create each year for nesting and roosting are used in subsequent years by cavity-nesting songbirds, small owls, ducks, and native squirrels that cannot fully excavate their own nest site.

Aesthetically, the value of woodpeckers is incalculable. The sound of their drumming is a sign that spring is approaching and bird-watchers eagerly seek the sight of any woodpecker.

Any annoyance woodpeckers may cause for homeowners is greatly outweighed by the large number of insect pests they eat, and the number of homes they create for other wildlife.

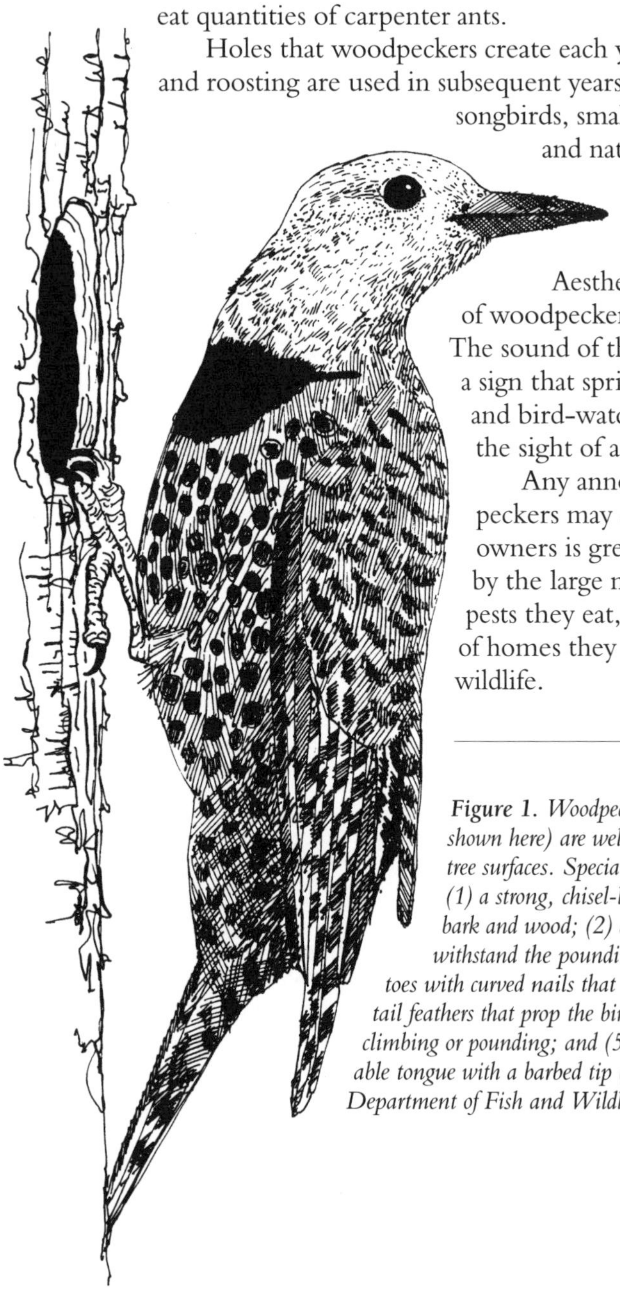

Figure 1. Woodpeckers (Northern flicker shown here) are well adapted for life on tree surfaces. Special adaptations include (1) a strong, chisel-like bill to hack into bark and wood; (2) a thick skull that can withstand the pounding; (3) long, strong toes with curved nails that can grab bark; (4) stiff tail feathers that prop the birds up while they are climbing or pounding; and (5) a very long, extendable tongue with a barbed tip (Fig. 5). (Washington Department of Fish and Wildlife.)

Facts about Woodpeckers

Food and Feeding Habits

- While most woodpeckers eat tree-dwelling and wood-boring insects, they also will eat sap, berries, fruits, nuts, acorns, and seeds.
- When searching for insects, woodpeckers tap on wooden surfaces and look and listen for insect movements. If they see or hear an insect, they will continue chiseling until the insects are caught. Hollow sounds also may indicate that insects are present, thus encouraging woodpeckers to continue chiseling away.
- Sapsuckers feed on tree sap and insects. They chisel small holes in live trees and then use their bristly tipped tongues to lap up the sap and insects (Fig. 11).
- Woodpeckers are attracted to suet feeders, especially in winter.

Nest Sites and Shelter

- Woodpeckers excavate nest sites in dead or dying trees, aging utility poles, fence posts, house siding, and earth banks. They will also use specially designed nest boxes.
- The entry hole is round or rectangular (pileated woodpeckers). From there the birds tunnel down 6 to 24 inches (15–61 cm), making a wide bottom for the egg chamber (Fig. 4).
- Nest holes may be started but never completed, possibly due to poor location or quality of the wood. Occasionally woodpeckers will re-use a nest hole after doing some minor work to it.
- Both male and female woodpeckers excavate the nest, the

Table 1. Pacific Northwest Woodpeckers

Thirteen species of woodpeckers live in the Pacific Northwest. A field guide is helpful for accurate identification. Illustrations should be in full color and show distinguishing traits. Popular field guides are listed under "Birds" in Appendix F. The following woodpecker species are seen on rural properties and in urban parks where large trees exist:

The **Northern flicker** (*Colaptes auratus*, Fig. 1 and back cover) is probably the most commonly observed and heard woodpecker species. It is also the species likely to arouse the tempers of homeowners when one of these large birds sets up a drumming station on their home. Two forms occur in the Pacific Northwest: the red-shafted flicker, and, less commonly, the yellow-shafted form.

Both species have a conspicuous white rump patch easily seen in flight. The red-shafted flicker has salmon-colored wing and tail undersides that are distinctly visible during its slow, undulating flight. The wing and tail undersides of yellow-shafted flickers are yellow, and this species has a red mark on the nape of its neck. Unlike most other woodpeckers, Northern flickers are principally ground feeders, though they also forage on tree trunks and limbs.

The **hairy woodpecker** (*Picoides villosus*) and **downy woodpecker** (*Picoides pubescens*, Fig. 2) live in areas with large deciduous and coniferous trees and orchards. Both species are found in suburbs and city parks. The hairy and downy woodpeckers look similar, being a striking black and white with a white stripe down the back. Size distinguishes these similar species, with the hairy woodpecker being larger overall and having a much larger bill than the downy. The downy woodpecker, which is only slightly larger than a sparrow, is more common than the larger hairy woodpecker. The males of both species have a small red patch on the back of the head.

The **pileated woodpecker** (*Dryocopus pileatus*, Figs. 5 and 6) lives in densely wooded areas and occasionally in forested urban areas. It is the largest woodpecker in North America (with the exception of the ivory-billed woodpecker, which is thought to be extinct in the United States). It is crow-sized, black and white, and has a bright red pointed crest.

The **acorn woodpecker** (*Melanerpes formicivorus*, Fig. 3) is seen in oak and oak-conifer woodlands throughout Oregon and in very local spots in southern Washington. It is a robin-sized woodpecker with a bright red cap that is more extensive on the male. The acorn woodpecker has a distinctive pattern of facial markings that give it the appearance of wearing a bandit mask.

The **red-breasted sapsucker** (*Sphyrapicus ruber*) occurs west of the Cascade mountains in Oregon and Washington and in the coastal areas of British Columbia. It is found in lowland parks and gardens and other areas with lots of trees, especially along streams. The adults are robin-sized and have a head and breast that is dull red. The **red-naped sapsucker** (*Sphyrapicus nuchalis*) occurs in similar habitats in eastern Oregon and Washington and south-central and southeastern British Columbia. It is black and white with yellow and red accents. The heads of both these sapsuckers are bright red and have a black stripe through and above the eyes.

Figure 2. The small downy woodpecker is a mix of black and white. Its wings, lower back, and tail are black with white spots; its upper back and outer tail feathers are white. Its underside is white, and its head is marked with wide alternating black-and-white stripes. Males have a red spot at the backs of their heads, which females lack. (Washington Department of Fish and Wildlife.)

Figure 3. Robin-sized acorn woodpeckers excavate shallow holes in dead trees and utility poles in which to store large quantities of acorns to feed their colony over the winter. These food-storage sites are called granaries, and are aggressively defended from scavengers. (Drawing by Elva Hamerstrom Paulson.)

Other Pacific Northwest woodpecker species include:

White-headed woodpecker
(Picoides albolarvatus)
Lewis's woodpecker
(Melanerpes lewis)
Three-toed woodpecker
(Picoides tridactylus)
Black-backed woodpecker
(Picoides arcticus)
Williamson's sapsucker
(Sphyrapicus thyroideus)
Yellow-bellied sapsucker
(Sphyrapicus varius)

male doing substantially more than the female. Complete excavation may take only a few days in soft wood, but averages 14 days.
• Eggs are laid on wood chips created during excavation of the nest.

Reproduction

• The breeding season for woodpeckers is from mid-February to July (depending on the species, elevation, and the north-south range), with young leaving the nest as late as mid-August.
• The time from start of nest building to departure of young is 50 to 70 days.
• Both male and female woodpeckers take turns incubating the four to eight glossy white eggs. Incubation time is around 11 days for the smaller woodpeckers, 18 days for pileated woodpeckers.
• Depending on species, the young leave the nest 20 to 30 days after hatching, but may stay with the parents for a month or more while they learn to forage.
• Individual woodpeckers return to the same area to breed year after year.

Viewing Woodpeckers

Even a novice bird-watcher can identify a woodpecker by its behavior. Only the much smaller brown creeper and red-breasted nuthatch spend so much time moving up and down on tree trunks. Woodpeckers can also be recognized by their undulating flight—wings flapping as the bird goes up and wings folded on the way down. Viewing opportunities are best when birds are given

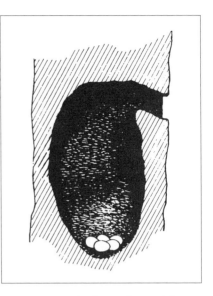

Figure 4. Woodpeckers will tunnel down 6 to 24 inches (15–61 cm) in a nest tree, making a wide bottom for the egg chamber. (Oregon Department of Fish and Wildlife.)

plenty of space. Observe woodpeckers from a distance and get your close-up by using binoculars, a spotting scope, a telephoto camera lens, or a strategically placed suet feeder.

Winter is a good time to watch woodpeckers. In the Northwest, most woodpeckers remain year-round, and when the leaves are off many trees the birds are more easily seen. Woodpeckers also cover more area in their search for food in winter.

The size of the nesting territory defended by a woodpecker is considerably smaller than the area it occupies at other times of the year. Still, the large pileated woodpecker may defend a nesting territory as large as 200 acres (80 h). The much smaller downy woodpecker defends an area of only 150 feet (46 m) around its nest tree.

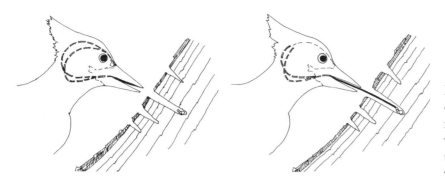

Figure 5. A woodpecker's tongue retracted (left) and extended (right). The exceptionally long tongue wraps around the skull and is anchored at the base of the bill. It is extended by a complex system that includes very long hyoid (tongue-base) bones. The tip of the tongue is barbed to help extract insects from holes, and the tongue is coated with sticky saliva. A pileated woodpecker is shown here. (Drawing by Jenifer Rees.)

Figure 6. Pileated woodpeckers excavate a rectangular hole that is 3½ inches (9 cm) wide and 5 inches (13 cm) tall. Nestling pileated woodpeckers are shown here. (Drawing by Elva Hamerstrom Paulson.)

Nesting and Roosting Sites

Look for signs of woodpeckers excavating areas for roosting and nesting sites in wooded areas where there are old, large trees that have some dead or rotting wood on them. Look for fresh wood chips on the ground below an excavation site.

The tapping involved in excavation consists of irregular pecking sounds that are not nearly as loud as territorial drumming. Once the hole is deep enough, the bird excavates from inside, and all you may notice are pecking sounds and bits of wood periodically being tossed out the entrance.

Small woodpeckers excavate nesting and roosting sites in tree trunks and limbs as small as 8 inches (20 cm) in diameter; the pileated woodpecker requires trees that are at least 12 inches (31 cm) in diameter. Depending on the species, entrance holes are from 1½ to 3½ inches (4–9 cm) in diameter and 6 to 70 feet (1.8–21 m) above the ground. Small entry holes may be camouflaged by surrounding lichens and moss. Pileated woodpeckers excavate a rectangular hole that is 3½ inches (9 cm) wide and 5 inches (13 cm) tall (Fig. 6).

Because the number of trees in urban areas that meet a woodpecker's requirements is often limited, they frequently excavate new holes in the same tree year after year. These trees are obvious when you come across them and are good places to look for woodpecker activity.

Feeding Sites

Dead and dying trees, known as snags, harbor many of the insects that woodpeckers eat. A popular feeding tree is obvious because of

Attracting Woodpeckers to Your Property

When managing space for woodpeckers around your property, the most important thing to do is to protect undisturbed wooded areas, particularly those that contain dead or dying trees. While larger trees may be more suitable as woodpecker housing, small trees rot faster and quickly attract insects that woodpeckers eat.

Provided they don't pose a hazard of falling on people, pets, or houses, leave any such trees for "woodpecker watching" whenever you can.

Other things you can do include:

• Install a suet feeder in winter (Fig. 8). Suet can be obtained neatly packaged from stores that cater to the bird-feeding public, and at farm supply centers and hardware stores.
• Plant sunflowers—popular seed sources in fall and winter.
• Install a nest box that is specially designed for woodpeckers (see "Nest Structures and Den Boxes" in Appendix F for resources and "Preventing Conflicts" in Chapter 38 for information on how to keep starlings out).
• Avoid using pesticides, especially insecticides.
• Leave ant colonies for flickers to find and harvest. (An Oregon biologist found over 2,000 ants in the stomachs of flickers.)
• Leave some fruit on orchard trees for woodpeckers to eat in late summer, fall, and winter.

Figure 7. Head-bobbing is the most common visual display of flickers, and is accompanied by a call. (From Stokes, A Guide to Bird Behavior.)

Figure 8. Suet or peanut butter can be placed in feeder holes that are one inch in diameter and one inch deep. Attach a tree limb to a lightweight skillet using outdoor wood screws and a threaded bolt. (From Link, Landscaping for Wildlife in the Pacific Northwest. Feeder design by Ken Short.)

the number of holes in it. On a snag, the outer surface of the bark is where bark beetles and ants are hunted; the underside of the bark is where the larvae and pupae of insects are found; the heartwood is where carpenter ants and termites are found and eaten by strong excavators such as pileated woodpeckers.

Pecking for insects consists of light taps in irregular rhythms—a sound entirely different from territorial drumming.

Offer suet at your feeder and woodpeckers can become regular visitors. In late winter, look for the male and female seen together. This indicates that courtship has begun. If you continue to offer suet through spring, the adults will bring their young to the feeder. (See "Preventing Conflicts" in Chapter 38 for information on how to prevent starlings from dominating a suet feeder.)

Drumming

Drumming refers to what happens when the tip of the woodpecker's bill rapidly hits any resonant surface, sounding much like an electronic drumroll. A typical drumming rate is 15 pecks per second. A woodpecker can peck faster at a hard tree trunk than a rotten one because the beak sinks more into soft wood, thus slowing down the attack.

Drumming serves roughly the same function as song does for other birds; lone males in late winter to midsummer may advertise for a female by drumming on posts around their range. Later, males and females will stay in close touch with each other by drumming. Woodpeckers' skulls are protected from the concussive force of pounding by a narrow space around the brain that functions as a shock-absorber.

Hearing drumming is often the best way to locate woodpeckers, since drumming is common and the sound carries well. Once you hear drumming, you know that their breeding season has begun.

Calls

A woodpecker's call is given during long-distance communication between a pair and during moments of alarm. The other member of the pair may answer it with the same call or by drumming. The following are common calls given by each species:

Northern flicker: a loud, high-pitched *wik-wik-wik-wik;* also a piercing, sharply descending *peeahr.*

Downy woodpecker: a high-pitched *pik* and a rattling *ki-ki-ki-k-ki.*

Hairy woodpecker: a loud, high-pitched *peek!*

Pileated woodpecker: a loud and penetrating *kuk-kuk-kuk-kuk-kuk-kuk.*

Acorn woodpecker: a loud *yak-up;* also a laughing *aka-aka-aka.*

Red-breasted sapsucker: a nasal, down-slurred *keerrr,* similar to that of the Eastern gray squirrel cry.

Visual Displays

Woodpeckers use various visual displays, including head-weaving and body-bobbing, during courtship and as signs of aggression toward intruders (Fig. 7).

The most active displaying occurs early in the breeding season, before nest-building, when the birds are pairing and there is competition for mates.

Preventing Conflicts

Woodpeckers that have been crowded out of their wooded territories will readily use alternative structures for **drumming,** seeking food, or excavating a cavity. So the flicker that awakens you in the spring, drumming loudly on a gutter or metal flashing, is making good use of the habitat you are providing. For this reason, don't remove dead or decaying

trees in the hope of driving woodpeckers away. That makes it more likely they may investigate your house for food or a nest site.

The following are descriptions of woodpecker activities and suggestions for ways to remedy problems. Success will depend on timing, the availability of food and shelter, and the woodpecker's previous exposure to the tactics suggested below. The homeowner will have to weigh the trouble and expense of control against the scope of the damage caused by the woodpeckers.

Drumming: A woodpecker drums to communicate with a mate or to proclaim its territory and attract a mate. This typically happens during the breeding season (mid-March to June), but may continue into July. For reasons that are not fully understood, drumming may also occur for a short time in the fall.

Drumming is the most common reason for woodpeckers to use buildings, and while it may be annoying, the bird's activity usually does not penetrate completely through wood siding.

Woodpeckers will return year after year to the same house because it works for them; they attract a mate this way. Thus, a woodpecker that has been using the same location for several years will be hard to move.

To discourage drumming, try a combination of the following strategies:

Scare the woodpecker by hanging strips of Mylar scare tape or floating Mylar party balloons in front of the area of woodpecker activity (Fig. 9). When using scare tape, strengthen each strip by attaching a piece of duct tape or

nylon packing tape to each end. Tack or nail one end to the outer end of the roof soffit, just under the gutter, and attach the other end to the side of the house. Before attaching the bottom, twist the tape 6 to 7 times and keep the tape loose enough to provide some slack. The slack and twisting are necessary to produce the shimmering effect. Apply these tape strands at parallel intervals of 2 to 3 feet (60–90 cm).

You can also try hanging aluminum pie pans horizontally along a rope or section of twine (Fig. 9). Run one end of the rope to a convenient window and fasten it to an object inside the house. Whenever you hear drumming, jerk on the string to make the pans move.

Some people have had success with stapling large rubber spiders in the vicinity of the drumming birds. Large, black rubber spiders are available from most party stores. The Birds Away Attack Spider® is vibration/sound activated and will respond to the drumming of woodpeckers, by dropping down on a "web" cord. Batteries then retract the spider back up the cord, where it waits for the next unsuspecting woodpecker to arrive.

Scaring the woodpecker by shouting and banging pans outside a nearby window may provide temporary relief. A squirt of water with a garden hose can have a similar effect. Again, woodpeckers living in urban areas likely will have grown accustomed to such noises and activity, and the results will be short lived. (***Note:*** Scaring woodpeckers away from an active nest is illegal.)

Create a barrier by covering or wrapping the gutter, down-

Figure 9. Scare the woodpeckers away from a drumming site by hanging strips of Mylar scare tape or aluminum pie pans or floating Mylar party balloons in front of the area of woodpecker activity. (Drawing by Jenifer Rees.)

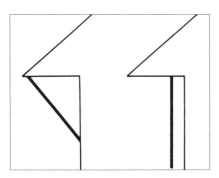

Figure 10. Prevent woodpeckers from accessing the side of a house by creating a barrier with a sheet, tarp, burlap, bird netting, or other material. (Drawing by Jenifer Rees.)

spout, or other drumming site with a sheet, tarp, burlap, or other material. A large area of siding can be protected by hanging a sheet, tarp, or bird netting from the roof gutter or eave (Fig. 10). Be sure to cover any ledges or cracks the bird uses as a foothold while drumming. If you cannot fasten the material to the gutter or eave, attach it to a board that has been temporarily fastened along the top of the wall.

If a single board on the house serves as a toehold, heavy monofilament fishing line or stainless steel wire can be tightly stretched approximately 2 inches (5 cm) above the landing site to prevent the woodpecker from perching.

Repel woodpeckers by applying a commercially available, nontoxic woodpecker-protective coating spray where activity is taking place. The spray exudes an aromatic and taste deterrent when pecked.

A note on where to get materials: Mylar scare tape, Mylar scare balloons, bird netting, and woodpecker repellents are available from farm supply centers, nurseries, and on the Web (see "Wildlife/Human Conflicts" in Appendix F or search for "bird control supplies.") Some pest-control companies sell a heavier netting with a larger mesh than common bird netting used to protect fruit. Such netting is not as likely to create problems for small songbirds, which sometimes get caught in the smaller mesh.

Seeking Food: Woodpeckers inspect tree trunks and branches for wood-boring beetles and other insects throughout the year. If a woodpecker's pecking is not restricted to one location on your house, and if it occurs any time of year, the bird is probably gathering insects, or their eggs or larva. Physical evidence of this behavior includes soft pecking in straight lines that result in dime-sized holes.

Once they have established a feeding pattern on a house, flickers, in particular, can be very persistent, and the holes they create may serve as visual attractants to other woodpeckers. So, it is important the get them to stop as soon as possible.

Note that the flickers may be doing you a favor by drawing attention to an insect infestation. As a temporary measure, you can create a barrier between the bird and food source by using one of the techniques described earlier.

For the long term, you'll need to control the insects if you have an infestation, and then make any necessary repairs or modifications with wood filler, caulk, or other materials. You may want to consult a licensed pest-control operator on how to remove the insects and eliminate future infestations.

Excavating a Nest or Roost Site: If you find a round opening—about the same width as the woodpecker that made it—in the siding or other boards, the woodpecker is probably excavating a cavity to nest or roost in. If an acorn woodpecker is excavating, it may be creating a food-storage site. Often the birds pull out insulation from between the walls and there may be evidence of this below the new hole.

In the spring or early summer, assume there is an active nest with eggs or chicks inside. Scaring woodpeckers away from an active nest is illegal. So, after the young birds have left the nest (generally by mid-June), immediately seal the opening to prevent starlings, house sparrows, squirrels, or other animals from using the cavity.

If a woodpecker has nested or attempted to nest in a wall, you might consider providing a nest box specially designed for a woodpecker as an alternative nest site. Flickers commonly use nest boxes. A nesting flicker may defend its territory and keep other woodpeckers away.

Public Health Concerns

Woodpeckers are not considered to be a significant source of any infectious disease transmittable to humans or domestic animals.

Legal Status

In **Oregon** and **Washington,** all woodpecker and sapsucker species are protected by the Federal Migratory Bird Treaty Act. A state permit and federal permit can be obtained to use lethal means to control woodpeckers and sapsuckers when extreme damage is occurring on private property. Such permits are only granted after all other nonlethal control techniques have proven unsuccessful. Contact your local state Fish and Wildlife for permit information.

In **British Columbia,** all woodpecker and sapsucker species are protected under the Federal Migratory Bird Convention Act and the provincial Wildlife Act. Any permit to lethally control woodpeckers or sapsuckers would need to be issued from the Canadian Wildlife Service and would likely only be issued in very extreme cases.

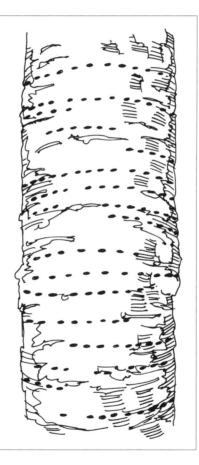

Figure 11. Sapsuckers get their name from their foraging strategy of drilling holes in tree trunks, and then coming back to those holes later to feed on the running sap and the insects attracted to that sap. (Drawing by Jenifer Rees.)

How about Them Sapsuckers

Unlike other woodpeckers, sapsuckers are not primarily interested in insects for food. Instead, they're looking for tree sap. They collect sap using their long, brush-tipped tongue as the sap flows out of the holes they've drilled.

The holes sapsuckers drill are about an eighth of an inch (3 mm) in diameter, and are evenly spaced up and down and around the trunk, appearing as if done by a machine (Fig. 11). Don't confuse sapsucker holes with holes created by insect borers. Borer holes are rarely as numerous as sapsucker holes and are randomly spaced.

In most cases, sapsuckers do not seriously harm trees. The holes are shallow and the wounds do not cause significant or permanent damage. But sometimes a particular tree becomes a favorite feeding place for an individual sapsucker. In this case, large areas on the trunk may be dotted with many holes. When this happens, the tree may be weakened and become more sensitive to other problems, such as disease or drought. The wounds themselves may attract harmful insects.

All the scare tactics, barriers, and repellents discussed above for woodpeckers will work to reduce or eliminate sapsucker damage to live trees. In addition, the affected area could be loosely wrapped with 1/4-inch (6 mm) hardware cloth. During winter months, sapsuckers may congregate in orchards and drill several trees. In these situations, it may be best to sacrifice one tree to the sapsuckers, to reduce damage to other unprotected trees.

PART 3

Reptiles and Amphibians

CHAPTER 41

Frogs and Toads

Frogs and toads, along with salamanders and newts, are members of the animal group called amphibians. Amphibians (from the Greek words *amphi,* meaning "both," and *bios* or "life") are fittingly named. Frogs and toads start their lives as totally aquatic animals with gills and a pronounced tail fin; this is familiar to many people as the tadpole stage. Over time, legs develop, the tail and gills are absorbed, and the frog or toad becomes a terrestrial, air-breathing animal.

The **Pacific treefrog** *(Pseudacris regilla)* is the smallest but most commonly seen and heard frog in the Pacific Northwest, and for that reason much of this chapter is devoted to it. For information on the other native species of frogs, see Table 1. For information on toads, see "Notes on Toads and Toadlike Creatures of the Pacific Northwest."

The Pacific treefrog is an adaptable species found from rain forests near sea level, to mountains at 11,000 feet (3352 m), and on into dry interior areas of the Pacific Northwest where water is available. Adults measure 2 inches (50 mm) in length and vary in color from a bronze brown to a light lime green, and from solid in color to intricate patterns. A sharply defined black mask extends from the tip of its snout to its shoulder (Fig. 1 and back cover).

The "song" or call of the male treefrog, designed to attract females, is a loud, two-part *kreck-ek,* or a *ribbit,* often repeated many times. This calling stimulates other males to join in, and large concentrations of these frogs can be heard far away, especially on nights when air temperatures remain above 45°F (7°C). Male treefrogs call mainly in the evening and at night, although they often call sporadically during the day at the height of the breeding season. A dry-land call made by male treefrogs away from their breeding ponds is a single-note *Krr-r-r-ek.*

When Hollywood moviemakers wanted frog calls to convey the feeling of nighttime outdoors, they recorded treefrogs. Consequently the "ribbit-ribbit" calls of this species have become the stereotypical frog call, even in areas where treefrogs don't occur.

Facts about Treefrogs
Habitat Needs of Adults

- Depending on location, treefrogs move into breeding sites from February (coastal areas) to July. Breeding sites include ponds, swamps, marshes, and roadside ditches—even puddles that dry up during the warm months.

- Outside of the breeding season, adult treefrogs inhabit a variety of habitats, including woodlands, meadows, pastures, and gardens—at times several hundred yards (meters) from water. *Note:* Ponds, swamps, marshes, and similar spots are used only a few weeks or months of the year; treefrogs spend the rest of the year in surrounding areas.

Figure 1. Individual treefrogs can change color between green and brown tones in a few minutes. This color change is related to the temperature and amount of moisture in the air, not to the background color as is the case for most reptiles. (Washington Department of Fish and Wildlife.)

- During dry periods and in arid areas, adult treefrogs are active only at night, spending the day in water or shaded vegetation, a rock or log crevice, rodent burrow, or other protected place.
- Treefrogs secrete a waxy coating from their skin glands that allows them to remain moist and travel far from water.

Food and Feeding Habits

- Toe pads on their front and hind toes enable treefrogs to climb in search of beetles, flies, spiders, ants, and leafhoppers. Adults have been seen and heard up in trees and outside windows two stories high.
- Adult treefrogs catch prey with their long, elastic-like, sticky-ended tongues.
- Treefrog tadpoles eat algae, decaying vegetation, and scavenge on dead earthworms, fish, or whatever else is available.

Reproduction and Life Cycle

- Male treefrogs are the first to move into their aquatic breeding areas and soon begin chorusing to attract females. Males chorus while floating at the surface or sitting partially submerged in shallow water.
- Females lay 400 to 750 eggs, which are externally fertilized by the male.
- Individual egg masses contain 10 to 75 eggs, measure 1 to 2 inches (2–5 cm) across, and contain a special jelly that swells up on contact with water.
- Egg masses are attached to sticks or grasslike vegetation below the surface, or may be on the bottom in shallow areas. Egg masses often become camouflaged with algae and sediment.

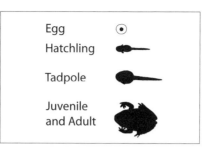

Figure 2. *The various life stages of a treefrog. (Adapted from Corkran,* Amphibians of Oregon, Washington, and British Columbia: A Field Identification Guide.)

- Eggs hatch more quickly in warmer water, with the time averaging two to three weeks.
- Tiny hatchlings soon turn into tadpoles with short, round bodies and eyes that bulge out at the sides of their heads (Fig. 2).
- In five to six weeks, tadpoles turn into ½-inch (13 mm) long, air-breathing juvenile frogs that climb onto land but eventually return, like their parents, to breed in water.
- Treefrog populations are notorious for dramatic year-to-year fluctuations. They may not breed at all if the rainy period of the year is too short, as happens in some droughty years.

Mortality and Longevity

- Treefrog eggs are eaten by caddisfly larvae and fish. Fungus and frost also kills some eggs.
- Treefrog tadpoles are eaten by dragonfly larvae, diving beetles, fish, long-toed salamander larvae, bullfrogs, garter snakes, and birds (herons, ducks, and jays).
- On land and at the water's edge, raccoons, foxes, coyotes, river otters, skunks, snakes, hawks, and owls eat adult treefrogs. Cats, children, lawn mowers, and vehicles all take their toll on adult treefrogs.

- The loss of wetlands and natural ponds eliminates breeding areas; chemicals from pesticides and runoff poison treefrogs and their food.
- Most treefrogs die at the egg or tadpole stage, hence having a life expectancy of only a few weeks. Treefrogs that reach adulthood live an average of two years in the wild.

Viewing Frogs and Toads

Frogs and toads tend to be more active at lower temperatures than snakes and lizards. As a result, they can be seen in the fall and early spring when most reptiles are in a hibernation-like state. On warm, rainy nights during spring and fall, search trails, roads, and other openings for adult frogs and toads on the way to or from breeding sites.

Examine ponds, swamps, marshes, and other bodies of fresh water from February through August for breeding adults, eggs, tadpoles, or juveniles. Slowly and quietly approach the area, using binoculars to detect animals visible along the waterline or the characteristic dual eye-bumps of frogs on the water surface.

Adult and juvenile frogs and toads are usually found at the surface, at the water's edge, or in moist vegetation along the shore. Recently metamorphosed juveniles often are found under objects around the edges of the water. Occasionally, after rains, they can be found moving into upland areas near breeding sites in large numbers. When examining areas under rocks, wood, and other material, always carefully replace the covering in its original position.

Table 1. Native Frogs of the Pacific Northwest

In the cooler, more moist areas of Oregon, Washington, and British Columbia, the frogs you are likely to hear or see are the treefrog (see text for information) and red-legged frog; in arid areas the dominant species are the treefrog, spotted frog, and Great Basin spadefoot (see "Notes on Toads and Toadlike Creatures of the Pacific Northwest"). The following are descriptions of these and other less common species:

The **red-legged frog** (*Rana aurora*) is fairly common west of the Cascade Range in Oregon and Washington and west of the Coast Range crest in British Columbia. It inhabits moist forests near cool ponds, lakes, and slow streams, especially where aquatic vegetation provides cover. During the nonbreeding season this frog may be found several hundred yards (meters) or more away from permanent water. It can be as much as 4 inches (10 cm) long and gets its name from the deep wine-red (burgundy) undersides of its legs, thighs, and portions of the belly.

The call is a series of five to seven quiet, low-pitched notes—*uh-uh-uh-uh-uh*. Males often call from under water and the call is barely audible when made above water.

The **foothill yellow-legged frog** (*Rana boylii*) is found in southwestern Oregon, where it prefers gravelly or sandy streams with sunny banks and open woodlands nearby. It is 3 inches (7.6 cm) long, gray to brown to olive above, with yellow under its legs and lower abdomen. Its raspy call of four to five rapid notes is rarely heard.

The **Columbia spotted frog** (*Rana luteiventris,* Fig. 3) occurs east of the Cascade mountains in Oregon and Washington and throughout most of British Columbia, except the coastal islands. It inhabits marshy edges of permanent ponds, lakes, and streams. Adults are 3 inches (7.5 cm) long and light to dark brown, gray, or olive green, with dark spots on their back, sides, and legs. The undersides of their legs are bright red, salmon, or orange. Their upturned bright-yellow eyes are characteristic of this frog.

It is uncommon to hear this frog call, but it may allow you to observe it closely. Its call is a rapid series of 5 to 30 faint, low-pitched, hollow notes that sound like a distant woodpecker tapping on hard, resonant wood.

The species formerly known as the "spotted frog" has recently been separated into this species and its near relative, the **Oregon spotted frog** (*Rana pretiosa*). The Oregon spotted frog is similar in appearance and occurs in a handful of localities in the Puget Sound lowlands and southern Cascade mountains through Oregon.

The **Cascades frog** (*Rana cascadae*) occupies mountain meadows and moist forests above 1,500 feet (457 m) in Oregon and Washington. This frog breeds in marshy bogs or ponds, often along the margins of slow-moving mountain-meltwater streams. It is 2½ inches (6.4 cm) long and olive to brown, usually with black spots on its back and legs. The undersides of the adults' legs are yellow. Its call is a series of low-pitched clucking noises.

The **Northern leopard frog** (*Rana pipiens,* Fig. 4) occurs only in a few areas in eastern Oregon, Washington, and southern British Columbia. Their populations have declined dramatically and they have been wiped out in much of the Pacific Northwest. This frog reaches 4 inches (10 cm) in length and is easily recognized by the large, dark spots with pale borders on its back, sides, and legs. Its call lasts several seconds and consists of a series of low-pitched snorts or grunts that have been compared to the puttering of a small motorboat.

Figure 3. The Columbia spotted frog occurs east of the Cascade mountains in Oregon and Washington, and throughout most of British Columbia except the coastal islands. (Washington Department of Fish and Wildlife.)

Figure 4. The native Northern leopard frog populations have declined dramatically and this frog is now absent in much of the Pacific Northwest. (Washington Department of Fish and Wildlife.)

Notes on Toads and Toadlike Creatures of the Pacific Northwest

Three species of toads occur in the Pacific Northwest, although none are very common:

The **Western toad** (*Bufo boreas*, Fig. 5) is warty and brown or green above, with a light back stripe down its middle. Females may reach 5 inches (13 cm) in length; males are slightly smaller.

Provided a source of water is nearby to breed in, the Western toad inhabits a variety of forested, brushy, desert, and meadow areas. It occurs throughout Oregon except for parts of the Coast Range and most of the Willamette Valley. In Washington it is found in all regions except for the most arid; the Western toad occurs throughout most of British Columbia. This toad is increasingly rare and completely gone from areas where it used to be common. The reasons for local declines remain unknown.

The Western toad may be active in the daytime when nights are cold, and during warm wet weather, but it is mostly nocturnal. Adults retreat from dry conditions or high temperatures to rodent burrows, spaces under logs and rocks, or by partially burying themselves in sand or soil.

Adult toads eat flying insects, spiders, earthworms, and other invertebrates. Tadpoles feed on algae, rotting vegetation, and occasionally animal matter.

Western toads breed later than do other native amphibians. (Only bullfrogs, which are not native, breed later than Western toads.) Their jelly-covered eggs are laid in pearl-like parallel strings. These strings are usually many feet long, contain thousands of eggs, and are threaded throughout vegetation in shallow ponds, irrigation canals, and similar places, and less often in deeper ponds.

The eggs hatch in three to ten days depending on water temperature. The resulting tadpoles are highly social and schools of a thousand or more may be observed swimming and feeding together. Juvenile toads live in marshes and forests where they can hide among logs, grasses, and shrubs; they live less often along the edges of streams.

Because Western toads lack the amplifying vocal sac of many of their toad relatives, the call of adult male toads sounds like the peeping of crickets or baby chicks.

The similar-looking **Woodhouse's toad** (*Bufo woodhousei*) is the most widely distributed toad in the United States; however in our area only isolated populations are found, primarily along the Columbia and Snake Rivers in Oregon and Washington and in eastern Morrow County in Oregon. There is little known about the life history or current status of this toad in the Pacific Northwest.

The **Great Basin spadefoot** (*Scaphiopus intermontanus*) is not considered a true toad. It inhabits sagebrush, bunchgrass prairie, and open forests, and breeds in a variety of temporary waters including rain pools, roadside ditches, and small ponds.

Spadefoots occur in eastern and central Oregon and Washington and south-central British Columbia. This 2- to 3-inch-long (5–8 cm) amphibian is gray or olive-gray, with dark brown or reddish spots and light stripes on its sides and back. It has a conspicuous black "nail" at the base of the first toe of each hind foot.

Spadefoots have extremely quick development times. Their eggs occur in soft, randomly shaped, grape- to plum-sized clusters and hatch in only two to three days in water that may reach 90°F (3°C).

Spadefoots are usually active above ground at night only in the moist spring months. They use their spade feet for digging into sand or soft soil, making hideouts that allow them to survive cold, hot, or dry periods. They eat ants, beetles, grasshoppers, and flies. Their skin secretions may cause humans to sneeze. Their ducklike calls, *waaah* and *kwaah-kwaah-kwaah,* can be heard from a great distance.

Horned toads are actually lizards; see Chapter 43.

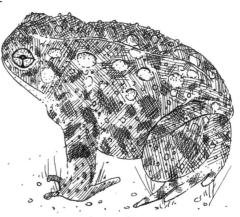

Figure 5. The Western toad is the largest and most widely distributed toad in the Pacific Northwest. (Washington Department of Fish and Wildlife.)

Tips for Frog and Toad Watchers

Here are some hints for frog and toad watchers heading into the field:

- If you frog-watch alone, let someone know where you are going and when you plan to be back.
- Don't wade out into ponds or marshes—you will scare the frogs, and there is the risk that you may slip and fall into the water.
- Keep a close eye on children.
- Have a change of clothes just in case you get wet.
- Watch for stinging insects. If you have an allergy to stings, be absolutely sure you have your sting kit with you.
- Use polarized sunglasses and binoculars for better viewing.
- If you are frog-watching on private land, ask permission from the landowner.
- Make sure dogs are leashed— better still, leave them at home.

Breeding sites for frogs and toads are often found in the warmest water locations, including water within or adjacent to thin-stemmed emergent plants such as rushes, sedges, and aquatic grasses. Tadpoles are found near large pieces of wood or rocks that store heat. Likewise, they tend to cluster in the warmest water they can find—often in shallow areas with lots of sun exposure.

Most frogs and toads start to call about half an hour after sunset. Calling also occurs during dark and rainy days. When listening to identify a frog or toad species, remember that the calls are generally slower in colder temperatures, and faster when it is warm. (See individual species descriptions for specific call information.)

Handling frogs and toads can be hazardous to their health and needs to be done carefully—or not at all. Their permeable skin could absorb harmful chemicals from your hands, such as lotion or bug repellent. Amphibians can also die from moisture loss, known as desiccation. Handling them increases this risk. (See "Collecting and Releasing Amphibians and Reptiles.")

Attracting and Maintaining Amphibians on Your Property

Because their total habitat requirements extend from water onto the land, it is difficult to maintain wild amphibians in most yards. If your yard is surrounded by concrete and highly maintained landscapes, chances are slim that these interesting animals will visit.

The chances of hosting frogs, toads, and other amphibians increase if your property adjoins an undeveloped area, such as a greenbelt or other wild area, or if it is next to a wetland, stormwater retention pond, or other freshwater area. Your chances further increase if you establish and maintain a natural landscape. Treefrogs are very good at colonizing new areas.

To attract and maintain amphibians on your property:

Protect existing natural areas to the greatest extent possible. Protect woodlands, wetlands, meadows, stream corridors, shorelines, and other wildlife habitat on your property; encourage your friends and neighbors to do the same. Support public acquisition of greenbelts, remnant forests, and other wild areas in your community. Write to legislators and attend public meetings when regulations are being considered.

Protect buffer areas next to streams, lakes, or ponds. The vegetated buffers surrounding these areas protect the ecological functions and value of the breeding habitat, and provide needed upland habitat for amphibians.

Wherever possible, protect migration paths between uplands and breeding sites. If amphibian migrations to breeding sites cross neighborhood roads, try placing signs to inform local drivers of this crossing. If a new road is to be constructed in migration areas, work for installation of amphibian crossing structures, such as small tunnels under the roadway. Amphibian movements can also be guided by means of drift fences and large logs. If you have an area on your property that is used by migrating amphibians, leave the area as natural as possible.

Leave a portion of your grass unmowed, especially in areas that

Figure 6. Retain stumps, logs, rootwads, rock piles, and other debris that provides a cool, moist habitat for amphibians. (From Link, Landscaping for Wildlife in the Pacific Northwest.*)*

adjoin a wet area, forest edge, or any other distinct habitat, as well as any area that is being used by migrating amphibians. If you must mow in these areas, mow at slower speeds and be ready to step on the clutch or brake. Set the mower blades as high as possible, or use a weed-whacker and leave grass 6 inches (15 cm) high.

Regularly mow any areas you want to keep as lawn to prevent longer grass developing where frogs may hide. Mowing in hot, dry weather will minimize the chances of finding amphibians, and making some disturbances before mowing will encourage frogs to hop out of the way. Don't mow or weed-whack when many amphibians are seen during breeding migrations or juvenile dispersal periods.

Preserve leaf litter under trees and shrubs. Such material provides cover and moisture; it also attracts organisms that amphibians eat.

Retain stumps, logs, rootwads, rock piles, and other debris that provides a cool, moist habitat for amphib-ians. Such habitat features provide much needed cover. All these can be strategically located as "stepping stones" across exposed areas, or to bridge gaps between breeding ponds and woods. To be effective in exposed areas, keep the structures within 15 feet (5 m) of each other.

With permission from landowners, you could salvage these materials from cleared or logged areas and install them in your landscape, preferably away from busy roads (Fig. 6).

Consider building a pond. Treefrogs will breed in almost any type or size pond. Toads do not seem to be directly affected by fish. If your pond is aimed at amphibians in general, however, it is critical not to stock fish, given the problems with predation and the increased nutrient load that may result. For fish lovers, one option is to build a separate pond specifically for fish. See "Ponds" in Appendix F for references on how to construct ponds.

Collecting and Releasing Amphibians and Reptiles

If amphibians or reptiles are not present in your yard, there is probably a good reason. Generally, the conditions surrounding your property are not right for them. Attempting to move an amphibian or reptile into your yard when conditions are not appropriate for it will probably result in killing the animal.

Another reason for not relocating these animals is their well-developed homing instinct. Many will immediately leave an unfamiliar area and, drawn by instinct, try to return to their place of origin. This usually results in their being killed on roads or by predators they are exposed to on their journey.

If you enhance your yard for reptiles and amphibians, and if they occur naturally in your general vicinity, sooner or later you will probably enjoy seeing them on your property. Meanwhile, work with the wildlife species that do visit your property and help preserve nearby wild areas.

Many species of reptiles and amphibians are popular as pets. Unfortunately, owners often tire of them and release them (illegally) to fend for themselves (see "Legal Status"). Many of the pet-store species cannot survive our climate and will die if released. On the other hand, some species, such as the red-eared slider turtle, the snapping turtle, and the bullfrog often do survive. They may introduce new diseases or compete with native reptiles and amphibians.

Figure 7. The male treefrog amplifies his voice with a resonating throat sac he blows up to three times the size of his head. (Washington Department of Fish and Wildlife.)

Fence large ponds to prevent access by livestock, to protect the water, and to allow a more diverse plant community to grow, providing cover for amphibians and their food source.

Avoid using pesticides and herbicides. Amphibians have highly permeable skin that can absorb toxic chemicals from your lawn, and they can be poisoned directly or indirectly through their food, such as slugs and snails. Moss-killers and roof treatment chemicals can also be toxic, and often such runoff is directly channeled into wetlands via pipes or sewer outflows.

Preventing Conflicts

The noise of one or more treefrogs may be annoying to some people, particularly light sleepers (Fig. 7). Complaints also arise when large numbers of young toads leave ponds en masse; however, the toadlets disperse quickly and the perceived problem takes care of itself in a few days.

If for some reason the presence of frogs or toads cannot be tolerated, or to exclude them from an area where they might be injured or killed, install a barrier. *Note:* A fence likely will not keep

treefrogs out. On land, create a 20-inch (50 cm) high barrier using ¼-inch (6 mm) mesh hardware cloth buried at least 4 inches (10 cm) in the ground. Secure the barrier to posts using wire or hog rings.

The same material (or better yet, aluminum hardware cloth) can be used to surround a pump or other mechanical device within the pond.

To prevent frogs (including treefrogs) and toads from breeding in a small pond, keep it empty, or keep a waterfall or other source of turbulence in place to discourage frogs from using it during the breeding season. Egg masses can also be removed, as can adults. Or just add fish! (See below for information on managing bullfrogs.)

Repellents and scare devices do not work on frogs or toads. No poisons should ever be added to

the water to keep these animals out.

Bullfrogs: The **bullfrog** (*Rana catesbeiana,* Fig. 8) has successfully spread throughout the low elevations of the Pacific Northwest. Large populations of this species are believed to have contributed both directly and indirectly to the drastic decline of native amphibians and reptiles.

The bullfrog is the largest true frog in North America. It can measure 8 inches (20 cm) in length, leap up to 3 feet (90 cm), and live nearly ten years. Bullfrogs are occasionally seen crossing roads, even during periods of dry weather, and may travel overland up to a mile (1.6 k). This movement allows them to expand their range from the source where they were introduced. The large number of eggs in each egg mass laid

Figure 8. Bullfrogs, and perhaps other non-native frogs, are expanding their range as individuals colonize suitable habitats—often using constructed stormwater ponds as stepping-stones between natural wetlands. Bullfrogs also spread to new habitats when released as unwanted pets and after people share frog eggs with their neighbors, not knowing the damage they can do. (Washington Department of Fish and Wildlife.)

by the females allows bullfrogs to quickly establish themselves within a new territory.

Bullfrogs get their name from the *bar-room* mating call made by the males. Juveniles and adults of both sexes emit a squeak just prior to jumping into water when avoiding an intruder.

The original native range of bullfrogs was the eastern United States, but they have been introduced to most of western North America, from southern British Columbia to Baja and on into parts of South America (e.g., Brazil), as well as Hawaii, the Caribbean Islands, Japan, England, Germany, and Italy. Bullfrogs were first introduced into the Pacific Northwest during the Great Depression (early 1930s) to provide opportunities for frog hunting, food (i.e., frog legs), and stock for frog farms, enterprises that rarely succeeded.

Bullfrogs thrive in the warm waters of natural and man-made ponds, marshes, sloughs, reservoirs, and sluggish irrigation ditches and streams. Bullfrogs tolerate polluted and muddy waters better than do most native frogs, and may be found within cities in wetlands, reservoirs, and stormwater ponds.

In their northern range and in cooler climates, bullfrogs persist only in year-round bodies of water because they require two years to develop from eggs into adult frogs. In their southern range and in warmer areas, they have been known to fully metamorphose in one year and colonize semipermanent and seasonal ponds. (Such cases have been documented by biologists in Oregon.)

Bullfrogs breed only after the nights warm up and reach the high 60s and 70s (Fahrenheit), generally

Figure 9. Largemouth bass (Micropterus salmoides)*, a species introduced from parts of the eastern half of the United States and Canada, is reported to prey on bullfrog tadpoles and adults. (Oregon Department of Fish and Wildlife.)*

June and July here in the Pacific Northwest.

Adult bullfrogs and tadpoles overwinter in mud on the bottom of ponds and other bodies of water. They hibernate by burying themselves in surface mud or by digging cavelike holes underwater. Adults also hibernate on land near ponds where they bury themselves within the soil. Their body temperature may drop virtually to the freezing point, and their hearts slow so drastically they seem to stop altogether. But they continue to absorb oxygen through their moist skin, and when their surroundings thaw, they emerge into the spring sunshine to resume their business of catching insects and other prey.

Adult bullfrogs usually are "sit and wait" predators that readily attack almost any live animal smaller than themselves—insects, frogs, tadpoles, fish, small snakes, turtle hatchlings, newts, salaman-

ders, bats, hummingbirds, and ducklings. Bullfrogs use their sticky tongues to subdue prey, but that's not their only method of securing food. Large frogs are more likely to lunge at their targets. Once they get a grip with their wide, sturdy jaws, they use their front feet to shove the items down their gullets.

Garter snakes regularly catch and eat bullfrog tadpoles and adults. Painted turtles also eat some in late summer, when adult and developing bullfrogs become sluggish for some unknown reason. Large bullfrogs also capture smaller ones and eat them.

The relatively unpalatable nature of bullfrog tadpoles may give them the ability to coexist with many otherwise potential fish predators. Largemouth bass, however, are reported to prey on tadpoles and adult bullfrogs (Fig. 9). Bullfrog eggs can be eaten by many predators (leeches, salaman-

Bullfrog Identification

Each female bullfrog creates one thin-jelly **egg mass** that may contain 6,000 to 20,000 very small eggs, which are black on top and white underneath. Egg masses are generally found in water that is less than 2 feet (60 cm) deep in mid to late summer. The eggs start out as a round, basketball-size mass (below or near the surface) that then rises, flattens out, and forms a 2- to 4-inch (5–10 cm) gelatinous mass 2 feet (60 cm) in diameter. The egg mass floats on the surface of the water or rests on the bottom within sparse vegetation. The mass remains attached to deep vegetation in some places and is often covered in algae.

The **tadpoles** are dark green with black dots, orange or bronze eyes, and opaque yellow underbellies. A two-year-old tadpole may be 4 to 6 inches (10–15 cm) long.

The **juveniles** are green to brown with a peppering of tiny black spots, and have orange or bronze eyes. A fold of skin extends from the eye around the eardrum.

Adult bullfrogs have thickset bodies, large, exposed eardrums, and are green, tan, or dark brown above (with dark spots). Male bullfrogs have a yellow throat. The eardrums on males are larger than the eyes, while the female's eardrums are the same size as the eyes (Fig. 10). The eyes of both sexes are gold.

Figure 10. The eardrums on male bullfrogs are larger than the eyes, while the female's eardrums are the same size as the eyes. (Washington Department of Fish and Wildlife.)

ders, fish) with no obvious detrimental effects to the predator.

Under no circumstances should you take or purchase bullfrog tadpoles for your home pond, transfer wild-caught bullfrogs, or in any way encourage them to expand their range. If you are adding plants or water to a small pond, make sure you are not also adding bullfrog eggs or tadpoles.

Managing Bullfrogs: The removal of bullfrogs is unlikely to be a viable management option in most wild or semiwild situations owing to the difficulty of removing all bullfrog eggs, tadpoles, and adults, and preventing surrounding bullfrogs from invading the water body. However, in a small wetland or pond it may be possible to eliminate the local bullfrog population by visiting the pond daily through the breeding season and removing all eggs, tadpoles, and adults. Long-term success depends on closely monitoring the pond to prevent other bullfrogs from breeding.

Bullfrog control techniques should be limited to those that cause the least harm to native amphibians. At least one person should be able to identify all stages of native amphibians when attempting to manage bullfrog populations. Foot traffic in areas where many juvenile toads, red legged frogs, or other species are about should be minimized.

Adult bullfrogs are difficult to gig or catch in nets because they are very wary and leap for the water at first approach. However, when they stare at a bright light at night, they seem unable to see a hand, net, or frog gig reaching

out to grab them. Wear a headlamp to keep your hands free, or have someone next to you spot the bullfrogs using a powerful flashlight. Some people have found capturing adult bullfrogs in turtle hoop-nets to be a useful management technique.

Shooting adult bullfrogs using a single-shot .410 shotgun has been successful. Alternatively, a pellet gun or a .22-caliber rim-fire rifle and dust-shot bullets can be used at close range. Use only dust shot in the .22 rifle, not conventional ammunition. Because all shot has the potential to ricochet, exercise extreme caution when discharging firearms on or near any water body.

Because adult bullfrogs are less active in cooler water, usually March and April are the best months for catching or shooting them.

Bullfrog tadpoles are equally difficult to catch because they quickly swim to deeper water to avoid capture. However, capture is made easier if the water body in which the tadpoles (and adults) live and breed is lowered with a pump or by another means to make them more accessible. Tadpoles are best captured using a long-handled dip net (not a fish net).

Tadpole collection should be done when nontarget species (especially native amphibians) are unlikely to be caught. For instance, in early September, when native frogs and salamanders occupying the site have turned into their juvenile stage and the only tadpoles left are bullfrog tadpoles.

Bullfrog eggs should be carefully collected in a large dip net or a 5-gallon bucket to avoid breaking up the egg masses. Clipping of vegetation may be necessary to dislodge the egg masses.

Euthanasia of adult bullfrogs and tadpoles can be by a blow to the head or the two-stage process of refrigerating the animals for an hour to slow down their respiratory system and then freezing them. The same can be done for the eggs, or they can be left a distance away from the water body to be eaten by a bird or other animal.

Since bullfrogs generally require two years to develop, it may be possible to rid an area of these animals by draining the water body they inhabit. This assumes that there isn't another nearby water body that can serve as temporary harborage. (Where ponds have dried up naturally, adult bullfrogs have been known to seek refuge in nearby wells, springs, animal burrows, and crevices in the ground.)

Public Health Concerns

Toads cannot jump as quickly as frogs do, so adult toads have evolved the defense of having toxic skin secretions that contain bufotoxins to protect themselves from predators. The toxins are also found in their eggs and tadpoles. The poison can be released during handling, and appears as a milky-white liquid coming from the glands behind the eyes or warts on the back of the toad. Handling toads and exposure to this skin secretion usually does not cause a problem in humans, but is an irritant to predators such as cats, raccoons, and garter snakes. As a general precautionary measure, however, wash your hands after handling any amphibian.

Toads do not give warts to people who touch them.

Legal Status

Because legal status about frogs and toads may soon change, contact your local, state, or provincial wildlife office, or visit their Web site for updates. (See Appendix E for contact information and where to access the state and provincial laws mentioned below.)

Oregon: All native frogs and toads are protected and may not be captured and/or held in possession (OAR 635-044-0130) unless a scientific collecting permit or other authorization has been obtained from Oregon Department of Fish and Wildlife (ODFW). Two exceptions include the Great Basin spadefoot and the Pacific treefrog. Although there are no restrictions on taking of these two species, they must be cared for in a humane manner (OAR 635-044-0132), and cannot be sold or purchased (ORS 498.022).

Bullfrogs may be hunted or angled without a license. No person may import, purchase, sell, barter, or exchange, or offer to import, purchase, sell, barter, or exchange live bullfrogs (OAR 635-056-0070 (1.c.)). Individual bullfrogs may be collected from the wild and held indoors in an escape-proof aquarium (OAR 635-044-0035), but they may not be released without authorization by ODFW (OAR 635-056-0070 (1.c.)).

It is illegal to release any non-native frog into the wild (ORS 498.052).

Washington: The Northern leopard frog and Oregon spotted frog are state endangered species (WAC 232-12-014) and cannot be

hunted or trapped. All other species of frogs and toads are unclassified and can be controlled without a permit.

The bullfrog is classified as a prohibited aquatic animal species (RCW 77.60). No license is required to hunt bullfrogs, there are no bag limits, and the season is open year round.

It is unlawful to import into the state, hold, possess, offer for sale, sell, or release all frog and toad species into the wild without the proper license to do so (WAC 232-12-064).

British Columbia: Red-listed frog species include the Rocky Mountain tailed frog, Northern leopard frog, and the Oregon spotted frog. Blue-listed frog species include the coastal tailed frog, Great Basin spadefoot, and the red-legged frog. (See "Legal Status" in Chapter 1 for listing information.)

The bullfrog and green frog are listed under Schedule "C" and may be captured or killed anywhere or at any time in the province. No permit is required, but a person must have land owner or occupier permission on private property.

Importation of all species of the family Ranidae (true frogs), and all species of toads is not permitted except under permit through an educational institution or a scientific organization, provided the Director of Wildlife is satisfied that the importation will not be detrimental to native wildlife or wildlife habitat (Wildlife Act, Permit Regulation, Section 7).

CHAPTER 42

Salamanders

Although the cool, moist areas of the Pacific Northwest may be home to few reptiles, they support nearly 20 species of salamanders (Table 1).

Because salamanders and lizards have similar body forms, they are often confused. However, salamanders are amphibians that have thin, moist skin, are slow moving, and are active at night in cool, moist places. Lizards are reptiles, with thick, dry skin covered with scales. They move quickly, and are active during the day mostly in warm, dry habitats.

Figure 1. Adult roughskin newts (a type of salamander) are active day and night, especially during rainy periods. Large numbers can be seen crossing roads to and from breeding ponds in the spring and fall. (Washington Department of Fish and Wildlife.)

Facts about Salamanders

Food and Feeding Habits

- Adult salamanders eat whatever prey they can overpower and fit in their mouths—slugs, worms, spiders, earwigs, crickets, ants, amphibian eggs, tadpoles, and small fish.
- Due to their size, adult Pacific giant salamanders are able to include frogs, shrews, mice, and other salamanders in their diet.
- A salamander's food must be swallowed whole, as they do not have cutting or crushing teeth.
- Salamander larvae (e.g., tadpoles) and juvenile salamanders feed on a variety of aquatic invertebrates. They also eat salamander larvae—including their own kind. Cannibalism may help to keep the population in check when food reserves are limited.

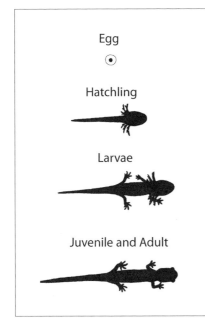

Figure 2. Following the hatching of the egg, the transition of the aquatic, swimming salamander larvae into the terrestrial form is called metamorphosis. (Adapted from Corkran, Amphibians of Oregon, Washington, and British Columbia: A Field Identification Guide.)

Figure 3. Land-breeding salamanders lay their eggs in the absence of free water. They complete their larval stage within the egg and hatch as miniature adults. (Adapted from Corkran, Amphibians of Oregon, Washington, and British Columbia: A Field Identification Guide.)

Reproduction of Water-breeding Salamanders

- Depending on the species and their location, water-breeding salamanders migrate from terrestrial areas to aquatic breeding sites from early winter to early spring.
- Breeding sites include swamps, marshes, lakes, ponds, puddles in meadows, and the shallow parts of streams with little or no flow.
- Typically, male salamanders deposit a sperm packet and females pick it up for internal fertilization. The female later attaches her eggs to submerged grasses and other aquatic plants, branches, or rocks.
- The eggs are surrounded by a special jelly that swells up on contact with water, and the egg mass may be larger than the salamanders themselves.
- Depending on the species and site conditions, the eggs hatch

after a few weeks, and metamorphose from the larvae into adults over a couple of weeks to several years (Fig. 2).

Reproduction of Land-Breeding Salamanders

- Land-breeding salamanders don't lay eggs in water; instead, the female lays her eggs in a rodent burrow or under a log, board, or stone, usually in spring.
- Females brood the cluster or strand of white eggs, keeping them moist with their skin secretions.
- The eggs hatch in late summer or early fall, and the juvenile salamanders emerge with the fall rains (Fig. 3).

Mortality and Longevity

- Many factors contribute to the overall decline of salamanders, some of which are still poorly understood. Loss of swamps, marshes, natural ponds—and the buffer areas surrounding them—eliminates breeding areas, while pesticides poison food supplies. Ozone depletion may also contribute to egg mortality.
- In water, snakes, fish, frogs, crayfish, and other salamanders eat salamander eggs and larvae.

Bullfrogs and bass—and other introduced fish species—are capable of eliminating an entire population of salamanders in a pond or small lake.
- On land, shrews, opossums, raccoons, and skunks eat salamanders. Towhees and other birds that scratch in the forest floor also eat them.
- Little is known about the longevity of salamanders. Ensatinas, a type of land-breeding salamander, have lived ten years in the wild.

Viewing Salamanders

Although newts and salamanders are often abundant, they are overlooked because of their slow speed and secretive habits. Because the life requirements of about half the Pacific Northwest species change with the seasons, you will have to search out both aquatic and terrestrial habitats to find them.

On warm nights throughout the rainy season, look along trails, roads, and other areas for adult salamanders on their way to or from their breeding sites.

During the day, look for egg masses of the aquatic breeding salamanders (see Table 1 for descriptions and times of year). Look for adult salamanders resting on the bottom of a pond or a clear stream pool, and under rocks and pieces of wood in the grass at the edge of a wetland. Avoid times of rain, high winds, or overcast conditions because these reduce visibility at and below the water surface.

Land-breeding salamanders conceal their eggs and the female often tends or guards them. The potential for accidentally destroy-

Table 1. Selected Pacific Northwest Salamanders

Aquatic Salamanders

Aquatic salamanders usually have a larval (tadpole-like) aquatic stage and an adult land stage (Fig. 2).

The **Northwestern salamander** *(Ambystoma gracile)* occurs in moist lowland forests westward from the crest of the Cascades in Oregon and Washington, and west of the Coast Range in British Columbia. Adults measure 6 to 9 inches (15–23 cm) in total length, and are gray-brown to chocolate-brown, with large swollen glands behind the eyes. Since they spend most of their lives in rodent burrows, under forest debris, and in rotting logs, adult Northwestern salamanders are seldom seen unless migrating to or from a breeding site. When threatened, they strike a warning posture (see "Salamander Defense Systems").

Northwestern salamanders attach their egg masses to thin-stemmed aquatic plants, 5 to 12 inches (13–30 cm) below the water surface. Egg masses have 40 to 150 visible brown eggs surrounded by thick Jell-O-like capsules. A single egg mass is approximately the size of an orange. Its greenish color is attributable to a symbiotic algae that lives with and provides oxygen to the developing embryos. Egg masses sink when removed from vegetation, so don't remove them!

Look for eggs and larvae in deeper water of permanent wetlands, areas they favor because larvae require a full year to metamorphose into terrestrial adults.

Both larvae and adult Northwestern salamanders are mildly toxic and presumably distasteful. Thus, they are one of the few salamander species that continue to be found in large ponds and medium-sized lakes that contain introduced bullfrogs and fishes.

The **long-toed salamander** *(Ambystoma macrodactylum)* is the widest-ranging salamander in the Pacific Northwest, occurring in lowland forests, cold mountain meadows, pastures, residential greenbelts, and sagebrush areas. Adults measure 5 to 6 inches (13–15 cm) in total length, are brownish-black above, and have a gold-yellow or green back stripe that may be made up of blotches. Adult long-toed salamanders are especially active above ground during rainy periods.

The long-toed salamander is an early breeder, and egg laying may occur in January and February in mild lowland areas. Look for eggs in large puddles or shallow ponds with no fish or frog predators. Eggs are attached to grasslike vegetation just off the bottom, or attached to the bark of submerged trees, fallen leaves, or other detritus along a pond bottom. The eggs can be difficult to find because they are laid singly, or in clutches of 3 to 25, and may be camouflaged by sediment.

Eggs hatch in a couple of weeks in favorable conditions, and larvae generally metamorphose the first year.

The **tiger salamander** *(Ambystoma tigrinum,* Fig. 4) occurs in scattered locations areas east of the Cascade mountains in Oregon and Washington, with the largest populations in northeast Washington and south-central British Columbia. They also occur in the Columbia Basin of central Oregon and Washington.

Adult tiger salamanders measure 13 inches (33 cm) in total length, are heavy bodied, and have olive or pale-yellow spots, bars, or blotches between black markings on their backs. They spend most of the year in underground burrows, and are active at the surface mainly at night and during or after spring rains when migrating. When threatened, an adult will arch its long tail over its head (see "Salamander Defense Systems").

Figure 4. Tiger salamanders are active at the surface mainly at night and during spring rains as they migrate to and from breeding grounds. (Washington Department of Fish and Wildlife.)

Table 1. Selected Pacific Northwest Salamanders *(continued)*

Tiger salamanders breed in late spring in warm ponds (either permanent ones or those that retain water until midsummer) or shallow lakes. Eggs are laid in water that is less than 3 feet (90 cm) deep, in clusters attached to a branch, stem, or rock. Most larval tiger salamanders metamorphose into adults the first year; in permanent water bodies free of predators, metamorphosis may be delayed for several years.

In the past, tiger salamander larvae were harvested and sold as fish bait called "water dogs."

The **roughskin newt** (*Taricha granulosa,* Fig. 1) occurs from coastal areas eastward to the foothills of the Cascade mountains in Oregon and Washington, and also in southwestern British Columbia. Adult newts measure 8 inches (20 cm) in total length; they are orange underneath and uniformly brown above, with rough skin (except for the breeding males).

Adult roughskin newts are commonly encountered. They are active day and night, especially during rainy periods. Large numbers can be seen crossing roads to and from breeding ponds in the spring and fall. Newts can also be seen gracefully swimming—undulating their bodies and large, paddle-like tails.

When threatened, the roughskin newt strikes a warning posture that presumably reminds predators of the newt's poisonous skin (see "Salamander Defense Systems"). The common garter snake *(Thamnophis sirtalis)* is apparently one predator that is able to eat roughskin newts without harm.

You are not likely to see the eggs of the roughskin newt. Females spawn in late spring and early summer, and the eggs are attached singly on the underside of vegetation. Eggs also are sometimes "glued" between leaves to keep them hidden from predators.

Land-Breeding Salamanders

Land-breeding salamanders lay their eggs in the absence of free water. They complete their larval stage within the egg and hatch as miniature adults (Fig. 3). The female usually exhibits parental care, and when the nests are found (a rare occurrence) the female is invariably with the clutch.

The **ensatina** *(Ensatina eschscholtzii)* occurs in moist lowland forests west of the Cascade mountains in Oregon and Washington, and the coastal areas in southern British Columbia. Adults measure 3 to 6 inches (8–15 cm) in total length, and display an amazing variety of colors and patterns. Coloring includes uniform brown or reddish-brown to dark brown or black above, with cream, yellow, or sometimes orange spots, blotches, or mottling. The belly is whitish or flesh-colored.

Ensatinas seem to adapt to disturbances, and can be found in pristine old-growth forests, logging slash piles, and urban wood-scrap heaps. They lay their eggs in rodent burrows, moist woody debris, and piles of firewood, old shingles, and plywood.

When threatened, an ensatina strikes a warning posture; if the tail is seized it easily snaps off, allowing the ensatina to escape (see "Salamander Defense Systems").

The **Western red-backed salamander** *(Plethodon vehiculum)* is one of the smallest and most abundant of the land-breeding salamanders in the Pacific Northwest. It is found under forest debris and wood and brush piles in damp coniferous forests and rocky areas. Its Pacific Northwest distribution includes areas west of the Cascade mountains in Oregon and Washington, and west of the Coast Range in British Columbia. Adults are up to 4 inches (10 cm) in total length, slate-colored underneath, with a back stripe that is usually red but sometimes tan, yellow, orange, or gray. Some populations are completely black.

Salamander Defense Systems

A variety of behavioral defense mechanisms exist among different salamander species. Some of the more agile species, such as the clouded salamander, merely flee when disturbed, seeking refuge under some sort of cover. Biting is another basic defense, and Pacific giant salamanders can deliver a formidable bite.

Another common behavioral ploy is camouflage or mimicry. Salamanders that have a color pattern similar to the background, such as the tiger salamander, remain immobile to go unnoticed. Others with dorsal stripes, such as the slender salamander, are thought to remain immobile in a coiled position to mimic millipedes, which are unpalatable to most predators.

Certain species, particularly the ensatina and the slender salamanders, readily lose their tails (as lizards do). The ensatina is particularly well adapted for this defensive strategy: The base of the tail has skeletal and muscular features that allow the tail to fall off easily. A predator may swallow the tail while the owner scurries off relatively unscathed. The tail regenerates over a year or two, but will not be as large as the original.

Some species, including newts, ensatinas, and Northwestern salamanders, often display themselves in a manner that makes them appear more intimidating or makes predators more aware of their poisonous nature. An adult Northwestern salamander will lift its back up and tip its snout down (like a bucking horse) to present its poison glands to an attacker. Large white drops will ooze out and run down the side of the salamander. The poison is not very toxic to humans, but it can kill small animals such as mice and shrews, and sicken larger animals such as skunks and raccoons. See "Public Health Concerns"!

ing a land nest is great, so if you find a nest, carefully replace its covering.

At night, use a flashlight to look for breeding adult salamanders and larvae where they may be swimming in shallow areas. Larvae may appear near surface sediments or move out from under rotting leaves, logs, or rocks in shallow water.

It can be difficult to tell the difference between hatchling salamanders and tiny hatchling tadpoles (larval frogs and toads). In each case, the gills are located behind the head and in front of the belly. However, hatching salamanders have slender bellies—so slender that they appear to be part of the tail (Fig. 5). Hatchling tadpoles have fat bellies. In addition, any tadpole-like creature with visible gills and four legs is a larval salamander.

In summer, when temporary ponds are drying up, recently metamorphosed juvenile salamanders sometimes remain under objects around the pond edges that still are wet, and will migrate to upland sites during or after a rain.

See "Collecting and Releasing Amphibians and Reptiles" in Chapter 41 for additional information.

Attracting and Maintaining Salamanders on Your Property

Because of their habitat requirements, it is difficult to maintain salamanders on land surrounded by concrete and highly maintained landscapes. The chances of seeing and hosting salamanders increase if your property adjoins an undeveloped area, such as a greenbelt or other wild

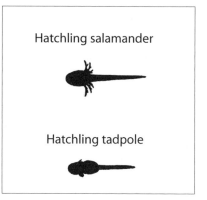

Figure 5. Hatching salamanders have slender bellies—so slender that they appear to be part of the tail. Hatchling tadpoles (larval frogs and toads) have fat bellies. (Adapted from Corkran, Amphibians of Oregon, Washington, and British Columbia: A Field Identification Guide.)

area, or if it is next to a wetland, stormwater retention pond, or other freshwater area. Your property may then become part of a salamander's home range if you establish and maintain a natural landscape.

Many salamander species breed in fishless ponds or temporary pools. Landowners may not think these small water holes are valuable, but salamanders and other species of amphibians may use them as breeding sites.

For additional information, see "Attracting and Maintaining Amphibians on Your Property" in Chapter 41.

Preventing Conflicts

Salamanders do not cause damage to people or property. Occasionally they frighten people who are not familiar with them. None have a poisonous bite.

Salamanders generally do not enter buildings, although they occasionally get trapped in window wells or find their way into damp basements and garages in search of food or shelter.

In both cases, they can be picked up with a gloved hand or scooped up with a piece of paper. Place the salamander outside in a cool, moist area away from human and pet traffic.

Salamanders enter at ground level, so sealing all ground-level holes or cracks can prevent their entry. Check for openings around the corners of doors and windows, water pipes, and electric service entrances. Holes in masonry foundations (poured concrete and concrete blocks or bricks) should be sealed with mortar. Openings in wood should be sealed with fine-mesh hardware cloth or aluminum flashing. Areas around window wells can be bordered with an 8-inch (20 cm) high barrier of hardware cloth, aluminum flashing, or a more decorative material.

Salamanders can find easy access to garage areas through poorly fitting garage doors. Cover door bottoms with sheet metal or a commercially available product for that purpose. Any weather-stripping along the garage and other outside doors should fit tightly.

All the above modifications will also help to exclude mice and other rodents.

To make the inside of a building less attractive to salamanders, steps should be taken to dry out the basement area (see a construction specialist for this) or to remove objects from the floor of a damp basement where salamanders can hide. Keep areas outside of the buildings free from objects that salamanders can hide under—including lumber, branches, old boards, or firewood. If such items

are off the ground (stacked on runners), the soil underneath can dry, making the area uninhabitable by salamanders.

Public Health Concerns

Certain species of salamanders emit toxic skin secretions. Handling salamanders and exposure to this skin secretion usually does not cause a problem in humans. As a general precautionary measure, however, wash your hands after handling any adult, juvenile, or adult salamander.

If a roughskin newt is ingested, get medical help immediately. If medical help isn't available, call the Poison Center at (800) 222-1222 or 911.

Legal Status

Because legal status and other information about salamanders and newts change, contact your local, state, or provincial wildlife office, or visit their Web site for updates. (See Appendix E for contact information and where to access the state and provincial laws mentioned below.)

Oregon: Seven species of salamanders found in Oregon are unprotected: the Northwestern salamander, long-toed salamander, Pacific giant salamander, ensatina, Dunn salamander, Western red-backed salamander, and the rough-skinned newt. Although there are no restrictions on the taking of these animals, wildlife taken from the wild and possessed must be maintained in a humane manner (OAR 635-044-0132), and cannot be sold or purchased (ORS 498.022).

All other species of salamanders are classified as State Sensitive species and may not be captured, killed, and/or held in possession (OAR 635-044-0130) unless a scientific collecting permit or other authorization has been obtained from the Oregon Department of Fish and Wildlife.

The tiger salamander is prohibited from being imported, possessed, or sold in Oregon (OAR 635-056-0050(1)(d)(A)), including being used as bait.

Washington: All salamanders existing in a wild state are considered wildlife (RCW 77.08.010). It is unlawful to remove or possess a newt or salamander from the wild without a permit to do so. It is unlawful to release newts and salamanders from captivity into the wild (WAC 232-12-064).

The Cascades torrent salamander, Columbia torrent salamander, Dunn's salamander, Larch Mountain salamander, and Van Dyke's salamander hold special state/federal status.

British Columbia: Importation of all salamander species is not permitted except under permit through an educational institution or a scientific organization. In such cases, the Director of Wildlife must be satisfied that the importation will not be detrimental to native wildlife or wildlife habitat (Wildlife Act, Permit Regulation, Sect. 7).

Red-listed species include the tiger salamander and the Pacific giant salamander. Blue-listed species include the Coeur d'Alene salamander (*Plethodon idahoensis*). See "Legal Status" in Chapter 1 for a definition of these listings.

CHAPTER 43

Snakes and Lizards

Snakes are among the most misunderstood of all animals. As a result, many harmless, beneficial snakes have met untimely deaths at the hands of shovel-wielding humans. Of the 15 species of snakes found in the Pacific Northwest, only the Western rattlesnake is capable of inflicting a poisonous bite, which it seldom does.

Snakes should be left alone, and, except for a rattlesnake that poses an immediate danger to people or pets, no snake should ever be killed. Observe snakes, like all wild animals, from a respectful distance. Don't attempt to capture them, and don't keep wild ones as pets.

All snakes are an important part of the natural food chain, eating a variety of prey—from mice and birds to frogs and insects. Besides their ecological value, snakes offer the careful wildlife viewer a chance to watch one of nature's most efficient predators.

See Table 1 for a list of snakes that are commonly seen in the Pacific Northwest.

For information on lizards, see "Notes on Pacific Northwest Lizards."

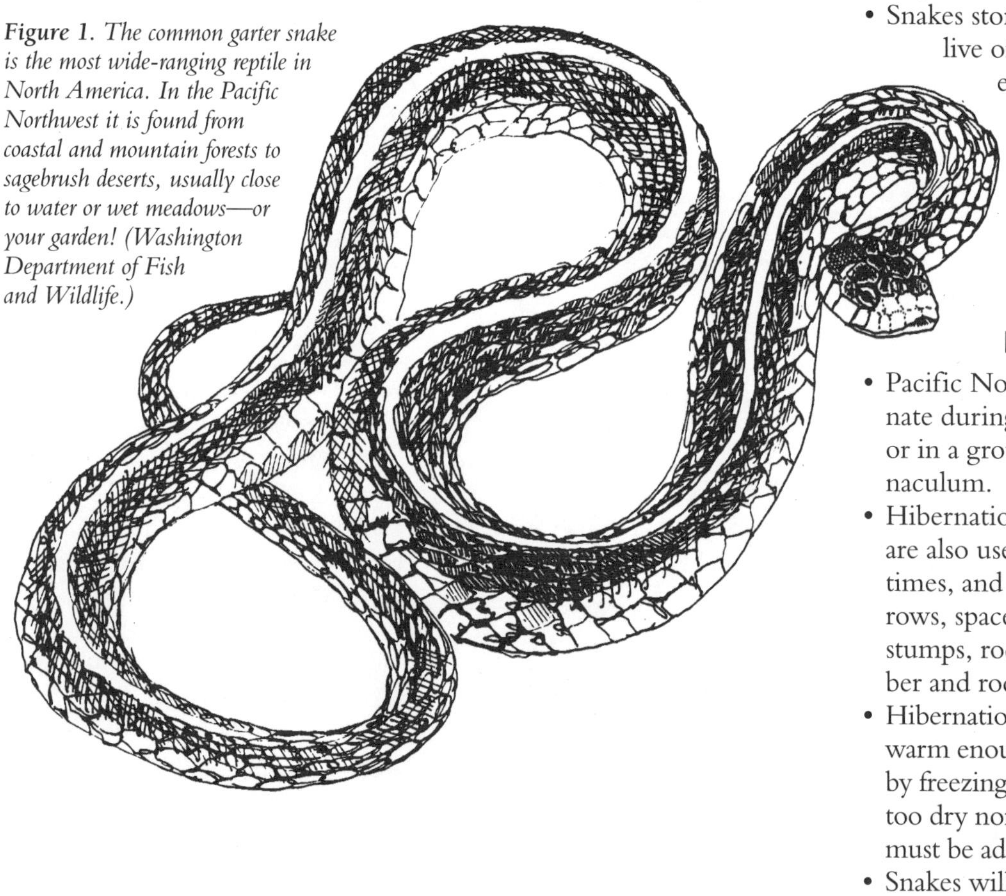

Figure 1. The common garter snake is the most wide-ranging reptile in North America. In the Pacific Northwest it is found from coastal and mountain forests to sagebrush deserts, usually close to water or wet meadows—or your garden! (Washington Department of Fish and Wildlife.)

Facts about Snakes
Food and Feeding Behavior

- Snakes are predators and eat many animals thought to be pests—mice, voles, snails, and slugs. Other prey items include insects, bird eggs and nestlings, fish, frogs, and lizards.
- Snakes have hinged jaws that allow them to consume food that is wider than their bodies. Even so, what a snake eats depends on its size; generally, larger snakes eat larger food items.
- Snakes have forked tongues that deposit air molecules on receptors in the mouth; thus, snakes "taste" the air, which helps them locate prey and sense their way in the dark.
- Snakes store food as fat, and can live off their fat reserve for extended periods of time.

Shelter and Hibernation Sites

- Pacific Northwest snakes hibernate during winter, either alone or in a group site called a hibernaculum.
- Hibernation sites (snake dens) are also used for shelter at other times, and include rodent burrows, spaces under logs and tree stumps, rock crevices, and lumber and rock piles.
- Hibernation sites must remain warm enough to prevent death by freezing, they must be neither too dry nor too wet, and they must be adequately ventilated.
- Snakes will use the same hibernaculum year after year; several

Table 1. Common Snakes of the Pacific Northwest

Four species of **garter snakes** occur in the Pacific Northwest. Small garter snakes eat earthworms and slugs; larger snakes include amphibians, small rodents, nestling birds, and fish in their diet.

Garter snakes survive in suburbia and towns because they give birth to live young, and so do not require safe places for their eggs. Their name comes from their alleged resemblance to the garters once worn by men to hold up their socks.

When disturbed, garter snakes will try to escape, but if threatened they may strike, bite, and smear foul-smelling anal secretions on your hands. A bite from one of these nonvenomous snakes may be alarming, but will rarely break the skin.

The **common garter snake** (*Thamnophis sirtalis,* Fig. 1) is found from coastal and mountain forests to sagebrush deserts, usually close to water or wet meadows—or your garden! Next to the Northwestern garter snake, this species is the most frequently encountered snake. It has brightly colored stripes (yellow, green, blue) that run lengthwise along its body, and a grayish-blue underside. It grows to 2 to 3 feet (60–90 cm).

The **Western terrestrial garter snake** (*Thamnophis elegans*) occurs in a wide variety of habitats and, despite its name, it spends a lot of time in water. This garter snake is usually gray-brown or black, with a dark, checkered pattern between yellow stripes. Identification difficult because there are four subspecies, all varying in coloration. Nearly black forms occur in some areas. It can grow to a length of 40 inches (100 cm).

The **Pacific coast aquatic garter snake** (*Thamnophis atratus*) occurs in southwest Oregon where it lives and hunts along bodies of water. It is extremely variable in color and markings; generally it has three dull-yellow stripes, or is spotted or blotched. Maximum length is about 33 inches (84 cm).

The **Northwestern garter snake** (*Thamnophis ordinoides*) is somewhat less widespread than its three cousins, preferring coastal and mountain forest habitats. However, it is commonly found in suburban areas and city parks. It's more slender than other garter snakes, reaching 2 feet (60 cm) at maturity. It is dark above and has stripes of varying colors, often red and orange.

The **gopher snake** (*Pituophis catenfer*), also known as the **bull snake,** is found in warm, dry habitats—deserts, grasslands, and open woodlands. It's a robust snake, measuring 3 to 4 feet (90–125 cm) in length, with dark blotches against tan along its back.

The gopher snake is often mistaken for a rattlesnake, owing to its coloration and its impressive display of coiling, striking, and loud hissing. It will also vibrate the tip of its tail in dry grass and leaves, further mimicking a rattlesnake. However, it is not poisonous. It is a constrictor, killing prey—mostly small rodents—by squeezing them until the prey suffocates.

The similar looking but rarely seen **night snake** (*Hypsiglena torquata*) occurs in similar habitats in eastern Oregon and Washington.

The **racer** (*Coluber constrictor*) occurs in warm, dry, open or brushy country where it is often observed streaking across roads. It is about 3 feet (90 cm) long, plain brown or olive above, with a pale yellow belly. It is thinner than a garter snake of comparable size. The racer is well named because it can move extremely fast. It holds its head and neck above the ground when hunting, and may climb into shrubs.

The **Western rattlesnake** (*Croatus viridis,* Fig. 2) is widespread and locally common in much of eastern and central Oregon, eastern Washington, and south-central British Columbia. It is distinguished by its broad, triangular head that is much wider than its neck, the diamond-shaped pattern along the middle of its back, and the rattles on the tip of its tail. Overall color patterns differ with habitat, ranging from olive to brown to gray. Black and white crossbars may occur on the tail. Western rattlesnakes measure 18 inches to 4 feet (45–120 cm) at maturity. Although many people talk of seeing "timber rattlers," "diamondbacks," and "sidewinders," none of these occur in Oregon, Washington, or British Columbia.

continued on next page

Table 1. *(continued)*

The number of segments on the rattle does not indicate the true age of the snake, since rattlesnakes lose portions of their rattles as they age.

Rattlesnakes are most common near their den areas, which are generally in rock crevices exposed to sunshine. They are most likely to be seen at night and dusk during the spring and fall when moving to and from hibernation sites.

Rattlesnake fangs are hollow and are used to inject the snakes' poisonous venom in order to stun or kill their prey—mice, woodrats, ground squirrels, and young rabbits and marmots. Their fangs are shed and replaced several times during their active season. Fangs may also be lost by becoming embedded in prey, or be broken off by other means.

Rattlesnakes cannot spit venom; however, venom may be squirted out when the snake strikes an object such as a wire fence. This venom is only dangerous if it gets into an open wound and has been used in the development of several human medications.

Rattlesnakes do not view humans as prey, and will not bite unless threatened. A rattlesnake bite seldom delivers enough venom to kill a human, although painful swelling and discoloration may occur. (For more information, see "Rattlesnakes.")

The **rubber boa** *(Charina bottae)* is a member of the same family as the world's largest snakes—including the boa constrictor, python, and anaconda. However, our local species only measures 14 to 33 inches (35–84 cm). It is olive-green, reddish-brown, or tan to chocolate-brown. It looks rubbery and has a short, broad snout and a short, blunt tail, giving it a two-headed appearance.

The rubber boa is found in damp wooded areas, large grassy areas, and moist sandy areas along rocky streams, being particularly fond of rotting stumps and logs. Although seldom encountered, this snake can be common in appropriate habitat.

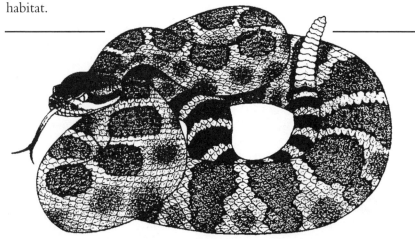

Figure 2. The number of segments on the rattle does not indicate the true age of the snake, since rattlesnakes lose portions of their rattles as they age. (Oregon Department of Fish and Wildlife.)

hundred snakes and different species may occupy the same hibernaculum.

- Emergence from hibernation can begin as early as March, depending on the species and location. Snakes may temporarily emerge from hibernation to feed and bask during warm periods in mild areas.

Reproduction

- Courtship and mating occurs shortly after snakes emerge from hibernation.
- Garter snakes, rubber boas, and Western rattlesnakes bear live young from eggs retained in the body until hatching. All other Pacific Northwest snakes lay eggs in protected areas where the eggs receive enough external heat to hatch.
- Young are born from July through September, and fend for themselves after hatching.
- Young snakes grow rapidly, and reach sexual maturity in two or three years.

Mortality and Longevity

- Snakes are preyed upon by badgers, coyotes, foxes, opossums, raccoons, skunks, weasels, great blue herons, hawks, eagles, and owls.
- Near human habitation, humans, domestic cats and dogs, lawn mowers, weed-whackers, and vehicles fatally wound or kill snakes.
- Garter snakes have lived as long as 18 years in captivity. Such ages might be exceptional for wild snakes, but little is known on this subject.

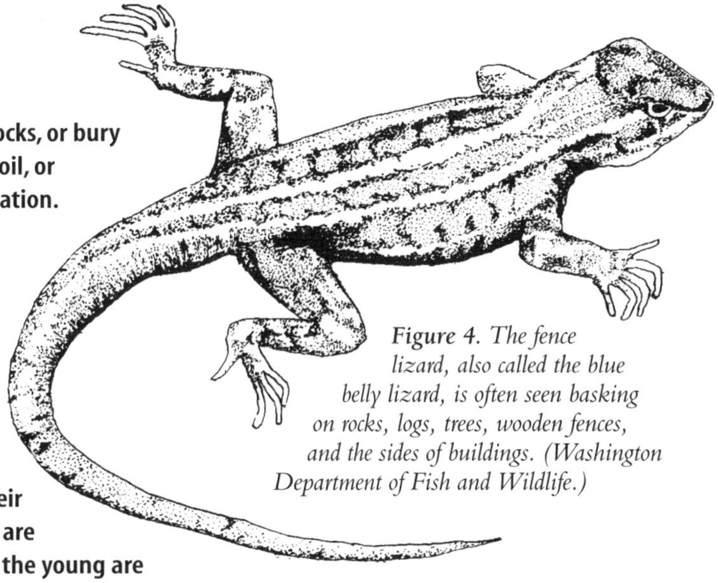

Notes on Pacific Northwest Lizards

Twelve species of lizards inhabit the Pacific Northwest's varied habitats. Four species occur fairly widely in areas with cool summers: the **Western fence lizard** and the **Southern alligator lizard** inhabit relatively open and dry areas, while the **Northern alligator lizard** and **Western skink** occur in forested areas. The other eight lizard species occur only in the dry, open habitats of interior Oregon, Washington, and British Columbia.

While snakes are easily recognized as being snakes, lizards are sometimes confused with salamanders because of their similar appearance. However, lizards have better adaptations for spending their lives on dry land: Protective scales cover their bodies, sharp claws help them climb rock and wood surfaces, and alertness allows them to rapidly seek food and flee from prey.

Lizards hibernate below ground from about October to March in rock fissures and rock piles, beneath the soil or surface debris, and in unoccupied rodent burrows.

Courtship begins after lizards emerge from hibernation. Most lizards lay eggs with leathery shell coverings under logs or rocks, or bury them in sand, soil, or decaying vegetation. The eggs hatch eight to ten weeks later. The female Northern alligator lizard and short-horned lizard retain their eggs until they are ready to hatch; the young are then born alive.

After birth, the hatchlings of all species fend for themselves and grow rapidly during the summer. Fall can be a stressful time for lizards hibernating for the first time, and they often suffer high mortality during the winter.

Lizards have numerous small, sharp teeth, relatively powerful jaw muscles, and are quick to seize and devour any available food item of the right size. Most lizards eat small invertebrates, including beetles, grasshoppers, crickets, aphids, spiders, and ants. Sagebrush lizards include wasps, black widow spiders, ticks, and scorpions in their diet.

Lizards are alert and can be very difficult to catch. However, they have

Figure 4. The fence lizard, also called the blue belly lizard, is often seen basking on rocks, logs, trees, wooden fences, and the sides of buildings. (Washington Department of Fish and Wildlife.)

many predators, including jays, crows, hawks, foxes, coyotes, and snakes.

Most lizards have "disposable" tails: When threatened, they can sever their tails by constricting the muscles at the base of the tail. The severed tail writhes and wiggles, thus distracting the predator while the lizard makes its escape. The lizard later generates a new tail.

Common lizards of the Pacific Northwest include:

The **Northern alligator lizard** (*Elgaria coerulea,* Fig. 3) is long-bodied, sometimes exceeding 10 inches (25 cm) in total length. It has short legs and wiggles as it walks. It resembles an alligator not only in shape but also in having large, platelike body scales. It will also bite if picked up or restrained. This lizard is active day and night and at lower temperatures than most lizard species.

It occurs from Vancouver Island and southern British Columbia south through most of western Washington and Oregon.

The similar-appearing **Southern alligator lizard** (*Elgaria multicarinata*) ranges as far north as south-central Washington, occupying oak woodlands and chaparral vegetation. Its preference for brushy areas and woodlands brings it into suburban yards that border wilder areas.

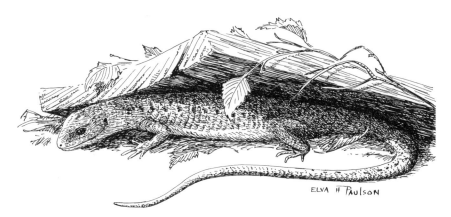

Figure 3. The Northern alligator lizard is thought to be the only the lizard that occurs along the coast north of Coos County, in Oregon. (Drawing by Elva Hamerstrom Paulson.)

The **Western fence lizard** (*Sceloporus occidentalis,* Fig. 4) measures 6 to 8 inches (15–20 cm) in total length. It is recognizable by its gray and brown topside and the blue patches on its belly. It occurs in desert canyons, grasslands, and open woodlands, and is often seen basking on rocks, logs, trees, wooden fences, and the sides of buildings.

The similar looking but slightly smaller **sagebrush lizard** (*Sceloporus graciosus*) is usually the most common lizard in the sagebrush habitat east of the Cascade mountains. It is also found in dunes and coastal open forests in southwest Oregon. It's not adept as a climber and is extremely wary when basking.

The **Western skink** (*Eumeces skiltonianus,* Fig.10) is a slender, shiny lizard, reaching 8 inches (20 cm) in total length. Young skinks often have bright blue tails, apparently to divert predators when they are trying to escape. The tail color changes to a blue-gray and then gets brownish with age. Rocky areas with some moisture, such as along streams and creeks, are its preferred habitat. This lizard may also be discovered in moist leaf-litter or under logs in pine or oak woodlands and in shrub-steppe habitat.

The **side-blotched lizard** (*Uta stansburiana*) is a slim lizard easily recognized by a blue-black spot behind each front leg. Its body becomes a golden-tan color during the breeding season in spring and early summer. At 5 inches (12 cm) long, it is the smallest lizard in the Pacific Northwest. The side-blotched lizard is found in rocky areas and in shrub-steppe flats in the dry interior.

The **short-horned lizard** (*Phrynosoma douglassii*) measures 4 inches (10 cm) in length and has a flat, round body with a basic color of gray or black. Its "horns" are stubby scales at the back of the head.

The short-horned lizard occurs in sagebrush deserts, juniper woodlands, and open coniferous forests east of the Cascade mountains and in extreme south-central British Columbia. Being quite cold tolerant, it can be common at elevations as high as the Cascade passes. It prefers sandy soil, but is also found in rocky areas.

The uncommon **desert horned lizard** (*Phrynosoma platyrhinos,* Fig. 5) occurs in southeastern Oregon. It has large pointed scales, or "horns," at the rear edge of its head.

Horned lizards are fairly easy to capture and are often taken home as pets. Unfortunately, they invariably die shortly thereafter. No lizard should ever be removed from the wild.

Viewing Snakes and Lizards

All snakes and lizards tend to be inconspicuous, preferring to move away and hide or lie still in the hope of being overlooked. Most encounters are momentary. Although snakes and some lizards are often seen as threatening, they will hiss, strike, or bite only if they are cornered or restrained.

Most of the time, snakes and lizards are slow moving, but they can make short dashes to chase prey or escape from predators. They are unable to sustain long-distance movement.

Because snakes and lizards are particularly active and less wary during the breeding season, begin to be on the lookout for them in spring.

Lizards are often inactive during the hottest part of the day, especially in mid- to late summer, and seek shelter or crawl underground to avoid overheating. In desert areas, snakes may become active at night when the air cools, and while the ground remains warm. Night snakes, as the name implies, are almost always nocturnal.

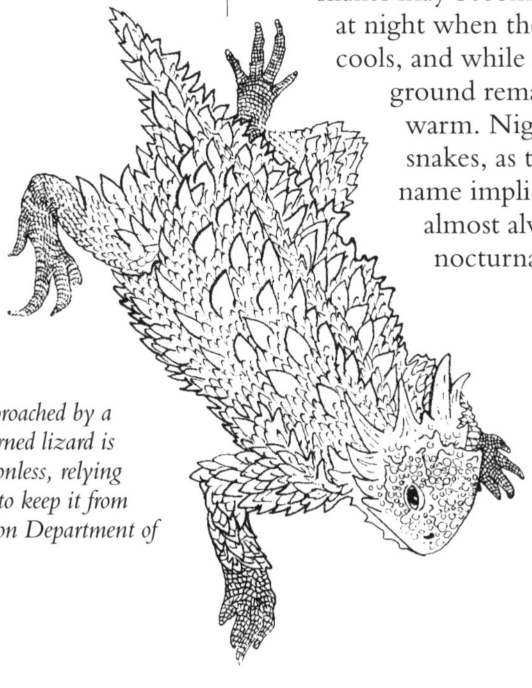

Figure 5. When approached by a human, the desert horned lizard is likely to remain motionless, relying on its protective color to keep it from being detected. (Oregon Department of Fish and Wildlife.)

It is environmentally unsound to capture snakes and lizards and try to relocate them on your property, or keep them as pets. Due to their well-developed homing instincts, most snakes will soon leave an unfamiliar area, which usually results in their being killed on the roads or by predators.

Basking Sites

Most snakes and lizards reach their preferred body temperature by basking on surfaces exposed to sun. They control their body temperature by moving in and out of the sunlight, and by changing their orientation to it (facing the sun, back to the sun, etc.). They also derive body heat by lying on or under warm surfaces.

In hot areas, look for snakes and lizards basking in the morning sun on asphalt, concrete, rocks, and wooden fences. In cooler regions, they can be seen basking throughout the day. Snakes, more than lizards, tend to bask on sun-warmed roads in the evening, a fact that often leads to them being run over.

Shed Snake Skin

A growing snake sheds its skin every four to five weeks. You can tell when it is ready to shed—its eyes look bluish-white and dull. Snakes may even become temporarily blinded until the old skin splits at the head and they are able to crawl out. Shed skin looks like thin, clear plastic, with every detail of the scales still visible, even the eyeball cover. Look for shed skin under boards, in rock piles, and other places where snakes congregate.

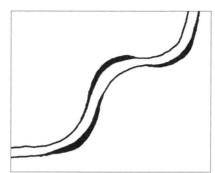

Figure 6. The trail of the Western rattlesnake is characterized by side-to-side undulations. The distance from one curve to the next varies according to the age of the snake and the speed at which it is traveling. (Drawing by Kim A. Cabrera.)

Trails

The trails of snakes and lizards are most easily seen in sandy or dusty areas in their preferred habitats. Snake tracks may be wavy or straight lines (Fig. 6). Surface material is usually pushed up at the outside of each curve.

Lizard tracks are difficult to distinguish from those made by invertebrates, particularly beetles.

Droppings

Snake and lizard droppings are interesting in that you will find a capping of white calcareous deposits at one end, as in the case of birds. The size of the droppings corresponds to the size of the animal. Snake droppings are cordlike, with constrictions and undulations.

Preventing Conflicts

Because of much false teaching, many people dread snakes and lizards, and consequently control often is practiced when it is not needed. The first thing you should do when encountering a snake or lizard is to leave it alone. Next, as long as it is not a rattlesnake or

inside a house or building where it is not wanted, continue to leave it alone. The chances that the animal will ever be seen again are fairly small.

If a snake or lizard gets into a house or other building, remain calm to avoid disturbing it and forcing it into hiding. Several methods are available to remove it. First, try opening a nearby door and using a broom to gently herd the animal out. You can also use a long pole, stick, or golf club to pick a snake up and place it in a box or wastebasket for transport outdoors.

If you are squeamish, the snake or lizard can be confined in a room or corner with barriers, such as boards or boxes, to be available for capture by a neighbor or an experienced handler. If possible, place an empty pail or wastebasket slowly over a lizard or a small or coiled snake until someone less squeamish arrives.

If someone else isn't available to remove the snake or lizard, you can hire a wildlife control company to do the job. To find such help, call your local wildlife office for a referral or look under "Animal" or "Wildlife" in your local phone directory. Police or fire departments sometimes remove rattlesnakes.

To prevent conflicts with snakes or lizards:

Prevent entry into buildings: Snakes and lizards in houses fall into two categories: those that entered accidentally and will be attempting to escape because they find the habitat unsuitable, and those that have entered to find prey or shelter and would take up permanent residence if allowed. The former include small snakes ands lizards that may

Tips for Attracting Snakes and Lizards

In addition to the persecution snakes experience routinely at human hands, they have suffered greatly from the habitat alterations we have created. Snakes and lizards fare poorly when we break up natural lands for urban and suburban development and isolate animals that cannot easily move across inhospitable terrain.

To help prevent this and to provide safe spaces on your property for snakes and lizards:

• Protect hibernation sites and other areas used by snakes and lizards.
• Mow at slow speeds and be ready to step on the clutch or brake. Leave grass unmowed in places that adjoin a wet area, sunny forest edge, or any other known snake habitat. If the grass has to be cut, survey the area and move or direct any snakes to a safe location prior to mowing. Set the mower blades as high as possible, or use a weed-whacker and leave grass 6 inches (15 cm) high.
• Build a small, fish-free (fish eat all stages of amphibians) pond for amphibians. Many snakes, and garter snakes in particular, feed on tadpoles, adult frogs, and invertebrates found in and around ponds (see Appendix F for sources of information on ponds).
• Build a rock wall or a rock pile with crevices for snakes and lizards to escape from severe weather and predators, to find food, and to give birth.
• Create a "snake board" by laying a sheet of plywood or corrugated sheet metal on the ground in a sunny location, propping it up on three sides with 3-inch (7.5 cm) rocks, lengths of plastic pipe, or similar objects (Fig. 7). In cool areas, paint the top black or cover it with dark asphalt shingles to increase the temperature below the board.
• Place habitat-enhancement features, such as reptile boards and rock piles, away from driveways or heavily traveled roads to avoid vehicle–reptile unpleasantries.
• Discourage cats and dogs from using your yard. They are effective hunters and can severely impact snake and lizard populations.
• Encourage your friends and neighbors to preserve wildlife habitat on their property, especially property that adjoins yours.
• Support public acquisition of greenbelts, remnant forests, and other wild areas in your community.
• Join a local conservation organization or a habitat enhancement project.

be considered trapped and will likely die from lack of food or moisture if not captured and removed. Some snakes may hibernate in older houses with leaky cellars or crawl spaces with dirt floors. The presence of shed skin usually indicates that a snake has been living in the house for some time.

Snakes and lizards usually enter at ground level, so sealing all ground-level holes or cracks can prevent their entry. Seal all cracks and holes in building foundations and exterior walls, including warped siding, where a small snake or lizard could enter. Use ¼-inch (6 mm) mesh hardware cloth, caulk, mortar, or a concrete patch to make the seal.

Snakes and lizards can find easy access to garage areas through open garage doors or under poorly fitting doors. Cover door bottoms with metal flashing or another material. Any weather-stripping along the garage and other outside doors should fit tightly. These modifications will also exclude mice and other rodents.

Modify the habitat: To limit the number of snakes and lizards around a living structure, reduce their food supply and shelter, and encourage their natural predators. The reduction of shelter

Figure 7. Snakes will seek out the shelter of a "snake board" in areas where similar types of shelter are lacking. (Drawing by Jenifer Rees.)

(rock piles, woodpiles, tall grass) not only limits hiding places, but also reduces the habitat used by mice and other rodents, which are a food source for snakes. (Snakes will also use holes made by mice or other rodents.) Mice and rat populations can be reduced by keeping food (including seed under bird feeders) inaccessible to these animals (see Chapter 37 for information on feeder management).

While snakes and lizards don't make up the majority of any predator's diet, hawks, owls, and a wide range of mammals eat them. See the appropriate chapters for information on attracting species that feed on snakes.

Note: As the number of snakes is decreased, the number of ground squirrels, gophers, and other rodents may increase, resulting in a different variety of problems.

Fences and repellents: In areas where rattlesnakes are commonly encountered, fences have been used to keep them away from buildings and out of yards (Fig. 8). This method is expensive, but if the yard is to be used as a playground by children the cost may be justified—if only to ease the worries of the parents.

A fence can be made from ¼-inch (6 mm) mesh galvanized hardware cloth, 30 inches (75 cm) high. The bottom edge should be buried 3 to 6 inches (8–15 cm) in the ground and the support stakes should be inside the fence to prevent snakes from crawling up them. Such a fence also can be added in front of an existing fence. Any gates should fit tightly and be kept closed. (For another fence design, see "Preventing Conflicts" in Chapter 14.)

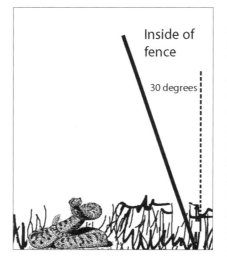

Figure 8. *A snake fence should slant outward at a 30-degree angle toward the area containing snakes. (Drawing by Jenifer Rees.)*

Regularly inspect the fence to be sure that holes haven't been opened under it, and that items have not been piled against the outside. Keep grass and weeds around the fence mowed.

Snake repellents, such as Snake-A-Way® and the stronger Doctor T's Snake Away® have produced mixed results. Snakes "smell" things via their tongues and the Jacobson's organ, which is located in the roof of the mouth. But, unlike mammals, snakes have no sense of smell associated with their breathing cycle. This means that unless the snake just happens to poke its tongue out at the precise moment that it is moving over the repellent, it will not notice a thing. Even if it does, the smell may not be noxious enough to drive the snake in another direction.

Trapping and snakes: As a last resort, a snake can be trapped and moved outside, or a one-way door can be installed that will allow the snake to exit but not reenter.

A live trap can be made from a 24- to 36-inch (60–90 cm) section of 4-inch (10 cm) PVC pipe. Temporarily cap one end and close off the other end with a cap that has a 1-inch (2.5 cm) hole drilled through the center. Place a hand warmer and a soft cotton rag in the far end of the tube. The hand warmer warms the tube up and a cold snake will crawl inside. (In hot settings, place an ice pack or a cool wet rag in the tube.) If the PVC trap feels too hot or cool, drill a few small air holes in the top of the tube to let some heat or coolness out.

Anchor the trap to prevent it from rolling. A tracking patch (a bit of flour) in front of the trap will confirm that a snake has entered. Place a piece of duct tape over the hole before moving the snake and trap outside.

To make another trap, attach three or four commercially available, rat-sized glue-boards to a piece of plywood. Place the glue-board trap along an inside wall or foundation. When a snake moves over the glue board trap it will get stuck. Once the snake is attached, the glue board trap can be removed. To avoid close contact with an agitated snake, consider fastening a wood extension handle to the glue board plywood base before placing the trap. *Note:* Use glue boards only indoors or under structures and only where children, pets, or non-targeted wildlife can't reach them.

To release a snake caught in the glue board trap, take it to a suitable area, place the trap on the ground, and pour vegetable oil on the snake. The oil will reduce the tackiness of the glue and allow the snake to free itself. *Note:* Again, glue boards should

Rattlesnake Research

Researchers in Eastern Washington are beginning to attach radio transmitters on rattlesnakes to discover how and where rattlesnakes hibernate, as little is known about their behavior. Researchers wonder if they hibernate in groups or separately, if they tend to go back to the same winter dens year after year, or if they hibernate with other snake species. Researchers are also unclear about how many rattlesnakes there are and whether the population is growing or shrinking. The research is not purely academic. The practical result is that if researchers know how rattlesnakes migrate, they can help identify where roads, trails, or other uses should be avoided.

not be used outdoors or at any location where they are likely to trap pets or other non-target animals.

To create a one-way door, seal all the openings except the suspected main entrance being used by the snake(s). On that opening, install a one-way door made from a piece of aluminum window screen rolled into a cylinder about 10 inches (25 cm) long and with a slightly larger diameter than the entrance hole. Suspend the outlet end of the tube off the ground to prevent the returning snake(s) from finding the entry. The device may be left in place for a month or longer to allow time for the snake(s) to leave.

Make any necessary repairs to the house or other structure to prevent the problem from reoccurring.

Rattlesnakes

If you live in or visit rattlesnake country, be alert and aware of this species in order to avoid threatening it.

Also know the recommended treatment steps in case a human or pet is bitten.

If you encounter a rattlesnake, move away: A rattlesnake will coil into a defensive posture if it cannot escape by crawling away. If you remain too close, the rattlesnake will usually warn you with its distinctive rattle. Its last defensive move is to strike. Remember, all of these warnings are meant to help avoid conflict. Rattlesnakes want to avoid you as much as you want to avoid them.

Prevent Problems While Hiking: To minimize conflicts with rattlesnakes while hiking:

- Stick to well-used, open trails. In brushy areas, use a walking stick to alert a snake of your approach.
- Avoid walking through thick brush and willow thickets.
- Do not step or put your hands where you cannot see.
- Wear over-the-ankle boots and loose-fitting long pants.
- Watch rattlesnakes from a distance, and be aware of defensive behaviors that let you know you are too close.

Rattlesnake Bites: All rattlesnake bites should be considered life threatening. When someone has been bitten, time is of the essence. If possible, call ahead to the emergency room so anti-venom can be ready when the victim arrives.

If a rattlesnake bites a person or a pet, do the following:

- Keep the victim calm, restrict movement, and keep the affected area below the heart level to reduce flow of venom toward the heart.
- Wash the bite area with soap and water.

Figure 9. Researchers are unclear about how many Western rattlesnakes there are and whether the population is growing or shrinking. (Oregon Department of Fish and Wildlife.)

- Remove any rings or constricting items; the affected area will swell.
- Cover the bite with a clean, moist dressing to reduce swelling and discomfort.
- Shock is responsible for more snakebite deaths than the actual venom is. To treat for shock, keep the victim quiet and maintain his or her body temperature. If the victim is cold, wrap them in a blanket; if hot, cool them off by fanning.
- Get medical help immediately. Make sure the doctor who treats the victim knows how to treat snakebites and, if not, call the Poison Center at (800) 222-1222.

Things NOT to do:

- Do not allow the person to engage in physical activity such as walking or running. Carry the victim if they need to be moved.
- Do not cut or suck the wound, do not apply ice or cold packs to the wound, and never use a tourniquet.
- Do not give the victim stimulants or pain medications, unless instructed by a physician.
- Do not give the victim anything by mouth.
- Do not raise the bite area above the level of the victim's heart.

Figure 10. Due to its secretive, burrowing lifestyle, the bright-blue-tailed Western skink is rarely seen. (Washington Department of Fish and Wildlife.)

Public Health Concerns

Nonvenomous snakebites are harmless. The only concern may be for potential infection. If bitten, clean and sterilize the wound as you would a cut or abrasion. A few people may be allergic to what are usually harmless bites, such as those from a garter snake. Contact your physician if a rash or a sign of infection appears.

Legal Status

Several snake and lizard species in **Oregon, Washington,** and **British Columbia** hold special state, provincial, and/or federal status. Because their legal status is currently undergoing change, contact your local, state, or provincial wildlife office, or visit their Web site for updated information. (See Appendix E for contact information.)

CHAPTER 44

Turtles

Turtles are among the most distinctive of animals, and are recognized at a glance by most people. Turtles kill diseased and weakened fish and other aquatic wildlife, clean up dead or decaying animal matter, and their eggs are a food source for a variety of predators. In addition, turtles offer the wildlife watcher the opportunity to see one of nature's most enduring living creatures (Fig. 1).

Most people know that frogs and salamanders are amphibians, and that snakes and lizards are reptiles. But what about turtles, which can be found on land or in water? Turtles are reptiles: Protective scales cover their bodies, sharp claws help them climb out of the water to bask on rocks and logs, and alertness allows them to rapidly flee from predators.

Both freshwater and marine turtles occur in the Pacific Northwest. This chapter focuses on freshwater turtles because they are encountered around wetlands, reservoirs, drainage ditches, slow-moving streams, ponds, and lakes.

No native tortoises (turtle relatives) occur in the Pacific Northwest. Those found wandering through backyards are pets that have escaped or been released (see "Finding a Wandering Turtle").

Facts about Freshwater Turtles

Food and Feeding Habits

- Turtles feed on whatever plants and animals are found in their waters.
- Plant foods include algae, moss, and the roots, shoots, and leaves of aquatic plants.
- Animal foods include spiders, beetles, insect larvae, tadpoles, and the eggs of frogs, toads, and salamanders, as well as adult frogs, small fish, and carrion (animal carcasses).
- Turtles lack true teeth, but the sharp-edged horny ridges on the jaw margins serve as teeth.

Figure 1. Turtles are reptiles: Protective scales cover their bodies, sharp claws help them climb out of the water to bask on rocks and logs, and alertness allows them to rapidly flee from predators. (Oregon Department of Fish and Wildlife.)

Figure 2. Turtles do not care for their eggs or young beyond providing a protective underground chamber (Western pond turtle shown here). (Washington Department of Fish and Wildlife.)

Nest Sites and Nesting Behavior

- Female turtles leave the water and may travel as far as 1,000 feet (300 m) before they find a suitable nest site.
- Turtles nest on beaches, floodplains, shrubby fields, gravel or soil roadsides, meadows, and pastures.
- Females dig 1- to 4-inch (2.5–10 cm) deep chambers by using their hind feet alternately. Nesting usually occurs in late afternoon or early evening, and the entire process takes two to eight hours.
- Female turtles don't set on the eggs, so for warmth the nest site is always open to direct sunlight for much of the day.

Reproduction

- Courtship and mating of turtles occurs in the water.
- Sexual maturity is more dependent on the size of the turtle than on its age. Slider males become sexually mature from two to five years of age; male Western pond turtles require ten years to attain sexual maturity.
- Depending on the size and species of the female turtle, 4 to 20 oval, hard-shelled eggs are laid in the nest in early summer.
- Hatchlings emerge in about 15 weeks, or overwinter in the nest and emerge the following spring (Fig. 2).
- Incubation temperature of the eggs determines the sex of the hatchlings in many species of turtles. For instance, more female painted turtles and pond turtles hatch out after a warmer summer than one with normal temperatures, and more males hatch out after a cooler than normal summer.
- Young turtles grow rapidly in their early years, but growth slows considerably as they get older.

Turtles in Winter

In late fall, turtles begin to look for a place to overwinter. Overwintering for turtles means periods of reduced or no activity that may include a hibernation-like state. Turtles remain in their overwintering site, which is in a protected location either on land or in water, from late fall to mid-spring.

On land, turtles will dig in under leaf litter and loose soil, often in dense shrub cover. Those in the water will find a space under a log, overhanging bank, or other crevice to wedge themselves into. They may also dig down under vegetation, mud, or debris on the bottom, leaving only the eyes and snout exposed.

Western pond turtles may overwinter under logs or in rodent burrows up to 1,500 feet (450 m) from the nearest water source, sometimes changing sites during the season. Stream-dwelling pond turtles are thought to be more likely to overwinter on land than are pond-dwellers.

Turtles, like all reptiles, have lungs. They have to breathe air. However, they can go a few seconds to several hours (when they sleep) to several months during winter dormancy without doing so.

In winter, the waters of their ice-covered ponds are generally about 40°F (4.4°C), and their metabolic rate is extremely low. This allows them to survive months of dormancy by relying on anaerobic metabolism (metabolism without oxygen), supplemented with some direct uptake of oxygen through specialized gill-like tissues around the cloaca (the common opening of the reproductive and digestive tracts).

Table 1. Pacific Northwest Turtles

Painted turtle *(Chrysemys picta)*
The painted turtle has an olive-gray or blackish-green shell, a reddish or orange-ish underside, and yellow or thin reddish stripes on the neck and legs. The shell of adults is 8 to 11 inches (20–28 cm) long. The painted turtle can be distinguished from the red-eared slider turtle by the absence of the bright-red eye stripe behind the eye. Sliders also have bright yellow undersides.

The painted turtle is found in northeastern Oregon and throughout the Columbia River Basin in both Oregon and Washington. It is also found in a few areas around the Willamette Valley in Oregon, Puget Sound in Washington, and in southern British Columbia. Often several can be observed basking on a log.

Western pond turtle *(Clemmys marmorata,* Figs. 2 and 3)
The Western pond turtle has a smooth, broad, low shell that is olive to dark brown. Its underside is pale yellow with dark-brown or black blotches. The shell of adults is 5 to 8 inches (13–20 cm) long.

Western pond turtles originally ranged from northern Baja California, Mexico, north to the Puget Sound region of Washington, but in recent years they have become rare or absent in the northernmost part of their former range. Few turtles now occur in the Seattle and general Puget Sound region. (An attempt is in progress to reintroduce them from captive-bred stock.) Only small populations are known in the Willamette Valley of Oregon, where the pond turtle population has declined by as much as 98 percent since the beginning of the century. Pond turtles are more common in large river basins in southern Oregon.

The initial cause of decline in Western pond turtle numbers may have been commercial exploitation for food. Pond turtles never recovered from this decline; and this is due, in part, to continued loss of habitat. Wetlands were filled for development, particularly in the Puget Sound region. Dam construction and water diversion projects also reduced available habitat and caused populations to become isolated and vulnerable. Human disturbance may have kept females from crossing overland to lay eggs, or may have reduced the amount of time the turtles were able to spend comfortably basking. Loss of wetland vegetation on lakesides may also have made habitat less suitable for hatchling and juvenile turtles.

Slider *(Trachemys scripta,* Fig. 4 and back cover)
Sliders occur naturally from eastern North America to northernmost South America. Millions have been raised on turtle farms and sold as pets, resulting in releases in the Pacific Northwest and elsewhere. This introduced species isn't very cold hardy, so most large populations occur in mild lowland areas west of the Cascade mountains in Washington and Oregon, and west of the Coast Range in British Columbia.

The slider has a prominent red (red-eared slider) or yellow (yellow-eared slider) stripe or blotch behind its eyes (this telltale marking is not always present, particularly in old individuals). The shell is olive to brown and streaked with yellow and black; its underside is bright yellow. An adult slider shell is 4 to 12 inches

---

Figure 3. Like most turtles, the Western pond turtle has a distinctive shape and shell, familiar to everyone. The top of the shell is domed and known as the carapace. *The carapace is fused to the backbone and to the ribs and is covered with large plates, or* scutes. *The bottom part, called the* plastron, *is flat. Most species can withdraw the head, tail, and all four legs into the shell for protection from predators. (Washington Department of Fish and Wildlife.)*

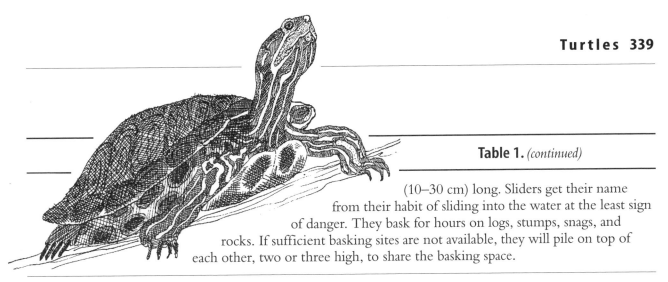

Table 1. *(continued)*

(10–30 cm) long. Sliders get their name from their habit of sliding into the water at the least sign of danger. They bask for hours on logs, stumps, snags, and rocks. If sufficient basking sites are not available, they will pile on top of each other, two or three high, to share the basking space.

Snapping turtle (*Chelydra serpentina*, Fig. 5)

Snapping turtles are native to eastern North America and have been intentionally introduced at various sites in the Pacific Northwest, primarily in western Oregon and Washington. Because snapping turtles are rarely sold in pet stores, how this species became so widespread is a puzzle. Perhaps because they grow quickly and tend to bite, owners release them into the wild after only a year or so.

Snapping turtles have large heads and limbs, causing their shells to look relatively small, and giving them a prehistoric appearance. The tan to dark brown shell of adults can reach 18 inches (46 cm) in length. The underside is yellow to tan.

Snapping turtles prefer quiet waters with muddy bottoms and much vegetation. They rarely leave the water to bask and often hide during daylight hours, which makes them difficult to detect.

Unlike other turtles, which retract into their shells when threatened, snapping turtles squarely face their attackers, lunge, and bite. They can bite very hard, and will not let go easily, so don't try to pick them up (even by the tail, as they have a long neck and can bite you from this position). Snapping turtles have evolved this behavior—defending themselves by attacking before they are attacked—because their large heads and broad necks are too big for complete retraction into their shells.

Other Introduced Turtles

Besides the slider and snapping turtle, at least nine other introduced species of turtles now occur in Pacific Northwest waters, including the Reeve's turtle *(Chinemys reevesi)* and the Malayan box turtle *(Cuora amboinensis)*, two Asian species that are sold in pet stores. These turtles are seldom detected because they rarely bask. Four turtle species from eastern North America are also occasionally found in our area: the mud turtle *(Kinosternon subrubrum)*, musk turtle *(Sternotherus odoratus)*, softshell turtle *(Apalone spiniferus)*, and the box turtle *(Terrapene* sp.*)*.

Pet turtles are released into nature when people tire of them, don't want to treat them when they become sick, or are bitten (e.g., sliders and snapping turtles have sharp beaks).

These introduced species may compete with or displace our native turtles. For example, the Western pond turtle is severely depleted in the Puget Sound region where the red-eared slider has become common.

See "Don't Let Your Pet Turtle Go" for more information.

Figure 4. The red-eared slider, introduced from eastern North America, is now widespread and common in the mild lowland areas of the Pacific Northwest.

Figure 5. Unlike other turtles, which retract into their shells when threatened, snapping turtles squarely face their attackers, lunge, and bite. (Drawings from Oregon Department of Fish and Wildlife.)

Mortality and Longevity

- Raccoons, river otters, coyotes, foxes, dogs, and eagles prey upon adult turtles.
- Turtle eggs are dug up and eaten by river otters, skunks, raccoons, opossums, mink, coyotes, foxes, and dogs. Predators may smell out turtle nests because female turtles urinate on the finished nest to help solidify the plug of soil tamped into the entrance. Nests are occasionally trampled by livestock or people.
- Herons, owls, crows, starlings, bullfrogs, and all the above mammals catch and eat the 1-inch (2.5 cm) long hatchling turtles (Fig. 2).
- Threats to native turtles include habitat alteration, predation on young turtles by exotic bullfrogs (see Chapter 41) and fishes, drought, local disease outbreaks, and fragmentation of remaining populations.
- Turtles have a well-deserved reputation for longevity. The maximum age achieved by Western pond turtles in the wild is at least 42 years.

Viewing Turtles

With turtles, as is true for all reptiles, the temperature of the surrounding air determines the animal's body temperature. As a result, turtles are most often seen when temperatures are warm, usually from late April through mid-October.

During hot weather, turtles conceal themselves in or under masses of floating vegetation or algae, occasionally sticking their heads up for air and to look around. Pond turtles will sometimes find a cool, shady area out of the water where they can avoid the heat. During winter, turtles overwinter in protected areas, occasionally coming out to bask on warm days.

Having no external ears, turtles cannot hear high-frequency sounds, but the shell serves as a sort of eardrum by conducting low-frequency vibrations to the middle ear. Turtles also have a well-developed sense of sight. In park and golf course ponds and drainage-ditch systems that run through towns and cities, turtles sometimes appear quite tame and do not move unless people go off the trails. In wilder places, turtles are wary and will seek safety when a perceived threat is 150 feet (45 m) or more away. Consequently, turtles are often overlooked in the wild.

The best way to observe turtles is by carefully stalking them—moving slowly and hiding behind shrubs and trees in areas where turtles are known to live. Use binoculars to look for their dark-colored shells on partially submerged logs, rocks, planks, barrels, and drainage pipes, and also on muddy or sandy banks.

Normally passive, turtles are aggressive toward one another when competing for basking sites, which may be limited in number. Aggressive behavior may involve biting, pushing, and open-mouthed threats, in which the bright edges of the mouth and reddish tissues inside the mouth are exposed and may serve as warning signals.

As a wildlife viewer, you can help the turtles a number of ways:

- Watch turtles from a distance with a good pair of binoculars or a spotting scope. Stay back so you don't scare the turtles off their basking sites.
- Do not create needless paths to the water's edge where turtles congregate. Raccoons, skunks, and other predators will use your trail to hunt for adult turtles and their nest sites.
- Keep track of the locations and numbers of native turtles (painted and Western pond turtles) you observe. In particular, watch for young turtles to determine if the population is successfully breeding. Keep land or wildlife managers informed of your observations.

Tips for Attracting Turtles

Because of their habitat requirements, it is difficult to maintain breeding populations of wild turtles in most yards and on most properties. If your yard is small and surrounded by concrete and highly maintained landscapes, chances are slim that wild turtles will find their way to it. The chance of seeing and hosting turtles increases if your property adjoins a wetland, stormwater retention pond, or other freshwater area. Your property may then become part of a turtle's home range if you establish and maintain a natural landscape.

Private landowners are key to the survival of western pond turtles because most of the best turtle habitat is privately owned. If you have western pond turtles on or near your land, your habitat improvement and management efforts can play a major role in conserving these turtles.

The following can help protect turtles and their habitat:

- Protect areas on your property used by turtles—streams, canals, lakes, and ponds—and the buffer areas next to them. Buffers protect the ecological functions of the freshwater area, and are used as escape cover, nest sites, and hibernation sites.

- Build one or more 4-foot (1.25 m) deep ponds for turtles and amphibians—turtles feed on tadpoles, frogs, and invertebrates found in and around year-round ponds (see Appendix E for resources on pond construction). Include aquatic plants for food and cover, and a mud or sandy bottom for overwintering turtles.
- Add basking sites, including logs, rocks, or artificial platforms, to existing ponds or lakes (Fig. 6).
- Create underwater refuges, such as rocks of various sizes, submerged logs or branches, submerged vegetation, or holes or undercut areas along the bank. Turtles use all these as hibernation sites and escape cover.
- To improve the survival of juvenile turtles, keep exotic predators such as bullfrogs and warm-water fish out of ponds or other bodies of water.
- To keep turtles from wandering onto a road, fence critical areas. The fence should reach at least 1 foot (30 cm) into the ground, since turtles dig.
- Discourage dogs from wandering your property. They are effective hunters and can have a severe impact on on turtle populations.
- Encourage your neighbors to preserve wildlife habitat on their properties.
- Support public acquisition of greenbelts, remnant forests, and other wild areas in your community.

Preventing Conflicts

Conflicts with turtles are rare, and every effort should be made to ensure that local populations are not harmed. Larger adult turtles, particularly introduced snapping turtles, can inflict a severe bite. Because of the potential severity of the bite, never attempt to feed or catch these animals.

If necessary, a snapping turtle can be captured with a large, long-handled fishnet. Watch for moving weeds or the turtle's bubble trail to see what direction the turtle is heading, then make the grab. There are several commercial turtle traps available to consumers (see Appendix F for resources). The hoop-net design is the most common.

Finding a wandering turtle: Turtles are somewhat sedentary, although they are capable of moving significant distances and have been known to travel a half-mile (0.80 km) in a few days. Most movements on land are associated with nesting, overwintering, or escaping hot temperatures. Some turtles move between ponds on an annual basis, moving to larger ponds as water levels recede in smaller ones.

If you find a turtle, do nothing if the turtle is in a safe place and healthy.

If it is found on a road, carry the turtle to the side toward which it was going, otherwise it may try crossing the road again. If the tur-

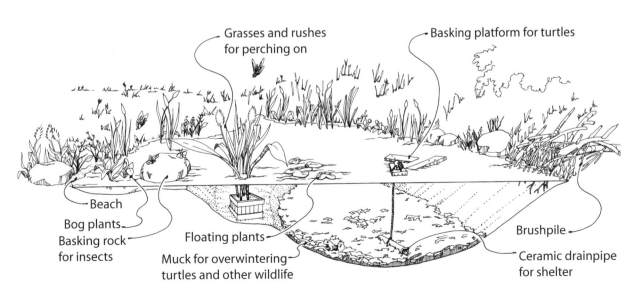

Grasses and rushes for perching on

Basking platform for turtles

Beach
Bog plants
Basking rock for insects
Floating plants
Muck for overwintering turtles and other wildlife
Brushpile
Ceramic drainpipe for shelter

Figure 6. Side view of a wildlife pond. (From Link, Landscaping for Wildlife in the Pacific Northwest.*)*

tle is in a place where it will be hurt, transport it out of danger.

Don't take the turtle home. It is not only environmentally unsound but also illegal to keep a wild turtle as a pet (see "Legal Status"). Due to its well-developed homing instinct, the turtle will soon leave an unfamiliar area, which usually results in its being killed on the road or by a predator. Most native turtles are rare, threatened, or of unknown status, and there is no justification for moving them to areas where they may well not survive. If turtles are not present in your yard, there is probably a good reason.

Attempting to move a turtle into your yard when conditions are not appropriate for it will probably just result in the death of the animal.

Finding a turtle nest: Female turtles leave the water and may travel as far as 1,000 feet (300 m) before they nest, although most nests are within 100 feet (30 m) of the water. If you find a turtle nest at a site where you know the nest will be destroyed, place a wire cage over the nest site to protect it from dogs, other predators, and trampling. One-inch (2.5 cm) mesh chicken wire attached to a wood frame that is secured to the ground with stakes should be effective. Make sure the hatchlings will be able to travel through or under the wire cage after hatching. Stay out of the area to prevent attracting attention to it.

Don't let your pet turtle go: Turtles are popular as pets. Unfortunately, owners often tire of them and release them to fend for themselves. Having been bred in captivity, fed, and well-protected,

most pet turtles die in the wild. On the other hand, some species, such as the red-eared slider turtle and snapping turtle, often do survive. This presents a particular risk when pet turtles are released into the wild, because often they are sick (and they may be at the most advanced stages of disease).

It is much more humane to find a home for your pet turtle and not release it into the wild. You can advertise on the Internet, in a local paper, or at a pet shop. (A pet shop may be willing to take the turtle and resell it.) You can also donate the turtle to a school for educational purposes.

Public Health Concerns

Most turtles harbor *Salmonella* organisms, which may be necessary for proper functioning of the turtle. The strains of *Salmonella* found in turtles may not be the ones that affect humans; however, turtles may also harbor the form that is infectious to humans. Wash your hands after handling turtles to avoid transferring this potential disease.

If a turtle bites you, clean the wound to avoid infection. Seek medical assistance for severe wounds.

Legal Status

Because legal status and other information about turtles change, contact your local, state, or provincial wildlife office, or visit their Web site for updates. (See Appendix D for contact information and where to access the state and provincial laws mentioned below.)

Oregon: The painted turtle and the Western pond turtle are

classified as State Sensitive species and are protected. These turtles may not be captured, killed, and/or held in possession unless a scientific collecting permit or other authorization has been obtained from the Oregon Department of Fish and Wildlife (OAR 635-044-0130).

All other freshwater turtles are unprotected. They cannot be possessed, transported, or sold alive (OAR 635-056-0050). If there are questions regarding which non-native turtle species may be kept as pets, refer to Oregon Administrative Rules 635–056.

Washington: All turtles existing in a wild state are considered wildlife (RCW 77.08.010). It is unlawful to remove or possess a turtle from the wild or release a turtle from captivity into the wild without a permit to do so (WAC 232-12-064). The Western pond turtle is state endangered (WAC 232-12-014); the painted turtle is classified as protected wildlife (WAC 232-12-011).

British Columbia: The painted turtle is a blue-listed species (see "Legal Status" in Chapter 1 for a definition of listings).

Importation of all turtle species of the family Emydidae (pond and river turtles), and all species of the family Chelydridae (snapping turtles) is not permitted except under permit through an educational institution or a scientific organization. In such cases, the Director of Wildlife must be satisfied that the importation will not be detrimental to native wildlife or wildlife habitat (Wildlife Act, Permit Regulation, Section 7).

Part 4

Appendices

Trapping Wildlife

If you cannot resolve a conflict with an animal by other means—such as removing the attractant, installing a barrier, applying a scare tactic—a final alternative is to trap it. Trapping is the last option because it presents many problems for both the animals and the trapper. Also, it rarely is a permanent solution if other animals are in the area, and food and/or shelter remain available to them.

An animal that is simply passing through the yard or living in a tree in the backyard should not be trapped. There is no guarantee that the next animal to move in won't be a problem. Living near animals is like having new neighbors—you never know what the situation will be until after the new family has already moved in.

Appropriate times to trap an animal in or around a home or property include emergency situations, the removal of a targeted problem animal, or when trapping is the only practical solution.

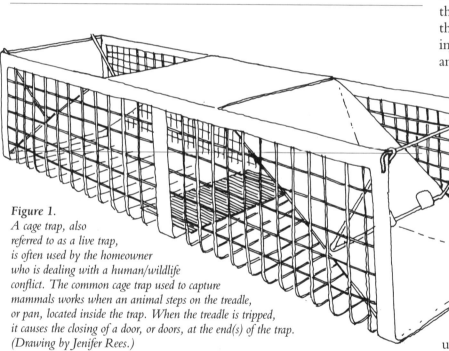

Figure 1.
A cage trap, also referred to as a live trap, is often used by the homeowner who is dealing with a human/wildlife conflict. The common cage trap used to capture mammals works when an animal steps on the treadle, or pan, located inside the trap. When the treadle is tripped, it causes the closing of a door, or doors, at the end(s) of the trap. (Drawing by Jenifer Rees.)

Basic Trap Designs

Modern traps fall into two main categories: quick-kill type traps and live-holding traps. Kill-type traps are designed to quickly kill the captured animal, much like a common snap-trap used on house mice. Live-holding traps can be separated into cage traps, foothold traps, and snares.

This appendix deals only with the use of cage traps used to capture mammals (Fig. 1). These are the traps most often used by people dealing with human/wildlife conflicts in yards, gardens, and houses. They come in a variety of designs; their sizes range from those that capture mice to those that capture large dogs. To avoid injuring people or trapping pets and other non-targeted animals in urban or suburban settings, cage traps often are the only traps permitted for use in these areas.

Except mice and rat snap-traps, quick-kill traps and other live-holding traps should be left to professionals and individuals who meet state or provincial requirements and are authorized to use these traps under permit.

Cage-Trapping Wildlife

Two questions to ask yourself before attempting to trap an animal are: (1) Can I do this legally (comply with state or provincial laws and regulations regarding trapping and transporting wildlife? and (2) Can I do this humanely? (see "Options for What to Do with the Trapped Animal"). If the answer to either of these is no, consider hiring a professional who is better able to meet these ends (see Appendix C, "Hiring a Wildlife Damage Control Company," for information).

When used properly, cage traps can offer nonlethal solutions to conflicts. However, despite the perception that live capture in cage traps is humane, animals often experience stress and physical damage during the capture. Captured animals may also suffer from exposure to extreme weather and lack of water. Such injuries, trauma, and disorientation may lead to the death of an animal days after it has been released.

For these reasons, it is important that the appropriate-size cage trap is used, that no sharp edges be exposed inside the trap that could injure an animal, and that all precautions described throughout this Appendix are carefully followed.

Cage traps can be purchased from hardware stores, farm supply centers, and over the Internet (see "Traps and Trapping" in Appendix F for resources). Some rental business and wildlife damage control companies rent them. Before using a trap, be sure it is clean, to prevent the animal coming in contact with or spreading potentially dangerous organisms. A dirty trap should be washed, disinfected with a bleach solution (1 part bleach to 9 parts of water and let it remain on for 20 minutes), and thoroughly rinsed. To protect yourself, always wear gloves when handling the trap.

When Not to Trap

Never trap an adult animal that is caring for dependent offspring. Look and listen for young—even outside the animal's known birthing season. If young are seen, heard, or you suspect they may be present, refer to Step 5 in Appendix B, "Evicting Animals from Buildings."

When an adult animal is trapped, look for enlarged teats that are relatively free of hair, which indicate it is a female nursing young. (By standing the trap up on end you can usually observe the underneath side of the animal.) In such a case, release the female on site so she can tend to her young.

In emergency situations, when the family needs to be removed, refer to Step 5 in Appendix B.

Permanently separating the nursing female from her young would likely cause the offspring to starve to death, which is inhumane. (Orphaned wildlife must be cared for by licensed professionals. Do not attempt to care for the animals yourself. Not only could you further harm the animals, it may be illegal for you to do so.)

Never trap an animal during poor weather with the intension of releasing it. Trapped animals expend energy that is normally used to cope with winter conditions and they may die soon after.

What to Do with the Trapped Animal

Before trapping an animal, you need to know what you are going to do with it after the animal has been captured. There are a variety of options. *Note:* State and provincial wildlife offices have requirements that you'll need to follow in certain areas, including mandatory euthanasia of certain species, such as raccoons, Eastern gray squirrels, and pigeons (rock doves).

The Professional Trapper

The fur-trapping industry has been important throughout Pacific Northwest history. The fur trade helped motivate the great sea voyages of Cook and Vancouver. On land, the Lewis and Clark Expedition (1804–1806) to the Columbia River also sought new fur resources.

Trapping has traditionally been the primary form of managing the populations of furbearers, including beaver, river otter, and muskrats, with quotas set by state or provincial wildlife agencies. Although today the price of pelts generally makes trapping nonlucrative, trapping continues to be a form of wildlife management and commercial recreation, where legal.

In the wildlife damage control industry, private companies made up of one or more people offer a variety of services, including trapping. (See Appendix C, "Hiring a Wildlife Damage Control Company," for information.)

Experienced trappers know the behavior of each species and the methods required to trap them. They also recognize signs of diseases and nursing females. Often their solution to a conflict will involve setting several cage traps to make sure the entire family of animals is caught—or as many as possible at one time. Multiple traps are not something most homeowners have on hand.

Note: Persons working at state or provincial wildlife offices do not provide trapping services, but they can provide names of individuals and companies that do.

Option 1. Release the animal at the site of capture. With this option, an animal is trapped and released on site after its reentry into a structure is prevented by physical exclusion. (See Appendix B for exclusion techniques.)

In such a case, the animal is evicted within its home range and because it is familiar with its surroundings, it can soon find suitable food and shelter. In the event young are present but were not noticed prior to trapping, allow the female back inside to tend to her dependent offspring.

A downside to this approach is the possibility that the animal may simply enter someone else's attic, chimney, or similar place. Then, if someone else has to trap the animal, they will be dealing with a trap-smart animal, making its capture difficult.

Option 2. Release the animal outside of its home range. The release of elk, bear, and other wildlife by professional wildlife managers to reintroduce or augment populations is a proven and valid technique for wildlife management. However, releases of this kind should not be confused with moving problem wildlife, which in most cases is also illegal (see "Legal Status" at the end of each chapter and contact your local wildlife office if you have any questions). For instance, in the State of Washington, it is unlawful to possess or transport live wildlife or wild birds (except starlings and house sparrows by falconers) without a permit (WAC 232-12-064). *Note:* **This includes Eastern gray squirrels, Eastern cottontail rabbits, raccoons, and opossums. They are considered wildlife because they occur in Washington in a wild state—which includes neighborhood parks and backyards.**

Many times, not enough consideration is given to the impact of the capture and release process on the animal, or to the animal's impact on the established wildlife populations at the release site. While wildlife released in a new location is an option often preferred by well-meaning people opposed to killing animals, this may be at the expense of the released animal or the animals at the release site.

Most biologists do not recommend releasing wildlife outside their home range for the following reasons:

- Mortality rates increase when animals are subjected to stress and trauma associated with capture, handling, transport, and being released into an unfamiliar territory.
- Animals that are released may harm or be harmed by resident animals (e.g., by territorial disputes, disease transmission, gene-pool disruptions, etc.).
- The same (or a competing) species may already be overabundant in the area. Excess animals have to move or die.
- Habitat conditions in the new area might not be suited to the animal being released.
- Many animal species have strong homing instincts and, upon release, they begin traveling in the direction of their capture sites, resulting in exposure to roads and other hazards.
- Animals may cause problems for humans in the vicinity of the release site.

Option 3. Euthanize the animal. The term euthanasia is derived from the Greek terms eu meaning good, and thanatos meaning death. A "good death" is one that occurs with minimal pain and distress.

Whether to use euthanasia is a personal question and a matter of an individual's perspective and values. People's opinion on the topic often depends on the experience they have with an animal. Opinions also depend on what will be killed—people are often less upset if it's a mouse or a snake that is killed, and more upset if it is a raccoon or a beaver.

The most widely accepted—but still disputed—guidelines for euthanasia practices follow the standards set by the American Veterinarian Medical Association (AVMA), which include:

- An injection of sodium pentobarbital or other pharmaceutical.
- Carbon monoxide (CO) or carbon dioxide (CO_2) supplied to a chamber from a compressed gas cylinder (small and medium sized animals).
- A gun shot to the head (small and medium sized animals. Check local firearm ordinances).
- Stunning, followed by decapitation (amphibians, reptiles, and birds only).
- Cervical dislocation by stretching the animal so the neck is hyperextended to separate the first vertebrae from the skull (birds, rabbits, and small rodents only).

Unfortunately, the majority of the above agents of euthanasia require training and care to administer. In addition, most are not available to, or do not appeal to, the do-it-yourselfer.

Care should always be taken to guarantee that the animal is euthanized humanely. If it cannot, or you do not have the necessary training, an alternative would be to contact your local wildlife damage control company, veterinarian, or animal shelter. They may euthanize the animal for a fee.

While drowning and freezing have long been considered a humane way to deal with problem wildlife, animal experts no longer generally accept these techniques, and they are not considered humane by the AVMA standards.

While shooting an animal may sound extreme, in many cases it is the best available method because of its quickness, and it may cause the least amount of stress and pain to the animal. If shooting is used, the operator and firearm must be capable of producing a quick death. To calm down an active animal, the trap can be covered with a dark towel or other cover.

Depending on the species and size of the animal, a .22-caliber rifle or revolver, or a high-velocity pellet gun should be used. A pellet gun fired to the head is capable of quickly killing tree squirrels, rabbits, and similar-size mammals. Local laws and regulations regarding the discharge of firearms must be followed. See Step 4 below for information on how to handle the dead animal.

Note: In order to properly check an animal for possible rabies, the animal must not be shot in the head; instead, aim for the lung area directly behind the front shoulder.

Human psychological responses to euthanasia of animals need to be considered, with grief at the loss of life as the most com-mon reaction. People who have to euthanize animals, especially under public pressure to save the animals rather than destroy them, can experience extreme distress and anxiety.

For additional information on euthanasia methods, see the references in Appendix F, "Wild-life/Human Conflicts," or contact a local veterinarian.

Cage-Trapping Wildlife: Steps to Take

If an animal needs to be trapped and you are uncomfortable or have no interest in doing the work yourself, contact a wildlife damage control company as described in Appendix C. If you are somewhat knowledgeable about wildlife, have identified the species of animal to be trapped, and feel you can handle the situation in a humane and legal way, follow the steps below.

See Table 1 for detailed trapping information for individual species.

Step 1.
Develop a Plan that Includes Options

- Contact your local wildlife office and municipality for current information on trapping restrictions (types of traps to be used, requirements for euthanization, species of biological concern in the area) and any required authorization.
- Decide if the animal will be released on site, euthanized, or moved somewhere else by someone who has a legal permit to transport it.
- If it is to be released on site, be ready to make all necessary construction repairs to ensure that the animal will not reenter the structure after being released.
- If the animal is to be euthanized, decide who, and if necessary, how it will be done. Note: Have a backup plan in case your original plan changes.
- If the animal is to be relocated off-site, decide where it will go and how it will get there. You must adhere to trespass laws at the release site and laws regarding transporting wildlife (see "Legal Status" at the end of each chapter).

Step 2.
Set the Trap

- Set the cage trap as near to the den as possible, in the animal's pathway, or in the area of damage (see Table 1 for specific recommendations). When locating the trap, consider the possibility of young children approaching the trapped animal, theft of the trap, or damage to the trap by vandals.
- If setting a trap on concrete or another hard surface, place the trap on plywood or some other protective surface to prevent the animal from damaging its paws when trying to dig its way out. To prevent raccoons and opossums from toppling the trap, make sure the protective material extends out from the trap at least 8 inches (20 cm) and locate the trap away from shrubs or other objects that they could grab.
- A captured animal often defecates in a trap. If the trap is set outdoors, biological risk is minimal but still real. Refrain from setting a trap near a shallow well, garden, playpen, or where a dog is tethered. Traps set inside the living area of the house should

be placed on top of at least ten sheets of newspaper.

- Place a tennis ball in the trap to give a large animal a way to release energy and frustration; a piece of wood will provide a small animal something to chew.
- Anchor the trap so it won't tip or misfire when the animal enters—**an animal will not enter a tipsy trap,** and misfires teach it not to try and enter again. Anchor the trap with a cinder block or other heavy object placed on top. You can also pound rebar stakes into the ground at the corners, or wire or clamp the trap to a stable object.
- Set the trap and then trip it several times to be sure the cage is steady and functioning properly. Trip the trap by using a pen or pencil, sticking an end through the side of the cage and pushing down on the treadle. If the doors do not work fast enough, place small stones or other weights on top of the door to make it drop faster.
- Use plenty of the suggested bait so it will be seen and its odor released into the air (see Table 1 and "Capturing a Wary or Trap-Smart Animal" for detailed information regarding baiting).

Step 3.
Monitor the Trap and Animal

- Be "on call" the entire time a trap is set.
- A trap set for a nocturnal animal should be set at or near dusk. The unset trap should be closed at dawn to avoid trapping a non-target animal during the day. Reverse this procedure when attempting to capture a diurnal (active during daylight) animal. Change the trap loca-

tion or try different bait if it doesn't produce a catch within three days.

- When an animal is captured, move the trap to a quiet, protected spot and cover it with a tarp until time of release or euthanasia. A captured animal should not remain in the trap longer than necessary. **Note:** In summer, a trap set where the sun can beat down on it can cause the animal to dehydrate rapidly, suffer a heat stroke, or die.
- If the captured animal appears injured or sick (i.e., having a discharge from eyes or nose, or a dull, sparse coat or scabby skin) and you don't want to euthanize it or have it euthanized, contact a wildlife rehabilitator (see "Wildlife Rehabilitators and Wildlife Rehabilitation" in Chapter 20). **Note:** Most vets and animal shelters won't accept a sick wild animal because of their concern for the spread of disease.

Step 4.
Remove the Animal from the Trap

- Release nocturnal animals at night and diurnal species during daylight.
- Point the opening of the trap toward escape cover, so the animal can see and move toward it. Stand at the opposite end of the trap, open the door, and tap the trap with your foot. If the animal is reluctant to leave, try placing the open trap on its side and moving away from the trap.
- When releasing an animal that offers the potential of a bite or a spray, attach a long string to the door of the trap prior to setting it, so the door can later be

opened from a distance. Place the trap under the driver's-side door of a truck, or a window on a house, lean out the window, and hold the door open with the string until the animal exits. **Note:** Skunks and opossums often take their time when leaving a cage trap.

- The carcasses of euthanized animals must be disposed of properly. To dispose of an animal on-site, the carcass must be covered by at least 2 feet (60 cm) of soil and located at least 200 feet (60 m) from any groundwater well that is used to supply drinking water. Cover the burial hole with rocks or strong wire screening to help prevent animals digging into it. Sprinkling a layer of garden lime on the carcass will also help reduce the odor, which attracts digging animals. If it is not feasible to dispose of a carcass on-site, contact a local veterinarian or wildlife damage control company for assistance with disposal. Animal carcasses should never be handled with bare hands.

Step 5.
Follow up

- A trap that contained a sick animal should be washed, disinfected with a bleach solution (1 part bleach to 9 parts of water and let it remain on for 20 minutes), and thoroughly rinsed after each capture so as to stop the spread of any potential disease.
- Immediately complete all repair work necessary to prevent another conflict.

Capturing a Wary or Trap-Smart Animal

Tips from trappers who capture wary or trap-smart animals include:

Entice an animal into a trap by sprinkling bits of bait leading from the travel route of the animal to the trap door and into the trap. To prevent filling the animal up with food before it enters the trap, use a small amount of bait, every 6 to 12 inches (15–30 cm).

Entice a young coyote or a hungry fox into a trap by taking advantage of their curiosity about food that has been hidden or stored by another animal. Dig a hole approximately 6 inches (15 cm) wide and deep. Bury some slightly spoiled meat in the hole, being certain to leave some of the bait exposed. Set and secure the rear of the trap over the hole to entice the animal into the trap for a look down the hole.

Funnel the animal into the trap using the funnel method (Fig. 2).

Wire bait to the back of a single-door trap to force the animal to step on the trip pan while reaching forward to get the bait. Wrap the back of the cage with small-mesh wire screen so the animal can't reach through the larger mesh wire to get the bait.

Put out an unset trap with the door(s) wired open for several nights. Offer some bait outside the trap the first night, at the trap's entrance the second night, and then inside the trap the third night—still without setting it. On the fourth night, place the bait inside and set the trap with an unwired door(s).

Camouflage the trap by covering the bottom with soil, leaf litter, grass clippings, or similar material, using enough to just hide the treadle. (A different material placed on the bottom of the same trap can sometimes be used to catch an animal a second time.) Also, place a few things like branches, boughs, or boards over or leaning against the trap to cover any glare and break up the outline of the trap. Make sure this camouflage does not interfere with the operation of the trap.

To help make the trap as "invisible" as possible, reduce the high glare of the cage steel with some earth-tone paint.

To deal with the animal's sensitive nose, wear old gloves when preparing the trap, and do not walk or linger around the site any longer than necessary. New cage-traps may need to be washed with water and vinegar to remove oils. Before leaving, mist the area around the trap with a spray bottle containing water and fir needles, or other local aromatic vegetation.

When using a two-door trap, securely fasten one door down to prevent an animal from backing out and getting away. With two-door traps, the trip pan to release the doors is in the middle of the trap, and only allows the animal to get halfway into the trap before it closes. The door may land on the back or tail of a large raccoon and not close completely.

To prevent this, use a trap with only one door, use a larger two-door trap, or secure one of the doors down so only one is in operation. This allows the bait to be placed in back of the trap, forcing the animal to go farther into it. You can also place a wedge-shaped piece of wood under the trip pan on the open-door side of the trap. This allows the animal to step on the front side, without anything happening. As it steps on the back side of the pan, the door closes.

To prevent capturing birds, mix peanut butter/oatmeal/sunflower seeds as bait. Place this on the inside roof of the trap above the pan. Spreading the bait above the trip pan keeps it out of sight of birds and causes the animal to stand on the pan to taste

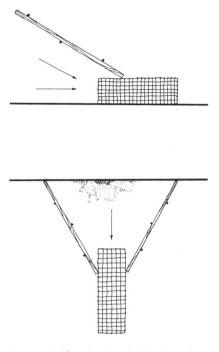

Figure 2. Place 2 x 6 inch (5 x 15 cm) boards, cinder blocks, or a similar barricade in such a way as to funnel raccoons, opossums, skunks, or other animals into a trap. (Drawing by Jenifer Rees.)

the bait. An option is to smear peanut butter on a pine cone or other inedible object—not bread, which is another bird attractant—and place it on the bottom of the trap.

To prevent catching a domestic cat or dog, don't bait with meat or fish products. However, if the target animal has been eating pet food, use pet food as bait.

To prevent bait from getting wet, place it in a light-colored, covered container with plenty of holes punched in the side.

Table 1. Cage-trapping Specifications
(Refer to "Traps and Trapping" in Appendix F for additional information)

Note: The below are minimum height, width, and length measurements for traps.

Animal Species	Trap Type, Size (height, width, length)	Bait	Notes
Badger	Single-door type, 10 x 12 x 42 in. (25 x 30 x 107 cm)	Chicken and attractors such as feathers and eggshells, cotton balls, or marshmallows.	Due to the strength and aggressiveness of badgers, it is recommended that a professional trapper trap them. Due to the badger's questionable population status, your local wildlife office should be informed of any trapping activity.
Bat	Trapping bats is not recommended. Traps can be fatal to bats if left unattended, or if they become overcrowded. In addition, bats have excellent homing instincts, making moving bats unlikely to succeed. Instead, use the exclusion methods described under "Bats Roosting in Buildings" in Chapter 2.		
Beaver	Hancock or Bailey suitcase-type trap	Freshly cut tree sprouts or branches, commercial scents and lures.	Due to the weight and dangers associated with suitcase traps, it is recommended that only people experienced with these traps use them. Some success has also come from using a 4 foot (1.2 m) long, single-door cage trap set right at the water's edge next to the beaver slide.
Bobcat	Single-door type, 24 x 24 x 48 in. (60 x 60 x 120 cm)	Poultry or rabbit carcass and feathers for a sight attractor.	Set the trap in the vicinity of an animal kill or a travel way to and from cover. Use brush or grass on the top and sides of the trap to give the appearance of a natural "cubby" or a recess in a rock outcrop or in brush. Cover the cage bottom with soil. (See "Capturing a Wary or Trap-Smart Animal" for detailed information.)
Cat (domestic cat)	Single-door type, 11 x 11 x 32 in. (28 x 28 x 81 cm). Double-door traps should be 42 in. (107 cm) long	Moist or dry cat food, tuna; also a fake nest with eggs and/or an electronic chirping device.	Set the trap in the area being frequented by the particular cat. Pre-baiting and laying a towel or something over the floor will help. (See "Capturing a Wary or Trap-Smart Animal" for detailed information.) A special notice may need to accompany the trap in urban areas. Contact the local jurisdiction for restrictions. *Note:* A cat that is assumed to be a docile pet can turn into a hostile animal when captured. Handle a caged domestic cat with the same respect you would any wild animal.

Animal Species	Trap Type, Size (height, width, length)	Bait	Notes
Coyote	Single-door type, 20 x 26 x 48 in. (51 x 66 x 122 cm)	Sight attractors like chicken feathers, eggshells, cotton balls. An auditory lure that "squeals" can be effective. Wrap it in paper towels and a baggie to muffle the volume.	Cage traps are rarely effective at capturing healthy adult coyotes and most effective at capturing young or sick coyotes living in urban areas or entering a chicken coop or other holding area for pets, live-stock, or birds. The trap should be thoroughly concealed with a tarp or other material, and extra precautions need to be taken to eliminate human scent from the area of the trap. (See "Capturing a Wary or Trap-Smart Animal" for detailed information.)
Chipmunk	Single- or double-door type, 5 x 5 x 16 in. (13 x 13 x 41 cm)	Unroasted peanuts, sunflower seeds, grain, popcorn, apple slices.	Place the trap where the chipmunk is active. Place a few sunflower seeds in front of the trap entrance.
Dog (domestic dog)	Single-door type, 12 x 12 x 36 to 20 x 28 x 72 in. (30 x 30 x 91 cm to 51 x 71 x 183 cm), depending on size of dog	Moist dog food.	Set the trap in the area being frequented by the particular dog. A special notice may need to accompany the trap in urban areas. Contact the local jurisdiction for restrictions. ***Note:*** A dog that is assumed to be a docile pet can turn into a hostile animal when captured. Handle a caged dog with the same respect you would any wild animal.
Fox (red and gray fox)	Single-door type, 15 x 15 x 48 in. (18 x 18 x 61 cm)	Tainted meat, eggs placed in a nest, marshmallows, cotton balls (they resemble eggs and have eye appeal).	Foxes are long-bodied animals, so the trap must be long. Take precautions to eliminate human scent from the trap and the area around the trap. Place bait in a hole dug under the rear of the trap. Cover all sides of the trap with a tarp or other material. Sift dirt onto the bottom of the cage to cover the wire bottom. (See "Capturing a Wary or Trap-Smart Animal" for detailed information.)
Gopher	See "Lethal Control" and "Legal Status" in Chapter 11.		
Ground squirrel	Single-door type, 5 x 5 x 15 in. to 7 x 7 x 24 in. (13 x 13 x 38 cm to 18 x 18 x 61 cm), longer if a double-door trap is used	Peanut butter, oats, barley, fresh fruit, vegetables, greens.	Set the trap near an active burrow with signs of recent diggings. Placing guide logs on either side of the path between the burrow opening and the trap will help funnel the animal into the trap. Cover the floor of the trap with soil and leave the bait highly visible. Cover the trap with a tarp to conceal the trap and provide an enticing nook for the animal to enter. (See "Capturing a Wary or Trap-Smart Animal" for additional information.)

Wildlife Species	Trap Type, Size (height, width, length)	Bait	Notes
Marmot (yellow-bellied marmot)	Single-door type, 12 x 12 x 36 in. (25 x 25 x 91 cm), longer if a double-door trap is used	Peanut butter, oats, barley, fresh fruit, vegetables, greens.	See Ground squirrels.
Mice	Single or double-door type, 3 x 3 x 10 in. (7 x 7 x 25 cm)	Peanut butter, grain.	See "Lethal Control" in Chapter 8 for information on trap placement.
Mink	Single-door type, 7 x 7 x 17 in. (18 x 18 x 43 cm), longer if a double-door trap is used	Cheese or fresh bloody meat such as chicken or rabbit; use sight attractors like feathers or fur.	Wrap the cage trap in something dark; mink like to investigate dark holes. Set the trap in the animal's line of travel.
Mole	See "Lethal Control" in Chapter 13.		
Mountain beaver	Single-door type, 7 x 7 x 17 in. (18 x 18 x 43 cm)	Piece of apple, sweet potato, or whatever is being eaten in the area.	Set trap directly in the entrance of an active tunnel. Alternatively, place a plastic laundry basket over the hole and cut out an opening just large enough to insert the door of a trap into the basket cut-out. The mountain beaver will search for an opening to go through and enter the trap. Stake the basket down so it cannot be moved. Mountain beavers are very prone to hypothermia, so wrap the trap with black plastic or burlap and cover it with soil. Trap when the weather is mild and check the trap early in the morning.
Muskrat	Single-door type, 6 x 6 x 20 in. (15 x 15 x 51 cm), longer for double-door traps	Corn, carrot greens, sweet apples, cattail roots.	Place the trap just outside the burrow and partially in the water, taking every precaution to make sure that the captured muskrat will not be under water and will be able to breathe. Conceal the cage trap well with grass or leaves. A short line of bait leading to the entrance of a trap will increase capture success.

Animal Species	Trap Type, Size (height, width, length)	Bait	Notes
Nutria	Single-door type, 9 x 9 x 45 in. (23 x 23 x 114 cm), longer if a double-door trap is used	Cantaloupe rind, ripe bananas, sweet potatoes.	Place the trap along an active trail or where nutria are seen. A short line of bait leading to the entrance of the trap will increase capture success. A trap placed on a floating raft will effectively catch nutria, but pre-baiting is necessary (see "Capturing a Wary or Trap-Smart Animal").
Opossum	Single- or double-door type, 11 x 11 x 36 in. (28 x 28 x 91 cm)	Dry or canned pet food, sardines, old meat, chicken entrails, bacon, fish, apples.	Place the trap where the animal, or evidence of it, has been seen, or at its den entrance (Fig. 2).
Porcupine	Double-door type, 10 x 12 x 42 x in. (25 x 30 x 106 cm)	A salt-soaked cloth, sponge, or piece of wood, also water softener tablets, sweet potatoes, apples, roasted peanuts.	Place the trap in the vicinity of damage or at the den entrance. To lure the porcupine, blend a cup of raw sweet potatoes and an apple, and dribble the puréed mixture at the opening of a single-door live trap.
Rabbits and Hares	Single- or double-door type, 9 x 9 x 26 in. (23 x 23 x 66 cm) See "Notes"	Fresh vegetables in summer; apples, carrots, or bread in winter.	Place the trap near cover where rabbits feed or rest, or where they gain entry under a fence. Place some bait just outside the trap and spray the inside with apple juice to increase effectiveness. To capture hares (jackrabbits) in open terrain, use a double-door trap with weighted doors to prevent escape. It is best to use a larger trap than used on rabbits.
Raccoon	Single-door type, 10 x 12 x 42 in. (25 x 30 x 107 cm)	Fish-flavored cat food, corn, ripe bananas, bacon, sardines, peanut butter, jelly, or marshmallows (resemble eggs and have eye appeal).	Place the trap where the animal, or evidence of it, has been seen, or at its den entrance (Fig. 2). (See "Capturing a Wary or Trap-Smart Animal" for additional information.)
Rats	Single- or double-door type, 5 x 5 x 18 in. (13 x 13 x 45 cm)	Peanut butter, grain.	See "Lethal Control" in Chapter 16 for information on trap placement.

Animal Species	Trap Type, Size (height, width, length)	Bait	Notes
River otter	Single-door type, 10 x 12 x 42 in. (25 x 30 x 107 cm), longer if a double-door trap is used	Fresh fish.	Cover the bottom of the trap with sand. (See "Capturing a Wary or Trap-Smart Animal" for detailed information.) River otters may be trapped in suitcase-type traps used to capture beavers. Modify the sides so the otters can't escape.
Skunk (spotted skunk)	Single-door type, 7 x 7 x 20 in. (18 x 18 x 51 cm)	Dry or canned pet food, sardines, old meat, chicken entrails, bacon, fish, apples.	See striped skunk.
Skunk (striped skunk)	Single-door type, 10 x 10 x 24 in. (25 x 25 x 61 cm) or longer if a double-door trap is used	Peanut butter, bananas, honey, or molasses spread on a piece of bread or dried fruit; also yogurt, cheese, raw egg (trail some through the trap and leave the rest in the back of the trap).	Place the trap along a travel route or immediately outside the den entrance, using the funnel method (Fig. 2). Alternatively, place a plastic laundry basket over the hole and cut out an opening just large enough to insert the door of a trap into the basket cut-out. The skunk will search for an opening to go through and enter the trap. Stake the basket down so it cannot be moved. Box traps designed specially for trapping skunks are available, or a cover can be made out of a dark-colored blanket, plywood, or cardboard fastened with bungee cords. If a skunk is accidentally caught, use a long stick or other device to **slowly** cover the trap with a towel or blanket before moving it. Avoid sudden, jarring movements or loud noises that may frighten the skunk.
Squirrel (Eastern gray and fox squirrel)	Single- or double-door type, 6 x 6 x 24 in. (15 x 15 x 61 cm), longer for double-door traps	Peanut butter, nuts, corn, sunflower seed, popcorn, bread.	A squirrel may not find a trap set in the dark, or it may bump the trap, causing it to close prematurely. A trap set on the roof is safe from theft, children, and pets, and offers a better chance for catching the squirrel. If possible, find a window adjacent to a roof that the squirrel is using and you won't need a ladder. To prevent catching birds, see "Capturing a Wary or Trap-Smart Animal."
Squirrel (Douglas and flying squirrel)	Single- or double-door type, 5 x 5 x 18 in. (13 x 13 x 46 cm)	Apples, sunflower seeds, roasted peanuts.	For Douglas squirrels, see above. For flying squirrels, set the trap inside a structure and near the animal's point of entry.

Animal Species	Trap Type, Size (height, width, length)	Bait	Notes
Weasel (long-tailed weasel)	Single- or double-door type, 5 x 5 x 24 in. (13 x 13 x 61 cm)	Fish, fresh chicken liver, chicken entrails.	Set the trap in an old brush pile, or under an outbuilding or fence, since the weasel is likely to investigate any small covered area.
Weasel (short-tailed weasel)	Single- or double-door type, 5 x 5 x 18 in. (13 x 13 x 46 cm)	Fish, fresh chicken liver, chicken entrails.	Set the trap in an old brush pile, or under an outbuilding or fence, since the weasel is likely to investigate any small covered area. Also see "Lethal Control" in Chapter 26.

Evicting Animals from Buildings

Occasionally a raccoon, skunk, tree squirrel, or other animal will find a suitable shelter in or under a house, shed, or other structure. These animals may occupy an area sporadically, using the site only two or three consecutive days or nights—usually until available food sources are exhausted. However, some may choose to overwinter there if the surroundings remain favorable. During the mating and nesting season, females attracted to warm, dry, easily defended areas may attempt to den or nest in these settings.

You may choose to let the animal use the area if it doesn't pose a direct problem to you, your family, or your pets and other animals. However, its discarded food, urine, or droppings may create odors and become a potential health hazard. Animals also may make considerable noise, chew on building parts, or destroy insulation during the nest-building process.

Should you choose to remove the animal, you can complete the process yourself or hire someone to do it. (See Appendix C, "Hiring a Wildlife Damage Control Company.") A wildlife damage control company is recommended for work that poses health or safety hazards. Examples include removing a large accumulation of droppings, removing a mother and/or her young in emergency situations, or working in a precarious location.

Note: State and provincial wildlife offices do not provide animal removal services, but they can provide names of individuals and companies that do.

If the animal you are trying to evict appears sick or injured, call a nearby wildlife rehabilitator for assistance (see "Wildlife Rehabilitators and Wildlife Rehabilitation" in Chapter 20 for information).

To encourage an animal to move on its own or to evict it from a place where it is undesired, follow the steps below. (For information on evicting bats, mice, or rats, see the appropriate chapters in this book.)

A Seven Step Strategy to Conflict Resolution

Before anything else occurs in a wildlife/human conflict in or around a structure, it is absolutely necessary to be sure of the following: the species involved, where the animal(s) are entering, and whether or not young animals are present. After that it is important to proceed in such a way that is humane and prevents the problem from reoccurring.

Step 1.
Try to identify the suspect

Identifying the species of wild animal causing the conflict is vital to resolving the problem. Note the time and location of calls, cries, or scampering noises heard coming from inside the structure, and

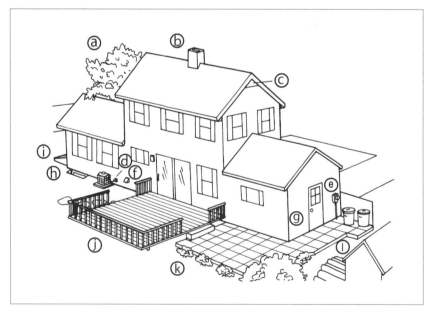

Figure 1. The house and yard can provide homes for wild animals in the form of shelter and cover. An overhanging branch (a) can provide access to a roof, while the tree itself may be used as a refuge. An uncapped chimney (b) or broken vent (c) can provide access to warm, dry living quarters. Entry for small mammals can occur where wiring or pipes enter the house (d) and (e), vents are uncapped (f), doors are improperly fitted (g), ground-level window sills, foundations (h) and bulkhead doors (i) have gaps. Shelter can be found under decks (j); burrowing animals may tunnel beneath patios (k) or wood piles (l). (Adapted from Hodge, Wild Neighbors: The Humane Approach to Living with Wildlife.*)*

have a look outside for the animals themselves. Tree squirrels (except flying squirrels) and marmots are heard exiting around sunrise and returning from late afternoon until dark. Both can be seen outside during the day. Flying squirrels, raccoons, river otters, skunks, and opossums are heard rummaging around shortly after dark until just before dawn, and are generally seen outside at night.

If the animal isn't seen, try to identify it from its method of entry, odor, tracks, droppings, or any damage it is causing. (Read the appropriate chapter for specifics on any suspects.) Always be cautious around animal droppings; they can contain organisms extremely harmful to people.

Step 2.
Do some detective work to locate the animal's method of entry

Inspect the outside of the structure for visible entrances. More than one entry may be used, and entry holes are often smaller than expected. (Small native squirrels enter holes 2 inches [5 cm] in diameter; Eastern gray and fox squirrels chew open baseball-size entries.)

Common points of entry are around utility cables and pipes that come into a structure, attic louvers, and roof vents, as well as holes in roofs, siding, soffits, and foundations (Fig. 1). Raccoons often leave scratches, tracks, and body oil stains where they shimmy up downspouts, trees, or the corners of buildings to access roofs. Rats, skunks, and marmots often dig under foundations or concrete slabs.

Use a bright flashlight to locate holes in shadowy areas,

and a ladder to search for holes high on a structure. You know there is a hole when you shine the light at an area and it remains black. (This is because the light is entering the hole, instead of reflecting back to you off the structure.)

(For information on raccoons or tree squirrels down chimneys, see "Preventing Conflicts" in the appropriate chapter.)

Step 3.
Determine the animal's main entry

Indications that you've found the animal's main entry include a newly dug hole or dirt stains, nest materials, and/or hairs stuck around a narrow hole in the roof, siding, or foundation. Fresh animal tracks may be found in dry soil near the entry.

To verify that an entry is being used, lightly stuff wadded-up newspaper, burlap, or dirt in the entry and watch daily to see if the material has been moved. (Don't use this technique if you think it may be bats or birds entering; they will get trapped inside.)

An alternative approach at ground level is to spread a tracking patch outside the entry, covering an area large enough to record footprints as the animal enters or exits. A tracking patch is a light layer of an inert material such as unscented baby powder, fine dirt, or sand. Don't use flour—it may attract a hungry animal.

If you can't find the entry, during daylight and with a strong flashlight or headlamp containing fresh batteries, very carefully enter the attic, crawlspace, or other area. Wear gloves and a dust mask or a respirator (see "Public Health Concerns" in Chapter 29), and

be on alert for animal life. From inside, you can better inspect the screening on the vents for signs of entry. Turn the light off to reveal light coming through any potential entry holes in the roof or walls. Securing something in these holes will make locating them from the outside easier when it comes time for repair.

Step 4.
Determine if young are involved

After finding the main entry you need to verify that no young are inside before proceeding with the eviction process. Because each situation and each animal is different, do this even if it seems early or late in the year for young to be present. Failing to do so can lead to major problems from an unhappy female animal separated from her young.

Reenter the attic, crawl space, or similar place and search for a nest or young. Focus on the area near the active entry or where you have been hearing noises. Squirrel nests are often made of insulation and other material that is torn up or piled within 20 feet (6 m) of the entry, and close to the outer edges of the attic. Raccoons, skunks, and river otters don't make an obvious nest.

To get the young to move or make noise and alert you to their presence, pound on a floor joist, ceiling joist, or wall. A stick may be used to search for babies in hard-to-reach places, such as in a wall between studs.

Note: Use care to prevent injuring any animals, and never approach a mother with her young; her protective instincts can make her very dangerous.

Step 5.
Evict the family only
when necessary

If young are present, the most humane thing to do is to leave the family alone until they move on their own. Squirrels, raccoons, opossums, and other young mammals generally leave the nest area eight to ten weeks after being born. Occasionally one of the young may stay behind, in which case the eviction methods described in Step 6 may be used.

If the young need to be moved, you will want to get the mother to move them on her own using one of the techniques described below. Even in an emergency, females can often be persuaded to move their young, thus avoiding the need to trap or euthanize families.

Note: Anytime you try to evict a mother animal and her young there is a chance that she may leave some or all of her young behind. If the young end up as orphans they will not survive in the wild without mom. In such a case, they should be taken to a local wildlife rehabilitator. Do not attempt to care for the animals yourself. Not only could you further harm the animals, it may be illegal for you to do so. (See "Wildlife Rehabilitators and Wildlife Rehabilitation" in Chapter 20 for information.)

Note: State and provincial wildlife offices have requirements that you'll need to follow in certain areas, including mandatory euthanasia of certain species, such as raccoons, Eastern gray squirrels, and opossums.

If you are lucky and the weather is fair, the mother may move her young, even newly born ones, to an alternate den within an hour or so after they have been disturbed. If the weather isn't favorable, or she has to find a new den or build a new nest, it may take a couple of days.

Each animal is different, and river otters in particular tend to be quite stubborn when they have young with them. To help the eviction process go smoothly, keep children and pets away from the animal's entry.

If steps taken to evict the family are unsuccessful and the young must be moved immediately, the female can be live-trapped and the dependent young placed in a weather-protected releasing box. Place the box outside and adjacent to the point of entry—after the entry has been sealed to prevent reentry. This will allow the mother to relocate her young at her own pace. It is recommended that a wildlife control company experienced with live trapping and releasing boxes complete this procedure.

Step 6.
Begin the removal process

If no young are present, an option at this point is to live-trap the animal (see Appendix A, "Trapping Wildlife").

Because trapping presents additional problems for both the trapper and the animal, the preferred option is to get the animal to leave on its own. This will require effort on your part in the form of encouraging the animal to leave, and then following up to make sure the animal doesn't return, or a different animal take its place.

Begin the removal process by sealing off all entries but the active one. First, carefully seal any potential entries, as the animal will seek other ways to get back inside. Use wood, ¼-inch (6 mm) mesh galvanized hardware cloth, sheet metal, aluminum flashing, or another sturdy material that will prevent the animal from entering. Small holes in hard-to-reach locations can be plugged with wadded-up wire, copper Stuf-fit®, or copper or stainless-steel mesh scouring pads (steel wool quickly corrodes after becoming wet). High-quality and reasonably priced bulk material is available for larger jobs. If necessary, foam or caulk the openings to seal them. Paint will help hide the repair job.

To create a barrier along a foundation to prevent skunks and other species from burrowing, refer to Figure 4a and "Preventing Conflicts" in Chapter 23.

After all entries except the active one are sealed, and during a period of fair weather, encourage the animal to leave using one or more of the following methods:

Seal the remaining entry hole while the animal is outside feeding.
Note: Do not do this if young are present; they will be separated from their mother, which will quickly create other problems.

First, have all materials ready that are needed to seal the entry. Next, place wadded-up newspaper in the entry or use a tracking patch as described in Step 3 to determine that the animal has gone outside. For squirrels and other species that are active during the day, look for the signs that they have exited early in the morning; for raccoons and other nocturnal species, begin the surveillance an hour after dark. Survey the entry frequently, as animals will return to rest or escape bad weather. When you are certain the animal is outside, seal the entry to prevent the animal from reentering.

An alternative approach—and one to use if mobile young are present—is to lightly pack the active entry hole with wadded-up newspaper, burlap, or dirt, and repack it whenever you see it open. Just block the hole enough so the animal must expend energy to reopen it, but not get trapped inside. When the barrier has not been removed for three days during fair weather, the animals have gone and repairs can be made to prevent reentry.

Harass the animal. Simply banging on the ceiling, wall, or floor in the vicinity of the animal may cause it to vacate; also, your initial search for young may have already made the animal uncomfortable enough to leave.

Alternatively, with a powerful flashlight or headlamp containing fresh batteries, and wearing gloves and a dust mask or respirator, carefully enter the area where you think the animal and/or its young are sleeping. Shine the light on the adult animal, bang on a rafter, clap your hands and tell the animal to leave, or do anything that doesn't put you or the animals in danger. If the adult is outside, gently tamper with the nest—pull off the top and/or slide it over a foot or so.

In addition, roll rags into tight balls and secure them with twine or tape. Sprinkle the rag balls with predator urine available from farm supply centers, hunting shops, or over the Internet (see "Wildlife/ Human Conflicts" Appendix F), and throw or place them near the nest. Sprinkling stinky kitty litter around the nest will also create an

unpleasant atmosphere; Raccoon Eviction Fluid® works well on raccoon families.

The animal(s) may leave within the hour or it may take a couple of days. Revisit the area to see if the young are gone, and to make sure the adult didn't move them elsewhere within the structure.

Use wadded-up newspaper as described above to verify that the animal is gone, and make the necessary repairs to prevent reentry.

Intensely harass the animal. Using a mechanic's bright droplight (grid enclosed bulb) or other portable light located away from burnable objects, light up the sleeping area being used by the animal. (A fluorescent light will conserve electricity and keep the heat level down.) In addition, put a radio in the area and play a talk station as loud as you can tolerate. If the animal moves to an unlit area, move the light and radio to that area, or install an additional light and radio.

(There is no scientific evidence that commercially available ultrasonic devices will drive animals from buildings. Animals quickly become accustomed to the noise or move to a noise-free area because the devices do not penetrate objects, and the sounds quickly lose their intensity with distance.)

Leave the lights and radio on 24 hours a day to interrupt the animals' sleep. Use a visual verification, a one-way door (see "One-Way Doors"), or the wadded-up newspaper or tracking-patch approach described above to verify that the animal has gone. Be patient—it may take several days for the animal to make the move, especially in

urban areas where animals are used to lights and noises.

Step 7.
Follow up

If you hear noise coming from inside the enclosure after sealing the entry, an animal may be inside. Reopen the area and repeat Step 6 until all the animals have departed. Then reseal the entry. If for some unknown reason an animal will not leave the area, it can be live-trapped (see Appendix A, "Trapping Wildlife").

Make frequent inspections for two weeks to make sure an animal hasn't tried to get back inside using the original entry or a new entry. Where one animal enters, a scent trail is left which others may find and use. This scent lasts for several months, sometimes longer. As a preventative measure, pepper spray or a commercial taste repellent such as Ropell® can be applied to the area. Applications will need to be repeated if the area is exposed to damp weather.

Consider hiring a wildlife damage control company to inspect for piles of droppings and other contaminants. If an animal has spent a lot of time in an area with exposed wiring, inspect the area for wire damage or have an electrician inspect it. (You should also inspect for damage done to insulation and heating ducts.) In the meantime, check your smoke detectors to make sure they are functioning in case of a fire.

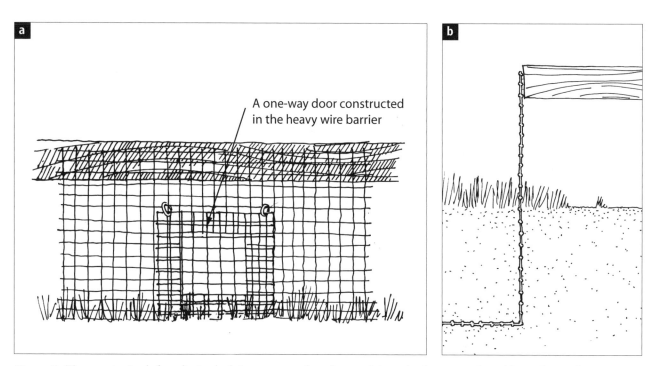

Figure 3. To prevent animals from digging back into an area where they are being evicted, one-way doors (a) are often used in conjunction with an L-shaped footer made of welded wire or hardware cloth (b). (Drawing by Jenifer Rees.)

One-Way Doors

An active entry can be fitted with a one-way door so an animal can exit but not reenter. A one-way door takes time and effort to install correctly, but is effective at evicting squirrels, raccoons and other animals above the ground on buildings where they have gained access. Ready-made one-way doors that trap burrowing animals

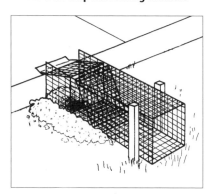

Figure 2. A ready-made one-way door can be set to evict animals that burrow under houses and concrete slabs. (From Hodge, Wild Neighbors: The Humane Approach to Living with Wildlife.)

(Fig. 2) are available from companies advertising over the Internet (search for "Animal Control" or "Animal Traps"); also see "Traps and Trapping" in Appendix F.

A one-way door must be used only when you can be sure that no young will be trapped inside after the adult is evicted. Thoroughly inspect the area for young prior to installation.

Leave the one-way door in place for seven days (longer during particularly cool or rainy weather). To verify the one-way door's success, look for scraping or digging on the outside of the door—this means the animal is out and can't get back in. For further proof, place a tracking patch on the outside of the one-way door, as described in Step 3, and keep an eye out for prints. After all animals have been excluded, remove the door and immediately seal up the exit.

A simple one-way door can be constructed from plywood, sheet metal, or ¼-inch (6 mm) mesh hard-

ware cloth. Attach the top to the structure with strap hinges (wood) or fence staples (sheet metal, hardware cloth) to create a flap door that opens easily and closes completely. The door should extend out at least 6 inches (15 cm) from all sides of the exit hole. To help prevent the animal from reentering, the bottom of the door can be weighted with a piece of rebar or a similar heavy object. To further help prevent an animal from trying to open a wire-mesh door, the wires around the edges can be bent out to create sharp points.

On angled areas (trim, eves, etc.) where gravity would keep the door open, use two small screw-eyes below the door, and run fish line from the bottom of the door through the screw eyes. Weight the ends of the line with a few metal nuts or whatever is needed to pull the door closed.

Hiring a Wildlife Damage Control Company

State and provincial wildlife departments are legislatively mandated to preserve, protect, and perpetuate wildlife. They also have the responsibility to assure that individual animals do not pose a threat to human safety or create unreasonable damage to crops, livestock, or property.

The expanding human population and the habitat alteration (or loss) accompanying it are resulting in a progressive increase in the frequency of wildlife/human conflicts. Although laws give citizens latitude to deal with problems, many people are either unwilling or unable to handle human/wildlife conflicts.

In addition to their official staff, the wildlife departments of Oregon, Washington, and British Columbia call on the skills of private citizens who have knowledge and training in the capture and handling of many wildlife species that commonly generate wildlife complaints. Typically these individuals are referred to as "wildlife control operators," "nuisance wildlife trappers," or simply "trappers." Although they must have permits from the wildlife department, and conform to its regulations, they are not state or provincial employees. They operate as private enterprises and normally charge a fee for their services.

Under the authority of their permit or letter of authorization, wildlife control operators are able to trap, capture, and transport animals for relocation year-round.

State and provincial wildlife offices continue to provide technical advice and/or informational pamphlets on request to citizens who are experiencing wildlife problems. Wildlife control operators, however, provide direct assistance to landowners who are willing to pay for the cost of trained individuals to resolve their wildlife problems.

While many conflicts can be solved with information about an animal's activities, or by adopting a more tolerant stance or doing some repair work, wildlife control operators are recommended for work that poses health or safety hazards. Examples include removing a large accumulation of droppings, removing a mother and/or her young from a precarious location, or installing a prevention device such as a chimney cap.

To find a wildlife control operator, contact your local wildlife office for names of companies or individuals that specialize in wildlife control work in your area. You may also look under "Animal Control," "Pest Control" or "Wildlife" in your local phone directory.

The wildlife control business is a new industry and companies vary widely in expertise and professionalism. How do you tell whom you're dealing with?

Follow these guidelines to choose a company that suits your needs:

- Does the representative listen to the description of the problem and ask relevant questions?
- Does the person appear to be professional and care about their work, the animals involved, and your concerns?
- Does the person appear knowledgeable and take the time to explain not only what the source of the problem is, but also its causes and potential solutions?
- Is the person licensed by the state or province, and bonded or insured against any incidental damage that might occur?
- Is the person willing to give you names and phone numbers of satisfied customers?
- Does the person try to scare you with talk about wildlife diseases or dangerous animals, or do they simply make you aware that you need to be cautious when dealing with wildlife to avoid the risk of infection?
- Are the procedures to be used simply and concisely explained, with a clear indication as to whether any procedure could harm the animal? (**Note:** The current laws may require mandatory euthanization of certain species. Call your local wildlife office for current requirements.)
- Does the person have more than one recommendation to resolve the problem—including nonlethal solutions?
- Does this person's approach to the problem include fixing it so that it does not reoccur? This should include discussion of needed structural repairs or

changes (such as installation of chimney caps or other exclusion devices to prevent animals from getting into areas where they are not wanted), and/or ways you should alter your own habits (e.g., birdfeeder or trash maintenance).

- Is any part of the work guaranteed? Although the kind of guarantee will vary depending upon the species involved and the type of work being performed, getting a guarantee suggests the person might be in business long enough to back it.
- Does the person offer a written contract? This is a must!
- Does the person provide a variety of pricing options to fit your budget? For example, can you share in the work by checking traps to save trips to your home?
- Ask the person who is responsible for checking the traps how

often the traps are to be checked. (The correct answer is that the traps must be checked daily, including weekends and holidays. If you must check the traps then the trapper must be available to remove the trapped animals.)

And, finally but importantly:

- Just because a company charges a lot of money for its services doesn't necessarily mean that it is better or more reputable than other companies. Be cautious of low quotes; you often get what you pay for.
- Discuss the situation with someone else and do the math to make your own estimate of what you are going to pay per hour for the job. Consider the following: How dangerous is the job? (Ladder work is always dangerous.) How much travel

and equipment is involved to resolve the problem? (If the person has to travel 20 miles one-way to reach your location, they will need to get paid for the time both ways.) Also consider how expensive it is to live in your area, and what kind of warranty or guarantee the company provides.

- Be wary of a company that requires all the money up front. Any reputable company should be satisfied with 50 percent down and the remaining amount due upon completion of the job.
- Be in control of all your negotiations and do not be pressured into buying the service. If it doesn't feel right, take your business elsewhere.

The Impact of Domestic Cats and Dogs on Wildlife

Often—unbeknownst to their loving owners—pet cats and dogs that are allowed to roam free represent a serious threat to wild animals. With forethought and preventive measures, however, people and their pets can coexist with wildlife.

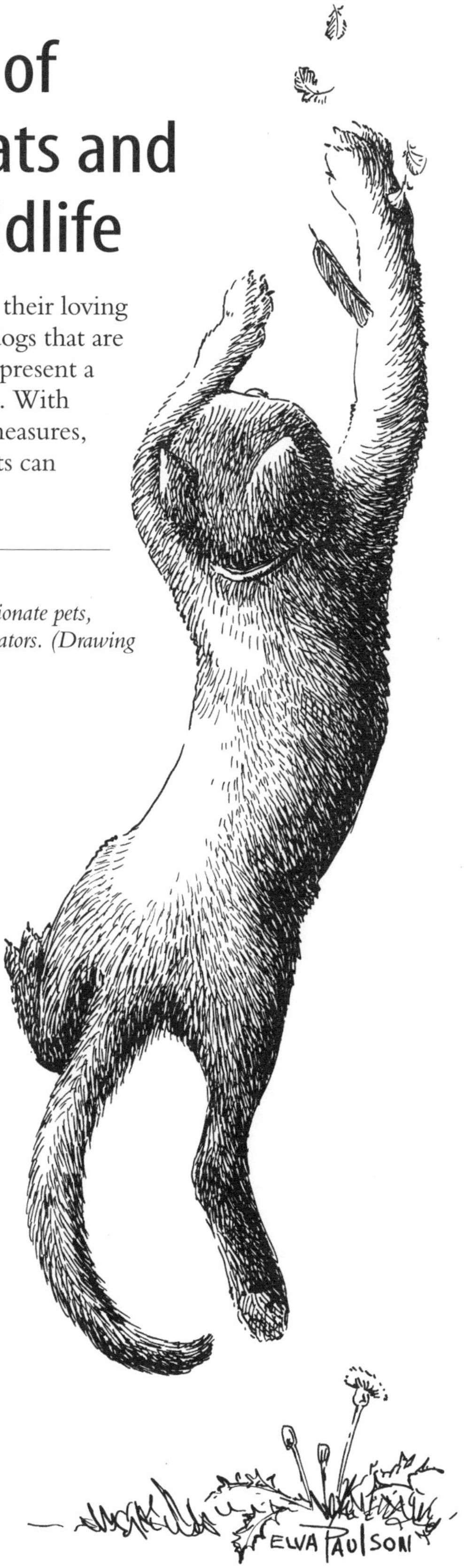

Figure 1. Although cats make affectionate pets, many hunt as effectively as wild predators. (Drawing by Elva Hamerstrom Paulson.)

Domestic Cats

The following questions and answers address some misunderstandings about free-ranging domestic cats:

Do cats really injure and kill a significant number of wild animals?
Yes. Although cats make affectionate pets, many hunt as effectively as wild predators (Fig. 1). Wildlife rehabilitators in Washington State have reported that, of all the animals treated in wildlife rehabilitation centers, nearly 20 percent were injured by cats. This number represents only those animals that were attacked by cats, rescued by humans, and transported to a wildlife rehabilitator. For every animal that makes it, there were likely countless others that were found dead or never found at all.

Studies in other states have shown that cats are a major threat to bird populations, especially ground-nesting species; a Wisconsin study tallied 19 million songbirds and 140,000 game birds killed by cats each year in that state (Fig. 2).

Note: Only about half of the animals killed by pet cats are brought home.

Can I teach my cats not to hunt wildlife?
No. The domestic cat, a descendant of the wild cat of Africa and southwestern Asia, instinctively hunts and captures prey. Cats cannot be trained to ignore these natural hunting instincts. In one study, six cats were presented with a live small rat while eating their preferred food. All six cats stopped eating the food, killed the rat, and then resumed eating the food.

Note: Even cats that have been declawed are still able to bat down and kill wildlife.

Figure 2. Quail are among the many species of ground-feeding birds that domestic cats prey on. (Washington Department of Fish and Wildlife.)

Since cats are hunters, don't they fill a niche in the food chain?

No. The cats that do the most damage to wildlife are usually well-fed, healthy, strong pets that do not need to hunt to survive—in fact such cats can focus hunting activities on a broader array of wild animals.

Cats are also nonselective predators and capture healthy as well as sick and injured wildlife. Some of these are house mice and rats, but many are native species including songbirds, snakes, lizards, squirrels, rabbits, and chipmunks, whose populations may already be stressed by other factors, such as habitat destruction.

Rural cats kill many more wild animals than do urban or suburban cats. Several studies found that up to 90 percent of the diet of free-ranging rural cats was wild animals, and less than 10 percent of rural cats did not kill wild animals.

Can cats live happily indoors?

Yes. Indoor cats lead long and healthy lives. The Humane Society of the United States reports that the expected life span of an indoor cat is at least triple that of cats that spend their lives outside.

Indoor cats avoid the dangers of traffic, diseases, fleas, parasites, poisoning, and confrontations with other cats, dogs, or wildlife. Pet owners benefit from lower health-care expenses and the knowledge that their cats are safe.

It is easier to start kittens on a completely indoor living regime, but you can change an outdoor cat into an indoor one (see "How to Have a Happy Indoor Cat").

Having your cat neutered or spayed reduces aggression, wandering, and territorial behavior (including the urge to spray).

Does putting a bell or bells on a cat really do any good?

No. Studies have shown that bells on collars are not effective in preventing cats from killing birds or other wildlife (Fig. 3). Birds do not necessarily associate the sound of a bell with danger. Cats with bells also can learn to silently stalk their prey. Even if the bell on the collar rings, ground-nesting baby birds, baby mammals, and slow-moving species have little chance of escape.

How do veterinarians feel about keeping cats indoors?

The American Veterinary Medical Association, the nation's largest professional veterinary group, passed a resolution on June 1, 2001, strongly encouraging cat owners in urban and suburban areas to keep their cats indoors. The Association of Avian Veterinarians and the Alliance of Veterinarians for the Environment also support keeping cats indoors.

My cat is clearly unhappy since I've started keeping it inside. What should I do?

Despite your best efforts, some cats may develop behavioral problems when they are no longer allowed outside. Most of these problems can be attributed to a change in routine that is too abrupt or to a lack of attention and stimulation indoors.

If your cat shows signs of extreme stress, you and your veterinarian might consider short-term drug therapy to relax the

Figure 3. Studies have shown that bells on collars are not effective in preventing cats from killing birds or other wildlife. (Drawing by Elva Hamerstrom Paulson.)

How to Have a Happy Indoor Cat

Although it takes patience, an outdoor cat can transition into a perfectly content indoor pet. The key is to make the conversion gradually and to provide lots of attention and stimulation while the cat is indoors.

Cats are creatures of habit, so you must be careful to slowly replace your cat's old routine of going outside with the new exciting routine of staying in. If your cat is outdoors most of the time, bring your cat inside for increasingly longer stays. Gradually shorten the length of time the cat is outside until you no longer let him or her out at all.

Cats need human companionship to be happy, and when they spend all their time outdoors, they get very little attention. An outdoor cat may welcome the indoors if he or she gets more love, attention, and play time.

To make life more enjoyable for your indoor cat, try the following:

Make life good.
- Spend quality time loving your cat: it is dependent on you for care and companionship.
- Build or buy resting shelves or high perches for your cat. The taller models, especially those with multiple perches, make the most of vertical space and appeal to cats' natural interest in heights.
- Cats love to sun themselves and enjoy looking outdoors. Install perches or shelves to provide your cat with more windows of opportunity.
- Provide a scratching post (one with different levels is good), corrugated cardboard, or sisal rope for your cat to scratch. Praise your cat for using it.
- Safely tether your cat outside on a "dog trolley" for part of the day (see "Domestic Dogs" for safety tips). Make sure coyotes, dogs, and other predators can't access the tethered animal.
- Consider adopting a playmate for your cat. Two cats are just as easy to care for as one and they keep each other company.

Make life interesting.
- Place a bird feeder near a window to attract a variety of wildlife and engage the interest of the indoor cat. (See Chapter 37 for detailed information on feeder management.)
- Amuse your cat with safe toys such as boxes, paper bags, crumpled paper, soft rope, or a ping-pong ball in a tub of water. Some cats enjoy searching for toys. Cats love to chase catnip toys, ribbons or strips of fabric, and feathers dangled in front of them on wands or short poles.
- If your cat likes to explore the house looking for "prey," hide its toys at various places so the cat can find them throughout the day. The key here is variety. Rotate favorites in and out of your cat's toy box.

Make life safe.
- Give your cat its own protected, screened outside enclosure that it can access through a window or pet door. Try to make the enclosure at least 5 feet (1.5 m) wide and 12 feet (3.6 m) long. Add a ceiling or make the sides high enough to prevent the cat from jumping out.
- Place identification and license tags on your cat's collar, or get the cat microchipped. If it accidentally gets outdoors, this may enable your cat to be returned home safely.

Make life healthy.
- When the weather allows, leave securely screened windows open so your cat can get fresh air.
- Bring in a pot of fresh green grass or catnip for the cat to chew. Garden centers and pet stores sell wheat or oat grass seed to be planted in small pots for indoor cats.
- Clean your cat's litter box regularly and have fresh drinking water available at all times.
- Feel good about your indoor arrangement. You're doing your cat—and wildlife—a favor by giving them the protection they deserve!

If you cannot or prefer not to offer your cat a run or enclosure, consider leash-training the cat so you can supervise its time outside. All you need is a leash and a sturdy figure-eight or figure-H style harness from which the cat cannot escape. Your cat may resist wearing one of these at first, so let him or her become accustomed to it gradually. Put it on for brief periods indoors, and then later attach the leash and walk him or her around the house. When the cat becomes comfortable with that, venture outdoors for short trips. Cats may resist leash-training at first, but most cats will eventually accept the leash.

Finally, remind family members, housemates, and visitors not to let the cat outside. Post signs near all doors, and if you live with children, teach them to close the door behind them. Be especially cautious with screen doors that may not latch tightly. See "How to Prevent Cat Escapes."

Figure 4. A low fence around a bird feeder won't keep cats out, but if ground-feeding birds are ambushed, it will give them a chance to escape. (Drawing by Jenifer Rees.)

cat during the transition. Another option is to explore homeopathic remedies, which can be effective calming agents in many animals. Consult a veterinarian for specific treatments, or find a pet supply store that sells safe and easily administered remedies that help to take the edge off the upset.

If you are absolutely convinced that the indoor life simply does not fit your cat's lifestyle, let your cat outside at a regular time each day, and try to keep the amount of outside time to a minimum. This is extremely important during the wildlife breeding season—March through June.

Pledge to make your next cat an indoor-only cat.

I keep my cats indoors but my neighbors let theirs roam. Can I do anything to keep their cats from killing animals in my yard?

Yes. First, try talking with your neighbors, sharing your concerns, and giving them a copy of this Appendix. Many people are unaware of the pressures their pets put on wildlife populations or of the benefits of indoor life for cats. If you are not a pet owner, keep in mind the emotional attachment your pet-owning neighbors have for their animals when sharing this information with them.

To discourage wandering cats:

- Sprinkle a dog and cat repellent, available from pet stores and garden centers, where cats like to lurk.
- Although it requires persistence, spraying cats with water from a hose or large squirt gun is effective for the short term. Short-term success has also come from laying spring mousetraps under several sheets of flat newspapers

where cats lurk or walk. Reset traps as needed.

- Install a commercially available motion-activated water sprayer, such as the Scarecrow®. When a cat comes into its adjustable, motion-detecting range, a sharp burst of water is sprayed in the direction of the cat.
- Keep your dog inside a fenced yard.

To prevent problems at bird feeders:

- Place your bird feeder at the top of a tall, nonclimbable pole, or one equipped with a squirrel baffle (see Chapter 24 for an example). Stop feeding birds until you take measures to prevent cats from successfully killing birds.
- Keep the area around bird feeders and birdbaths free of shrubbery. Absence of cover prevents cats from ambushing birds at these locations.

- A 2-foot (60 cm) high wire fence can be placed around a feeder (Fig. 4), a birdbath, or other areas where there is a concentration of birds at ground level. A low fence won't keep cats out, but will give birds the opportunity to escape. A higher fence will eliminate the problem. (See Chapter 37 for an example of a fence around a birdbath.)

To deter cats from walking on or climbing over a fence:

- Apply a thin layer of Tanglefoot® to the top of the fence. More is not better in this case.
- Install Kitty Klips® or anther commercially available product on top or inside the fence (Fig. 5). The pipe is too big and smooth for the cat to get a grip on. (See "Wildlife/Human Conflicts" in Appendix F for contact information.)
- To keep cats inside a yard, add a barrier made from 1-inch

How to Prevent Cat Escapes

Our feline friends are sneaky, quick, and curious—attributes that make them exceptional escape artists if given the opportunity. Here's how to keep your indoor-only cat from answering the call of the wild, along with important steps to take if he or she manages to make a break for the not-always-so-great outdoors.

- Keep doors and windows shut and make sure screen doors latch properly. If you do partially open windows, check screens often to see they're secure and intact.
- Pay attention to your cat's whereabouts whenever entering or leaving the house.
- Be particularly careful when trying to open, go through, and shut the door while carrying something like an armload of groceries.
- Make sure the whole family understands and adheres to rules for the cat's safety.
- Alert visitors and repairmen of your indoor-only cat rules. Close your cats in a bedroom when workers, guests, or family members will be going in and out.
- Train your cat to avoid doors that lead outside. Teach cats to shun these areas by setting them up for some negative reinforcement. For example, stand outside with a spray bottle while the door is cracked open enough to attract the curious feline.
- To handle a "bolter" that waits near the door so he or she can dart outside as you're juggling groceries, try enlisting a friend to make a loud noise outside the door when you open it, such as banging a pan with a spoon or shaking a pop can with pennies in it.
- Spaying or neutering your pet will help curb its roaming tendencies, along with providing a stimulating indoor environment that includes plenty of toys, climbing opportunities, and playtime.
- Training an indoor cat to come when called should be a priority, in case it does get outside. To help train the animal, call it by name each time you feed it or summon it for playtime.

If the cat is missing, act fast. If you're not positive your cat went outside, first conduct a thorough search of your house and check all those hidden spots cats love to squeeze into. Next, scour the immediate area around your home. See if you can persuade the escapee to remain in the area by setting out some fragrant food, his or her litter box, scratching post, and anything else that carries the animal's scent.

Here's what to do to recover the cat as quickly as possible:

- Take a casual approach if your cat is still in sight. Try not to panic and run shouting after your possibly spooked pet; this could chase him or her even farther from home.
- Try using a lure toy or another item that interests your cat to playfully tempt him back into the house.
- Food is another great motivator. A cat that has been trained to eagerly respond when you tap a spoon against a cat food can or shake a bag of treats might be lured within reach by one of these sounds.
- If your cat escapes and you can't catch it, consider renting a cat trapping cage from the local Humane Society (see Table 1 in Appendix A for detailed information).
- Spread the word. Alert your neighbors that your cat is missing and ask them to tell their children, too— kids tend to get around the neighborhood more than adults. Post signs and pass out flyers that include a clear photo and description of your pet, its name, your contact information, and "Reward" on them (you don't need to specify an amount).
- Don't give up too soon.

Adapted from: *Housecat: How to Keep Your Indoor Cat Sane & Sound* by Christine Church.

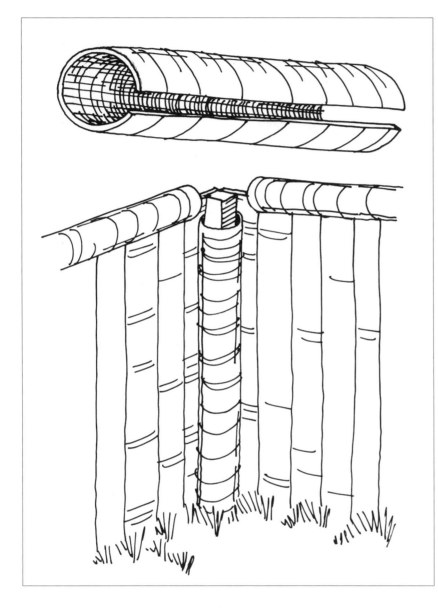

Figure 5. The Kitty Klip® system works by preventing cats (and dogs) from gaining hold of the fence. To use the Kitty Klip® system on the top of a wooden fence, take 4-inch (10 cm) PVC pipe and saw a line down the length. Then simply pry open the "klip" and slip it onto the top of the fence. If the length of this is fairly long, you may need a second person to assist in this step. It is also helpful to wear gloves in order to prevent your fingers from getting pinched. If you can see the 2 x 4 and 4 x 4 inch support posts on a wooden fence, you will need to cut enough pipe to fit both the upper horizontal 2 x 4s, as well as the upper part of the vertical 4 x 4s. You will also need to make another cut to remove 1 inch (2.5 cm) of pipe to make "klipping" the pipe over the thicker boards easier. (See top of illustration.) (Drawing by Jenifer Rees.)

(2.5 cm) chicken wire slanted at about 45 degrees (or horizontal) toward the cat (Fig. 6). Use this large mesh wire so cats can't get their feet caught.

- Be careful about placing chairs or planters too close to walls and fences. If the cats have a leg up, they may be able to simply clear the fence or wall.

Are there any laws regulating free-roaming cats?

The answer depends on where you live. Because laws vary, one should check local ordinances for the appropriate way to deal with free-roaming cats. The answer may be that there is no clear legal way to deal with the problem if animal control laws are murky or your town does not fund cat control.

In most areas, if cats wander onto other people's property, they can be live-trapped in a cage trap and either returned to the owner or turned over to authorities. If the cat has proper identification on its collar, it can be returned unharmed to its owner by you, an animal control agent, or another professional hired to trap the cat. In some cases, a letter delivered to the neighbor describing the trapping process that is about to begin is all that is necessary to curtail the problem—especially if the cat's owner is informed of the fees involved in trapping the cat. Some areas have "Cat at Large" fines and if the animal is not neutered or its shots up-to-date, additional fines may be charged. (See Appendix A for information on trapping domestic cats.)

What can I do if I find wildlife that has been injured by a cat?

A bird or other wild animal that seems to have escaped a cat attack has often not escaped uninjured. The animal may be doomed, either from bacteria or viruses in cat saliva, or from internal injuries. Because birds have high metabolisms, they are particularly susceptible to bacteria from cat bites and scratches, and infections set in quickly.

Any wild animal that has been caught by a cat probably needs medical care to survive. Do not attempt to care for the animals yourself. Not only could you further harm the animal, it may be illegal for you to do so. Instead, call your closest wildlife rehabilitation center for assistance. (For detailed information on wildlife rehabilitators, see Chapter 20.)

What about unwanted or stray cats?
Finally, don't dispose of unwanted cats by releasing them in rural areas, parks, or anywhere else. This practice enlarges feral cat populations and is an inhumane way of dealing with unwanted cats (see "Feral Cats are Not Wildlife in Need of Help"). Cats suffer in an unfamiliar setting, even if they are good predators. Contact your local animal welfare organization for help.

Don't feed stray cats. Feeding strays maintains high densities of cats that kill and compete with native wildlife populations.

Domestic Dogs

The following questions and answers address some misunderstandings about free-ranging domestic dogs:

Do dogs really injure and kill a significant number of wild animals?
Yes. Dogs that are allowed to roam free will often harass, injure, or kill deer, raccoons, skunks, squirrels, chipmunks, rabbits, grouse, quail, and other wildlife. Even the most obedient dog may harass deer or other wildlife if given the opportunity (Fig. 7).

Two or more dogs working together can relentlessly pursue an animal, causing a loss of valuable energy that is especially detrimental in cold weather. In deep snow crusted over with ice, deer and elk

punch through with their small, sharp hooves, while dogs race along the top on their big paws, snowshoe-style. A dog-chased deer or elk can be rapidly stressed in these conditions, allowing the dog to catch it and slowly tear it apart.

A nationwide survey of state departments of agriculture, wildlife conservation agencies, and related agencies was done to determine problems caused by unconfined dogs. Damage to wildlife, especially deer, small game, and birds was considered the primary problem caused by dogs. The second most serious problem reported was damage to livestock.

Washington Department of Fish and Wildlife offices, particularly on the east side of the state, receive hundreds of complaints about such situations every severe winter.

What can happen to my dog if it is allowed to run free?
Dogs running at large are in danger of being hit by a car or of being bitten or attacked by another dog or by wild animals, including coyotes, raccoons, and foxes. Although rabies is rare in the Pacific Northwest, dogs can contract rabies from an infected animal. They can also develop bacterial infections from bite or scratch wounds.

Many local jurisdictions consider free-ranging dogs to be feral and subject to being shot, especially if they are harassing livestock or otherwise being a nuisance or threat. Dogs chasing deer or elk in Washington are legally considered a nuisance under wildlife law and owners can be cited with a misdemeanor, subject to minimum fines of about $150. In severe winter conditions, wildlife law enforce-

What Did It— Coyote or Dog?

"Hobby farms" and "ranchettes" are more common today than ever before. Many people are enjoying a return to the rural lifestyle, living on a few acres and keeping livestock. Many of these new "farmers" and "ranchers" are quick to blame coyotes for killing and eating their animals. To help distinguish coyote predation from dog predation on lambs, calves, and similar-size animals, look for the following telltale signs:

COYOTE
- Usually attacks the front of a victim
- Bite marks found on throat of victim
- Seldom chews on ears and tail of victim
- Usually kills one animal each day
- May drag the remains of victim away
- Other nearby animals slightly stressed
- Feeds upon the animal it attacks

DOMESTIC DOG
- Usually attacks the rear of a victim
- Bite marks found everywhere on victim
- Often chews on ears and tail of victim
- Often kills more than one animal each day
- Seldom drags the remains of victim away
- Other nearby animals severely stressed
- Seldom feeds upon the animal it attacks

Note: Sometimes dogs kill the way coyotes do, and young and inexperienced coyotes may attack any part of the body of their prey as dogs would.

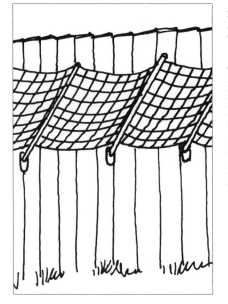

Figure 6. A barrier can keep cats (and dogs) within an area surrounded by an existing 5-foot (1.5 m) fence. Make sure there aren't places along the bottom of the fence that cats (and dogs) can crawl through, or overhanging branches or buildings, such as garages, that cats can climb over. (Drawing by Jenifer Rees.)

ment officers are often authorized to remove or even destroy offend-

ing dogs. For information on trapping dogs, see Appendix A.

What can be done to prevent dog/wildlife conflicts?

Keep dogs under restraint, either kenneled or leashed, whether you live in town or in the country. Suburban or rural living does not entitle your dog to roam free; in fact, the greater your proximity to wildlife, the greater your responsibility to keep dogs confined.

Neuter or spay your dog to help the animal's overall health and decrease the desire to roam. Neutering also reduces aggression and territorial behavior, which result in injuries from fighting with other dogs and other animals.

Keeping vaccinations current will protect both wildlife and your pet by preventing the spread of viruses and infectious disease. Both pet cats and dogs should be closely monitored during the nesting seasons of wildlife, generally in the spring and summer months.

Tethering Dogs and Cats

The following are guidelines for tethering a dog (or house cat):

- Purchase or construct a "dog trolley." To build one, slide a metal ring onto a length of heavy wire or rope. Fasten each end securely to suspend the wire between two posts. Connect the dog's lead to the ring. Make sure the lead is short enough to avoid tangling yet ensures enough length to allow the dog to lie down. The trolley may be either overhead or on the ground.
- Avoid chaining a dog near stairways and fences. Choose a safe area away from all possible escape routes.
- Provide at least 15 feet (4.5 m) of clear space for your dog to move around in. Remove all debris that could cause your pet to become tangled or injured.
- Do not use a choke chain. Use a harness or a leather or nylon collar instead, and check it regularly for proper fit.
- Provide shade, water, and shelter for the dog at all times.

ELVA PAULSON

Figure 7. Dogs that are allowed to roam free will often harass, injure, or kill deer, squirrels, rabbits, quail, snakes, and other wildlife. (Drawing by Elva Hamerstrom Paulson.)

Feral Cats are NOT Wildlife in Need of Support

As a backyard wildlife sanctuary manager, you most likely keep your cat confined and talk to cat-owning neighbors about doing the same.

But what about homeless cats?

"Feral" cats, which are usually strays that are untamed or wild, are estimated to range from 60 to 100 million throughout the United States. They are NOT wildlife. Feral cats are non-native predators that can, and have, seriously damaged wild bird and other wildlife populations.

The last known individuals of the Tacoma pocket gopher (*T. mazama tacomensis*) were killed by cats.

While domestic cats are solitary animals, colonies of feral cats often form around food sources like bird feeding stations, garbage dumps, or places where people deliberately leave food for them. In fact, many colonies of feral cats are supported by well-meaning but misinformed advocates of what's become known as "TNR" management: Trap, Neuter, Release.

This wrong solution to a tragic problem works this way: Feral cats are trapped and taken to a clinic or veterinarian for disease testing. Those that are seriously ill or test positive for contagious diseases are usually euthanized, otherwise they are simply spayed or neutered. Then the feral cats are released back to where they were trapped and where they are supplied with food and water daily.

The theory behind TNR programs, which are funded by both private and public entities across the country, is eventual reduction of feral cat colonies. But sadly, such claims are not substantiated.

Cat colonies often serve as dumping grounds for other unwanted cats.

The food provided usually attracts more cats. Contrary to TNR proponent beliefs, colony cats do NOT keep other cats from joining the colony.

As time goes on, some colony cats become too wary to be caught, so rarely are all spayed or neutered. With females capable of producing up to three litters of four to six kittens every year, it doesn't take long to grow a feral cat colony.

In addition to their threats to wildlife, feral cat colonies pose human health risks. Even TNR-managed colonies can spread disease such as ringworm, toxoplasmosis, cat scratch fever, and rabies, since every cat is not captured regularly for health care. *Note:* Cat feces has been implicated in infecting sea otters off the California coast with toxoplasmosis. Toxoplasmosis is thought to be a factor in the recent population crash of this species.

Free-roaming cats of any kind are their own worst enemy, too. They usually have short, miserable lives, due to collisions with motor vehicles, attacks by other domestic and wild animals, accidental poisoning or trapping, and parasites and diseases.

TNR management of feral cats is clearly not in the best interests of anyone, and it often overwhelms the ability of well-meaning people who genuinely want to help animals. It also undermines efforts of responsible pet owners who keep their cats indoors.

The National Association of State Public Health Veterinarians, American Association of Wildlife Veterinarians, American Bird Conservancy, American Ornithologists' Union, and Cooper Ornithological Society oppose TNR practices. In addition, the Humane Society of the United States, People for the Ethical Treatment of Animals, and American Society for the Prevention of Cruelty to Animals have expressed concerns about this practice.

So what should be done to protect wildlife and treat all animals humanely?

First and foremost, spay or neuter your own cats and help promote community-wide spay/neuter information campaigns and low or no-cost spay/neuter clinics. The fewer kittens produced and possibly abandoned, the smaller feral cat colonies will be.

Second, keep your cat indoors. Spread the word to other cat owners that indoor cats live longer lives and avoid harassing wildlife. Unspayed or non-neutered cats kept indoors also won't add to the feral cat population explosion.

Help inform and educate others that practicing TNR is not the solution for feral cat management. Initiate or support local ordinances that prohibit cat abandonment and feral cat feeding. Humanely trap and remove feral cats, especially in public areas that provide habitat for wildlife, and take them to an animal shelter for possible adoption or humane euthanasia.

Feral Dogs and Wolf Hybrids

Feral dogs may occur wherever people are present and permit dogs to roam free, or where people abandon unwanted dogs. In appearance, most feral dogs (sometimes referred to as wild or free-ranging dogs) are difficult, if not impossible, to distinguish from domestic dogs.

The primary feature that distinguishes feral from domestic dogs is the degree of dependence on humans, and, in some respects, their behavior toward people. Feral dogs survive and reproduce independently of human intervention or assistance. While it is true that some feral dogs use human garbage for food, others acquire their primary subsistence by hunting and scavenging like other wild canids (coyotes, foxes, wolves). In addition, feral dogs are usually secretive and wary of people, growling, barking, and attempting to bite if approached.

Feral dogs are best described as opportunistic feeders. They can be efficient predators, preying on small and large animals, including domestic livestock. Many rely on carrion, particularly road-killed animals, crippled waterfowl, green vegetation, berries and other fruits, and refuse at garbage dumps. They often travel in packs or groups.

To protect livestock and poultry from feral and domestic dogs, refer to "Preventing Conflicts" in Chapter 6.

Hybridization between feral dogs and wolves and coyotes can occur in nature, but nonsynchronous estrus periods and pack behavior (that is, habits that exclude nonresident canids from membership in the pack) may preclude much interbreeding.

The wolf is the ancestor of all dog breeds that exist today. The term "wolf hybrid" refers to an animal whose ancestry contains a high genetic percentage of wolf. While genetics are the only way to determine how much wolf and dog is in a hybrid, at this time our genetic tests are not sophisticated enough to make this determination. A true wolf hybrid will display primary wolf appearance and behaviors (Fig. 8).

Unless they are part of a special government program, where they are taught wild survival skills and are kept from becoming socialized to humans, captive wolf hybrids cannot humanly (nor in most cases legally) be released into the wild. They do not have the hunting skills necessary for survival, and because they are socialized to humans they will seek food near human habitations.

"Released" wolf hybrids slowly starve to death and/or create problems that may be blamed on wild wolves. A released wolf hybrid also is much more likely to be killed by a wild wolf than to mate with it. However if, due to highly unusual circumstances, a hybrid bred with a wild wolf, the resulting hybrid offspring would compromise the genetic soundness of its wild wolf pack.

(See "Wildlife/Human Conflicts" in Appendix F for additional information on wolf-hybrids.)

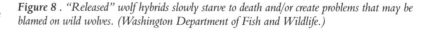

Figure 8. "Released" wolf hybrids slowly starve to death and/or create problems that may be blamed on wild wolves. (Washington Department of Fish and Wildlife.)

State, Provincial, and Federal Contact Information

State of Oregon

Oregon Department of Fish and Wildlife (ODFW)

ODFW is responsible for managing fish and wildlife resources, and regulating commercial and recreational harvests.

ODFW Headquarters

3406 Cherry Avenue NE
Salem, OR 97303-4924
Phone: (503) 947-6000
E-mail Odfw.Info@state.or.us
Web site: www.dfw.state.or.us/

ODFW Regional Offices
High Desert Region
61374 Parrell Road
Bend, OR 97702
Bend phone: (541) 388-6363
Hines phone: (541) 573-6582

Southwest Region
4192 N Umpqua Highway
Roseburg, OR 97470
Phone: (541) 440-3353

Northwest Region
17330 SE Evelyn Street
Clackamas, OR 97015
Phone: (503) 657-2000

Northeast Region
107 - 20th Street
La Grande, OR 97850
Phone: (541) 963-2138

Oregon State Laws

State laws provide for the protection of virtually every resident animal species. Where state law duplicates federal law, state law standards are often more restrictive than their federal counterparts. Local laws are sometimes more restrictive than state laws.

Oregon Administrative Rules:
http://arcweb.sos.state.or.us/
banners/rules.htm

Oregon Revised Statutes:
www.leg.state.or.us/ors/home.html

Oregon Department of Human Services (Public Health)

The Department of Human Services sets public health policy and provides administrative and technical assistance to county health departments and other local partners who deliver services in their communities.

ODHS Headquarters
500 Summer Street NE, E94
Salem, OR 97301-1097
Phone (voice): (503) 945-5733
TTY: (503) 945-6214
Fax: (503) 378-2897

Web site:
www.dhs.state.or.us/publichealth/

County health departments:
www.dhs.state.or.us/publichealth/
lhd/index.cfm

Oregon State University Extension Service

Oregon State University Extension Service provides education and information based on timely research to help Oregonians solve problems and develop skills related to youth, family, community, farm, forest, energy, and marine resources.

Headquarters
1849 NW Ninth Street
Corvallis, OR 97330-2144
Phone: (541) 766-6750
Web site: http://extension.oregon-state.edu/

County offices:
http://extension.oregonstate.edu/
locations.php

State of Washington

Washington Department of Fish and Wildlife (WDFW)

WDFW's mission is to provide sound stewardship of fish and wildlife.

Headquarters
Natural Resources Building
1111 Washington Street SE
Olympia, WA 98501

Mailing address:
600 Capitol Way N
Olympia, WA, 98501-1091
Telephone: (360) 902-2200
Fax: (360) 902-2230
Web site: www.wa.gov/wdfw/

WDFW Regional Offices

Eastern Washington – Region 1
8702 N Division Street
Spokane, WA 99218
Phone: (509) 456-4082

North Central Washington – Region 2
1550 Alder Street NW
Ephrata, WA 98823-9699
Phone: (509) 754-4624

South Central Washington – Region 3
1701 S 24th Avenue
Yakima, WA 98902-5720
Phone: (509) 575-2740

North Puget Sound – Region 4
16018 Mill Creek Boulevard
Mill Creek, WA 98012-1296
Phone: (425) 775-1311

Southwest Washington – Region 5
2108 Grand Boulevard
Vancouver, WA 98661
Phone: (360) 696-6211

Coastal Washington – Region 6
48 Devonshire Road
Montesano, WA 98563
Phone: (360) 249-4628

Washington State Laws

State laws provide for the protection of virtually every resident animal species. Where state law duplicates federal law, state law standards are often more restrictive than their federal counterparts. Local laws are sometimes more restrictive than state laws.

Revised Code of Washington (RCW) provisions:
www.leg.wa.gov/rcw/index.cfm

Washington Administration Code (WAC) provisions:
www.leg.wa.gov/wac/

Washington State University Cooperative Extension

Cooperative Extension Agents have a wide variety of home, garden, forestry, 4-H, and wildlife information. For your local county office, look under "Government," then county, in your phone directory. Publications are available from the WSU Web site, local county offices, or by mail.

Mailing address:
Cooperative Extension
Publications Building
Bulletin Department
Washington State University
PO 645912
Pullman, WA 99164-5912
Phone: (800) 723-1763 and (509) 335-2857
Web site: http://ext.wsu.edu/

County offices:
http://ext.wsu.edu/locations/

Washington Department of Health

State hotline: (800) 525-0127

For your county office, look under "County Health" or "Health" in a local phone directory or check the Web site below.

Web site: www.doh.wa.gov/

Local Health Department Web sites:
www.doh.wa.gov/LHJMap/LHJMap.htm

British Columbia

British Columbia Ministry of Water, Land, and Air Protection Offices

General responsibilities include environmental stewardship of biodiversity, including wildlife, fish, and protected areas; also park and wildlife recreation management, including hunting, angling, park recreation, and wildlife viewing.

Ministry Office
PO Box 9339 Stn Prov Govt
Victoria, BC
V8W 9M1 Canada
Phone: (250) 387-1161
Fax: (250) 387-5669
E-mail:
www.wlapmail@gems5.gov.bc.ca
Web site: www.gov.bc.ca/wlap/

Local Ministry Offices
Vancouver Island Region
2080 Labieux Road
Nanaimo, BC
V9T 6J9 Canada
Phone: (250) 751-3100

Lower Mainland Region
#10470 152nd Street, 2nd Floor
Surrey, BC
V3R 0Y3 Canada
Phone: (604) 582-5200

Thompson and Okanagan Regions
970-A Camosun Crescent
Kamloops, BC
V2C 6G2 Canada
Phone: (250) 371-6200

Kootenay Region
#401 333 Victoria Street
Nelson, BC
V1L 4K3 Canada
Phone: (250) 354-6333

Cariboo Region
#400 640 Borland Street
Williams Lake, BC
V2G 4T1 Canada
Phone: (250) 398-4530

Skeena Region
PO Box #5000
3726 Alfred Avenue
Smithers, BC
V0J 2N0 Canada
Phone: (250) 847-7260

Omineca and Peace Regions
Fourth Avenue, 3rd Floor
Prince George, BC
V2L 3H9 Canada
Phone: (250) 565-6135

Provincial Laws

Revised Statutes and Consolidated Regulations of British Columbia:
www.qp.gov.bc.ca/statreg/default.htm

Wildlife Act:
www.qp.gov.bc.ca/statreg/stat/W/96488_01.htm

Canadian Wildlife Service

The Canadian Wildlife Service of Environment Canada handles wildlife matters that are the responsibility of the federal government. These include the protection and management of migratory birds as well as nationally significant wildlife habitat. Other responsibilities are the identification of endangered species, control of international trade in endangered species, research on wildlife issues of national importance, and international wildlife treaties and issues.

Canadian Wildlife Service Environment Canada
Ottawa, Ontario
K1A 0H3 Canada
Telephone: (819) 997-1095
Fax: (819) 997-2756
E-mail: cws-scf@ec.gc.ca
Web site: www.cws-scf.ec.gc.ca/

Other Contacts

U.S. Department of Agriculture (Wildlife Services)

The Wildlife Services program, part of the U.S. Department of Agriculture's Animal and Plant Health Inspection Service (APHIS), helps alleviate wildlife damage to agricultural, urban, and natural resources. Wildlife Services also addresses wildlife threats to public health and safety and protects endangered and threatened species from predators.

Oregon phone:
(503) 326-2346

Washington phone:
(360) 753-9884

Web site: www.aphis.usda.gov/ws

U.S. Fish and Wildlife Service (USFWS Pacific Region)

USFWS works with others to preserve, protect and enhance fish, wildlife, and plants and their habitats for the continuing benefit of the American people.

Oregon phone:
(503) 231-6158

Washington phone:
(360) 753-9440

Web site: www.fws.gov

Pacific Region office directory: http://offices.fws.gov/directory/ListOffices.cfm

The Center for Disease Control and Prevention (CDC)

The CDC is recognized as the lead U.S. federal agency for protecting the health and safety of people, both at home and abroad, providing credible information to enhance health decisions, and promoting health through strong partnerships.

Public Inquiries

Phone: (800) 311-3435

Web site: http://www.cdc.gov/

Books, Organizations, and Internet Resources

Pacific Northwest Natural History (General)

Books

Csuti, Blair, et al. *Atlas of Oregon Wildlife: Distribution, Habitat, and Natural History*. Corvalis: Oregon State University Press, 1997.

Kozloff, Eugene N. *Plants and Animals of the Pacific Northwest*. Seattle: University of Washington Press, 1976.

Kruckeberg, Arthur R. *The Natural History of Puget Sound Country*. Seattle: University of Washington Press, 1991.

Mathews, Daniel. *Cascade-Olympic Natural History: A Trailside Reference*. Portland, OR: Raven Editions, 1999.

Schwartz, Susan H. *Nature in the Northwest: An Introduction to the Natural History and Ecology of the Northwest United States from the Rockies to the Pacific*. Englewood Cliffs, NJ: Prentice-Hall, 1983.

Mammals

Books

Christensen, James R., and Earl J. Larrison. *Mammals of the Pacific Northwest: A Pictorial Introduction*. Moscow, ID: University of Idaho Press, 1982.

Ingles, L. G. *Mammals of the Pacific States*. Stanford, CA: Stanford University Press, 1965.

Larrison, Earl J. *Mammals of the Northwest: Washington, Oregon, Idaho, and British Columbia*. Seattle: Seattle Audubon Society, 1976.

Maser, Chris. *Mammals of the Pacific Northwest: From the Coast to the High Cascades*. Corvalis: Oregon State University Press, 1998.

Muller-Schwarze, D., and Lixing Sun. *The Beaver: Natural History of a Wetland Engineer*. Ithaca, NY: Cornell University Press; and London: Comstock Publishing, 2003.

Nagorsen, David W., and Mark R. Brigham. *Bats of British Columbia*. Vancouver, BC: University of British Columbia Press, 1993.

Nagorsen, David W. *Royal British Columbia Museum Handbook: Opossums, Shrews, and Moles of British Columbia*. Vancouver, BC: University of British Columbia Press, 1996.

Shackleton, David. *Royal British Columbia Museum Handbook: Hoofed Mammals of British Columbia*. Vancouver, BC: University of British Columbia Press, 1999.

Verts, B. J., and Leslie N. Carraway. *Land Mammals of Oregon*. Los Angeles: University of California Press, 1998.

Internet Sites

Alaska Department of Fish and Game's Wildlife Notebook Series: www.state.ak.us/local/akpages/FISH.GAME/notebook/nothome.htm

Burke Museum's Mammals of Washington: www.washington.edu/burkemuseum/mammalogy/mamwash2.html

Canadian Wildlife Service's Hinterland Who's Who: www.cws-scf.ec.gc.ca/hwwfap/index_e.cfm

Mountain Beaver Journal: http://infowright.com/mtbeaver/

U.S. Forest Service Wildlife Species Life Form Information: www.fs.fed.us/database/feis/

Woodland Fish and Wildlife Project. Publications may be ordered from Washington State University Cooperative Extension, 800-723-1763. You also may download copies from their Web site: www.woodlandfishand-wildlife.org/
Beaver, Muskrat and Nutria on Small Woodlands
Coastal Douglas Fir Forests and Wildlife
Managing Deer on Small Woodlands
Managing Ponderosa Pine Woodlands for Fish and Wildlife
Managing Small Woodlands for Elk
Managing Western Juniper for Wildlife

Birds

Books

Baicich, Paul J., and Colin J. O. Harrison. *A Guide to the Nests, Eggs, and Nestlings of North American Birds.* New York: Academic Press, 1997.

Bosakowski, Thomas, and Dwight G. Smith. *Raptors of the Pacific Northwest.* Portland, OR: Frank Amato Publications, 2002.

Campbell, R. Wayne, et al. *Wildlife Habitat Handbooks for the Southern Interior Ecoprovince. Vol. 2: Species Notes for Selected Birds.* Victoria, BC: Wildlife Report No. R16, Ministry of the Environment, 1988.

Ehrlich, Paul R., et al. *The Birder's Handbook: A Field Guide to the Natural History of North American Birds.* New York: Simon & Schuster, 1988.

Gaussoin, Bret, and Janice Lapsansky. *The Barn Owl and the Pellet.* 2nd Edition. Bellingham, WA: Pellets (3004 Pinewood), 1994.

Harrison, Hal H. *Western Bird Nests.* Boston: Houghton Mifflin, A Peterson Field Guide, 1979.

Nehls, Harry B. *Familiar Birds of the Northwest: Covering Birds Commonly found in Oregon, Washington, Idaho, Northern California, and Western Canada.* Portland, OR: Audubon Society of Portland, 1989.

Peterson, Roger Tory. *A Field Guide to Western Birds.* Boston: Houghton Mifflin, 1990.

Proctor, Noble S., and Patrick J. Lynch. *Manual of Ornithology.* New Haven, CT: Yale University Press, 1993.

Scott, Shirley L., ed. *National Geographic Society Field Guide to the Birds of North America.* Washington, DC: National Geographic Society, 2002.

Sibley, David Allen. *The Sibley Guide to Birds.* New York: Alfred A. Knopf, 2000.

Sibley, David Allen. *The Sibley Guide to Bird Life and Behavior.* New York: Alfred A. Knopf, 2001.

Stokes, Donald W., and Lillian Stokes. *A Guide to Bird Behavior.* Volumes 1–3, Boston, MA: Little, Brown, Stokes Nature Guides, 1979, 1983, and 1989.

Udvardy, Miklos D. F. *Audubon Society Field Guide to North American Birds—Western Region.* New York: Alfred A. Knopf, 1977.

Organizations

National Audubon Society chapters: For local chapter information look in your phone directory under "Audubon Society." Also see below.

Internet Resources

Alaska Department of Fish and Game's Wildlife Notebook Series: www.state.ak.us/local/akpages/FISH.GAME/notebook/nothome.htm

Canadian Wildlife Service's Hinterland Who's Who: www.cws-scf.ec.gc.ca/hww-fap/index_e.cfm

Cornell Lab of Ornithology: www.birdsource.org/

eNature.com: a searchable nature database www.enature.com

National Audubon Society: www.audubon.org/

Seattle Audubon's Birds of Washington State: www.birdweb.org/birdweb/

The Original Owl Pellet People: www.pelletsinc.com/index.html

U.S. Forest Service Wildlife Species Life Form Information: www.fs.fed.us/database/feis/

Reptiles and Amphibians

Books

Corkran, Charlotte C., and Chris Thoms. *Amphibians of Oregon, Washington, and British Columbia: A Field Identification Guide.* Vancouver, BC, and Redmond, WA: Lone Pine, 1996.

Leonard, William P., et al. *Amphibians of Washington and Oregon.* Seattle: Seattle Audubon Society, 1993.

Nussbaum, Ronald A., et al. *Amphibians and Reptiles of the Pacific Northwest.* Moscow, ID: University of Idaho Press, 1983.

Storm, R. M., and W. P. Leonard, eds. *Reptiles of Washington and Oregon.* Seattle: Seattle Audubon Society, 1995.

Internet Sites

U.S. Forest Service Wildlife Species Life Form Information: www.fs.fed.us/database/feis/

Pacific Northwest Herpetological Society: www.pnhs.net/

Woodland Fish and Wildlife Project. Publications may be ordered from Washington State University Cooperative Extension; phone: 800-723-1763. You also may download copies from their

Web site: www.woodlandfishand
wildlife.org/

*Hawk, Eagle and Osprey
Management on Small
Woodlands*
*Managing Small Woodlands for
Cavity Nesting Birds*
*Managing Western Juniper for
Wildlife*

Wildlife/Human Conflicts

Books and Pamphlets

Bird, D. M. *City Critters: How to
Live with Urban Wildlife.*
Montreal, Quebec, Canada:
Eden Press, 1986.

Coleman, John S., S. Temple, and
S. Craven. *Cats and Wildlife: A
Conservation Dilemma.* (Available
from: Publications Room 170,
630 W. Miffin Street, Madison,
WI 53703; phone: 608-262-
3346.)

Conover, Michael. *Resolving
Human–Wildlife Conflicts: The
Science of Wildlife Damage
Management.* Boca Raton, FL:
Lewis Publishers, 2002.

D'eon, R., et al. *The Beaver
Handbook: A Guide to Understanding and Coping with Beaver
Activity.* Timmins, Ontario,
Canada: Ontario Ministry
of Natural Resources, Northeast Science and technology,
Technology Fieldguide FG-
006, NEST, 1995.

Hodge, G. R., ed. *Wild Neighbors:
The Humane Approach to Living
with Wildlife.* Golden, CO:
Fulcrum Publishing;
Washington, DC: Humane
Society of the United States,
1997. (Available from: Urban
Wildlife Resources, 5130 W.
Running Brook Road,
Columbia, MD 21044.)

Hygnstrom, Scott E., et al.
*Prevention and Control of Wildlife
Damage.* Lincoln, NE:
University of Nebraska-
Lincoln, Institute of Agriculture
and Natural Resources, 1994.
(Available from: University of
Nebraska Cooperative
Extension, 202 Natural
Resources Hall, Lincoln, NE
68583-0819; phone: 402-472-
2188; also see Internet Sites
below.)

Juhre, Robert G. *Preventing Deer
Damage.* 3rd edition. Kettle
Falls, WA, 1998. (Available
from: Robert G. Juhre, 1723
Mountain Garden Way, Kettle
Falls, WA 99141-9771.)

Lawrence, Fred A. *Mole Control:
A "How To" Handbook for
Homeowners Who Want to Solve
Their Own Mole Problems.* The
Washington State Trappers
Association Inc., 1985.

Smith, Arthur E., Scott R.
Craven, and Paul D. Curtis.
*Managing Geese in Urban
Environments.* (For ordering
information, see www.wildlife
damagecontrol.com)

Trout, John. *Solving Coyote
Problems: How to Outsmart North
America's Most Persistent Predator.*
New York: Lyons Press, 2001.

Vantassel, Stephen. *Wildlife
Removal Handbook: A Guide for
the Control and Capture of Wild
Urban Animals,* 1999. (Available
from: Wildlife Damage
Control, 340 Cooley Street,
Springfield, MA 01128;
admin@wildlifedamagecontrol.
com)

Vantassel, Stephen, and Tom
Olander. *Wildlife Damage
Inspection Handbook.* 1998.
Available from: Wildlife
Damage Control, 340 Cooley
Street, Springfield, MA 01128;
admin@wildlifedamagecontrol.
com

Internet Resources

American Bird Conservancy (Cats
Indoors and other programs):
www.abcbirds.org

Coexisting with Coyotes (from
the Stanley Park Ecology Society):
www.vcn.bc.ca/spes/urban
wildlife/cocoyote.htm

Habitat Modification and Canada
Geese: Techniques for mitigating
human/goose conflict in urban and
suburban environments:
www.canadageese.org/nlcon-
trol.html

Centers for Disease Control and
Prevention:
www.cdc.gov/

Government of British Columbia
Publications:
http://wlapwww.gov.bc.ca/wld/
pub/pub.htm

How to Avoid Conflicts with
Wildlife (Oregon State Fish and
Wildlife):
www.dfw.state.or.us/spring
field/wildlifeandpeople.html

Internet IPM Resources on
Vertebrate Pests (Oregon State
University):
www.ippc.orst.edu/cicp/Pests/
vertebrate.htm

Prevention and Control of
Wildlife Damage:
wildlifedamage.unl.edu/hand
book/handbook/

The Berryman Institute for
Wildlife Damage Management:
www.berrymaninstitute.org/

The Internet Center for Wildlife Damage Management:
http://wildlifedamage.unl.edu/

The Wildlife Management Web Site:
http://www.wildlifemanagement.info/home.htm

UC Davis Pest Management Guidelines:
www.ipm.ucdavis.edu/PMG/selectnewpest.home.html

Vertebrate Management Links:
www.snohomish.wsu.edu/verturl.htm

Veterinarian's Guide to Managing Poisoning by Anticoagulant Rodenticides:
www.liphatech.com/vetguide.html

Wildlife Rehabilitation: International Wildlife Rehabilitation Council
www.iwrc-online.org

Wildlife Rehabilitation Information Directory
www.tc.umn.edu/~devo0028/

Wildlife Solutions Online:
www.wildlifesolutionsonline.com/home.htm

2000 Report of the American Veterinarian Medical Association Panel on Euthanasia:
www.avma.org/resources/euthanasia.pdf

Additional Internet Resources

Benner's Deer Fencing®:
www.bennersgardens.com/bg/

Birds-Away Attack Spider®:
www.birds-away.com

Cat Containment Systems:
www.feralcat.com/fence.html
www.corporatevideo.com/klips/
www.friendlyfence.com/ff/prod_fence_cat_houdini.asp

Fickle Hill Deer Fence Supply:
www.ficklehillfence.com

Flood Control Devices (Beaver):
www.na.fs.fed.us/spfo/pubs/stewardship/accessroads/beavers.htm
www.co.snohomish.wa.us/publicwk/swm/steward/beavers/
www.fsiculvert.com
http://www.clemson.edu/psapublishing/PAGES/AFW/AFW1.PDF

Procedures for Evaluating Predation on Livestock and Wildlife:
http://texnat.tamu.edu/ranchref/predator/pred.htm

Roll Guard, Inc. (Coyote Roller®):
www.coyoteroller.com/Products/flash_demo.htm

Scarecrow® (The motion-activated sprinkler by Contech):
www.scatmat.com

Wildlife Control Supplies:
www.wildlifecontrolsupplies.com

Traps and Trapping
Internet Resources

Eco Traps:
www.ecotrap.com/mainindex.html
Bird Traps:
www.audubon-omaha.org/bbbox/nestbox/bauldry.htm
www.vanerttraps.com/urban.htm
www.purplemartin.org

Tomahawk Live Traps:
www.livetrap.com

Wildlife Viewing/Tracking
Books

Brown, Tom, Jr. *Tom Brown's Field Guide to Nature Observation and Tracking.* New York: Berkeley Books, 1983.
Halfpenny, James A. *Field Guide to Mammal Tracking in North America.* Boulder, CO: Johnson Books, 1986.
Hudson, Wendy, ed. *Naturewatch – A Resource for Enhancing Wildlife Viewing Areas.* Helena, MT: Falcon Press, 1992.
La Tourrette, Joe. *The National Wildlife Federation's Wildlife Watcher's Handbook: A Guide to Observing Animals in the Wild.* New York: Henry Holt, 1997.
McDougall, Len. *The Complete Tracker: Tracks, Signs, and Habits of North American Wildlife.* New York: Lyons Press, 1997.
Pandell, Karen, and Chris Stall. *Animal Tracks of the Pacific Northwest.* Seattle: The Mountaineers Press, 1992.
Rezendes, Paul. *Tracking & the Art of Seeing: How to Read Animal Tracks & Sign.* Charlotte, VT: Camden House, 1992.
Stokes, Donald W., and Lillian Q. Stokes. *Guide to Animal Tracking and Behavior.* Boston: Little, Brown, 1986.

Organizations and Web Sites

Animal Tracks (Kim A. Cabrera):
www.beartrackersden.com

Hawk Watch International:
www.hawkwatch.org

Wilderness Awareness School:
www.natureoutlet.com

Nest Structures and Den Boxes

Books

Campbell, Scott. *The Complete Book of Birdhouse Construction for Woodworkers*. New York: Dover, 1984.

Corkran, C. *Birds in Nest Boxes: How to Help, Study, and Enjoy Birds when Snags Are Scarce*. Happy Camp, CA: Naturegraph Publishers (in press).

Henderson, Carrol. *Woodworking for Wildlife: Homes for Birds and Mammals*. 2nd edition. St. Paul, MN: Nongame Wildlife Program, Section of Wildlife, Department of Natural Resources, 1992. (Available from: Minnesota's Bookstore, 117 University Avenue, St. Paul, MN 55155; phone 800-657-375.)

Link, Russell. *Landscaping for Wildlife in the Pacific Northwest*. Seattle: University of Washington Press and the Washington Department of Fish and Wildlife, 1999.

McNeil, Don, and Betsy Rupp Fulwiler. *The Birdhouse Book: Building Houses, Feeders, and Baths*. Seattle: Pacific Search Press, 1979.

Organizations and Web Sites

Den Boxes for Raccoons
Urbanwildliferehab.netfirms.com/2002/raccoons2k2_pg3.htm

Predator Guards for Nest Boxes:
www.dnr.state.md.us/wildlife/wapredguard.html
www.virginiabluebirds.org/pages/box_grd.htm

www.mts.net/~gbdsnell/plans-shield.html

Washington Department of Fish and Wildlife:
www.wa.gov/wdfw

Bird Feeders and Feeder Management

Books

Alder, Bill, Jr. *Impeccable Birdfeeding*. Chicago: Chicago Review Press, 1992.

Burton, Robert. *National Audubon Society North American Birdfeeder Handbook*. London: Dorling Kindersley; Boston: Houghton Mifflin, 1992.

Campbell, Scott D. *Easy-to-Make Bird Feeders for Woodworkers*. New York: Dover, 1989.

Henderson, Carrol L. *Wild About Birds: The DNR Bird Feeding Guide*. St. Paul, MN: Minnesota's Bookstore, 1985. (Available from: State of Minnesota, Department of Natural Resources; to order call 800-657-3757.)

Kress, Stephen W. *The Audubon Society Guide to Attracting Birds*. New York: Scribner's, 1985.

Waldon, Bob. *Feeding Winter Birds in the Pacific Northwest*. Seattle: The Mountaineers Press, 1994.

Organizations

Canada: Project Feeder Watch
Bird Studies Canada,
PO Box 160, Port Rowan
Ontario, Canada N0E 1M0
Phone: 1-888-448-BIRD
www.bsc-eoc.org/national/pfw.html

Project Feeder Watch
Cornell Laboratory of Ornithology
PO Box 11, Ithaca, NY 14850-0011

Phone: 800-843-2473
http://birds.cornell.edu/pfw/

Bat Houses and Bat Detectors

Books

Tuttle, Merlin D., and Donna L. Hensley. *The Bat House Builder's Book*. Austin, TX: Bat Conservation International, 1995; University of Texas Press, 2001.

Also see "Mammals."

Organizations

Bat Conservation International,
PO Box 162603
Austin, TX 78716-2603
Phone: 800-538-BATS
www.batcon.org

Bats Northwest
PO Box 18735, Seattle, WA 98118
Phone: 206-256-0406
www.batsnorthwest.org

Internet Resources

All about bats:
www.nyx.net/~jbuzbee/bat_house.html

Bat house enthusiasts:
www.batcon.org/bhra/bhratop.html

Mayberry Bat Homes:
www.maberrybat.com

Ponds

Books

Allison, James. *Water in the Garden: A Complete Guide to the Design and Installation of Ponds, Fountains, Streams, and Waterfalls*. Boston: Little, Brown, 1991.

Clafin, Edward B. *Garden Pools and Fountains*. San Ramon, CA: Ortho Books, 1988.

Grinstein, Dawn Tucker. *For Your Garden: Pools, Ponds, and Waterways*. New York: Grove Weidenfeld, 1991.

Link, Russell. *Landscaping for Wildlife in the Pacific Northwest*. Seattle: University of Washington Press and the Washington Department of Fish and Wildlife, 1999.

Matson, Tim. *Earth Ponds: The Country Pond Maker's Guide to Building, Maintenance, and Restoration*. Woodstock, VT: Countryman Press, 1991.

Nash, Helen. *The Pond Doctor: Planning and Maintaining a Healthy Water Garden*. New York: Sterling, 1994.

Gardening and Landscaping for Wildlife

Books

Cox, Jeff. *Landscaping with Nature*. Emmaus, PA: Rodale Press, 1991.

Kress, Stephen W. *The Bird Garden*. National Audubon Society; London: Dorling Kindersley; Boston: Houghton Mifflin, 1995.

Link, Russell. *Landscaping for Wildlife in the Pacific Northwest*. Seattle: University of Washington Press and the Washington Department of Fish and Wildlife, 1999.

Merilees, Bill. *Attracting Backyard Wildlife*. Stillwater, MN: Voyageur Press, 1989; North Vancouver, BC, Canada: Whitecap Books, 1990.

Stein, Sara. *Planting Noah's Garden: Further Adventures in Backyard*

Ecology. Boston: Houghton Mifflin, 1997.

Organizations with Wildlife Habitat Certification Programs

Backyard Wildlife Sanctuary Program
Department of Fish and Wildlife
16018 Mill Creek Boulevard
Mill Creek, WA 98012
www.wa.gov/wdfw

Backyard Wildlife Habitat Program
National Wildlife Federation
11100 Wildlife Center Drive
Reston, VA 20190-5362
www.nwf.org

Naturescape British Columbia,
PO Box 9354, STN PROV GOV
Victoria, BC, Canada, V8W 8M1
www.hctf.ca/nature.htm

Wildlife Plants

Books

Alexander, Martin C., Herbert S. Zim, and Arnold L. Nelson. *American Wildlife and Plants: A Guide to Wildlife Food Habitats*. New York: Dover, 1951.

Benyus, Janine M. *The Field Guide to Wildlife Habitats of the Western United States*. New York: Fireside Books, 1989.

Kruckeberg, Arthur. *Gardening with Native Plants of the Pacific Northwest*. 2nd Edition. Seattle: University of Washington Press, 1996.

Link, Russell. *Landscaping for Wildlife in the Pacific Northwest*. Seattle: University of Washington Press and the Washington Department of Fish and Wildlife, 1999.

Parish, Roberta, Ray Coupé, and Dennis Lloyd, eds. *Plants of*

Southern Interior British Columbia. Vancouver, BC, and Redmond, WA: Lone Pine Publishing, 1996.

Pettinger, April. *Native Plants in the Coastal Garden: A Guide for Gardeners in the Pacific Northwest*. Portland, OR: Timber Press, 2001.

Pojar, Jim, and Andy MacKinnon. *Plants of the Pacific Northwest Coast*. British Columbia Ministry of Forests, Vancouver and Edmonton, BC. Redmond, WA: Lone Pine, 1994.

Organizations

British Columbia Native Plant Society
Contact: Erin Skelton
3860 West 19th Avenue
Vancouver, BC V65 1C8, Canada
email: feskelton@telus.net
www.npsbc.org

Oregon Native Plant Society
PO Box 902
Eugene, OR 97440-0902
www.npsoregon.org

Washington Native Plant Society
6310 NE 74th Street, Suite 215E
Seattle, WA 98115
Phone: 206-527-3210 or
1-888-288-8022
www.wnps.org

Internet Resources

Pacific Northwest Native Wildlife Gardening:
www.tardigrade.org/natives/

Identifying, Propagating and Landscaping with Native Plants:
gardening.wsu.edu/text/nwnative.htm

Landscaping with Native Plants:
www.epa.gov/greenacres

Beaver Deceiver

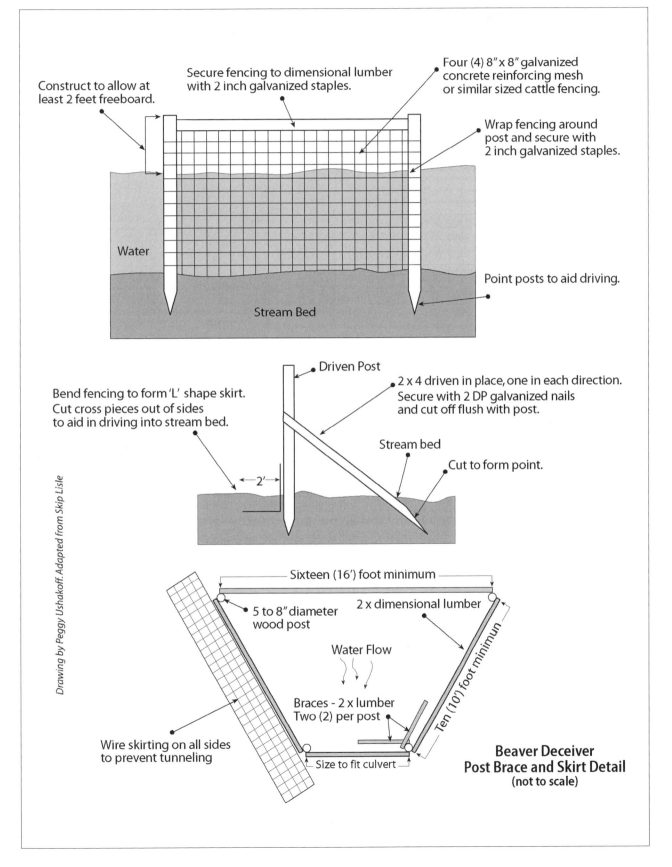

Construct to allow at least 2 feet freeboard.

Secure fencing to dimensional lumber with 2 inch galvanized staples.

Four (4) 8" x 8" galvanized concrete reinforcing mesh or similar sized cattle fencing.

Wrap fencing around post and secure with 2 inch galvanized staples.

Water

Point posts to aid driving.

Stream Bed

Bend fencing to form 'L' shape skirt. Cut cross pieces out of sides to aid in driving into stream bed.

Driven Post

2 x 4 driven in place, one in each direction. Secure with 2 DP galvanized nails and cut off flush with post.

Stream bed

Cut to form point.

2'

Drawing by Peggy Ushakoff. Adapted from Skip Lisle

Sixteen (16') foot minimum

5 to 8" diameter wood post

2 x dimensional lumber

Water Flow

Ten (10') foot minimun

Braces - 2 x lumber Two (2) per post

Wire skirting on all sides to prevent tunneling

Size to fit culvert

Beaver Deceiver Post Brace and Skirt Detail (not to scale)

Flexible Leveler

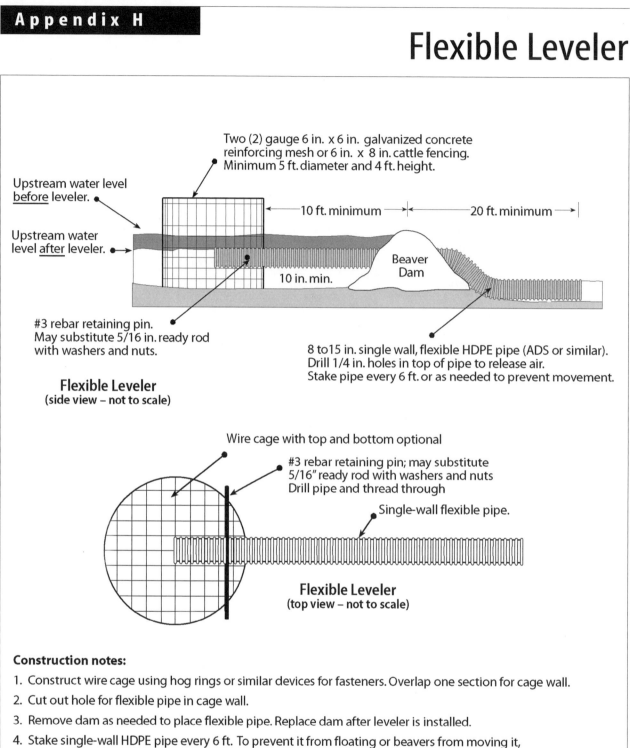

Two (2) gauge 6 in. x 6 in. galvanized concrete reinforcing mesh or 6 in. x 8 in. cattle fencing. Minimum 5 ft. diameter and 4 ft. height.

Upstream water level <u>before</u> leveler.

Upstream water level <u>after</u> leveler.

← 10 ft. minimum → | ← 20 ft. minimum →

Beaver Dam

10 in. min.

#3 rebar retaining pin. May substitute 5/16 in. ready rod with washers and nuts.

8 to 15 in. single wall, flexible HDPE pipe (ADS or similar). Drill 1/4 in. holes in top of pipe to release air. Stake pipe every 6 ft. or as needed to prevent movement.

Flexible Leveler
(side view – not to scale)

Wire cage with top and bottom optional

#3 rebar retaining pin; may substitute 5/16" ready rod with washers and nuts Drill pipe and thread through

Single-wall flexible pipe.

Flexible Leveler
(top view – not to scale)

Construction notes:

1. Construct wire cage using hog rings or similar devices for fasteners. Overlap one section for cage wall.
2. Cut out hole for flexible pipe in cage wall.
3. Remove dam as needed to place flexible pipe. Replace dam after leveler is installed.
4. Stake single-wall HDPE pipe every 6 ft. To prevent it from floating or beavers from moving it, use two T-posts and wire between them and over the top of the pipe to secure the pipe.
5. Drill 3/8th in. hole in culvert for rebar to allow for friction fit. If ready rod is used, place washers next to pipe and secure with double nuts.
6. One (1) 16 foot section of fencing will construct a cage wall approximately 5 feet in diameter. An additional section is needed to construct the top and bottom of each cage.
7. Pipe diameter should be sized to pass the stream base flow.
8. Final layout of the pipe should allow for a shallow gradient to facilitate fish passage.

Drawing by Peggy Ushakoff

Predator Guard–Birds and Mammals

Materials

½" hardware cloth (6½" x 18")
Tire wire

4 – # 8 x ½" sheet metal screws
4 – # 10 flat washers

Bend hardware cloth into shape and
lace edges together using tie wire.

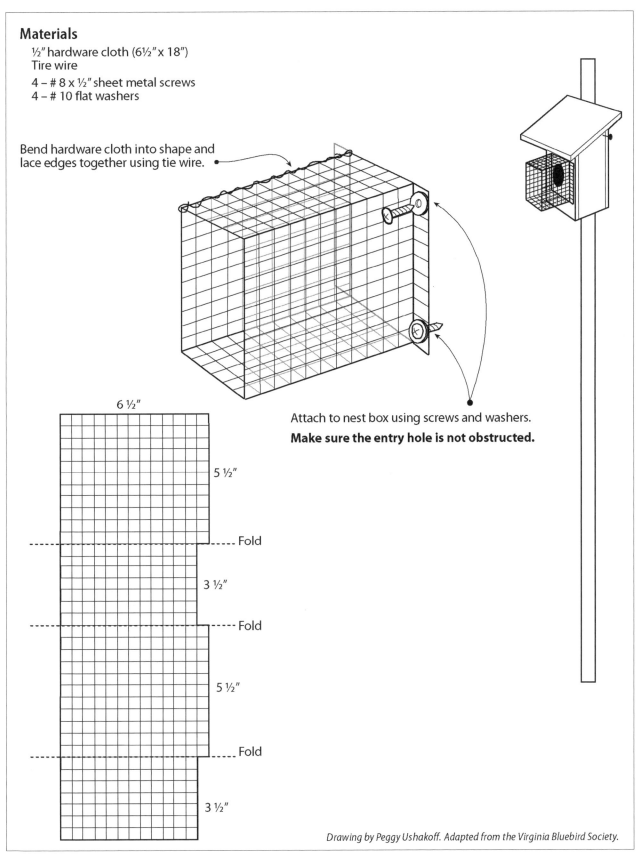

6 ½"

5 ½"

---- Fold

3 ½"

---- Fold

5 ½"

---- Fold

3 ½"

Attach to nest box using screws and washers.
Make sure the entry hole is not obstructed.

Drawing by Peggy Ushakoff. Adapted from the Virginia Bluebird Society.

Appendix J

Predator Guard–Mammals

Materials
- 1– 6 x 24" round duct or PVC pipe.
- 1– 6" round duct cap or PVC cap.
- 2 # 8 x ½" sheet metal screws.
- 1– 3" long ¼" bolt with two nuts and locking washers.
- 1" EMT metal electical conduit (or other 1 inch metal pole) buried in ground.

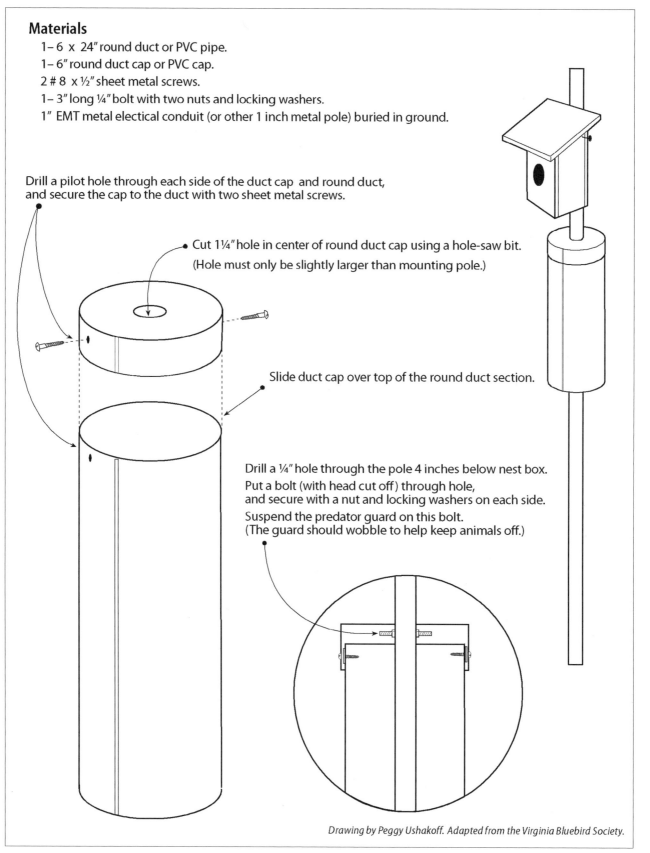

Drill a pilot hole through each side of the duct cap and round duct, and secure the cap to the duct with two sheet metal screws.

Cut 1¼" hole in center of round duct cap using a hole-saw bit.
(Hole must only be slightly larger than mounting pole.)

Slide duct cap over top of the round duct section.

Drill a ¼" hole through the pole 4 inches below nest box.
Put a bolt (with head cut off) through hole, and secure with a nut and locking washers on each side.
Suspend the predator guard on this bolt.
(The guard should wobble to help keep animals off.)

Drawing by Peggy Ushakoff. Adapted from the Virginia Bluebird Society.

Squirrel and Chipmunk Nest Box

Materials

One 1 x 8" x 6' rough cedar board

Twenty 1¼" outdoor wood screws or # 7 galvanized nails

7¼"

Floor	5¾"
Baffle	5½" / 3"
Top	10"
Front	9"
Side	9"
Side	9"
Back	14"
Support	2½"
Ledge	4½"

1½"

2¼"

Holes for attaching to a tree or post.

4"

The baffle

Pivot screws work as hinges. To allow the side to open easily, the pivot screw on the opposite side needs to be level with the one in front.

The ledge support.

Floor is ½" above bottom of front board.

Back

Front

Top view of nest box showing baffle and location of triangular cutout.

The Baffle

The baffle creates a protected nesting spot for squirrels. It also allows the box to have a dark interior and helps prevent heat loss.

Locate the baffle flush with the bottom of the entry hole (see above drawing). Orient the baffle with the triangular cutout in the far corner opposite the entry hole. (see drawing).

Drawing by Peggy Ushakoff.

Index